Plant Function and Structure

The Macmillan Core Series in Biology

General Editors: **Norman H. Giles and John G. Torrey**

Plant Function and Structure

Victor A. Greulach
University of North Carolina, Chapel Hill

The Macmillan Company, New York
Collier-Macmillan Publishers, London

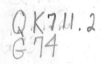
THE MACMILLAN COMPANY
866 THIRD AVENUE, NEW YORK, NEW YORK 10022
COLLIER-MACMILLAN CANADA, LTD., TORONTO, ONTARIO

Library of Congress catalog card number: 75-190153

Printing: 1 2 3 4 5 6 7 8   Year: 3 4 5 6 7 8

PREFACE

The writing of this textbook was stimulated by several marked trends in undergraduate biological education that have been in progress for the last decade or so and promise to continue in the future. There has been a trend away from general botany or general zoology courses toward general biology courses and from majors in botany, zoology, or microbiology toward a biology major. In many colleges and universities there has been a substantial reduction in the number and variety of undergraduate courses offered in the biological sciences, accompanied by the establishment of a limited number of core courses. These are designed to provide the biology major with a well-rounded foundation in the various biological disciplines that will prepare him effectively for graduate school, medical school, high school biology teaching, or other biological endeavors. Such core curricula generally include courses in biochemistry, cell biology, ecology, genetics, development, and evolution. The courses are usually designed to deal with organisms in general, rather than with specific groups such as plants, animals, or microorganisms.

Desirable as such undergraduate curricula may be, they tend to have certain defects. Although they deal well with most of the important biological principles and concepts, they often give inadequate attention to the structural aspects of organisms, including anatomy and the comparative morphology of the great diversity of organisms. Such neglect may extend even to the general biology course. Furthermore, despite the fact that the core courses are presumably designed to treat organisms in general, they often give little or no consideration to plants, particularly as regards their structure, function, development and morphogenesis, and comparative morphology. Plant structure is obviously quite different from that of other organisms. It is equally true, although not as generally recognized, that the physiology and development of plants differ substantially from those of animals.

The author, as well as many others, feels that any biologist regardless of his field of specialization can profit by some basic knowledge of all important groups of organisms and that the undergraduate biology curriculum should include a reasonable consideration of plants. This textbook has been designed to embody the important facts, principles, and concepts about plant structure and function that are often inadequately covered in the usual core courses. The treatment of structure and function was merged into one book and one course primarily because there is really not room in an undergraduate core curriculum for separate courses in plant physiology, plant anatomy, and plant development or morphogenesis. Furthermore, the intimate interrelations between structure and function seem to make such a merger desirable. In this book the emphasis is on vascular plants, although the nonvascular plants are by no means ignored. Originally much more consideration of nonvascular plants was intended, but their great structural and functional diversity made this plan impractical in a textbook designed for a one-semester course.

In addition to a course based on this book, the undergraduate curriculum should also include a course dealing with a comparative survey of the major groups of plants. Such a combination would provide (along with the usual core courses) an adequate basic knowledge of plants, not only for those who will specialize in biological disciplines other than botany but also for those who plan graduate study in botany.

This book has not been designed for students who have had a good strong course in general botany, particularly a two-semester course. Such students would find a considerable amount of repetition. However, since many of the discussions extend beyond those of most general botany textbooks, some instructors may wish to consider this book for a second-level course in plant structure and function, plant physiology, or plant development.

Chapter 1 is a modification of an article by the author under the pen name of V. A. Eulach published in the October 1956 issue of *Astounding Science Fiction* (now *ANALOG Science Fiction—Science Fact*), Copyright 1956 by Smith & Street Publications, Inc., and reprinted in the December 1956 issue of *Science Digest*. Permission for its inclusion in this book has been obtained from the present copyright holder, Conde Nast Publications, Inc.

Although the source of the illustrations is credited in the captions, I wish to express appreciation here to all those who cooperated by supplying photographs or by giving permission to reproduce illustrations from their publications. In particular, thanks are due to the Carolina Biological Supply Company, Ripon Microslides, and Triarch Products for numerous photomicrographs and photographs, including many good ones that could not be included in the book, and to McGraw-Hill Book Company, John Wiley & Sons, Inc., and The Macmillan Company for giving permission to reproduce many illustrations from their books without charge. The drawings reprinted from *Plants: An Introduction to Modern Botany* by Victor A. Greulach and J. Edison Adams, John Wiley & Sons, Inc., were made by Peggy-Ann Kessler Duke.

I also wish to express great appreciation to the botanists who reviewed the manuscript or parts of it. Their numerous criticisms and suggestions have made this a far better book than it would have been otherwise. However, I take full responsibility for any remaining errors, misinterpretations, or obscurities.

The entire manuscript was reviewed by A. E. DeMaggio, Douglas Pratt, and John G. Torrey. Aubrey W. Naylor reviewed the first ten chapters. The following reviewed certain chapters that related to their fields of specialization: Orlin Biddulph, R. Malcolm Brown, Jr., Harold J. Evans, Charles Heimsch, Paul J. Kramer, Tom K. Scott, and N. E. Tolbert. I also wish to thank several reviewers who remained anonymous to me. One reviewed the entire manuscript, one the first half, and several did individual chapters. Their comments and suggestions added greatly to the quality of the book.

Chapel Hill, North Carolina   V. A. G.

Brief Contents

1 Plant Structure As an Effective Autotrophic Design   1

2 Plant Cells   6

3 Photosynthesis   58

4 Respiration   94

5 Some Other Biochemical Activities of Plants   114

6 Mineral Nutrition and Metabolism   161

7 Cellular Traffic   173

8 The Vascular System   198

9 Absorption, Translocation, and Loss of Water   224

10 Translocation and Mobilization of Solutes   267

11 Introduction to Plant Growth and Development   294

12 Auxins and Ethylene   325

13 Other Phytohormones   355

14 Plant Reproduction   384

15 Reproductive Development of Angiosperms   435

16 Vegetative Development of Vascular Plants   488

17 Some Developmental Phenomena   520

18 Phases in the Life of Plants   553

Appendix A Metric Units and Equivalents   565

Appendix B Major Groups of Plants   569

Index   571

Detailed Contents

1 PLANT STRUCTURE AS AN EFFECTIVE
AUTOTROPHIC DESIGN 1

2 PLANT CELLS 6
Some General Characteristics of Plant Cells 6
The Cells and Tissues of Vascular Plants 7
 Meristematic cells 10
 Parenchyma cells 11
 Collenchyma cells 13
 Epidermal cells 13
 Endodermal cells 15
 Sieve cells and sieve-tube elements 16
 Laticifers 18
 Sclerenchyma cells 18
 Tracheids and vessel elements 19
 Cork cells 24
Organelles of Plant Cells 25
 The nucleus 26
 Ribosomes 29
 Microtubules 31
 Membranes 32
 Golgi apparatus 36
 Mitochondria 38
 Plastids 40
 Other membranous organelles 46
 Cell walls 48
Procaryotic Cells 54

3 PHOTOSYNTHESIS 58
Historical Development of Concepts About Photosynthesis 60
 Prechemical concepts and notions 60
 Pioneering basic discoveries 61
 The research plateau 62
 The modern period of research 64
The Photosynthetic Reactions 67
 The photochemical reactions 68
 The reduction of carbon 76

Efficiency of Photosynthesis in Energy Conversion 81
Factors That Influence the Rate of Photosynthesis 82
   Apparent and true photosynthesis 82
   Internal factors affecting photosynthesis 84
   The principle of limiting factors 84
   Carbon dioxide 86
   Water 88
   Light 89
   Temperature 90
   Oxygen 92

4 RESPIRATION 94
Aerobic Respiration 95
   Glycolysis 96
   Citric acid cycle 97
   Oxidative phosphorylation 98
   ATP energy release 101
   Hexose monophosphate shunt 101
   Comparison of respiration and photosynthesis 102
Anaerobic Respiration 104
Other Oxidation Systems in Plants 106
The Respiratory Quotient (RQ) 107
Rates of Respiration in Plants 108
Factors Influencing the Rate of Respiration 109
   Internal factors 110
   Water 111
   Oxygen 111
   Temperature 112

5 SOME OTHER BIOCHEMICAL ACTIVITIES OF
PLANTS 114
Carbohydrate Metabolism 115
   Monosaccharides 116
   Disaccharides 120
   More complex sugars 122
   Polysaccharides 122
Lipid Metabolism 130
   The fats 130
   Phospholipids 135
   Waxes 135
Nitrogen Metabolism 136
   The nitrogen cycle 136
   Amino acid synthesis 142
   Purines and pyrimidines 144
   Nucleic acids 145
   Proteins 148
   Other important nitrogenous compounds 151
Some Other Classes of Compounds Synthesized by Plants 155

	Terpenes	155
	Anthocyanins	156
	Tannins	158
	Sterols	158
	Lignins	159

6   MINERAL NUTRITION AND METABOLISM   161

Development of Concepts About Mineral Nutrition   161

Roles of the Mineral Elements   164

Mineral elements as constituents of organic compounds   164

Mineral elements as enzyme activators   165

Other roles of mineral elements   165

Mineral Deficiency Symptoms   166

Nitrogen   167

Phosphorus   168

Potassium   169

Sulfur   169

Calcium   169

Iron   169

Magnesium   169

Copper   170

Zinc   170

Boron   171

Manganese   171

Molybdenum   171

7   CELLULAR TRAFFIC   173

Membrane Permeability   173

Permeability to various substances   174

Basis of differential permeability   174

Factors influencing membrane permeability   175

The significance of differential permeability   175

Plasmodesmata   175

Gas Exchanges of Plant Cells   176

Diffusion of gases   176

Factors influencing the rate of diffusion   177

Movement of Ions Into and Out of Cells   177

Diffusion   178

Ion exchange   179

Mass flow   180

Active transport and accumulation of ions   180

Diffusion of Solutes Into and Out of Cells   183

Movement of Water Into and Out of Plant Cells   183

Influence of solutes on water potential   184

Influence of imposed pressure on water potential   185

Osmosis in plant cells   188

Some quantitative examples of osmosis in plant cells   190

Imbibition   193

Significance of Osmosis and Imbibition in Plants   195

8   THE VASCULAR SYSTEM   198
  The Vascular System of Stems   199
    Herbaceous monocotyledons   199
    Herbaceous dicotyledons   202
    Woody dicotyledons   204
    Conifer stems   210
  The Vascular System of Roots   212
    Herbaceous roots   212
    Woody roots   214
  The Vascular System of Leaves   216
    The vascular tissue of petioles   217
    The vascular tissue of leaf blades   218
  The Vascular System of Reproductive Organs   221

9   THE ABSORPTION, TRANSLOCATION, AND LOSS OF
        WATER   224
  The Loss of Water from Plants   226
    Leaf structure in relation to transpiration   226
    Structure and distribution of stomata   228
    Stomatal opening and closing   229
    Diffusion through stomata   231
    Factors influencing the rate of transpiration   233
    Wilting   241
    Periodicity of transpiration   244
    Magnitude of transpiration   246
    Significance of transpiration   247
  The Translocation of Water Through the Xylem   250
    Root pressure   250
    The cohesion mechanism   252
  The Absorption of Water   256
    Soil water   257
    Active absorption   261
    Passive absorption   262
    Factors influencing the rate of water absorption   263
    Mycorrhizae and absorption   264

10 TRANSLOCATION AND MOBILIZATION OF SOLUTES   267
  Principal Substances Translocated   267
  Pathways of Solute Translocation   269
    Upward translocation of solutes   270
    Downward translocation of solutes   273
  Xylem Translocation of Solutes   274
  Phloem Translocation of Solutes   274
    Some facts about phloem translocation   274
    The cytoplasmic streaming hypothesis   281
    The pressure flow hypothesis   281
    The activated diffusion hypothesis   284

The interfacial flow hypothesis   284
Bioelectrical potential hypotheses   285
Summary   285
Mobilization of Solutes   285
Circulation of Solutes in Plants   291

11   INTRODUCTION TO PLANT GROWTH AND DEVELOPMENT   294
Growth Versus Development   295
Growth and Development at the Cellular Level   296
Cell division   296
Cell enlargement   298
Cell differentiation   298
Interaction of Factors in Growth and Development   299
Hereditary Potentialities   301
Environmental Factors   303
Biological factors   303
Chemical factors   304
Physical factors: General   305
Physical factors: Temperature   308
Physical factors: Electromagnetic radiation   310
Internal Conditions and Processes   314
Observable Growth and Development   315
Growth Rates and Periodicity   315
Kinetics of growth   315
Range of growth rates   318
Determinate and indeterminate growth   318
Differential growth   319
Daily periodicity of growth   320
Seasonal periodicity of growth   321

12   AUXINS AND ETHYLENE   325
Chemistry of Auxin   328
Chemical nature   328
Synthesis and degradation of IAA   329
Quantitative Aspects of Auxin   329
Effective auxin concentrations   329
Quantitative determination of auxin   330
Occurrence of Auxin in Plants   331
Species distribution of auxin   331
Regions of auxin synthesis   332
Translocation of auxin   332
Distribution of auxin in plants   333
Effects of Auxin at the Cellular Level   334
Cell elongation   334
Cell division   334
Cell differentiation   335

Effects on Observable Growth and Development   335
    Organ elongation   336
    Tropisms   337
    Apical dominance   342
    Inhibition of abscission   344
    Adventitious root development   346
    Flower initiation   346
    Fruit development   347
The Biochemistry of Auxin Action   348
Synthetic Growth Regulators with Auxin Activity   350
Ethylene   352

13   OTHER PHYTOHORMONES   355
Gibberellins (GA)   356
    Species distribution of gibberellins   357
    Chemical nature of the gibberellins   357
    Biosynthesis of gibberellins   358
    Distribution of gibberellins in vascular plants   359
    Effective concentrations of gibberellins   359
    Bioassays for gibberellins   360
    Effects of gibberellins on observable growth and development   362
    Effects of gibberellins at the cellular level   365
    Biochemistry of gibberellin action   367
Cytokinins   367
    Chemical nature of the cytokinins   369
    Occurrence and distribution of cytokinins in plants   369
    Effects of cytokinins at the cellular level   370
    Effects of cytokinins on growth and development   371
    Bioassays for cytokinins   374
    Biochemistry of cytokinin action   375
Abscisic Acid (ABA)   375
    Chemical nature of abscisic acid   377
    Occurrence and distribution of abscisic acid   378
    Effects at the cellular level   378
    Effects on plant growth and development   378
    Assays for abscisic acid   380
    Biochemistry of abscisic acid action   380
Summary   380

14   PLANT REPRODUCTION   384
Asexual Reproduction of Nonvascular Plants   385
    Binary fission   385
    Sporulation   386
    Vegetative propagation   386
Asexual Reproduction of Vascular Plants   387
    Propagation by stems and buds   388
    Propagation by leaves   390

Propagation by roots	391
Vegetative propagation by flowers	391
Sexual Reproduction of Plants	392
Selected Plant Life Cycles	392
Ulothrix	393
Ulva	394
Oedogonium	396
A moss: Polytrichum	397
A fern: Polypodium	399
A clubmoss: Selaginella	403
A conifer: Pine	406
An angiosperm	411
Comparison and summary of sexual life cycles	421
Angiosperm Reproductive Structures	423
Flowers	423
Inflorescences	427
Fruits	429
Seeds	431

15 REPRODUCTIVE DEVELOPMENT OF ANGIOSPERMS 435

Initiation of Flowering	435
Ripeness to flower	436
Photoperiodism	437
Vernalization	456
Flower Physiology	461
Sex expression	461
Flower development	463
Pollen physiology	464
Flower movements	465
Senescence and abscission of flower parts	467
Fruit Physiology	467
Fruit set	467
Fruit growth	470
Fruit ripening	474
Seed physiology	477
Seed development	477
Seed viability	478
Seed dormancy	479
Seed germination	483

16 VEGETATIVE DEVELOPMENT OF VASCULAR
PLANTS 488

The Growth and Development of Shoots	489
The shoot apex	489
Differentiation of the vascular tissues	496
Differentiation of the fundamental tissues	500
Differentiation of the dermal tissues	501
Stem elongation	502

	Secondary growth of stems	503
	The growth and development of leaves	504
	The Growth and Development of Roots	508
	The root apex	509
	Differentiation of the primary root tissues	513
	The development of lateral roots	515
	Secondary growth of roots	517

17 SOME DEVELOPMENTAL PHENOMENA 520

Some Aspects of Plant Morphogenesis 520
Polarity 520
Symmetry 525
Spirality 528
Differentiation 530
Correlation 531
Regeneration 532
The Development of Isolated Plant Parts 535
Organ culture 537
Tissue culture 538
Cell cultures 538
Abnormal Plant Development 543
Calluses and tumors 543
Galls 545
Other abnormal organs 549
Abnormalities induced by ionizing radiation 549
Abnormalities induced by applications of growth substances 551

18 PHASES IN THE LIFE OF PLANTS 553
Juvenility 554
Maturity 559
Senescence 560

Appendix A METRIC UNITS AND EQUIVALENTS 565
Appendix B MAJOR GROUPS OF PLANTS 569
Index 571

Plant Structure As an Effective Autotrophic Design

Every now and then in science fiction the more or less human beings arriving from outer space turn out to be green. Like plants they contain chlorophyll, and like plants they are independent of outside sources of food—that is, they are what biologists call autotrophic organisms. Unless we want to let our science fiction roam into the realm of the completely fantastic and impossible, the question arises as to just how probable such human beings or other animals could be. Could there be human beings with chlorophyll in their skin? If so, would they be able to carry on sufficient photosynthesis to supply all the foods they would need? And if they could, might there not be some planet on which all life could consist of self-sufficient animals with none of the plants on which the human beings and other animals of the Earth are so dependent for their food, their oxygen, and indeed their lives?

There are, of course, on Earth a few autotrophic animals. A minority of the single-celled animals, or protozoa, contain chlorophyll and live, at least largely, on the food they produce by photosynthesis. Some more complex animals, like the green hydra and several marine gastropod mollusks, have single-celled algae living symbiotically within their tissues and so are at least partly independent of outside sources of food. Also, there are a few species of bacteria—really plants rather than animals, if either—that can synthesize their own food. Some contain chlorophyll and other pigments by means of which they trap light energy for their photosynthesis. Others use energy secured from the oxidation of inorganic substances such as sulfur or mineral salts, rather than light energy, for synthesizing their food. However, with these few exceptions all animals and all nongreen plants such as the bacteria, molds, and mushrooms are entirely dependent on green plants as the ultimate source of their food. No very large, very advanced, or very complex animal is in any way autotrophic.

Now it is by no means implausible that a human being or other animal might have mutations that would result in the ability to produce chlorophyll, because the prosthetic group of hemoglobin is chemically very similar to chlorophyll. It differs chemically from chlorophyll only in several details, one of which is that it contains iron whereas chlorophyll contains magnesium. It would not take much of a change in synthetic abilities for a human being to produce chlorophyll, though apparently such a mutation has never occurred. In such a green animal the chlorophyll would probably be restricted to the skin, in a layer not much thicker than our ordinary leaf. The question is, could a human being or other large animal with such a chlorophyll-containing skin carry on enough photosynthesis so that sufficient food would be produced to provide all needed in respiration and assimilation, thus making the animal independent of outside food supplies?

We can easily provide the answer to this question by taking a quantitative look at the problem. In the first place, we must consider the maximum rate at which plants can produce sugar by photosynthesis under the most favorable natural conditions. The figure is about 20 milligrams per square decimeter (mg/dm^2) of leaf area per hour. Actually, the usual photosynthetic productivity of most plants, even under favorable conditions, is only a fraction of this. The highest rate of photosynthesis ever measured was 52 $mg/(dm^2)$ (hr) of sugar, but this was achieved by supplying the plants with 5% carbon dioxide, rather than the 0.03% present in the air. Suppose that we grant generously that animals might be able to produce sugar at the 20-mg rate and that we are similarly generous in our other estimates.

An outside estimate of the amount of human skin suitably exposed to light would be 170 dm^2. Accepting this figure, the photosynthetic production would be 3.4 grams (g) of sugar per hour, or 41.8 g/12-hr day. This is considerably more than might be reasonably expected, for we are assuming uniformly favorable conditions of light and other factors throughout the day. Though the light period is greater than 12 hr in the summer, it is less than that in the winter; the 12-hr figure is a rough annual average. However, to get a round figure and one as optimistic as possible suppose we accept 42 g as the average daily sugar production.

This 42 g of sugar contains about 157 kilogram calories (kcal) of energy. How far will this go toward supplying the food requirements of our green man for both energy and body-building material? If you have ever been concerned with your daily calorie consumption, you have already noted that this food supply is completely inadequate. A person who weighs 160 pounds (lb) uses about 1800 kcal/day in his basal metabolism alone, that is, when he is completely at rest. This is the energy required to operate such automatic body processes such as the beating of the heart, peristaltic movements, and breathing. In addition, normal activity will require about 500 kcal more of energy, whereas a man doing heavy work would require a total of around 5000 kcal/day. In other words, a man's daily energy expenditure will range from about 2000 to about 5000 kcal.

Let us accept the minimum figure as the energy requirements of our photosynthetic man, because he would not need to be very active. He would be relieved of the necessity of securing food, assuming for the moment that he could synthesize all he needed; that is, he would neither have to obtain his food by his own direct efforts nor would he have to work to get money with which to buy food. Indeed, he probably would not have to work at all. To take advantage of his photosynthetic skin he would have to shun any clothing, and he would have to live in a tropical climate where he could comfortably expose himself to the sun the year around. Thus there would be no need for shelter or fuel for heating.

His only needs external to himself would be those required by plants—carbon dioxide and oxygen from the air, water, and mineral salts. The latter would perhaps be present in the water he would drink, though he might have difficulty in securing all the essential minerals in sufficient quantity. However, in addition to photosynthesis, our green men would also have to acquire several other synthetic abilities restricted to plants. Most important of these would be the ability to synthesize vitamins and amino acids, the latter for use in building up proteins. Although animals are able to make some amino acids from others, only plants can synthesize amino acids from sugar and inorganic salts of nitrogen and sulfur. However, there is no reason why animals could not carry on these processes if they had the proper enzymes.

We have been assuming that the green men could produce enough sugar by photosynthesis to meet their daily energy needs. The assumption is, of course, not valid, as our figures indicate. The maximum of 42 g of sugar produced daily by photosynthesis would provide only 157 of the basic 2000 kcal needed. Conceivably, a photosynthetic mechanism much more efficient and productive than that of plants might evolve. There might be more chloroplasts or more lamellae per chloroplast. There might be two photosynthetic layers—a superficial one in which chlorophyll is the principal pigment and a somewhat deeper layer with a light-absorbing pigment that could absorb wavelengths of light transmitted through the superficial layer, similar to the phycoerythrin of red algae. However, if such photosynthetic systems were even three or four times as productive as plant photosynthesis, there would still be a severe calorie deficit. The green men would rapidly starve to death if they attempted to rely solely on photosynthesis for their entire food supply.

Now suppose we attempt to propose structural and physiological modifications of the green men which would make them self-sufficient as far as food is concerned. The first obvious suggestion would be an increase in surface area so that more photosynthesis could be carried on. An increase in size would not help at all, for as size increases volume increases faster than surface area. A decrease in size would help, but the decrease would have to be so drastic that we would end by having an animal no larger than the small photosynthetic animals which actually exist.

A more fruitful possibility would be thin tissue extensions from the body, which would greatly add to the photosynthetic surface without adding too much tissue which would itself require additional energy for its respiration. In other words, the green men would have what would really amount to leaves. Considering the fact that our 2000-kcal goal represents only the energy used in respiration and does not include the food required for building of the body tissues in growth and repair, the surface area would have to be at least 20 times that of the present skin, or 3400 dm^2 rather than 170 dm^2. This extensive display of "leaves" would greatly hamper such movements as the green men would have to make, and there would be constant danger that they would be injured or broken off. The result would be that those men who, like plants, engaged in the minimum amount of movement would be in the best position to survive.

Another approach to the problem of making the green men self-sufficient would be to reduce the rate of basal metabolism and, therefore, the daily calorie requirement. One logical place to start would be the digestive tract, which no longer would be needed for taking in, digesting, and absorbing, food. However, some provision would have to be made for means of securing water and mineral salts. Perhaps the most logical proposal would be a system of projections extending into the soil—roots, if you will. This would, of course, anchor the organism to the soil and make the muscles used in moving from place to place unnecessary, saving more energy expenditure. The energy used in supplying the nervous system which controls these muscles would also be saved. The heart would still be expending considerable energy pumping blood; it might be well to replace it with the type of system by which water and food are transported by plants.

By this time we would have an organism which could easily be able to produce all its needed food, and then some, by photosynthesis. But we no longer would have an animal, let alone a man. The result would be what anyone would call a plant—something not too different from plants as they exist on Earth. In attempting to devise a workable photosynthetic animal we have lost the animal, and have gotten right back to a plant. The moral is, of course, that, in any conceivable biological scheme even remotely resembling that on Earth, plants are an absolutely essential component. There must be a group of organisms that have the ability to synthesize foods from things that are not foods. This group of organisms must also have a low enough energy requirement of their own so that they can make sufficient food to meet their own requirements and to meet those of the animal population as well.

Just how good a job are the plants of the Earth doing along this line? A few figures will tell the story. It is estimated that in a year the plants of the Earth produce some 87×10^9 *tons* of sugar by photosynthesis, that is, 87 billion tons. That photosynthesis is by far the greatest production process on Earth is indicated by the fact that the annual mining and manufacturing product of the Earth is only about 1 billion tons, and the agricultural productivity is another billion. Despite this immense photosynthetic productivity, plants use in photosynthesis only about 0.1 or 0.2% of the total light energy of the sun falling on the earth in a year.

If plants make 87 billion tons of sugar in a year, they are also producing 93 billion tons of oxygen, which are added to the air, and they are using 128 billion tons of carbon dioxide and 52 billion tons of water in photosynthesis. Without this continued addition of oxygen to the air, the atmosphere would eventually become devoid of oxygen because it would be used up in various oxidation processes. This is why astronomers consider the presence of oxygen in the atmosphere of a planet to be an indication of the presence of plant life. Since carbon dioxide constitutes only 0.03% of the air, it might seem that photosynthesis would soon exhaust it. However, the atmosphere contains about 600 billion tons of carbon dioxide.

More important, there are 50,000 billion tons of carbon dioxide dissolved in the waters of the oceans and 66 million billion tons tied up in limestone rocks. Since these supplies are in chemical equilibrium with the carbon dioxide of the air, the potential supply is inexhaustible. The amount of carbon dioxide added to the air annually by the respiration of plants as well as animals and by combustion is relatively insignificant, totaling altogether only about 33 billion tons. It is obvious that animals are not essential for maintaining the carbon dioxide level of the air for plants, as is sometimes assumed, when we realize that only 11 of these 33 billion tons of carbon dioxide are added to the air by animals, including humans.

The 87 billion tons of sugar produced annually by plants is a net figure, or what is left after plants have used perhaps a tenth of this amount in their own respiration. The 87 billion tons is used in making the cell walls and protoplasm of the plant cells; what is left from this accumulates in the plant, principally in the form of carbohydrates and fats. Animals consume only about 1 billion tons of these plant tissues as food each year; therefore, it is obvious that the photosynthetic productivity of plants is keeping well ahead of the food requirements of animals, despite the fact that some plant tissues such as wood are not fit food for most animals.

One other fact of interest in this connection is that perhaps as much as 50% of all photosynthesis is carried on by plants of the ocean. The bulk of this great ocean food production is carried on by microscopic plants.

We have wandered away from our green men and have become involved with green plants, but this is a necessity in a discourse such as this, just as it is in any workable biological scheme. If there is life on other planets, the organisms may be quite different from those on Earth. There could well be planets with no form of life comparable with our animals but, if there is life at all beyond the simplest sort of thing, like viruses or bacteria, we can be pretty sure that there will be something comparable to our plants, even though they might have a quite different anatomical structure. We can also be sure that there is no planet populated with photosynthetic animals of any size or complexity and devoid of plant-type organisms.

Reference

[1] Norman, A. G. "The uniqueness of plants," *Amer. Sci.,* **50**:436–449 (1962).

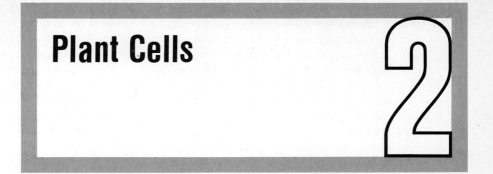

Plant Cells

Because cells are the site of metabolic processes and may be considered as the basic structural unit of organisms, it is appropriate at this point to consider the structure of plant cells and the metabolic roles of the various cell organelles. Subsequent chapters will deal with the metabolic processes of plant cells, the movement of substances into and out of cells, and the roles of cell division, cell enlargement, and cell differentiation in growth, development, and reproduction.

Some General Characteristics of Plant Cells

Except for bacteria and blue-green algae, which are procaryotes, all plants have **eucaryotic** cells, that is, cells that have discrete nuclei enclosed within a nuclear membrane. Plant cells are characterized by the presence of cell walls with cellulose as the usual fundamental structural component. In eucaryotic plant cells plastids are usually present, and in most mature living cells a large central vacuole occupies a major part of the total cell volume.

Although plant cells vary greatly in shape and size, they are generally considerably larger than animal cells. The dimensions of the cells of higher plants usually range between 10 and 200 microns (μ) although cotton fibers may be as long as 5 centimeters (cm) and the phloem fibers of ramie are as long as 55 cm. The pith cells of some species and the cells of watermelon flesh are unusually large, each dimension ranging from 100 to 200 μ. Such cells can be seen with a low-power magnifying glass or even by the unaided eye. However, most cells of higher plants are microscopic in width if not always in length. Some comprehension of the size of plant cells can be gained from the fact that an apple leaf, with an area of about 25 cm^2, contains on the order of 50,000,000 cells.

The Cells and Tissues of Vascular Plants

Vascular plants are composed of more than a dozen distinct types of cells, which are organized into tissues, tissue systems, and organs. The three major organs are the roots, stems, and leaves. Flowers, fruits, and seeds are sometimes designated as reproductive organs, but flowers (and cones) are considered to be composed of modified leaves attached to modified stems. Roots, stems, and leaves are all composed of the same three tissue systems: dermal, vascular, and fundamental.

The **dermal** system includes only the epidermis, except in roots and stems with secondary growth, where the epidermis is replaced by the periderm or corky bark. The **vascular** system is composed of two tissues, the **xylem** and the **phloem,** through which the long-distance transport of water and solutes occurs. This system is discussed in some detail in Chapter 8. The **fundamental** system is composed predominantly of parenchyma tissue, but it also includes collenchyma and sclerenchyma tissues. The fundamental tissue system of leaves is generally referred to as the **mesophyll** (Figure 2-1). In roots and stems the fundamental tissue system is commonly divided into two morphologically distinct regions: the **cortex** outside the vascular system and the **pith** inside it, although the roots of many species lack pith (Figure 2-1). In addition, in roots the part of the fundamental tissue system adjacent to the vascular system is separated from the cortex by a cylinder of specialized cells (the endodermis) and is referred to as the **pericycle.**

The **primary** tissues of vascular plants are derived from the apical meristems of the stems and roots (Figure 2-2). Many herbaceous (nonwoody) plants consist entirely of primary tissues. Some herbaceous plants and all woody plants also have **secondary** tissues derived from other meristems: secondary xylem, phloem, and ray parenchyma from the vascular cambium, and phellem (cork) and phelloderm from the phellogen (cork cambium). For our present purpose of charting the location of the various types of plant cells in the different tissues, the diagram of the primary tissues of a flowering plant (Figure 2-3) should be adequate, at least in conjunction with Table 2-1.

Before considering the principal types of cells of which vascular plants are composed, one unique characteristic should be noted. Among all organisms vascular plants are unique in having substantial numbers of cells that are functional after they are dead and consist only of cell walls. These include the tracheids and vessel elements of the xylem, through which water flows, the sclerenchyma cells that may provide mechanical support, and the cork cells that are impermeable to water.

Most tissues of vascular plants are complex, that is, they are composed of more than one type of cell, in contrast with simple tissues that are composed entirely of one kind of cell. At first glance the organization of plant cells into tissues, tissue systems, and organs may appear to be rather haphazard, but it should become clear that the arrangement of cells and tissues is by no means random and that the structural organization provides for the efficient and coordinated functioning of the plant.

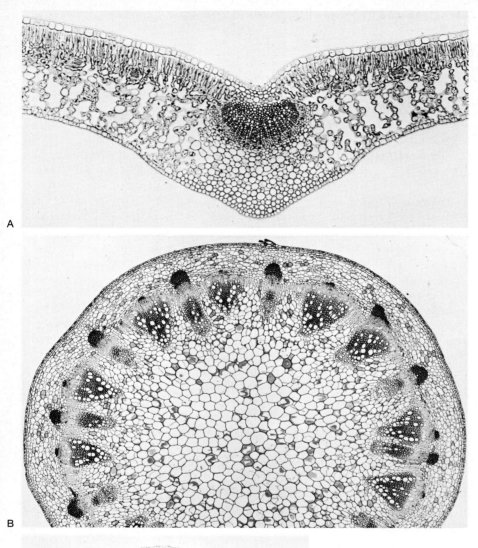

Figure 2-1. Photomicrographs of sections through **(A)** a Ligustrum (privet) leaf, **(B)** a Helianthus (sunflower) stem, and **(C)** a Ranunculus (buttercup) root showing the arrangement of their tissues. [**A** and **B** courtesy of Carolina Biological Supply Company; **C** courtesy of Ripon Microslides]

Table 2-1 Principal types of cells present in the various tissue systems and regions of vascular plants

Tissue systems and regions	Cell types												
	Meristematic	Parenchyma	Collenchyma	Epidermal	Guard	Endodermal	Sieve	Companion	Fibers	Sclerids	Tracheids	Vessel elements	Cork
Dermal													
Epidermis				x	x								
Phelloderm		x											x
Vascular													
Phloem		x					x	x	x				
Xylem		x							x		x	x	
Fundamental													
Cortex		x	x						x	x			
Endodermis		x				x							
Pericycle		x											
Pith		x											
Mesophyll		x								x			
Meristematic													
Apical meristems	x												
Vascular cambium	x												
Phellogen	x												

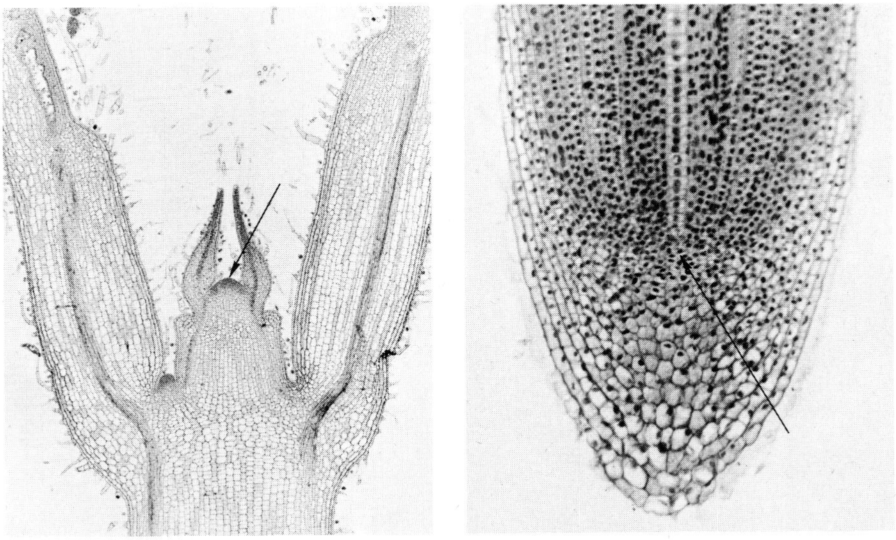

A   B

Figure 2-2. Photomicrographs of median longitudinal sections of **(A)** a Coleus stem tip and **(B)** an Allium (onion) root tip showing the apical meristems (arrows). In the stem tip note the two pairs of young leaves and (left) the primordium of an axillary bud. The tissue below the apical meristem of the root is that of the root cap. [Courtesy of the Carolina Biological Supply Company]

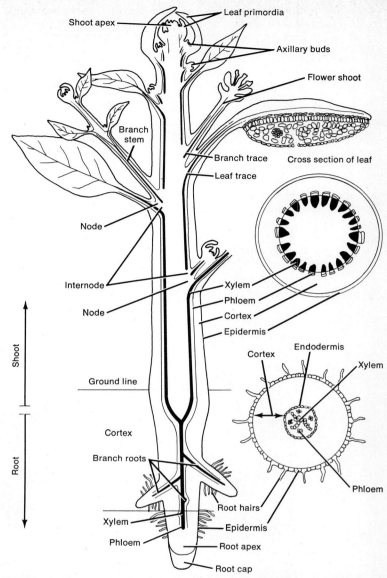

Figure 2-3. Diagrammatic longitudinal and cross sections of a young angiosperm plant showing the principal organs and primary tissues. [After W. W. Robbins, T. E. Weier, and C. R. Stocking, *Botany: An Introduction to Plant Science,* 3rd ed., John Wiley & Sons., Inc., New York, 1964]

Meristematic Cells

Meristematic cells give rise to all other cells of the plant by cell division and subsequent cell enlargement and differentiation. Embryo plants within seeds are composed predominantly, if not entirely, of meristematic cells. As a plant grows and differentiates, meristematic cells persist only in the root and stem apices, in very young leaves, and (in species that have secondary growth) in the vascular cambium. In addition, meristems may be derived from parenchyma, as in the formation of a cork cambium, the initiation of branch roots, the healing of wounds, or the development of a gall. The apical meristems and vascular cambium generally persist and continue to be sources of new cells as long as the plant lives, because not all of their meristematic cells differentiate into other kinds of cells.

Although the meristematic cells of root and stem apices vary considerably in size, shape, and structural details, they are in general approximately cubical with a width of around 10 μ. They have thin cell walls and numerous small vacuoles, which are sometimes visible only with an electron microscope. Thus, the protoplast occupies most of the cell volume, a considerable portion of it being taken up by the relatively large nucleus (Figure 2-4). Mitotic figures are often abundant in the apical meristems, but it should be noted that apices have quiescent regions where cell divisions are rare. Although the meristematic cells of the vascular cambium have many of the structural features of those in the apical meristems, they are generally long and pointed rather than cubical.

Parenchyma Cells

Parenchyma cells are the least highly specialized, the most ubiquitous (Table 2-1), and in herbaceous plants the most numerous cells derived from meristematic cells. Because of the abundance of parenchyma cells and the fact that

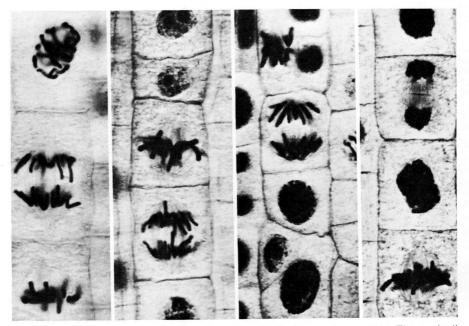

Figure 2-4. Photomicrographs of meristematic cells from an onion root tip. The nucleoli are visible in several of the nuclei (for example, first two cells at top of second row). Several of the cells are in various stages of mitosis. [Courtesy of Carolina Biological Supply Company]

many of them have chloroplasts, they are the site of the major portion of the metabolic activity of a plant, even though other types of living cells can carry on most or all of the metabolic processes. In some cases the parenchyma cells make up a simple tissue, such as most pith, the cortex of many species, the pericycle, the flesh of many fruits, and the leaf mesophyll of many species. Parenchyma cells are also constituents of most complex tissues such as xylem and phloem. Also, most of the cells of bryophytes are similar to the parenchyma cells of vascular plants.

Parenchyma cells have a considerable range of sizes and shapes. In pith and the flesh of fruits they tend to be isodiametrical and polyhedral, with an average of 14 faces. These cells often have a dimension of around 100μ or more. In the cortex, parenchyma cells are often six-sided and elongate, with proportions somewhat similar to those of a shoe box. Such cells generally range in width and breadth from 30 to 80μ and in length from 80 to 300μ. The palisade cells of leaf mesophyll are also elongate, whereas the spongy parenchyma are more nearly isodiametrical and often rather irregular in shape. Parenchyma cells have from 50 to 1000 or more times the volume of the meristematic cells from which they developed. As a somewhat extreme example, a cell in the flesh of a watermelon has a volume of about 350,000 times that of a meristematic cell.

Despite the variation in size and shape of parenchyma cells, their structural

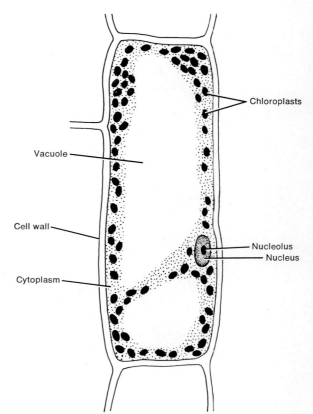

Figure 2-5. Drawing of a parenchyma cell as seen under a microscope.

features are quite uniform. Their walls are thicker than those of meristematic cells, and they have large central vacuoles that occupy a major portion of the cell volume, the protoplast thus being limited to a thin peripheral layer (Figure 2-5). However, strands of cytoplasm may traverse the vacuole. Although the nucleus of a parenchyma cell is as large or larger than that of a meristematic cell, it occupies a much smaller fraction of the cell volume. Since the nuclei are compressed within the thin layer of cytoplasm they are ellipsoidal in a side view, but in a face view they appear circular. Plastids are generally a conspicuous structural feature of parenchyma cells and are distributed throughout the cytoplasm. Those parenchyma cells that contain chloroplasts are often called **chlorenchyma** cells. The term is useful from a physiological standpoint, since it identifies cells capable of carrying on photosynthesis.

Collenchyma Cells

Collenchyma cells differ from parenchyma cells in having a substantial amount of wall thickening, principally in the corners of the walls. Collenchyma tissues occur as strands or cylinders in the outer region of the cortex of stems and petioles and along the larger veins of leaves. Some species of plants, in particular many monocotyledons, have no collenchyma. The thickened walls of collenchyma cells provide mechanical support. The protoplasts are essentially the same as those of parenchyma cells and may, or may not, contain chloroplasts.

Epidermal Cells

The epidermis is a tissue that is only one cell thick and constitutes the surface layer of all of the organs of vascular plants, except for woody stems and roots that have bark. The epidermis is composed of several types of cells, the most numerous being the epidermal cell proper. This type of cell is generally tabular, that is, considerably thinner than its length and breadth. In face view the epidermal cells of the shoot (the stems with their attached leaves and flowers) vary greatly in size and shape from species to species, for example, they may be roughly hexagonal, elongated, or irregular in outline like the pieces of a jigsaw puzzle (Figure 2-6). A characteristic structural feature is the **cuticle,** a layer of waxes and other lipids secreted by the cells on their outer surfaces. The mixture of substances making up the cuticle is called **cutin** [8]. There is considerable species variation in the thickness, compactness, and structure of the cuticle and thus in its permeability to water and other substances. However, cuticles in general are quite impermeable to water and are, therefore, a most important factor in the survival of plants with aerial shoots. The epidermal cells of most species lack chloroplasts.

The epidermal cells of roots do not have a cuticle and are generally regular in outline and elongated in the direction of the root axis. Some of them have long projections, the **root hairs,** that greatly increase their absorbing surface. Root hairs range between 5 and $17\,\mu$ in width and between $80\,\mu$ and 15 millimeters (mm) in length when mature, depending on the species. They are generally limited to the younger fully-elongated epidermal cells, and thus form

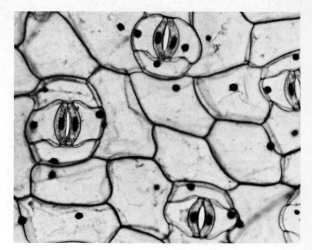

Figure 2-6. Photomicrograph of face view of epidermis of Tradescantia, showing open stomata and the subsidiary cells surrounding the guard cells. [Courtesy of Carolina Biological Supply Company]

a zone about 1 to 3 cm long beginning about 0.5 cm behind the root tip. Since the root hairs continue to elongate for some time, there is a progressive increase in length from the youngest to oldest part of the root hair zone. Also, there is continuing death of the older root hairs and formation of new ones; therefore, the root hair zone remains about the same length and distance from the root

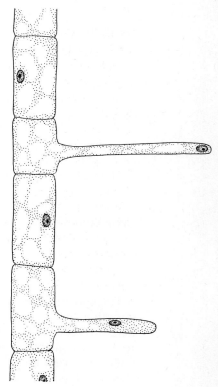

Figure 2-7. Diagrammatic drawing of a radial longitudinal section through the epidermis of a root showing cells with root hairs alternating with cells that lack root hairs.

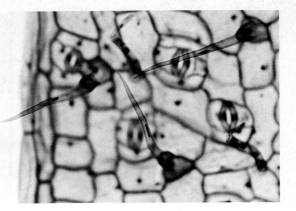

Figure 2-8. Photomicrograph of Tradescantia epidermis showing several trichomes. [Courtesy of Carolina Biological Supply Company]

tip as a root grows. Not every epidermal cell in the root hair zone develops a root hair (Figure 2-7). In some species every other epidermal cell in a longitudinal series has a hair, whereas in other species it is every third cell.

Some epidermal cells of stems and leaves also have hairs **(trichomes).** Trichomes are particularly abundant in some species, giving them a hairy appearance. Trichomes may be either unicellular or multicellular; the multicellular type are often complex and highly branched (Figure 2-8). Trichomes may also take the form of stalked discs. Some trichomes are glandular, and they secrete substances characteristic of a species, including essential oils, gums, mucilages, or resins.

Guard cells are important components of the epidermis of stems, leaves, and flower parts. They occur in pairs, with a **stomatal pore** between them, the complex constituting a **stomatal apparatus** (Figure 2-6). The guard cells of various species have different shapes, the most common being like a bean seed or a kidney. Their concave sides are adjacent, thus forming the stomatal pore or stoma (stomata, plural). In some species the guard cells are surrounded by epidermal cells that are generally somewhat smaller and different in shape from the other epidermal cells. These are called **subsidiary cells.** When the turgor pressure of guard cells decreases their walls come together, thus closing the stoma. Stomata are generally closed at night and when there is a water deficit. If the stomata are closed, there is little or no loss of water vapor from the plant by transpiration. Thus, the guard cells as well as the cutinized epidermal cells make an important contribution toward preventing the desiccation of land plants and so favoring their survival. The stomata are also important as pathways through which carbon dioxide and oxygen can diffuse readily into and out of the plant.

Endodermal Cells

In roots the innermost cylinder of cells of the cortex (or perhaps it is the outermost cells of the pericycle) constitutes a structurally distinct layer known as the endodermis. The endodermal cells differ from ordinary parenchyma cells in having their radial and tangential walls impregnated in a band or strip (the

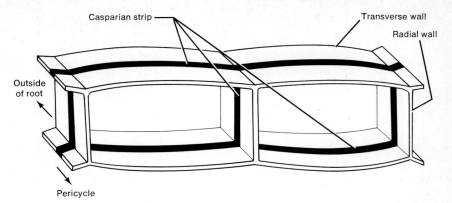

Figure 2-9. Three-dimensional drawing of the walls of two endodermal cells showing the location of the suberized Casparian strip, which is impermeable to water. As a result, the water and its solutes moving from the cortex to the stele must pass through the protoplasts of the endodermal cells. [After T. Weier, C. R. Stocking, and M. G. Barbour, *Botany,* 4th ed., John Wiley & Sons, Inc., New York, 1970]

Casparian strip) with lignin and suberin (Figure 2-9). This strip is essentially impermeable to water and prevents the movement of water through the walls but not through the protoplasts. Many botanists believe that the endodermis plays an essential role in the movement of water and solutes from the cortex into the vascular tissues.

Sieve Cells and Sieve-Tube Elements

From the standpoint of both structure and function, the sieve cells and sieve-tube elements can be regarded as the most highly specialized living cells of plants. These cells are found only in the phloem and function in the rapid, long-distance translocation of solutes. Sieve-tube elements are found in angiosperms. These cells are arranged end to end, thus forming a long, multicellular sieve tube that is easily recognizable. Sieve cells are found in gymnosperms and lower vascular plants. Although they adjoin one another, they do not form a definite sieve tube.

Sieve cells are long (generally 1 to 2 mm) and narrow with pointed ends that dovetail with those of other sieve cells. In these dovetailed end walls, and to a lesser extent in the side walls, there are depressed areas that are roughly circular in outline, the **sieve areas.** Each of these contains numerous pores through which strands of cytoplasm connect the cytoplasm of adjacent cells. Toward the ends of the cells the sieve areas occur in compact groups, each group constituting a compound **sieve plate.**

The sieve-tube elements of angiosperms are generally shorter than sieve cells, ranging from somewhat less than 1 mm to as little as 0.13 mm in length. They are generally broader than sieve cells, and their end walls are more nearly at right angles to their side walls (Figure 2-10). Their sieve plates are usually restricted to the end walls. In some species the sieve plates are compound, but in most species the sieve plate consists of a single sieve area that occupies most of the end wall.

The protoplasts of sieve cells and sieve-tube elements are essentially similar. The very young cells have the structural features of parenchyma cells but, as they differentiate and mature, numerous changes occur. The cells become wider and longer, the end walls become perforated and thus the sieve plates differentiate, and the mitochondria, endoplasmic reticulum, vacuolar membrane, and plastids become reduced and may even disappear. Perhaps the most striking event is the complete disintegration and disappearance of the nucleus, although isolated nucleoli may persist for a time. Sieve cells and sieve-tube elements are the only cells known to remain alive and active for several months or even a year or so without a nucleus. The young cells contain proteinaceous slime bodies that fuse into poorly defined longitudinal strands as the cells mature.

Each sieve-tube element of angiosperms is intimately associated with one to three parenchymatous cells located in a longitudinal series on one side of the sieve-tube element (Figure 2-10). These are called **companion cells.** The sieve-tube element and a companion cell are derived from the same parent cell. As the sieve-tube element elongates, its companion cell may divide transversely into two, or more rarely three, companion cells. The companion cells have the usual organelles of parenchyma cells, including a nucleus (which is usually polyploid). The role of companion cells is not definitely known, but it seems probable that the nucleus of a companion cell may serve also the associated enucleate sieve-tube element; this could explain the rather long life of a sieve-tube element after the loss of its nucleus. Sieve cells do not have companion cells derived from the same parent cell, but it has been suggested that some of the phloem parenchyma cells could play a comparable role here.

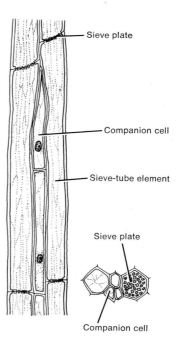

Figure 2-10. Drawings of longitudinal and cross sections of sieve-tube elements and companion cells. [After V. A. Greulach and J. E. Adams, *Plants: An Introduction to Modern Botany,* 2nd ed., John Wiley & Sons, Inc., New York, 1967]

Laticifers

The laticifer is another highly specialized type of living cell found in some species of vascular plants, principally in the composite, poppy, spurge, morning glory, lily, milkweed, and mulberry families. Laticifers produce and accumulate **latex,** a milky or clear liquid consisting of a high concentration of various substances dispersed in water. The substances present in latex vary with the species. Rubber latex is best known, but other terpenes, essential oils, resins, camphor, tannins, sugars, or proteins may also be present. Poppy latex contains alkaloids, whereas that of the tropical pawpaw contains papain, a proteolytic enzyme used as a meat tenderizer.

The laticifers form laticiferous ducts that may be either unicellular or multicellular in origin. The laticifer of a unicellular duct generally elongates greatly, forming a long tube that may be branched. The multicellular ducts are composed of long chains of cells that may or may not branch. In some species the end walls of these cells remain intact, whereas in others the end walls disintegrate, forming a long, continuous tube. The unicellular laticifers may undergo mitosis without cytokinesis, thus forming a multinucleate cell. The resin ducts of pines and other gymnosperms are not laticifers but rather are tubular intercellular spaces lined with secretory parenchyma cells.

Sclerenchyma Cells

The remaining types of specialized cells that we will consider are all dead at the time they are functioning. They consist only of thickened cell walls. These include sclerenchyma, tracheids, vessel elements, and cork cells.

The various types of sclerenchyma cells differ greatly in size and shape, but all have very thick cell walls that presumably provide mechanical support or protection. The two principal types are fibers and sclerids.

Fibers are long, narrow cells with pointed ends (Figure 2-11). They are most abundant in the xylem and phloem, although they are also present in the cortex of some species. Wood fibers are generally abundant in the xylem of angiosperms but are lacking in the wood of gymnosperms and the lower vascular plants. Phloem fibers and cortical fibers are often referred to as bast fibers, particularly when they are utilized commercially for making ropes or textile fibers such as linen. The long axis of a fiber is oriented along the length of a vascular bundle or cortex.

Sclerids (Figure 2-12) vary greatly in size and shape but are more nearly isodiametrical than fibers. Sclerids are abundant in some species and rare or absent in others. They may be found in almost any tissue of vascular plants, perhaps most often in the cortex, the mesophyll or epidermis of leaves, seed coats, and some of the tissues of fruits. Except for wood, essentially every hard tissue of plants is composed partly or entirely of sclerids. Examples are hard seed coats, nut shells, and other dry fruits, and the stone or pit of fruits such as cherry and peach. The grains of pear fruits are compact clusters of isodiametric sclerids called **stone cells.**

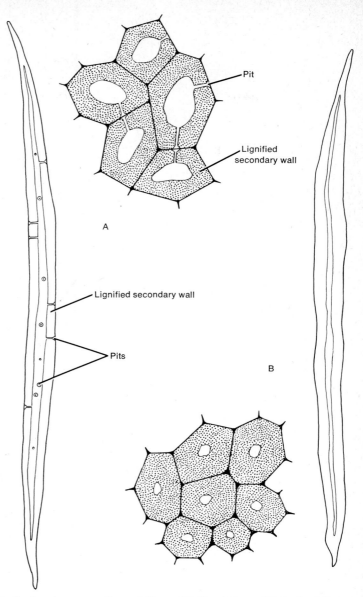

Figure 2-11. Longitudinal and cross sections of fibers. **(A)** Wood fibers. **(B)** Cortical fibers. [After V. A. Greulach and J. E. Adams, *Plants: An Introduction to Modern Botany,* 2nd ed., John Wiley & Sons, Inc., New York, 1967]

Tracheids and Vessel Elements

Tracheids and vessel elements are restricted to the xylem; their relatively large lumens (the space formerly occupied by the protoplasts) serve as tubes through which water flows. Vessel elements are not present in gymnosperms or the lower vascular plants. The great bulk of their xylem or wood is com-

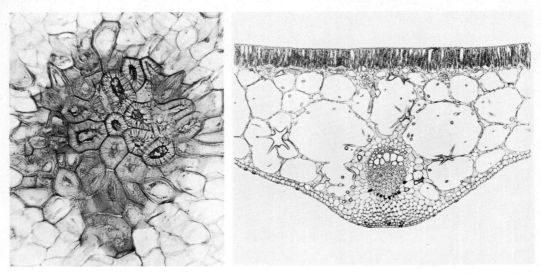

Figure 2-12. Two of the various types of sclerids. **Left:** A cluster of stone cells in the flesh of a pear fruit, surrounded by parenchyma cells. **Right:** Stellate sclerids in the mesophyll of a Nymphaea (pond lily) leaf. Note also the network of cells and the large intercellular spaces in the spongy mesophyll. [Courtesy of Ripon Microslides]

posed of tracheids. Fibers are also absent. Although tracheids may be present in some angiosperms, their predominant water-conducting cells are vessel elements.

Like fibers, tracheids are elongated cells with pointed ends, but they are generally wider and have thinner walls and larger lumens (Figure 2-13) than fibers. The walls generally contain numerous **pits,** small circular areas without secondary wall thickening. The pits in many species are of the complex bordered type (Figure 2-14). The pits of adjacent tracheids are opposite one another, forming a pit pair. Instead of pits, some tracheids have wall thickenings restricted to rings or spirals (Figure 2-13). As in fibers, the pointed ends of tracheids dovetail with one another.

Vessel elements (Figure 2-15) are usually much shorter and wider than tracheids and have truncate ends. They are arranged end-to-end in a longitudinal series. As they differentiate from their parenchymalike precursors, vessel elements lose their protoplasts and in most cases their end walls, thus forming a long hollow tube, the **vessel.** In some species the vessels (from one remaining end wall to another) are a meter or more in length, whereas in other species they range down to only 2 or 3 cm in length. Vessel elements have a considerable variety of secondary wall thickening patterns (Figure 2-16).

It is believed that both vessel elements and wood fibers have evolved from tracheids, each assuming one of the two functional roles of tracheids: a pathway through which water flows and mechanical support. This hypothesis is supported by the fact that the vessel elements of the more primitive angiosperms are more similar to tracheids than those of the more advanced ones.

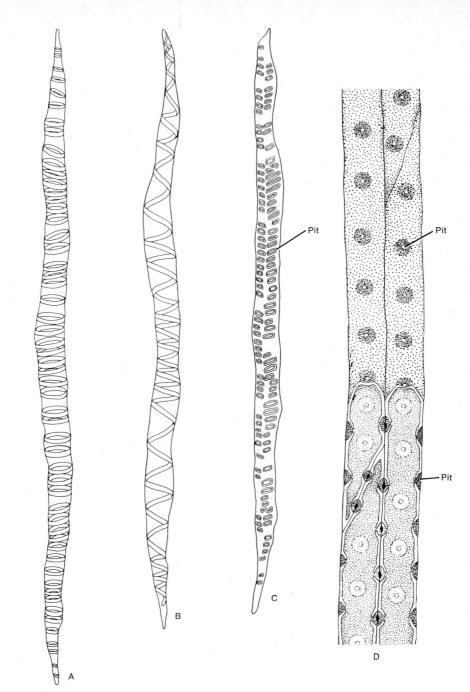

Figure 2-13. Tracheids with several types of secondary wall thickening: **(A)** annular, **(B)** spiral, **(C)** scalariform, and **(D)** pitted. [After V. A. Greulach and J. E. Adams, *Plants: An Introduction to Modern Botany,* 2nd ed., John Wiley & Sons, Inc., New York, 1967]

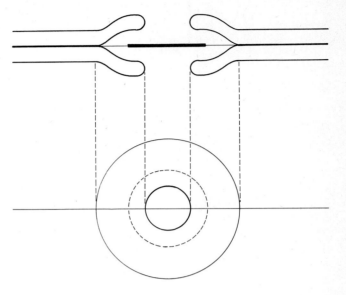

Figure 2-14. Diagrammatic representation of a bordered pit in sectional view (above) and face view (below). Note the torus, which is the thickened central portion of the closing membrane.

Figure 2-15. Photomicrograph of vessel elements and fibers from macerated wood. [Courtesy of the copyright holder, the General Biological Supply House, Chicago]

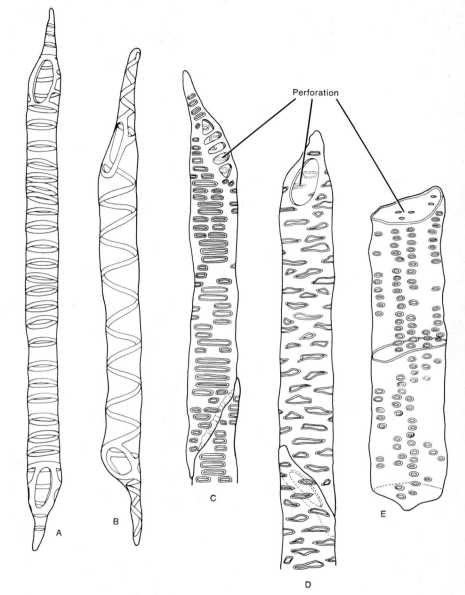

Perforation

Figure 2-16. Vessel elements, showing various shapes, sizes, and types of secondary wall thickening: **(A)** annular, **(B)** spiral, **(C)** scalariform, **(D)** reticulate, **(E)** pitted. [After V. A. Greulach and J. E. Adams, *Plants: An Introduction to Modern Botany,* 2nd ed., John Wiley & Sons, Inc., New York, 1967]

Cork Cells

Cork tissue (Figure 2-17) is an important component of the outer bark of the stems and roots of woody dicotyledons and gymnosperms. This tissue may also be present in the older parts of the stems of herbaceous dicotyledons that have secondary growth in diameter. The relatively few species of woody monocotyledons may also have cork, but its origin in these plants is different from that in other woody plants as described below [9]. Only the cork oak tree has sufficiently thick layers of pure, high-quality cork to be commercially valuable.

Cork (or **phellem**) is derived from a meristem (the **phellogen**) that differentiates first from cortical cells (Figure 2-17) and later from parenchyma cells of the phloem. The phellogen develops in patches that later coalesce and, as growth in diameter continues, new phellogens arise progressively deeper in the cortex and phloem. In addition to producing multiple layers of cork cells toward the outside, the phellogen commonly cuts off from one to several layers of parenchyma cells toward the inside. This tissue is known as the **phelloderm.** The phellogen and the phellem and phelloderm it produces are referred to collectively as the **periderm.**

Cork cells are compactly arranged, with essentially no intercellular spaces except in the lenticels (Figure 2-17). **Lenticels** are specialized regions where the cells are loosely arranged and provide a pathway through which gases can diffuse. As cork cells mature they impregnate their walls with **suberin**, a complex

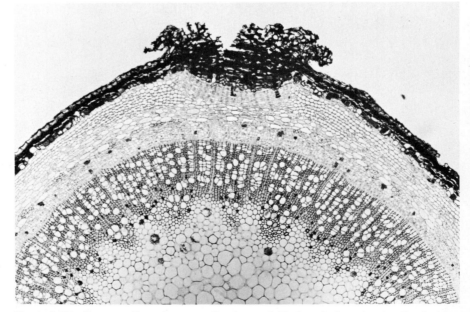

Figure 2-17. Cross section of a young Sambucus (elderberry) stem showing the first few layers of cork cells outside the cortex and a lenticel (top center). [Courtesy of Carolina Biological Supply Company]

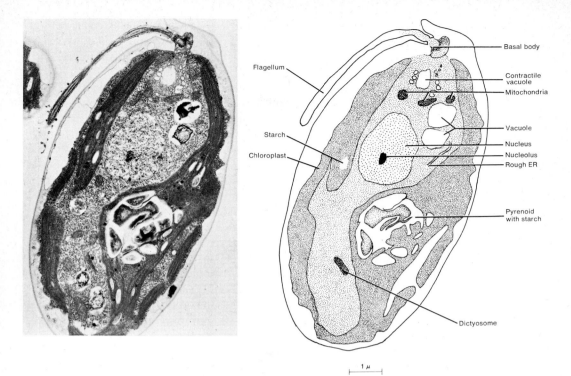

Labels in figure: Basal body, Contractile vacuole, Mitochondria, Vacuole, Nucleus, Nucleolus, Rough ER, Pyrenoid with starch, Dictyosome, Flagellum, Starch, Chloroplast

1 μ

Figure 2-18. Electronmicrograph of a Chlamydomonas (a green alga) cell, showing the principal organelles characteristic of eucaryotic plant cells. [Courtesy of R. Malcolm Brown, Jr.]

mixture of waxes and other lipids, and then their protoplasts die. The suberized walls are impermeable to water and, essentially, to all other substances. Thus, cork prevents excessive water loss from woody stems.

Organelles of Plant Cells

We now turn to a consideration of the various organelles of plant cells. Although the discussion centers on meristematic and parenchyma cells, the protoplasmic organelles described are also generally characteristic of other living cells of vascular plants and, to a considerable extent, of the eucaryotic cells of nonvascular plants. Indeed, with several exceptions such as plastids and cell walls, the organelles are characteristic of eucaryotic cells in general (Figure 2-18).

A major portion of our knowledge of the structure of cell organelles has been accumulated during the past 30 years by the use of electron microscopes and also light microscopes with improved optical systems, including interference contrast and phase contrast. In association with these studies, biochemical investigations have provided much information regarding the chemical nature and the functions of various organelles. It has become increasingly evident that many metabolic processes of cells are localized in certain specific organelles. We shall consider both the structure and the known functions of each organelle. More detailed information than can be presented here can be found elsewhere [3, 4].

A plant cell consists basically of the enclosing cell wall and the living protoplast. The protoplast is commonly considered to be composed of the nucleus and the cytoplasm, although the nucleus could just as logically be regarded as coordinate with the other protoplasmic organelles. The part of the cytoplasm surrounding the various organelles is known as the cytoplasmic matrix or **hyaloplasm.** This is basically a complex of proteinaceous hydrophilic colloids that can undergo phase reversals from a sol to a gel or a gel to a sol, with both phases often present in adjacent regions. Numerous solutes are also present. In the sol phase the hyaloplasm is often flowing, most conspicuously in the cytoplasmic streaming **(cyclosis)** in cells with large vacuoles. Cytoplasmic inclusions and organelles such as plastids are carried along by the flowing cytoplasm. In addition to the protoplast and the cell wall, a plant cell contains inclusions such as starch grains, crystals of proteins, crystals of salts (notably calcium oxalate), oil globules, and granules of fat-soluble substances. The **cell sap** of vacuoles, a water solution of sugars, salts, and often anthocyanins and other substances, is also an inclusion. These inclusions, along with the cell wall, are referred to as **paraplasmic** (or **ergastic**) substances.

The principal structural components of the protoplast, in addition to the water that usually constitutes over 90% of its weight, are proteins, phospholipids, and nucleic acids. Chlorophyll and other pigments are also important constituents of some cells. In addition, there is an immense variety of other substances present, including hormones, vitamins, coenzymes, and many different metabolic substrates and products.

The Nucleus

The nuclei of meristematic cells generally range between 7 and $10\,\mu$ in diameter, whereas those in parenchyma and other enlarged cells are usually 35 to $50\,\mu$ in diameter. However, the nuclei of some plant cells are much larger, notably those of Cycad zygotes which may reach $1000\,\mu$ (1 mm) in diameter, and so are visible without magnification. At the other extreme are the nuclei of fungi, which may be as small as $1\,\mu$ in diameter.

A nucleus (Figure 2-19) is enclosed within its nuclear membranes (or **nuclear envelope**) and contains **chromatin,** which appears to be in the form of threads and blobs, and one or more **nucleoli.** The rest of the nuclear volume is occupied by the nuclear sap or nucleoplasm, a rather viscous sol. In most eucaryotic plant cells the nuclear membrane and nucleoli disintegrate during mitosis (Figure 2-4) and meiosis, spindle fibers form, and the chromatin condenses into discrete chromosomes. Since a basic knowledge of mitosis is assumed it will not be described in detail, but the changes in nuclear structure during mitosis should not be ignored. Note that, except in plant cells with flagella, most dividing plant cells do not have centrioles as do dividing animal cells. Also, dividing plant cells lack the astral rays that are a conspicuous feature of mitosis in animal cells.

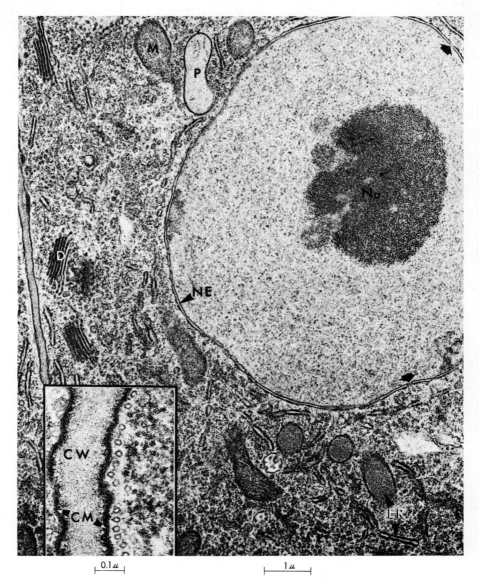

├ 0.1µ ┤   ├ 1µ ┤

Figure 2-19. Electron micrograph showing a section through the nucleus of a root tip cell of *Arabdopsis thaliana* (mouse ear cress). Note the nucleolus (Nu) and the nuclear envelop (NE) or membrane, with pores in the membrane at the points of the arrows. In the cytoplasm there are mitochondria (M), a proplastid (P), rough endoplasmic reticulum (ER), and dictysomes (D). The insert (lower left) shows a section through cell walls (CW) and cell membranes (CM) of root tip cells of *Phleum pratense* (timothy). The circles are cross sections of microtubules. [Electron micrographs by M. C. Ledbetter. Reprinted with permission of The Macmillan Company from *Plant Structure and Development* by T. P. O'Brien and Margaret E. McCully. Copyright © 1969 by The Macmillan Company]

CHROMATIN AND CHROMOSOMES. Chromatin is composed of nucleoproteins. Analysis of chromatin from pea embryos showed that it contained 36.5% deoxyribonucleic acid (DNA), 9.6% ribonucleic acid (RNA), 37.5% histone protein, and 10.4% other proteins. As is now well known, the DNA carries the genetic codes that constitute the genes or hereditary potentialities of an organism. Although DNA is also present in the cytoplasm, particularly in chloroplasts and mitochondria, the great bulk of the hereditary potentialities are in the chromosomal DNA.

DNA plays two essential roles. First, prior to each mitotic division it is replicated precisely, thus providing each daughter cell with the same complete set of genes present in the parent cell. DNA replication may occur even when mitosis or meiosis does not. Second, the DNA serves as a template for the synthesis of the RNA of the cell, thus transcribing the DNA codes into complementary RNA codes. In turn, the messenger RNA (mRNA) code determines the sequence in which various amino acids are linked together during protein synthesis. There will be more about this in Chapter 5. For the present, we merely wish to emphasize the role of the chromatin as the carrier and transcriber of genetic information.

During mitosis and meiosis the chromatin is organized into discrete chromosomes. Four distinct parts of chromosomes are visible under light microscopes: centromeres, euchromatin, heterochromatin, and nucleolar organizers. The **centromere** is the point of attachment of a spindle fiber and is constricted. The **euchromatin** constitutes the bulk of the chromosome and is apparently the site of most of the active genes. **Heterochromatin** remains condensed during the interphase, uncoils briefly near the beginning of mitosis, and then condenses before the euchromatin does. It synthesizes DNA at a different time than the euchromatin. It is often localized in small blocks along the length of the chromosome, particularly near the centromere, nucleolar organizers, and the ends. Few genes are found in the heterochromatin. Some of the constrictions in metaphase chromosomes are **nucleolar organizers** and, as the name implies, are evidently involved in formation of nucleoli.

Although the euchromatin is unraveled during interphase and distinct chromosomes are not visible, the chromosomes presumably retain their integrity during interphase. The blobs of heterochromatin in interphase nuclei are called **chromocenters.** These are absent in some species. The pattern of chromatin filaments and chromocenters varies considerably from species to species [3].

NUCLEOLI. The nucleoli (Figure 2-19) consist of about 11% RNA, 84% protein, and 5% DNA. Much of the DNA is in chromocenters associated with the nucleoli. The nucleoli have two components immersed in a protein matrix: convoluted filaments and granules. Both contain RNA. Although the roles of nucleoli are not known with any great degree of certainty, there is a variety of evidence to the effect that ribosomes may be assembled there and subsequently transported to the cytoplasm. Presumably the ribosomal RNA is made by the chromatin, but the ribosomal protein may be synthesized in the nucleoli. It has also been proposed that most of the nuclear protein synthesis occurs in the nucleoli. The disappearance and subsequent reappearance of nucleoli during mitosis may well reflect periods of their metabolic activity.

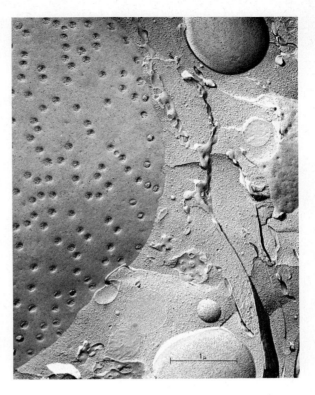

Figure 2-20. Freeze-etch preparation of an onion root tip cell showing the pores in the nuclear membrane (left). To the right of the nucleus is a strip of endoplasmic reticulum. [Courtesy of Daniel Branton]

THE NUCLEAR ENVELOPE. The nuclear envelope (Figure 2-19) is composed of two separate membranes. Each membrane is a bimolecular layer of phospholipids and proteins. There may be evaginations of the outer membrane alone or of both membranes. The outer membrane is continuous at many places with the endoplasmic reticulum (ER), which is also composed of two membranes and permeates the cytoplasm. The nuclear membrane has numerous circular pores (Figure 2-20). The inner membrane joins the outer one at the circumference of a pore. In onion root tip cells, for example, there are 35 to 65 pores/μ^2, and their diameters range from 80 to 200 nanometers (nm),* although those in any one nucleus are about the same size. The pores are not just holes in the membrane; they have a structure of their own. A circle of eight globular subunits is located adjacent to the circumference of the pore (Figure 2-21) and often a central dot is present. This is probably mRNA passing through the pore on its way from the nucleus to the cytoplasm, as indicated by the fact that RNAase removes the central dots.

Ribosomes

Ribosomes [19] are very small nonmembranous particles (15 to 30 nm in diameter) (Figure 2-22). They are found in nuclei and are probably produced

* 1 nanometer = 10^{-9} meter = 10^{-3} microns = 1 millimicron = 10 angstrom units. See appendix.

Figure 2-21. Electronmicrograph of a freeze-etch preparation of Chlamydomonas showing the pores in the nuclear membranes (viewed from inside the nucleus as indicated by the shadowing). A fragment of the inner membrane (below) is superimposed on the outer membrane. Note the structural details of the pores. Also shown are a part of a chloroplast with its lamellae (upper right), vacuoles (below and right of nucleus), plasmalemma, and separated cell wall (right). At the extreme lower right (arrow) is a cross fracture of the flagellum, showing the microtubules within it. [Courtesy of R. Malcolm Brown, Jr.]

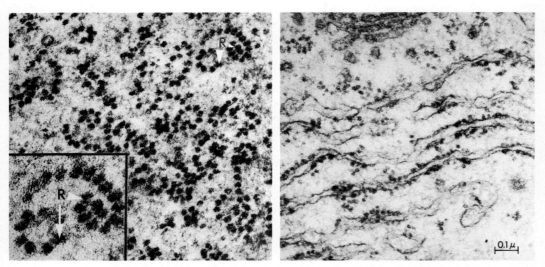

Figure 2-22. Rough endoplasmic reticulum in an epidermal cell of an oat coleoptile, showing ribosomes (R). **Left:** Surface view. **Insert:** Higher magnification of one of the polyribosomal aggregates. **Right:** Profile view. [Reprinted with the permission of The Macmillan Company from *Plant Structure and Development* by T. P. O'Brien and Margaret E. McCully. Copyright © 1969 by The Macmillan Company. Originally reproduced with permission from *Protoplasma,* **63:**405 (1967)]

there, but they are more abundant in the cytoplasm where they may be free or attached to the ER. Ribosomes are also present in mitochondria and chloroplasts, such ribosomes being produced within these organelles.

Ribosomes are composed of two subunits of unequal size. These subunits can separate from one another and then reattach. The ribosomes of mitochondria and chloroplasts are somewhat smaller than the other ribosomes of plant cells and are the same size as those of bacteria and blue-green algae. They generally have a sedimentation coefficient of 70 S* with subunits of 30 S and 50 S, whereas the ribosomes on the ER have about 80 S with 40 S and 60 S subunits. (The sedimentation coefficients of the subunits total more than the entire ribosome because shape, as well as weight, influences the rate of sedimentation.) Ribosomes are composed of about 50 to 60% proteins and 40 to 50% RNA. In 30 S subunits there is one RNA molecule and some 20 different protein molecules, whereas in 50 S there are two RNA molecules and over 30 proteins.

Ribosomes are essential constituents of the protein synthesis mechanism. Both the mRNA molecule coding for a protein and the transfer RNA (tRNA)–amino acid complexes must be attached to a ribosome for protein synthesis to occur, amino acids being incorporated in the growing polypeptide chain only at the point where the mRNA is attached to a ribosome. One mRNA molecule is generally attached to a linear series of 4 to 20 or more ribosomes with polypeptide synthesis occurring at each attachment. The series of ribosomes is referred to as a **polyribosome.**

Microtubules

Cells contain numerous microtubules [15], which are elongated, proteinaceous, nonmembranous cylinders (Figure 2-24). Microtubules range from 20 to 27 nm in diameter and are of undetermined length, although they have been traced through a distance of several microns. There are at least three different kinds of microtubules that differ from one another in size, structure, and function: cytoplasmic, spindle, and flagellar. In cross section a microtubule appears as a circle with a hollow center and with a circle of protein subunits (usually 13) around the periphery (Figure 2-19). Microtubules are generally straight and appear to be rather rigid and brittle. When they bend it is in an arc of large radius.

In plant cells the cytoplasmic microtubules are usually most abundant near the plasma membrane, and these are usually parallel with the cellulose fibers in the inner layer of the wall. It has been proposed that the cytoplasmic microtubules play a role in cytoplasmic streaming, particularly as regards its direction. The spindle microtubules are abundant in the nuclear region of dividing cells and play several roles in cell division. The spindle fibers are evidently composed of microtubules, and microtubules are also found perpendicular to the equatorial plane prior to cell plate formation. They apparently

---

* S is the abbreviation for Svedberg unit, measure of the rate of sedimentation of molecules and other small particles in a centrifugal field. $1 S = 10^{-13}$ cm/sec. The sedimentation rate is proportional to particle weight and shape.

play a role in the formation of the cell plate and perhaps of the middle lamella. Flagella and cilia [27] have circles of flagellar microtubules in them, and these function in the movement of the flagella and cilia. Flagellar movement and spindle fiber contraction is believed to result from the sliding of one micro-tubule along another, thus decreasing the length of the fibril.

Membranes

Every protoplast is limited by a membrane that makes an essential contri-bution toward maintaining the structural integrity of the cell (Figure 2-18). This outer cytoplasmic membrane is referred to as the **plasmalemma** or **plasma membrane.** In plant cells it is just inside of the cell wall. A second membrane, the **tonoplast** or **vacuolar membrane,** surrounds each vacuole. The third mem-brane, or more properly system of membranes, that we will consider in this section is the **endoplasmic reticulum.** The endoplasmic reticulum is a complex, convoluted, and branched system of double membranes that permeates the cytoplasm in the form of flat sacs or tubes. In addition, the Golgi apparatus, the plastids, the mitochondria, and the other cytoplasmic organelles that we will soon consider are all membranous.

Membranes are not only a pervasive structural component of a protoplast but are also the site of several essential functions. They control the movement of various substances into and out of the cell and from one part of the cell to another, a matter of critical importance in the maintenance of life. This control results partly from the fact that membranes are differentially permeable, permitting some substances but not others to diffuse through them. Also, some substances are carried across membranes by active transport systems in the membranes. Such active transport requires energy, which must be provided by respiration. Plasma membranes may have small invaginations (resolved only by the electron microscope) that engulf viruses and macromolecules (along with other particles and water) and then pinch off as a membrane-bounded vesicle, thus introducing into the cytoplasm particles that could otherwise not penetrate the membrane. This process **(pinocytosis)** was once thought not to occur in plant cells, but there is increasing evidence that it may. A similar but reverse process results in the delivery of cell wall components from the cytoplasm to the wall. In addition to controlling the movement of substances, membranes are the site of various biochemical processes. For example, many of the steps in photosynthesis and respiration are associated with the membranes of chloro-plasts and mitochondria respectively.

Membranes are composed predominantly of phospholipids and proteins. Calcium is also an essential constituent and, in calcium-deficient plants, the membranes disintegrate. The structural organization of the phospholipid and protein components of membranes is not known, but several hypothetical models of membrane structure have been proposed [14, 28]. One is a three-layered unit-membrane model. In this each of the two outer layers consists of proteins and the polar (water soluble) portion of the phospholipid molecules. The nonpolar (fat soluble) fatty acid portion of the phospholipid molecules makes up the middle layer. This model has been supported by the fact that, in electron micrographs, cross sections of membranes show a three-layered structure, with two electron-dense layers separated by a middle electron-transparent layer. Although this model gained wide acceptance, there is now increasing evidence in support of the mosaic model of membrane structure (one of the earliest models proposed). The current fluid mosaic model of membrane structure also proposes a phospholipid bilayer with the polar portion of each molecule at the membrane surface and the nonpolar fatty acid portions in the middle of the membrane. However, the protein molecules are located at intervals in the continuous phospholipid portion of the membrane, thus forming a mosaic, rather than being located in the two outer layers (Figure 2-23). The globular protein molecules project beyond the phospholipid surface, either on both sides or on only one side. The differential permeability of cell membranes can be explained much more readily on the basis of the mosaic model than the unit-membrane model. The three-layered membranes visible in electron micrographs are also in accord with the mosaic model.

PLASMALEMMA.   Since much of the above discussion relates primarily to the plasmalemma, further consideration of it here will be limited. It should be noted, however, that because of its peripheral location the plasmalemma is the membrane that influences the movement of substances into and out of the cell. The plasmalemma of a plant cell is connected with that of adjacent cells by the **plasmodesmata,** fine strands of cytoplasm that extend from one cell to another through numerous canals that penetrate the walls (Figure 2-24). These canals have an average diameter of less than 0.05 μ. The plasmodesmata also contain cytoplasmic matrix and endoplasmic reticulum. The protoplasmic continuity of plant cells provided by the plasmodesmata, despite the fact that each cell is enclosed by its wall, is probably a factor of considerable physiological significance.

TONOPLAST AND VACUOLE.   The tonoplast is generally somewhat thinner than the plasmalemma, having a thickness of 5 to 7 nm in fixed preparations. Its permeability may differ from that of the plasmalemma, either quantitatively or qualitatively. The anthocyanin, flavone, and flavonol pigments are generally

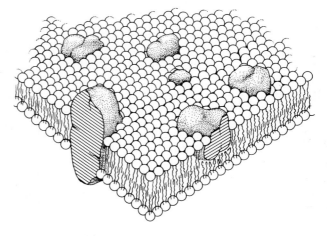

Figure 2-23. The lipid-globular protein fluid mosaic model of cell membranes. The spheres represent the polar portions of the phospholipid molecules that constitute the lipid matrix whereas the wavy tails represent the nonpolar fatty acid residues of the molecules. Note the precise orientation of the molecules in the lipid bilayer. The solid bodies represent the globular protein molecules of the membrane. [From S. J. Singer and G. L. Nicholson, Science 175:723, 18 February 1972, with the permission of Science and the authors. Copyright 1972 by the American Association for the Advancement of Science]

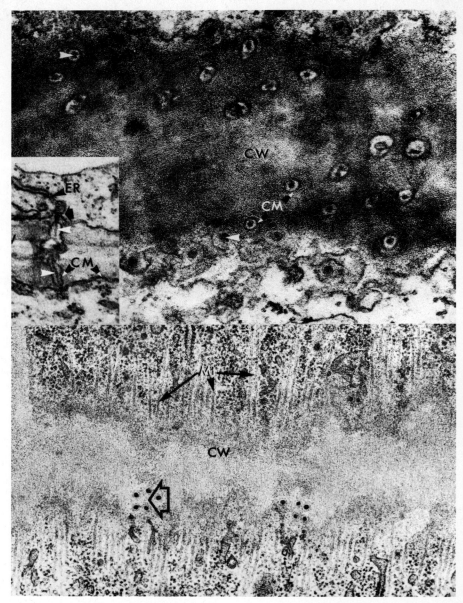

Figure 2-24. **Above:** Plasmodesmata in the walls (CW) of parenchyma cells from an oat coleoptile. In the insert the plasmodesmata are shown in longitudinal section, whereas in the larger picture they are shown in cross section. Note the cell membranes (CM) lining the plasmodesmata and the core of dense material (white arrows) that is continuous with the plugs that close the canal at the wall surfaces. The endoplasmic reticulum (ER) appears to touch these plugs. **Below:** A grazing section of the cell wall/cytoplasm boundry in cells from a root tip of *Arabidopsis thaliana.* The open arrow shows a primary pit field with plasmodesmata. Note the numerous microtubules (Mt) perpendicular to the cell wall. [Reprinted with the permission of The Macmillan Company from *Plant Structure and Development* by T. P. O'Brien and Margaret E. McCully. Copyright © 1969 by The Macmillan Company]

restricted to vacuoles and are not able to diffuse out through the tonoplast. The tonoplast is particularly effective in the active transport of ions into the vacuole and in holding them there. As a result the cell sap usually contains a much higher concentration of any particular ion than either the surrounding medium or the cytoplasm.

The vacuole plays several important roles in plant cells. Some of the substances that accumulate in vacuoles, such as ions of salts, sugars, amides, amino acids (in low concentrations), and proteins, are metabolites that may be utilized later on. Other substances that accumulate in vacuoles, notably calcium oxalate, are highly toxic, and the removal of such substances from the protoplast could be regarded as a substitute for excretion. Since calcium oxalate is not very soluble, it often forms crystals in the vacuoles. Still other substances that are often present in vacuoles, including pigments, tannins, alkaloids, and the latex emulsions of laticifers, are metabolic products. These may or may not be toxic or of any significance in plant metabolism, although some are of considerable economic importance. The proteins in vacuoles are generally present as colloidal sols, but in maturing seeds the proteins may aggregate into globular or crystalline aleurone grains as the vacuoles become dehydrated and often fragmented. During germination the vacuoles become rehydrated, and the proteins revert to colloidal sols that may be hydrolyzed and used.

A quite different role of vacuoles, but one related to the high solute concentration of the cell sap, is the development of turgor pressure as water diffuses into them. The differential permeability of the tonoplast plays an essential part in this process. Turgor pressure plays an important role in the support of organs with little wood or sclerenchyma, as evidenced by the wilting of herbaceous plants when their cells lose turgor, and is also essential for growth. The metabolic activity of cells that are partially dehydrated and lack turgor is at a low level.

Vacuoles can aggregate into a larger vacuole, as in the development of a parenchyma cell from a meristematic cell, and a large vacuole may also fragment into numerous small vacuoles. In the autumn the large vacuoles of the cambial cells of woody perennials fragment into numerous small vacuoles; in the spring these aggregate once more into the large central vacuoles.

It is interesting and perhaps significant that not all vacuoles in a tissue or even in a single cell contain the same substances, probably because of differences in their tonoplasts. Tannins may be present in some vacuoles of a cell and not others. Some vacuoles of the petal cells of *Hibiscus syriacus* contain a red anthocyanin while others contain a violet one, although this may result only from a pH difference between the vacuoles.

ENDOPLASMIC RETICULUM. The endoplasmic reticulum (ER) is a double membrane structure; the space between the two membranes ranges from 15 nm to over 100 nm. This space appears transparent in electron micrographs. It is difficult to interpret the three-dimensional structure of the ER from the extremely thin sectional views in electron micrographs (Figure 2-22), but most of the ER consists of membranous sheets that are highly branched and often wavy. In addition, one of the two membranes may have evaginated saccules or tubules, the latter connecting two of the sheet membranes. As has been noted the ER is continuous with the outer nuclear membrane at various places and, therefore, the inner space of the two double-membrane structures is also continuous. Some investigators have reported ER connections with other membranes also, but this appears dubious. It is thought that the entire ER of a cell may be continuous. However, much of the ER appears to be in the form of cavities or saccules that are much wider and longer than their thickness, their inner space being largely (or perhaps sometimes entirely) isolated from the cytoplasmic matrix. The ER is very flexible and apparently undergoes more or less continual changes in contour and position from one part of the cytoplasm to another as well as with movements of the cytoplasm. Although the ER generally permeates the cytoplasm, it is often more dense in some regions than others; for example, it is particularly dense parallel to cell walls and near the nucleus. Stacks of ER profiles are often present in cells active in protein synthesis.

The ER with ribosomes attached to it is called **rough ER,** in contrast with the **smooth ER** that lacks ribosomes. In animal cells the sheets of ER are apparently always rough whereas the ER tubules are smooth, but in plant cells the sheets of ER may be either rough or smooth. In companion cells and some other plant cells one outer side of an ER sheet may be rough and the other outer side smooth. Cambial cells have been reported to have smooth ER in the winter and rough ER in the summer. Ribosomes on the rough ER of plant cells do not seem to be as numerous or as orderly in arrangement as those in animal cells.

The principal role of ER so far identified for plant cells is the participation of rough ER in protein synthesis. Although polyribosomes may not have to be attached to a membrane for protein synthesis, their attachment to ER apparently greatly promotes the process. It has been proposed that the inner space of the ER, as well as that of the nuclear membrane, may provide channels through which intracellular transport of substances occurs. That proteins synthesized on the ER may be transported in this way is supported by the presence of protein crystals in the inner space of the ER in a few electron micrographs of both plant and animal cells. It seems likely that the ER plays other roles than these and that many other functions may be revealed by future investigations.

Golgi Apparatus

A Golgi apparatus (Figure 2-25) is composed of one to several **dictyosomes,** which are stacks of membranous, hollow discs called **cisternae.** A cisterna is generally 1 to 3 μ in diameter and 14 to 20 nm thick, 6 to 10 nm of this being the inner space. The cisternae are spaced 10 to 12 nm apart. There are usually 3 to 8 cisternae per dictyosome, but there may be as many as 20, particularly in lower plants (Figure 2-26). At their edges the cisternae pinch off **vesicles** that are approximately spherical. The vesicles move to other parts of the cell, particularly the plasma membrane and wall.

In both plant and animal cells the Golgi apparatus polymerizes various substances into macromolecules or even molecular aggregates, which are then transported to the sites where they are used in the vesicles. The substances

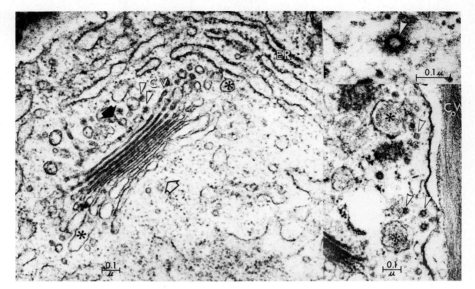

Figure 2-25. **Left:** A dictyosome in a cell of a hair of a bean leaf. The forming face is shown by the large black arrow and the secretory face by the open arrow. The small black arrows point to intercisternal elements. Several coated vesicles (CV) show near the forming face. The asterisks have been placed in two dictyosome vesicles. Note also the endoplasmic reticulum (ER). **Upper right:** A coated vesicle in an epidermal cell of an oat coleoptile. **Lower right:** Dictyosome vesicles in an epidermal cell of an oat coleoptile. The vesicles' contents resemble the materials of the cell wall (CW) in staining properties. [Reprinted with the permission of The Macmillan Company from *Plant Structure and Development* by T. P. O'Brien and Margaret E. McCully. Copyright © 1969 by The Macmillan Company]

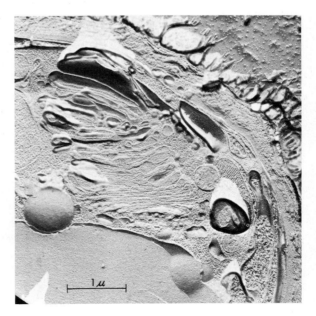

Figure 2-26. A freeze-etch preparation showing a Golgi apparatus of Pleurochrysis, an alga. The nucleus is at the left, with a mitochondrion above it. At the bottom is part of a vacuole. [Courtesy of R. Malcolm Brown, Jr.]

produced by the Golgi apparatus in plants are, to a large extent, different from those produced in animals. The Golgi vesicles of plants have been found to contain many substances including proteins, cellulose, hemicelluloses, pectic compounds, gums, and resins.

One of the most important roles of the Golgi apparatus of plants is the assembly of cell wall components. In several genera of algae the walls include highly structured discs or scales of cellulose, hemicelluloses, and pectic compounds, and Brown and his coworkers [2] have found these in the dictyosomes and Golgi vesicles. Dictyosomes are abundant in differentiating cells, particularly where secondary walls are being formed, and near the tips of pollen tubes and root hairs, where growth in length occurs. In diatoms the Golgi vesicles contain silicon compounds with which their walls are impregnated.

The dictyosomes of plant cells produce a number of substances other than cell wall constituents. Dodder (Cuscuta) is a genus of vascular plants parasitic on other vascular plants. It forms projections called haustoria that penetrate the tissues of the host plants and absorb water and nutrients from them. The dictyosomes in the haustoria are larger and more numerous than elsewhere in the plant and are thought to secrete the enzymes that break down the host tissues as the haustoria penetrate them. The sticky mucus on the upper leaves of the insectivorous sundew plants appears to be produced and secreted by the Golgi apparatus. In several genera of algae including Glaucocystis and Vacuolaria the Golgi apparatus plays the unusual and interesting role of secreting water. The water-filled vesicles enlarge into vacuoles that move to the plasmalemma and burst, discharging their contents, while their membranes are incorporated into the plasmalemma. It has been calculated that Vacuolaria can discharge a volume of water equal to the volume of the cell in a half hour. It seems likely that a good many more secretory roles of the Golgi apparatus of plant cells will be discovered. However, it should not be assumed that all plant secretions are products of the Golgi apparatus. For example, the endoplasmic reticulum seems to be the source of lignin and also the resin secreted by pines.

Mitochondria

Mitochondria (Figure 2-27) are the organelles in which all but the preliminary steps of aerobic respiration occur. They are usually either spherical or bacilluslike in shape, although they are sometimes sinuous filaments that are branched toward the ends. They range from 0.5 to 1.5 μ in width and from 3 to 10 μ in length if elongated. Most cells have 100 to 3000 or so mitochondria, although some lower organisms have only a few. Mitochondria are composed of two membranes. The inner membrane is convoluted into tubular projections or shelflike folds called **cristae** (Figure 2-28).

The interior of a mitochondrion is occupied by a liquid stroma containing proteins, RNA, DNA strands, ribosomes, and various solutes. The stroma surrounds the cristae.

In addition to carrying on respiration, mitochondria synthesize proteins and possibly complex lipids. The DNA of mitochondria carries genetic codes [10] for the production of at least some of the mitochondrial proteins, and presum-

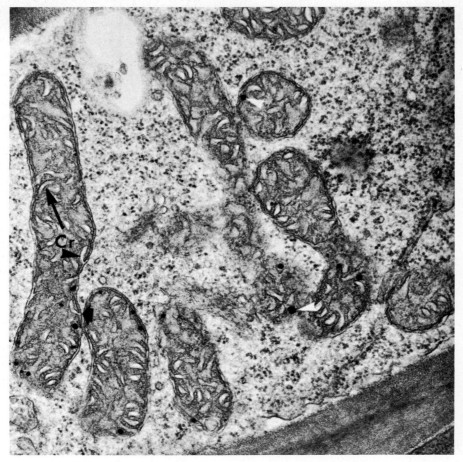

Figure 2-27. Mitochondria in a xylem parenchyma cell of an oat coleoptile. Note the cristae (Cr). The white arrows point to osmophilic granules, believed to be sites of calcium phosphate deposition. [Reprinted with the permission of The Macmillan Company from *Plant Structure and Development* by T. P. O'Brien and Margaret E. McCully. Copyright ⓒ 1969 by The Macmillan Company]

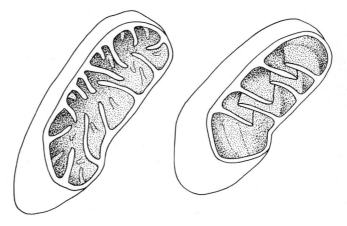

Figure 2-28. Semidiagrammatic drawings of mitochondria with villuslike cristae (left) and bafflelike cristae (right).

ably the mitochondrial RNA and ribosomes participate in protein synthesis in the usual way. This means that DNA replication and RNA synthesis also occur in mitochondria. Of course, the mitochondria contain all the enzymes, coenzymes, cofactors, and substrates essential for their respiratory and synthetic processes.

Plastids

Plastids, along with vacuoles and cell walls, are structures particularly characteristic of plant cells [13]. Plastids have a double limiting membrane and, except for proplastids from which mature plastids develop, they are much larger than mitochondria. They are also less numerous. Plastids are generally classified on the basis of the pigments they contain, and therefore on the basis of their colors. **Chloroplasts** contain chlorophylls and carotenoids, and they are green because the chlorophyll masks the yellow and orange colors of the carotenoids. **Chromoplasts** contain only carotenoids, and thus are orange or yellow. **Leucoplasts** contain no pigments and are colorless. However, because one kind of plastid may convert to another and because all arise from **proplastids,** it seems best to consider plastids in general as one kind of organelle. In the absence of light, proplastids that would otherwise be chloroplasts generally develop into leucoplasts or chromoplasts. They may later become chloroplasts when light is restored. In ripening fruits and autumn leaves the chloroplasts lose their chlorophyll and become chromoplasts. Although the three kinds of plastids have structural differences, particularly as regards the considerably more complex structure of chloroplasts, the classification is perhaps more important from a functional than a structural standpoint.

LEUCOPLASTS. Like other plastids of higher plants, leucoplasts have dimensions on the order of $2 \times 5 \mu$. They are generally lenticular or ellipsoidal. Like the proplastids from which they develop, leucoplasts have cristae similar to those of mitochondria. The differentiation of leucoplasts in principally a matter of enlargement. Some contain membranous internal lamellae, but these are less developed than those of chloroplasts.

Leucoplasts can be subdivided on the basis of their functions. Those that synthesize starch are called **amyloplasts,** and each one contains from one to a few starch grains. Amyloplasts can also hydrolyze starch. If more than one starch grain is present, each one appears to be enclosed in a membranous vesicle. Amyloplasts are the most abundant kind of leucoplast. **Aleuroplasts** contain protein crystals that apparently provide a protein reserve. These crystals may later be hydrolyzed and used in the synthesis of enzyme or structural proteins, as in the case of aleurone grains of vacuoles. The protein crystals may be assembled from polypeptides synthesized elsewhere, although aleuroplasts have been reported to contain ribosomes. The protein crystals are enclosed by a double membrane. **Elaioplasts** contain oil globules and are presumably capable of fat synthesis. However, some elaioplasts may be degenerate chloroplasts with the oil derived from chloroplast lipids.

Some leucoplasts are capable of producing a variety of substances. For example, those of dormant potato tuber buds contain starch, oil globules, proteins, nucleic acids, and phytoferritin. This variety of function, along with the fact that chloroplasts (and to a lesser degree chromoplasts) can also produce starch, proteins, and lipids, explains the difficulty of classifying plastids in any unequivocal way.

CHROMOPLASTS.   Chromoplasts are frequently present in the cells of flower petals and ripe fruits and are also found in other organs such as carrot roots and senescent leaves. Chromoplasts have a great diversity of shapes, although those in any one tissue are usually of some specific shape; they may be globular, ellipsoidal, filamentous, fusiform, crescent, rhomboid, or quite irregular with jagged points. The latter contain large crystals of carotenoids that influence their shape. The more regularly shaped ones either contain small crystals or granules of the pigments or may have the carotenoids dissolved in oil globules. The pigments may occur as crystalline rods, ribbons, plates, or spirals.

Chromoplasts generally differentiate from either chloroplasts or leucoplasts, rather than directly from proplastids. Those derived from chloroplasts generally contain remnants of the chloroplast lamellae, whereas those derived from leucoplasts have cristae. Developing chromoplasts frequently contain starch grains, although (except in carrot roots) these disappear before maturity. Oil globules are often present, so chromoplasts seem to be capable of synthesizing starch and lipids as well as carotenoids.

The carotenoid pigments of chromoplasts have no known metabolic role. It has been proposed that the colors they impart to petals and fruits may have some survival value by attracting pollinating or dispersing animals, but they obviously play no such role in carrot roots or senescent leaves.

CHLOROPLASTS.   Because they are the site of photosynthesis, chloroplasts are extremely important organelles for the entire biosphere as well as for the plants that contain them. The chloroplasts of vascular plants are lenticular or ellipsoidal and are about $2\,\mu$ wide and 3 to $6\,\mu$ long. There are usually 15 to 50 per cell, although both limits may be exceeded in some species. The chloroplasts of bryophytes are generally similar to those of vascular plants, except that in the genus Anthoceros each cell contains only a single, large, cup-shaped chloroplast. The various kinds of eucaryotic algae have a great diversity of chloroplasts as regards size, shape, and number per cell (Figure 2-29). In some genera algal chloroplasts do not differ greatly from those of higher plants, but many algae have chloroplasts of widely varying and complex shapes; these may be circular bands, flat plates extending most of the length of the cells, either solid or reticulate, cup shaped, or helical ribbons, as well as others. These kinds of chloroplasts are usually large, and there are generally only one or two per cell, depending on the species. They usually have associated with them one or two protein centers, called **pyrenoids,** covered with starch grains that they have synthesized. Some pyrenoids produce lipids instead of starch. The fine internal structure of these chloroplasts generally differs substantially from that of the chloroplasts of higher plants (Figure 2-18).

Several major groups of algae have accessory pigments in their chloroplasts that mask or modify the green color of the chlorophyll. For example, brown algae contain fucoxanthin and red algae contain phycoerythrin. Although such

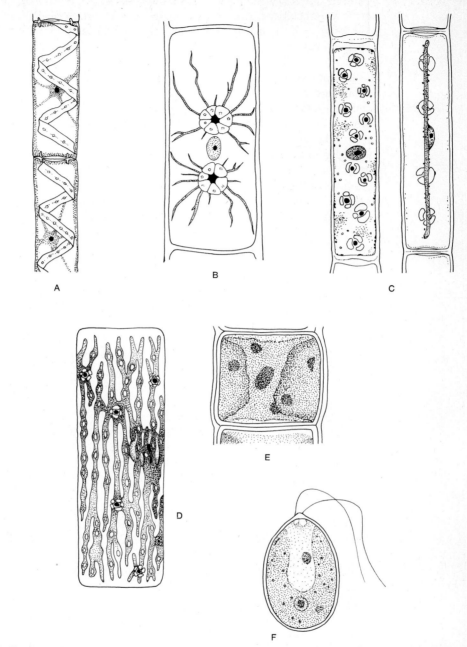

Figure 2-29. Six of the many varied shapes of chloroplasts found in the green algae. **(A)** Helical, in Spirogyra. Note the nuclei. **(B)** Stellate, in Zygnema. There is a pyrenoid in each chloroplast. **(C)** Lamellar (a flat plate), in Mougeotia, shown in face view (left) and from one side (right). Note the nucleus and the several pyrenoids. **(D)** Reticulate, in Oedogonium. **(E)** Open cylinder (like a snap-on bracelet), in Ulothrix. **(F)** Cup shaped, in Chlamydomonas.

pigments participate in light absorption, chlorophyll is essential for the photosynthetic reactions.

The fine internal structure of chloroplasts is considerably more complex than that of other plastids and, indeed, of most organelles of other kinds. The following descriptions of this internal structure apply specifically to the chloroplasts of vascular plants unless otherwise noted. Like other plastids, chloroplasts are limited by a double membrane. Each membrane is about 5 nm thick and the space between the membranes is about 2 to 3 nm thick. The inner membrane may have a number of cristalike evaginations into the body of the chloroplast. Extending through the liquid stroma of the chloroplast, almost from end to end or for a shorter distance, is a more or less parallel stack of pigmented **lamellae,** each lamella **(thylakoid)** being composed of a double membrane (Figure 2-30). The lamellae connect with one another by branching,

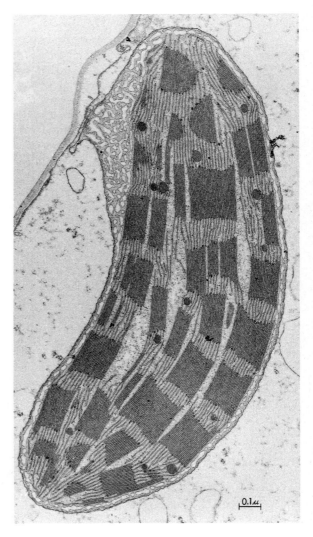

Figure 2-30. Electronmicrograph of a section through a corn chloroplast. Note the numerous thylakoids of the chloroplast extending lengthwise and the grana stacks (dark). The large globules are osmophilic granules; the small dots in the stroma are ribosomes. [Courtesy of L. K. Shumway]

and it is believed that usually all the lamellae of a chloroplast are a single interconnected system. It is also thought that a lamella may be irregularly perforated by relatively large spaces, making it a network of ribbons rather than a solid sheet [3]. Circular stalked tongues extend from the network at various points; in these the double membrane is forked or folded over, the space between the membranes being occupied by a disc.

The folded membrane with its disc is referred to as a **granal thylakoid.** These are found in stacks of 10 to 100, each stack being known as a **granum** (Figure 2-30). There is a 2 nm space between each thylakoid and its neighbors. The discs are 0.3 to 2 μ in diameter, and the thickness of a granum ranges from 150 nm to 1.5 μ, depending on the number of thylakoids. There are usually from 40 to 60 grana per chloroplast. Most of the photosynthetic pigments are in the grana. In a few species of higher plants and many eucaryotic algae there are no grana. Chloroplasts also contain 70 S ribosomes, strands of naked DNA, lipid globules [17], and often starch grains (Figure 2-18).

The dry weight of chloroplasts consists of about 50% protein (including both enzyme and structural protein), 35% lipids, and 7% pigments, the remainder being nucleic acids, photosynthetic products, and the numerous coenzymes involved in photosynthesis and other chloroplast processes, which include starch, lipid, protein, and pigment syntheses.

The detailed structural organization of the thylakoid membranes is still open to question; however, it appears evident that, even at the molecular level, there is a high degree of organization and a precise and orderly arrangement of both the membrane components and the chlorophyll and carotenoid molecules. The carotenes and the hydrocarbon tails of the chlorophyll molecules are fat soluble and can be expected to be in the phospholipid layer of the membrane, whereas the water-soluble porphyrin heads of the chlorophyll molecules are in the protein layers. As we shall see in Chapter 3, the orderly arrangement of the pigment molecules and their precise spacing are essential for the light reactions of photosynthesis.

Two different pigment systems with different light-absorption maxima are involved in photosynthesis, and it is believed that a pigment-system unit consists of about 300 chlorophyll molecules and their associated carotenoid molecules. There is increasing evidence for particles of two size classes embedded in the membranes [1], and these may be the structural manifestations of the pigment systems (Figure 2-31). There is considerable doubt that the quantasomes described by Park [21] are the basic photosynthetic units. There are also particles on the membrane surface that contain phosphorylation-coupling factors and show ATPase* activity.

PROPLASTIDS. Proplastids (Figure 2-19) are small (1 to 3 μ in diameter) plastids that enlarge and differentiate into the various kinds of mature plastids we have considered. They are present in the meristematic and reproductive cells of higher plants. The number per cell is species-dependent, but is usually 20 or more. Proplastids are generally spherical or ellipsoidal but may at times become amoeboid. Like mature plastids and mitochondria, they are bounded by a

---

* The enzyme or enzyme system that breaks down ATP (adenosine triphosphate).

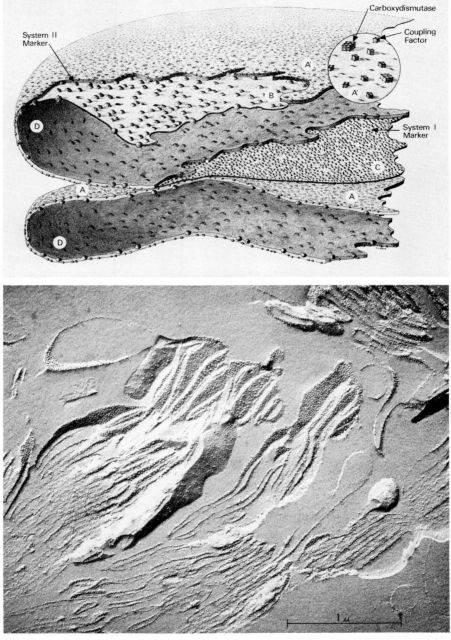

Figure 2-31. **Above:** Semidiagrammatic drawing of two granum discs showing particles embedded in the membranes. A, A': Carboxydismutase and phosphorylation coupling enzymes on the outer surface. B: Outer surface of inner membrane with System II markers. D: System II markers on the inner surface of the membrane. C: The smaller and more numerous System I markers. **Below:** Freeze-etch preparation of a corn chloroplast showing thylakoids in both sectional and face views. Note the System I and System II markers on the membranes. [Courtesy of Charles J. Arntzen. Drawing from C. J. Arntzen, R. A. Dilley, and F. L. Crane, *J. Cell Biol.,* **43:**16–31 (1969)]

double membrane. They contain only a few short primary thylakoids similar to the cristae of mitochondria, and it is sometimes difficult to distinguish them from mitochondria. However, proplastids are generally somewhat larger and often contain lipid globules, which are not present in mitochondria. Proplastids divide by fission, their divisions keeping pace with cell division. Most algae and some bryophytes do not have proplastids; it is their chloroplasts that divide during cell division. The division of chloroplasts has also been reported for a few vascular plants, but proplastid division appears to be the predominant means of maintaining the plastid population of cells.

The differentiation of proplastids into leucoplasts is principally a matter of enlargement and, in amyloplasts, the elaboration of enzyme systems essential for starch synthesis and hydrolysis. There is relatively minor development of the internal membrane systems. Proplastids may differentiate directly into chromoplasts, but these are usually derived from either leucoplasts or chloroplasts.

The differentiation of proplastids into chloroplasts involves the synthesis of many complex enzyme systems, the synthesis of the chloroplast pigments, and extensive differentiation of the internal membranes, as well as an increase in size [29]. In the light, the cristae and other invaginations of the inner limiting membrane form vesicles that detach, become aligned, and proliferate into the complex thylakoid system with its grana (Figure 2-30), and pigmentation occurs. In the absence of light normal thylakoid and pigment development are arrested, but tubular vesicles aggregate into one to three prolamellar bodies. The tubules fuse at each point where they touch. This structure persists until such a time as light may be available; then thylakoid differentiation and pigment formation occur. However, light is not essential for chloroplast development in most of the lower plants. Chloroplast differentiation is also impeded or prevented by deficiencies of oxygen or essential mineral elements or by genetic deficiencies in albino plants or tissues. Cells in the nongreen areas of variegated leaves, the ordinary epidermal cells of many species, and cells of the pith and roots generally do not contain mature pigmented chloroplasts even though all the essential environmental factors are present. This is good evidence for the genetic control of chloroplast development by cells.

In some cases, particularly where there is active secondary proliferation of cells as in the development of adventitious roots on cuttings, chloroplasts may revert to a proplastid condition. This is accomplished by repeated chloroplast divisions that result in progressively smaller plastids which lose their thylakoid systems and pigments. Such plastids generally do not redifferentiate into chloroplasts even if the environment is suitable. Leucoplasts may also dedifferentiate into proplastids, but chromoplasts rarely if ever do so. The developmental interrelations among the various kinds of plastids are summarized in Figure 2-32.

Other Membranous Organelles

Several other kinds of membranous organelles have been reported to occur in plant cells. These include three that we will consider briefly: spherosomes, lysosomes, and lomasomes. All three kinds have single limiting membranes,

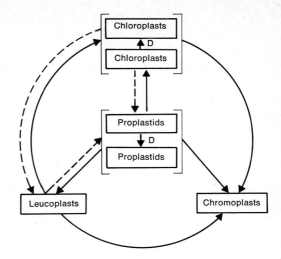

Figure 2-32. Diagram illustrating patterns of differentiation (solid arrows) and dedifferentiation (broken arrows) of plastids. The short arrows (D) indicate proliferation by division.

in contrast with the double membranes of plastids and mitochondria. These organelles are spherical and about the same size as mitochondria and proplastids.

SPHEROSOMES.   The spherosomes are abundant in most plant cells; many contain lipids and proteins, although they may lose their proteins and become membrane-enclosed oil globules. Thus spherosomes are centers of lipid accumulation and probably lipid synthesis. It has been proposed that spherosomes originate by being pinched off from the endoplasmic reticulum. Some spherosomes play other roles. They may have peroxidase activity or carry on glyoxalate metabolism. The latter are often called **glyoxysomes.**

LYSOSOMES.   The lysosomes are on the order of 0.4 μ in diameter and contain hydrolytic enzymes including DNAase, RNAase, protease, and a variety of carbohydrate-hydrolyzing enzymes. Their membranes are not permeable to these enzymes, but if the membranes are disrupted the enzymes will hydrolyze many important cell constituents and cause disruption and perhaps death of the cell. The specific roles of lysosomes are much better understood in animal cells [6] where they are also involved in phagocytosis by white blood cells. The function of lysosomes in plants are less evident, although they may well be involved in such things as the disintegration of the protoplasts of sclerenchyma, tracheids, and vessel elements and of the end walls of the latter. They could also play a role in the maturation of sieve-tube elements and laticifers and in the senescence of plant organs.

LOMASOMES.   The lomasomes are associated with the plasmalemma and cell wall and are thought to function in wall synthesis, particularly the wall expansion of enlarging cells. They contain bundles of tubules, but the relationship of these to microtubules is not known. Lomasomes were first identified in fungi.

Lomasomes, as well as spherosomes and lysosomes, are still not well understood, and there are still unanswered questions regarding their roles and the extent to which they occur in the various species of plants.

Cell Walls

The presence of cell walls that enclose each protoplast is one of the distinctive characteristics of plant cells. Despite the fact that cell walls are not considered part of the protoplast, they play a number of important and essential roles. Because of their great tensile strength and limited elasticity, they restrain the ballooning and possible rupture of the cell as water diffuses into it and high turgor pressure develops. The back pressure of the walls against the cell contents has a major influence on the water potential in the cells and therefore on osmosis. The turgor pressure of cells provides support of the nonwoody tissues and organs of plants, whereas the thick walls of cells in wood and sclerenchyma tissues provide mechanical support. The walls of tracheids and vessels enclose tubes through which water flows from one part of the plant to another. The cutinized walls of epidermal cells and the suberized walls of cork cells prevent excessive water loss and desiccation in plants. In addition, cell walls provide the protoplasts with at least some degree of mechanical protection and a hydrated surface.

A cell wall may have three more or less distinct layers. The outer one, called the **middle lamella,** is shared by adjacent cells and consists principally of pectic compounds. Each cell has its own **primary wall** that encloses it. The primary wall contains cellulose fibrils in a matrix made up principally of pectic compounds and hemicelluloses. After cell enlargement is complete, or essentially so, a multilayered secondary wall may be laid down inside the primary wall. This is thicker than the primary wall and, in some kinds of cells, may be very thick in proportion to the size of the cell.

Before we consider each of these wall layers in somewhat greater detail we will describe the structure of the cellulose fibrils. These fibrils may be regarded as the basic structural components of the wall and provide the tensile strength of the wall. Also, it should be noted that the walls of living cells, particularly the primary walls, are not just inert containers. Rather, they should be considered as functional parts of the cell. Primary cell walls are enzymatically active, containing up to 80% of the phosphatase activity of a cell for example. Some of the steps in the synthesis of cell wall components may occur within the wall itself. In addition, the wall is penetrated by numerous (1000 to 100,000 per cell) plasmodesmata that provide protoplasmic connections between adjacent cells. Both the formation and structure of cell walls are extremely complex phenomena that have been investigated extensively but are still incompletely understood. Much more information about them than can be presented below can be found elsewhere [4].

THE STRUCTURE OF CELLULOSE FIBRILS. Cellulose is a polysaccharide polymer composed of from 3000 to 10,000 β-D-glucose residues with 1,4 linkages, resulting in molecules that are extremely long in comparison to their width. Cellulose is a universal wall component of higher plants; however, in some fungi the fibrillar component of the wall is chitin, whereas in some algae it is a polysaccharide called xylan, which is composed of β-1,4 linked xylose residues [23].

The cellulose molecules are arranged parallel with one another and may be crosslinked by hydrogen bonds between the carbon 3 of one molecule and the ring oxygen of another. At intervals the molecules are arranged into highly ordered three-dimensional lattices, thus forming crystalline **micelles.** Each micelle is composed of about 100 cellulose molecules and has the shape of a thick ribbon. About 20 micelles are bound together into a **microfibril** that is 25 to 30 nm wide and is also ribbon shaped. It has been proposed that the micellar ribbons may not simply run lengthwise in the microfibril, as has been thought, but that each one may be twisted into a helix 3.5 nm wide. In the helix the cellulose molecules would run lengthwise, but in an unwound ribbon they would be essentially at right angles to the length of the ribbon. Some 250 microfibrils are bound into a **macrofibril,** which has a width of up to 0.5 μ and consists of a bundle of some 500,000 cellulose molecules. This binding together of cellulose molecules into progressively larger bundles results in fibrils that are very strong and can withstand great stress without rupturing.

MIDDLE LAMELLA. The middle lamella is an intercellular layer composed principally of pectic compounds, although some protein is present. The pectic compounds form plastic, hydrophilic gels that cement the cells to one another and provide coherent tissues. They are polymers of D-galacturonic acid (a derivative of galactose) residues with α-1,4 linkages. The principal pectic compounds in the middle lamella are pectic acid, its calcium and magnesium salts, and pectin. Pectic acid molecules have about 100 galacturonic acid residues and pectin molecules are about twice as large. Most of the carboxyl groups in pectin are esterified by the addition of methyl groups. Pectic acid is water soluble, but its salts are insoluble and pectin is soluble only in hot water. In senescent tissues, notably the flesh of overripe fruits, the pectic compounds are hydrolyzed and the cells are held together only loosely. In lignified cells the middle lamella as well as the primary and secondary walls may become impregnated with lignin.

PRIMARY WALL. In contrast with the middle lamella, the primary cell wall has cellulose microfibrils. In meristematic cells these may be more or less transverse, but as a cell enlarges they become arranged at numerous different angles to one another, giving an interwoven or matted appearance (Figure 2-33). Face-view electronmicrographs of walls give the impression that the microfibrils may be in contact with one another, but this may be an artifact of fixation. Other evidence indicates that microfibrils are separated from one another by some four times their own width—50 to 100 nm in fresh walls. Cellulose constitutes as little as 10% of the fresh weight of primary walls.

The matrix of primary walls is composed principally of pectic compounds and hemicelluloses. The pectic compounds include pectins and protopectins; the latter have higher molecular weights than the pectins and are less highly methylated. Protopectins are insoluble even in hot water. The hemicelluloses include a variety of polysaccharides with linear or branched molecules derived

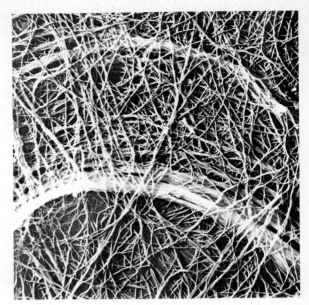

Figure 2-33. Electronmicrograph showing the matted cellulose fibers in a primary cell wall. [Courtesy of R. W. G. Wycoff]

from sugars such as D-xylose, L-arabinose, and D-mannose. Like the pectic compounds, they form hydrophilic gels that are highly hydrated. They do not form fibrils, but it has been reported that xylan chains may lie parallel to the cellulose microfibrils. Primary cell walls also contain proteins, particularly those that have a high content of hydroxyproline. This amino acid is an unusual protein constituent as it is restricted essentially to cell wall proteins and to collagen, a structural protein of vertebrate connective tissues. The framework of cellulose microfibrils is embedded in the gelatinous matrix, and is crosslinked with at least the proteins and pectic compounds of the matrix.

Unlike secondary walls, primary walls do not have pits but they do have circular depressed areas known as primary pit fields. Plasmodesmata may be more numerous in the pit fields than in the other areas of the wall.

SECONDARY WALLS. After the enlargement of a cell is complete, or nearly so, secondary cell walls may be laid down against the inner surface of the primary wall. Parenchyma cells and some of the specialized types of living cells generally do not have secondary walls, although these may be present in wood parenchyma and xylem ray cells. All cells that function after death, including sclerenchyma, tracheids, and vessel elements, have thick secondary walls. However, as noted previously, the secondary wall is not laid down over the entire primary wall. It is interrupted by pits and in some vessel elements is restricted to a helix, a series of rings, or a reticulum (Figure 2-16).

Cellulose is a more abundant component of secondary walls than primary walls, and the microfibrils of any particular layer of the secondary wall are essentially parallel with one another and are bound into macrofibrils (Figure 2-34). Secondary walls are generally formed in three distinct layers, designated as S_1, S_2, and S_3. S_1, which is next to the primary wall and is laid down first, consists of four submicroscopic lamellae, each with its cellulose microfibrils in a slow helix at a large angle to the axis of the cell. The direction of the helix is reversed in alternate lamellae. S_2 is generally the thickest part of the cell wall and may be composed of numerous lamellae. The S_2 microfibrils are in a steep helix at a small angle to the cell axis and, as in S_1, the direction of the helix reverses in alternate lamellae. S_3 is generally thin, consisting of only a few lamellae, or it may be absent. When present the microfibril helices of S_3 are at a large angle to the cell axis. Inside of S_3 in dead cells there is often a thin, noncellulose warty layer that may consist of remnants of the protoplast. The varying pitch of the cellulose microfibrils in the various layers and lamellae of the secondary wall results in an extremely strong wall. The structural organization is comparable with, but much more complex than, the alternating direction of the different layers of cords in an automobile tire.

Initially, the matrix of secondary walls consists primarily of hemicelluloses, but in some cells lignin is soon added. The matrix of secondary walls is generally more rigid and less highly hydrated than that of primary walls. The lignins have extremely large and complex molecules built up from numerous phenyl-propane molecules of different kinds. They are rigid substances that are not hydrophilic and contribute greatly to the stiffness and compressional strength of the walls. They are resistant to decay by most microorganisms that hydrolyze cellulose, although other microorganisms may be able to hydrolyze and use them, thus contributing toward the complete decay of wood. Lignins are also resistant to most chemical reagents and create a disposal problem when removed from wood pulp in the making of high grade paper.

The walls of sclerenchyma cells, tracheids, and vessel elements are always lignified, as are the Casparian strips of endodermal cells. Up to 30% of the dry weight of walls may consist of lignin. In secondary walls it is most abundant

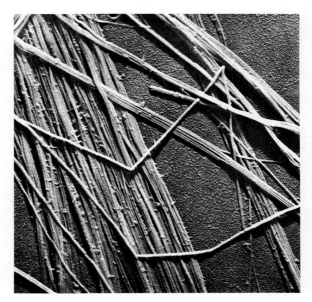

Figure 2-34. Electronmicrograph showing parallel cellulose fibers in a secondary cell wall. [Courtesy of R. W. G. Wycoff]

in the S_2 layer. In some species such as *Picea abies,* however, as much as 90% of the lignin is in the middle lamella and primary walls and 70% of their weight is lignin. In the cells of primary xylem, lignification of the middle lamella and primary walls is restricted to the areas with secondary thickening; however, in the cells of secondary xylem, the middle lamella and primary walls may be lignified throughout. Except for cellulose, woody plants synthesize more lignin than any other polymer.

In addition to cellulose, hemicelluloses, and lignin, secondary walls may include a variety of other substances such as tannins, suberin, gums, and mucilages.

The pits of secondary walls are often superimposed on the pit fields of the primary wall but may also form elsewhere. There are usually one or two pits per pit field. The middle lamella and the primary walls of the adjacent cells are continuous through a pit, constituting what may be called a pit membrane. Generally the pits of adjacent cells are opposite one another, forming a pit pair. The difference between simple and bordered pits has already been considered, but it should be noted that in adjacent cells of different kinds one pit of a pair may be bordered and the other one simple.

SURFACE LAYERS. External plant surfaces commonly have cells with walls impregnated with or covered with substances that reduce water loss and that may play other protective roles. The suberized walls of cork cells and the cuticle that covers epidermal cells have already been mentioned. The cutin of the cuticle, like suberin, is composed of long-chain hydroxy fatty acids, but the cuticle may also contain waxes and in some species extrusions of waxes occur from the cuticle. These extrusions may assume various complex shapes (Figure 2-35). The outer wall of pollen grains and pteridophyte spores (the exine) is made up of an extremely complex polymer of cellulose, xylans, lignins, and lipids that is called **sporopollenin.** It is highly sculptured in various patterns that are species-characteristic (Figure 2-36). Sporopollenin is highly resistant

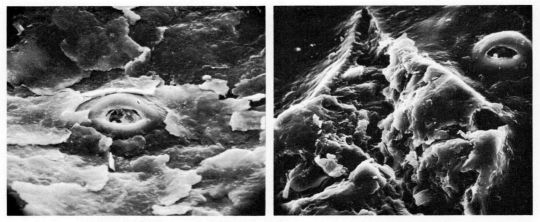

Figure 2-35. Scanning electronmicrographs showing wax platelets on the cuticle of orange epidermis ($\times$650). Note the stomata. The one on the left is plugged with wax. [Courtesy of L. Gene Albrigo]

Figure 2-36. Scanning electronmicrograph of *Polymnia uvedalia* pollen. (×725). [Courtesy of James R. Wells. From J. R. Wells, *Amer. J. Bot.,* **58:**124 (1971). Electronmicrograph by Thomas P. Schreiber]

to decay by microorganisms and may remain intact for many thousands of years in bogs and similar habitats. This makes pollen analysis useful for determining the past vegetation of an area and dating of geological core samples.

CELL WALL FORMATION. The formation of cell walls is an extremely complex phenomenon that cannot be discussed in detail here. Even though there is an impressive mass of information about it, much remains to be clarified. Cell wall formation has several rather distinct aspects: (1) the formation of a new middle lamella and new primary walls of the two cells resulting from cell division; (2) the incorporation of new wall components in and on the primary walls and middle lamella as a cell is enlarging; and (3) the deposition of secondary walls after cell enlargement is complete.

Following mitosis a cell plate forms between the two new nuclei, apparently by the concentration and coagulation of numerous 100-nm vesicles, which may be Golgi vesicles. The cell plate generally originates at the center of the cells and extends toward the side walls. The cell plate then gives rise to the middle lamella and primary walls. However, wall formation does not always follow mitosis immediately. For example, in the development of the embryo sac and endosperm, free nuclei may exist for some time before walls segmenting the cytoplasm around them are finally laid down.

Cell enlargement involves plasticization of the primary wall, which can then be stretched by turgor pressure, a process that requires the presence of the plant hormone, auxin. It is thought that the auxin acts by causing breakage of the crosslinkages between the cellulose microfibrils and either the pectic compounds or the wall proteins. Since the primary wall increases in thickness as well as area during cell enlargement, it is obvious that there must be deposition of additional wall materials, including the substances of the matrix and the middle lamella as well as cellulose.

Although the addition of new materials to enlarging primary walls may involve both intussusception of these among the previous wall components and apposition of new materials to the wall surface, laying down secondary walls is entirely a matter of apposition of successive layers of secondary wall, initially against the primary wall.

As has been mentioned earlier, the Golgi apparatus plays a major role in the prefabrication of cell wall components which are then transported to the walls in Golgi vesicles. Most of the wall components have been found in Golgi vesicles, with the exception of lignin, which appears to be produced by the endoplasmic reticulum. The lomasomes may also function in cell wall formation as may microtubules. As previously noted, the microtubules may influence the orientation of the cellulose microfibrils. It has also been suggested that the endoplasmic reticulum may possibly function in wall formation. Although the protoplast thus plays some major roles in wall formation, it seems likely that the final assembly of the wall structure occurs within the wall itself.

Procaryotic Cells

The cells of eucaryotic algae and fungi have such a great diversity of structural detail from species to species that it is not feasible to consider them further in this book. However, the procaryotic cells of bacteria and blue-green algae [7] deserve brief consideration because they differ substantially from eucaryotic cells and because we shall be referring to these organisms in connection with such topics as photosynthesis and the nitrogen cycle.

Procaryotic cells lack nuclear membranes and the wide variety of membranous organelles of eucaryotic cells, such as endoplasmic reticulum, the Golgi apparatus, plastids, mitochondria, and spherosomes. They do have a plasmalemma that bounds the protoplast. In blue-green algae and photosynthetic bacteria, invaginations of the plasmalemma give rise to internal membranes that contain the photosynthetic pigments (Figure 2-37). There are no grana. Often several of these membranes extend through most of the cytoplasm and are roughly parallel with one another and with the cell surface (Figure 2-38). Nonphotosynthetic bacteria either have no internal membranes or have only a few poorly developed ones. Procaryotic cells have 70 S ribosomes, similar to those of mitochondria and chloroplasts. Their nuclei consist of nothing more than a few chromosomes, which are strands of naked (nonhistone) DNA, that are commonly circular. There are no typical vacuoles.

The photosynthetic pigments of blue-green algae include phycocyanin and phycoerythrin in addition to chlorophyll a, carotene, and xanthophyll. The photosynthetic bacteria all have one or more of the several bacteriochlorophylls, and these differ from one another and from all of the chlorophylls of green plants as regards certain radicals of the porphyrin ring [22]. The green photosynthetic bacteria also have yellow carotenoids, whereas the purple and red photosynthetic bacteria contain red as well as yellow carotenoids. These pigments, along with the pale blue-gray color of the bacteriochlorophyll present, are responsible for the red to purple colors of the bacteria.

The walls of the procaryotic cells also differ substantially from those of eucaroytic plant cells. They contain neither cellulose nor chitin; the structural fibers are predominantly polysaccharides composed of derivatives of amino

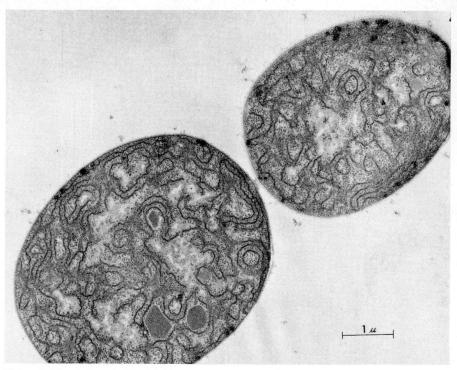

Figure 2-37. Electronmicrograph of two cells of Anabaena, a unicellular blue-green alga. Note the photosynthetic thylakoids, which extend through most of the protoplast. The lighter areas are nucleoplasmic regions, but note the absence of nuclear membranes. The larger dark bodies near the plasmalemma are lipid globules and the more numerous smaller ones are glycogen granules. Toward the bottom of the larger cell are two polyhedral bodies. [Courtesy of Norma J. Lang]

sugars (acetylglucoseamine and muramic acid) [26]. The substantial polypeptide component of the walls is built up of amino acids quite different from those of other organisms, including diaminopimelic acid and the D isomers of several common amino acids rather than the L isomers that are essentially universal in other organisms. Some lipids may also be present in the walls, particularly of gram-negative bacteria where they constitute up to 20% of the wall components.

The old hypothesis that mitochondria and plastids originated early in evolutionary history from primative procaryotes, which became symbyotic within eucaryotic cells, has in recent years attracted increasing attention and interest and a considerable number of proponents [5, 18, 25]. They cite the general resemblance of chloroplasts to blue-green algae and of mitochondria to nonphotosynthetic bacteria as regards size, structure, size of the ribosomes, and the presence of naked DNA strands. Also, the DNA of chloroplasts and mitochondria differs in density and base composition (AT/CG ratio)* from

---

*A, adenine; T, thymine; C, cytosine; G, guanine. These are the nitrogenous bases that provide the genetic code of DNA. See Chapter 5.

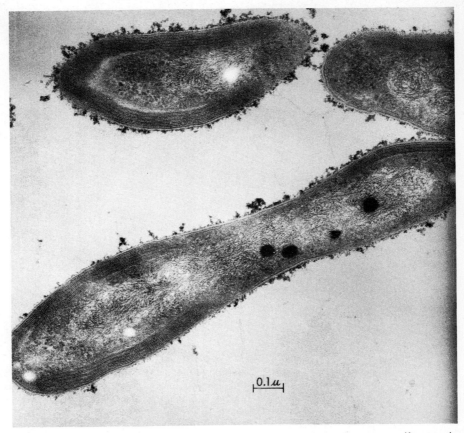

Figure 2-38. Electronmicrograph of *Rhodopseudomonas palustris,* a nonsulfur purple photosynthetic bacterium. Note the parallel photosynthetic thylakoids adjacent to the plasmalemma and the lighter nucleoplasmic regions with strands of DNA in them. [Courtesy of Germaine Cohen-Bazire]

the DNA of the nuclei of the cells in which they are located. Of course, chloroplasts and mitochondria do not have walls, as do the bacteria and blue-green algae.

References

[1] Arntzen, C. J., R. A. Dilley, and F. L. Crane. "A comparison of chloroplast membrane surfaces visualized by freeze etch and negative staining techniques; and ultrastructural characterization of membrane fractions obtained from digitonin-treated spinach chloroplasts," *J. Cell Biol.,* **43:**16–31 (1969).

[2] Brown, R. M., Jr. *et al.* "Cellulosic wall component produced by the Golgi apparatus of *Pleurochrysis scherffelii,*" *Science,* **166:**894–896 (1969).

[3] Buvat, Roger. *Plant Cells.* McGraw-Hill, New York, 1969.

[4] Clowes, F. A. L., and B. E. Juniper. *Plant Cells,* Blackwell Scientific Publications, Oxford, 1968.

[5] Coehn, S. S. "Are/were mitochondria and chloroplasts microorganisms?" *Amer. Scientist,* **58**:281–289 (1970).

[6] de Duve, Christian. "The lysosome," *Sci. Amer.,* **208**(5):64–72 (1963).

[7] Echlin, Patrick. "The blue-green algae," *Sci. Amer.,* **214**(6):74–81 (June 1966).

[8] Eglinton, G., and R. J. Hamilton. "Leaf epicuticular waxes," *Science,* **156**:1322–1335 (1967).

[9] Esau, Katherine. *Plant Anatomy,* 2nd ed., John Wiley & Sons, New York 1965.

[10] Goodenough, U. W., and R. P. Levine. "The genetic activity of mitochondria and chloroplasts," *Sci. Amer.* **223**(5):22–29 (Nov. 1970).

[11] Green, D. E., and J. H. Young. "Energy transduction in membrane systems," *Amer. Sci.,* **59**:92–100 (1971).

[12] Jensen, W. A., and R. B. Park. *Cell Ultrastructure,* Wadsworth Publishing Co., Belmont, Calif., 1967.

[13] Kirk, J. T. O., and R. A. E. Tilney-Bassett. *The Plastids,* Freeman, London, 1967.

[14] Korn, E. D. "Structure of biological membranes," *Science,* **153**:1491–1498 (1966).

[15] Ledbetter, M. C., and K. R. Porter. "Morphology of microtubules in plant cells," *Science,* **144**:872–874 (1964).

[16] Ledbetter, M. C., and K. R. Porter. *Introduction to the Fine Structure of Plant Cells,* Springer-Verlag, New York, 1970.

[17] Lichtenthaler, H. K. "Plastoglobuli and the fine structure of plastids," *Endeavour,* **27**:144–149 (1968).

[18] Margulis, Lynn. "The origin of plant and animal cells," *Amer. Sci.,* **59**:230–235 (1971).

[19] Nomura, M. "Ribosomes," *Sci. Amer.,* **221**(4):28–35 (Oct. 1969).

[20] Northcote, D. H. "The Golgi apparatus," *Endeavour,* **30**:26–33 (1971).

[21] Park, R. B., and John Biggins. "Quantasome: size and composition," *Science,* **144**:1009–1011 (1964).

[22] Pfenning, N. "Photosynthetic bacteria," *Ann. Rev. Microbiol.,* **21**:285–324 (1967).

[23] Preston, R. D. "Plants without cellulose," *Sci. Amer.,* **218**(6):102–108 (June 1968).

[24] Racker, E. "The membrane of the mitochondrion," *Sci. Amer.,* **218**(2): 32–39 (Feb. 1968).

[25] Raven, P. H. "A multiple origin for plastids and mitochondria," *Science,* **169**:641–646 (1970).

[26] Sharon, N. "The bacterial cell wall," *Sci. Amer.,* **220**(5):92–98 (May 1969).

[27] Sleigh, M. A. "Cilia," *Endeavour,* **30**:11–17 (1971).

[28] Singer, S. J., and G. L. Nicolson. "The fluid mosaic model of the structure of cell membranes," *Science,* **175**:720–731 (1972). See also Vanderkooi, G., and D. E. Green. *BioScience,* **21**:409–415 (1971); and D. E. Green, and R. F. Brucker. *BioScience,* **22**:13–19 (1972).

[29] von Wettstein, D. "The formation of plastic structures," *Brookhaven Symp. Biol.,* **II**:138–159 (1958).

Photosynthesis

3

As noted in Chapter 1, green plants are an essential component of the biosphere because they carry on photosynthesis. This is a complex series of reactions, driven by light energy absorbed by chlorophyll and other pigments, that results in the synthesis of organic compounds from carbon dioxide and water. Much of the absorbed light energy is eventually incorporated in the chemical bonds of the organic compounds (mostly sugars) produced, thus adding a vast amount of energy to the biosphere. This is essential for the maintenance of life because the life processes of all organisms require a continual expenditure of energy, most of it derived from the oxidation of sugars and other foods by respiration. All of this energy is eventually converted to heat, which is dissipated and lost from the biosphere. Energy flow through every ecosystem is one way: from light energy, to chemical bond energy, to free energy of various kinds that drives life processes, and finally to heat energy. Without the continued input of energy by photosynthesis, life would soon cease.

Furthermore, all the many and varied organic compounds that are essential constituents of cell structures are derived, in the final analysis, from the sugars or other organic compounds produced by photosynthesis. Thus, all organisms are dependent upon photosynthesis for their organic structural constituents as well as their energy.

The oxygen produced by photosynthesis in green plants (algae, bryophytes, and vascular plants) is also vital to all aerobic organisms, that is, to those organisms that require oxygen for respiration. It is generally considered that all the oxygen of the atmosphere has been produced by photosynthesis. Aerobic respiration, combustion, and various nonbiological oxidations would deplete the atmospheric oxygen if it were not replenished by photosynthesis. Thus, all aerobic organisms are dependent upon photosynthesis for oxygen, for the organic compounds that are essential to all organisms as cell constituents and as substrates for respiration, and for energy.

In addition to green plants there are a few other autotrophic organisms, that is, organisms that can synthesize their basic supply of organic compounds from carbon dioxide and a hydrogen donor rather than being dependent on an external source of food as are animals, fungi, and most bacteria. These autotrophs are mostly photosynthetic and chemosynthetic bacteria. There are also a few photosynthetic protozoa like Euglena, but the vast majority of animals is heterotrophic. Animals such as green hydra and a few gastropods are at least partially autotrophic, but the photosynthesis is carried on by symbiotic algae. All these autotrophs combined make a very minor contribution of organic compounds to the biosphere as a whole. Since the photosynthetic bacteria use hydrogen donors other than water, they produce no oxygen. The chemosynthetic bacteria secure the energy needed for synthesis of organic compounds from carbon dioxide by the oxidation of some inorganic substance like hydrogen, hydrogen sulfide, sulfur, iron compounds, ammonia, or nitrites, rather than from light. Although a few species use nitrates, sulphates, or carbonates as the oxidant, most species use oxygen. The latter are thus also dependent upon photosynthesis by green plants. There is no question about the essential role of photosynthesis in the maintenance of life on Earth.

However, if the current theory of the origin of life on Earth [22] is valid, the first organisms were heterotrophic rather than autotrophic. They used nonbiologically synthesized organic compounds as food. Such nonbiological synthesis was possible because, at that time, the atmosphere presumably contained no oxygen, perhaps little carbon dioxide and nitrogen, but an abundance of water vapor, ammonia, methane, hydrogen, cyanide, and perhaps other gases such as hydrogen sulfide from which organic compounds could be synthesized—and have been synthesized in laboratories. The energy required for the syntheses could have been provided by electrical discharges or by a much higher level of ionizing radiation (ultraviolet and radioactive) that may have then existed. The assembly of the resulting organic compounds, which included such essential substances as proteins and nucleic acids, into primitive cells is considered to be theoretically plausible. These first organisms were obviously anaerobic as well as heterotrophic. As they increased in number it seems likely that they used the organic compounds faster than they were being synthesized and, if this was the case, it would have led to the limitation or even the termination of life.

The eventual evolution of photosynthetic organisms avoided such a fate by making possible an abundant supply of organic compounds synthesized at the expense of light energy. The oxygen added to the atmosphere made possible the evolution of aerobic organisms. This was another major evolutionary advance, because anaerobic respiration frees only a small fraction of the chemical bond energy from sugar and is inadequate for supplying the energy requirement of any very large or very active organism. As the oxygen produced by photosynthesis accumulated in the atmosphere, it made continued nonbiological synthesis of organic compounds impossible by oxidizing the components of the primitive atmosphere as well as the organic compounds derived from them. From this time on life became completely dependent on the organic compounds and oxygen produced by photosynthesis.

Photosynthesis is not only a process essential for life, but also a process of immense magnitude. Estimates of the total annual production of organic matter by photosynthesis of all plants on Earth have been made in a variety of ways. Although the estimates differ from one another, they are all in billions of tons. One recent estimate is that photosynthesis produces 87 billion tons of organic matter per year, above that used by the plants in respiration. On this basis there is a net consumption of 128 billion tons of carbon dioxide and 52 billion tons of water and a net production of 52 billion tons of oxygen by photosynthesis. Earlier estimates were about five times as high, principally because the photosynthetic production in the oceans had been overestimated. However, it is still estimated that marine algae carry on from 45 to 60% of all photosynthesis.

Historical Development of Concepts About Photosynthesis

The emergence and evolution of man's knowledge of photosynthesis has been a slow and piecemeal process that has extended over more than 200 years [16]. What is generally regarded as the first bit of experimental evidence for photosynthesis was presented in 1772, but it was not until 1845 that a summary equation for photosynthesis, essentially the same as the present one, could be written. Most of the detailed knowledge of photosynthetic reactions has been acquired since 1940. It is obvious that any real understanding of photosynthesis could not be attained before the emergence of chemical concepts and techniques, but what is striking is that both began at essentially the same time and have subsequently kept pace with one another. There were a few experiments and insights that presaged the discovery of photosynthesis before modern chemistry began developing, but these were all directed toward the problem of the source of the material of which plants are composed.

Prechemical Concepts and Notions

Aristotle and other ancient Greeks believed that the solid components of plants were derived from the soil (earth), a reasonable hypothesis when everything was thought to be derived from earth, air, fire, or water, and earth was the only solid. As far as we know this notion was not challenged until, about 1450, Nicolaus of Cusa (Cusanius), a bishop and cardinal with a wide range of interests, described an experiment leading to the conclusion that the substance of plants was derived from water rather than earth. It seems likely that either he or someone else had actually conducted this experiment. In 1648 Jean-Baptiste Van Helmont, a Dutch investigator, published an account of an experiment that led him to the conclusion that plants were derived entirely from water [12]. He placed 200 lb of dry soil in an earthenware pot, then saturated it with water and planted a willow shoot in it. He covered the soil with a perforated metal plate to prevent accidental loss or contamination of the soil. After 5 years the willow had gained 164 lb, while the soil had lost only 2 ounces (oz). Whether or not Van Helmont knew about the Cusanius report is unknown; however, his experiment was essentially similar to that described by Cusanius, and Van Helmont has probably received too much credit for priority [12].

The belief that plants were derived entirely from water was evidently held for some time, although in 1656 J. R. Glauber reported that a substance now known as potassium nitrate improved plant growth; in fact he considered it "the essential principle of vegetation." In 1699 John Woodward reported to the Royal Society that the growth of spearmint was proportional to the quantity of dissolved material in water and concluded that an unknown earthy substance as well as water contributed to the substance of plants, a conclusion Van Helmont could have reached if he had not regarded the small loss in soil weight as insignificant.

More important from the standpoint of anticipating knowledge of photosynthesis was the 1687 report of Thomas Bortherton to the Royal Society of evidence that plants were nourished partly by air and partly by water. In his *Vegetable Staticks* (1727) Stephen Hales, the English clergyman who is frequently regarded as the father of plant physiology, described experiments that led him to conclude that "plants very probably draw through their leaves some part of their nourishment," and that they imbibe and fix air in a solid form. The conclusions were reached from experiments in which mint sprigs were maintained in glass jars inverted over water. Hales added, "May not light also, by freely entering the expanded surfaces of leaves and flowers, contribute much to the ennobling principles of vegetation." In a vague way Hales had identified several factors involved in photosynthesis, and he probably had a broader general preconception of the process than the early investigators whose experiments contributed directly to its discovery.

Pioneering Basic Discoveries

In 1772 Joseph Priestley, an English nonconformist minister and part-time scientist, reported to the Royal Society that plants restore air injured by the burning of candles [6]. He stated in part, "Finding that candles would burn very well in air in which plants had grown a long time . . . I thought it was possible that plants might also restore the air injured by the burning of candles. Accordingly, on the 17th of August, 1771, I put a sprig of mint into a quantity of air in which a wax candle had burned out and found that on the 27th of the same month another candle burnt perfectly well in it." He also reported, "This remarkable effect does not depend upon any thing peculiar to mint," because he later obtained the same results using balm, groundsel, and spinach. Priestley had discovered the production of oxygen by plants, although he did not announce his discovery of oxygen until 1774 and was still thinking in terms the dephlogistication of air that had been phlogisticated by combustion or animals.

In 1773 Jan Ingenhousz, a Dutchman who was a court physician to Empress Maria Theresa, learned of Priestley's experiments and was greatly impressed. He spent the three summer months of 1778 performing over 500 experiments on the effects of plants on air. These confirmed Priestley's results and, in addition, Ingenhousz found that the process was rapid rather than slow, that light and green tissues were essential, and that by night plants "contaminated" the air. His 1779 report stated in part,"that this operation of plants is more or less brisk in proportion to the clearness of the day and the exposition of the plants; that this operation of plants diminishes toward the close of the day, and ceases entirely at sunset; that this office is not performed by the whole plant, but only by the leaves and green stalks; that all plants contaminate the surrounding air at night. . . ." Ingenhousz thus identified the necessity of green tissue and light for oxygen production by plants, but there was still no awareness of photosynthesis and its important biological roles except for "purification" of the air.

In 1782 Jean Senebier, a Swiss pastor, reported the discovery of another aspect of the process: that the air-restoring activity of plants depends on the presence of fixed air (carbon dioxide). He stated, "fixed air dissolved in water is the nourishment that plants extract from the air which surrounds them and the source of pure air which they provide by the transformation to which they submit the fixed air." In 1796 Ingenhousz also reported on the use of carbon dioxide by plants and, in addition, recognized that it was the source of carbon in plants.

In 1804 Nicolas Théodore de Saussure, a Swiss scholar, published *Recherches chimiques sur la végétation,* an account of numerous experiments he had conducted. These experiments confirmed and extended previous discoveries and added a missing piece to the photosynthetic jigsaw puzzle: water. He showed that the weight of the oxygen and organic matter produced totaled considerably more than the weight of the carbon dioxide used. Since his experimental plants were provided with only water and air, he concluded correctly that water as well as carbon dioxide was used. Thus, the general nature of photosynthesis finally emerged, and it was possible for de Saussure to write a summary equation for the process essentially similar to the present one.

However, one important concept about photosynthesis was still missing, and this was added by Julius Robert Mayer in 1845. He clearly recognized that, in addition to providing the organic matter of organisms, photosynthesis also provided the energy used by organisms by converting light energy to the chemical energy of the organic substance produced. The concept of the importance of photosynthesis for life on Earth was now clear and complete, as was the overall nature of the process.

The Research Plateau

For almost a century after Mayer's contribution relatively little was added to the understanding of the photosynthetic reactions. In 1864 the French plant physiologist T. B. Boussingault put photosynthesis on a quantitative basis and found that the oxygen to carbon dioxide (O_2/CO_2) ratio was 1. In 1862 Julius von Sachs, a German plant physiologist, reported that the organic substance produced was starch. He reached this conclusion after finding that, if part of a leaf was darkened, the iodine test showed starch only in the lighted portion of the leaf.

Although it was generally assumed that the photosynthetic reactions were more complex than indicated by the summary equation, the only theory that received any consideration until after 1930 was the 1918 formaldehyde theory of Willstätter and Stoll. They proposed that carbonic acid (H_2CO_3) attached to chlorophyll and that the chlorophyll bicarbonate was then converted to chlorophyll, formaldehyde (CH_2O), and oxygen. The formaldehyde then presumably polymerized into sugar. Although there was never any real evidence to support this theory, it survived for lack of anything better.

One of the most important contributions to an understanding of the photosynthetic reactions during this period came from the English plant physiologist F. F. Blackman. In 1905 Blackman demonstrated that at least one step in the process was not directly dependent on light. In 1914 G. Bredig proposed that the oxygen produced in photosynthesis came from water, rather than from the carbon dioxide as demanded by the formaldehyde theory, but he had essentially no evidence to support his proposal and it made little impact. Nine years later Thunberg proposed that water was the reducing agent in photosynthesis, but the formaldehyde theory still survived.

Although little progress was made in understanding the photosynthetic reactions, a substantial amount of information about other aspects of photosynthesis was accumulated during this 90-year period. Much was learned about the influence of various environmental factors on the rate of photosynthesis and about limiting factors.

In 1882 the German botanist T. W. Engelmann [6] made a rough determination of the action spectrum of photosynthesis in a most ingenious manner. He placed an algal filament and a culture of motile aerobic bacteria under a microscope and used a prism to expose the filament to the spectrum of light. The bacteria then congregated at the portions of the filament where the oxygen was most concentrated, that is, in the portion exposed to red light and to a lesser extent in the portion exposed to blue light (Figure 3-1).

There were also several other aspects of photosynthesis that received considerable attention before 1930. For example, between 1906 and 1922 the German chemists Willstätter and Stoll worked extensively on determination of the structure of chlorophyll molecules. Also, beginning in 1922 Otto Warburg and

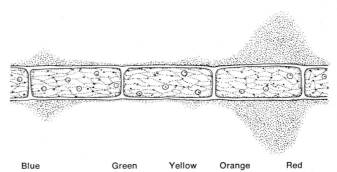

| Blue | | Green | Yellow | Orange | | Red |

Figure 3-1. Engelmann's pioneering demonstration of the action spectrum of photosynthesis. A filamentous alga, in a culture also containing motile aerobic bacteria, was exposed to a microspectrum. The bacteria moved to the regions of higher oxygen concentration, that is, the regions of more rapid photosynthesis. [Redrawn from Engelmann (1882). After V. A. Greulach and J. E. Adams, *Plants: An Introduction to Modern Botany*, 2nd ed., John Wiley & Sons, Inc., New York, 1967]

his associates conducted numerous experiments designed to determine the quantum efficiency of photosynthesis. They and other investigators continued research on quantum efficiency for many years. Their conclusions will be discussed later in this chapter.

The Modern Period of Research

It was during the 1930s that the first major breakthroughs in understanding the photosynthetic reactions occurred, although really substantial progress was not made until after World War II. One of the first important clues came from the investigations of bacterial photosynthesis in the late 1920s and early 1930s by C. B. van Niel [21] and others. Bacterial photosynthesis differs from green plant photosynthesis primarily in the nature of the light-absorbing pigments and in the fact that substances such as hydrogen or hydrogen sulfide (H_2S) rather than water serve as the hydrogen donors [20]. For example, the summary equation for photosynthesis by the green sulfur bacteria would be

$$CO_2 + 2\ H_2S \longrightarrow (CH_2O) + H_2O + 2\ S$$

(Note that if the product were a hexose sugar the entire equation would have to be multiplied by 6.) The comparable summary equation for green plant photosynthesis would be

$$CO_2 + 2\ H_2O \longrightarrow (CH_2O) + H_2O + O_2$$

Comparisons such as these led van Niel to conclude that the oxygen was derived from the water, rather than from the carbon dioxide, just as the sulfur was derived from the hydrogen sulfide. This supported Bredig's 1914 proposal and made it appear that the formaldehyde theory was untenable.

Further evidence that all the oxygen is derived from water came from at least two other sources. In 1937 the English biochemist Robert Hill found that, although isolated chloroplasts or chloroplast fragments did not take up carbon dioxide, they would evolve oxygen if provided with ferric salts, which were reduced to ferrous salts by accepting electrons. He later found that many other oxidants could be reduced, including quinone (Q) to hydroquinone (QH_2) and various dyes (D) to their reduced colorless forms (DH_2).

$$2\ H_2O + 2\ D \longrightarrow 2\ DH_2 + O_2$$

In 1941 Samuel Ruben and his coworkers used the stable isotope oxygen-18 (^{18}O) to trace the origin of oxygen from plants. If ^{18}O was used to label water, all the oxygen produced was ^{18}O; whereas, if the carbon dioxide was labeled with ^{18}O, all the oxygen produced was the ordinary isotope ^{16}O. The validity of these tracer experiments has since been questioned by some investigators, but it is now certain that oxygen is derived from water and not from carbon dioxide.

Further evidence that carbon dioxide was not involved in the photochemical reactions was provided in 1936 by Wood and Werkman. They found that heterotrophic bacteria could fix carbon dioxide by carboxylation, and it was later discovered that organisms in general can fix carbon dioxide whether or not they are photosynthetic [23]. Carboxylation in general can be represented in general as RH + CO$_2$ $\longrightarrow$ RCOOH, where RH is an organic compound and RCOOH is the organic acid derived from it by carboxylation. If RH is already an organic acid, carboxylation results in an additional COOH in the molecule.

Tracing the pathway of carbon through the photosynthetic reactions now became a discrete and soluble problem, depending only on the availability of suitable chemical tools and techniques and investigators competent to use them. The use of radioisotopes of carbon as tracers was attempted by Ruben and his associates in the late 1930s, but the only radioisotope available at that time was carbon-11 (^{11}C) and it had such a short half-life that it provided little information. When quantities of carbon-14 (^{14}C), a radioisotope with a half-life of around 6000 years, became available after World War II the stage was set for productive research on the problem. Several groups of investigators began work, the most productive being Melvin Calvin and his associates [4] at the University of California at Berkeley. Since 1950 they have published a series of papers providing more and more information on the sequence of compounds in which the carbon from radioactive carbon dioxide appeared. This work would have been difficult if not impossible without another relatively new chemical tool: two-dimensional paper chromatography.

In brief, the experiments were conducted by providing cultures of unicellular algae with radioactive carbon dioxide for short periods of time, killing them promptly, extracting the solutes, and separating these by chromatography (Figures 3-2 and 3-3). The substances present in the chromatograms were

Figure 3-2. Apparatus used by Melvin Calvin for supplying unicellular algae (in flat flask) with ^{14}CO$_2$ for brief periods of photosynthesis, ranging from a few seconds to a minute or so. At the end of each selected time period, the algae were suddenly killed by opening the stopcock and dropping them into the beaker of organic solvent. The algal extract was then chromatographed and radioautographs of the chromatographs were made. [Courtesy of Melvin Calvin, University of California, Berkeley]

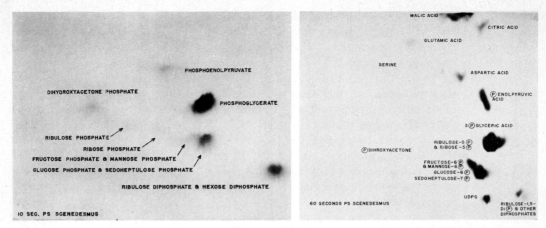

Figure 3-3. Two of the radioautographs of chromatographs made by Calvin and his co-workers. Note the marked radioactivity of phosphoglyceric acid (PGA) after 10 sec photosynthesis (phosphoglycerate) and also after 60 sec (3 Ⓟ glyceric acid), as well as the increased number of labeled compounds after 60 sec. [Courtesy of Melvin Calvin, University of California, Berkeley]

determined and the radioactive ones were identified by making radioautographs. By exposing the algae to $^{14}CO_2$ for various periods of time, it was possible to determine the time sequence in which the compounds became radioactive and reach conclusions about the sequence of reactions.

At the same time that these studies of the pathway of carbon were in progress other investigators, including Robert Emerson, William Arnold, Hans Graffon, James Franck, Daniel Arnon, Bessel Kok, and Eugene Rabinowitch, were working on the even more difficult problem of clarifying the nature of the photochemical reactions. The questions involved biophysical as well as biochemical phenomena, and the investigators were predominantly biochemists and biophysicists. Some of their contributions will be mentioned later.

As a result of the extensive and intensive research since about 1940, knowledge of the details of the photosynthetic reactions has increased from very little to almost complete step-by-step reactions. However, the present state of knowledge is not complete nor necessarily correct in every detail, and continuing research is providing more and more information that is extending and modifying the present concepts.

Before describing the photosynthetic reactions in greater detail, it may be well to note that our knowledge of photosynthesis has been provided by the work of numerous investigators who were predominantly not plant physiologists, although plant physiologists have contributed most of our knowledge of the factors that influence the rate of photosynthesis. In the early period the basic contributions were made mostly by clergymen, physicians, and others whose research was a sideline or hobby. Later on many of the important contributions were made by chemists, physicists, biochemists, and biophysicists. It is not surprising that a process as important and complex as photosynthesis has attracted the interest of a wide variety of scientists.

The Photosynthetic Reactions

It is now clear that photosynthesis consists of two quite distinct but closely related sets of reactions [15]. These are often referred to as the light reactions and dark reactions, but we shall call them the photochemical and carbon reduction reactions. In the photochemical reactions, light energy is absorbed by the chloroplast pigments and converted into chemical bond energy in adenosine triphosphate (ATP) and nicotinamide adenine dinucleotide phosphate ($NADP^+$) that has been reduced to NADPH (Figure 3-4). Water provides the electrons (e^-) and protons (H^+) used in the reductions, and its oxygen is the source of the O_2 produced by photosynthesis.

The reduction of the CO_2 that has been fixed to carbohydrate requires the

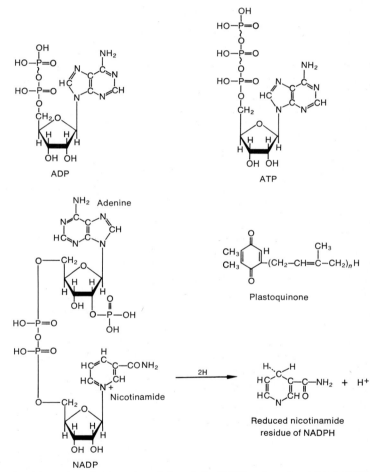

Figure 3-4. Structural formulas of adenosine diphosphate (ADP), adenosine triphosphate (ATP), plastoquinone, and nicotinamide adenine dinucleotide phosphate ($NADP^+$). At the right is the nicotinamide residue after the reduction of $NADP^+$ to NADPH. Nicotinamide adenine dinucleotide (NAD^+) differs from $NADP^+$ in that an —OH is present in place of the phosphate indicated by the arrow.

energy incorporated in ATP and NADPH. This reduction is therefore dependent upon the photochemical reactions. Although CO_2 fixation and reduction to not use light directly, they continue only momentarily after the photochemical reactions cease because the ATP and NADPH are quickly exhausted. Thus the term "dark reactions" is not too appropriate.

The summary equations for the photochemical and carbon reduction reactions per molecule of O_2 produced and CO_2 consumed are as follows.

$$2\,H_2O + 2\,NADP^+ + 3\,ADP + 3\,Pi \longrightarrow$$
$$O_2 + 2\,NADPH + 2\,H^+ + 3\,ATP$$

$$CO_2 + 2\,NADPH + 2\,H^+ + 3\,ATP \longrightarrow$$
$$(CH_2O) + 2\,NADP^+ + 3\,ADP + 3\,Pi + H_2O$$

Pi represents inorganic phosphate. The water released when ATP is synthesized and used when it is hydrolyzed to adenosine diphosphate (ADP) and Pi are not shown, nor is the required light energy. If all substances that appear on the left side of one equation and the right side of the other, and vice versa, are cancelled out (taking into consideration the number of molecules of each substance) and the remainder is then multiplied by 6 (to provide for the synthesis of a molecule of hexose sugar), the old standard summary equation for photosynthesis results.

$$6\,CO_2 + 6\,H_2O \longrightarrow C_6H_{12}O_6 + 6\,O_2$$

The following description of the photosynthetic reactions relates to eucaryotic plants. It will be recalled that the blue-green algae and photosynthetic bacteria do not have chloroplasts, that the latter have bacteriochlorophyll, which differs from the chlorophylls of green plants, and that hydrogen donors other than water are used so that no oxygen is produced.

The Photochemical Reactions

LIGHT ABSORPTION.   Photosynthesis begins with the absorption of light energy by the chloroplast pigments [17]. All green plants contain chlorophyll a and the vascular plants, bryophytes, and green algae also have chlorophyll b. The brown algae, diatoms, and some other algae have chlorophyll c instead of b, whereas the red algae have chlorophyll a and in some species d. Blue-green algae have only chlorophyll a, but both the blue-green and the red algae contain phycocyanin and phycoerythrin, which are blue and red in color. All green plants also have carotenes and xanthophylls in their chloroplasts. The structural formulas of a number of these pigments are shown in Figure 3-5.

Although all the chlorophylls absorb some light of all wavelengths and have their absorption peaks in the red and blue regions of the spectrum, the absorption peaks of the different chlorophylls are at different wavelengths. The carotenes and xanthophylls have their absorption peaks in the blue region between 420 and 480 nm. The absorption peaks of phycoerythrins are in the green and yellow regions and those of phycocyanins are in the orange region (Table 3-1, Figure 3-6).

The action spectrum of photosynthesis parallels the total absorption spectrum of the plant much more closely than it does the absorption spectrum of the

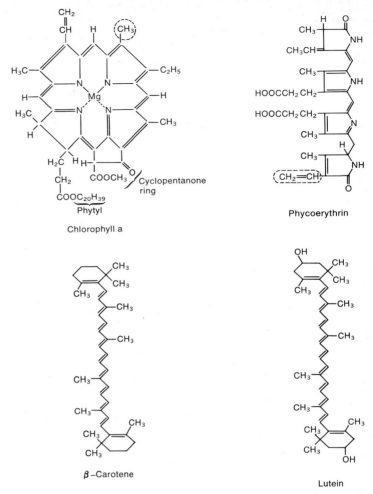

Figure 3-5. Structural formulas of several of the photosynthetic pigments. In chlorophyll b the circled —CH_3 of chlorophyll a is replaced by —CHO. In phycocyanin the circled CH_2=CH— of phycoerythrin is replaced by CH_3CH_2—. Note the absence of O in the carotene molecule, in contrast with the two —OH in lutein (a xanthophyll).

chlorophylls (Figure 3-7). This is interpreted to mean that light absorbed by the carotenoids, phycocyanins, and phycoerythrins as well as that absorbed by the chlorophylls is used in photosynthesis. However, only chlorophyll a can donate light-energized electrons to the electron acceptors involved in the subsequent reductions. The energy absorbed by the other pigments is effective in photosynthesis only when it is transferred to chlorophyll a by resonance [18]. Energy can also be transferred from one molecule of chlorophyll a to another (or to other specific pigments). Energy transfer from chlorophyll b to chlorophyll a or from one molecule to another of either one is highly efficient. By comparison, the efficiency of energy transfer from phycocyanin or phycoerythrin to chlorophyll a ranges between 60 and 100%, from fucoxanthin to chlorophyll a about 80%, and from carotenes to chlorophyll a 20 to 50%. Energy

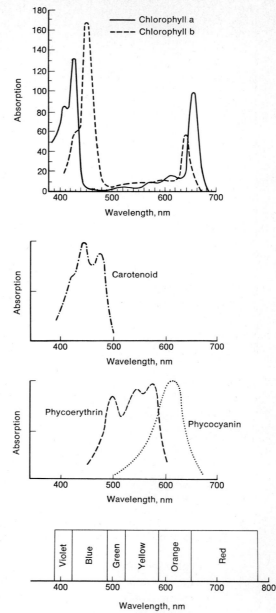

Figure 3-6. Absorption spectra of several of the photosynthetic pigments. The spectral scale at the bottom shows the colors visualized by the human eye in the various regions of the spectrum.

transfer from one pigment molecule to another can occur only when the molecules are closely packed. It is believed that the energy transfer from molecule to molecule is actually a migration of high-energy electrons or, viewed from the reverse standpoint, a migration of electron holes created by the emission of electrons.

The energy of light (and other electromagnetic radiation) can be exchanged only in discrete packets called **quanta** or **photons** [5]. The energy of a quantum

Table 3-1 Characteristic absorption peaks of photosynthetic pigments, exclusive of photosynthetic bacteria

Pigment	Absorption peaks, nm		Occurrence
	In organic solvents	*In cells*	
Chlorophylls			
Chlorophyll a	420, 660	435, between 670 and 700 in different forms	Universal
Chlorophyll b	453, 643	480, 650	Higher plants, green algae
Chlorophyll c	425, 625	645	Brown algae, diatoms
Chlorophyll d	450, 690	740	Some red algae
Carotenes			
α-Carotene	420, 440, 470		Higher plants, red algae, some other algae
β-Carotene	425, 450, 480		Principal carotene of most plants
Xanthophylls			
Luteol	425, 445, 475		Higher plants, green and red algae
Violaxanthol	425, 450, 475		Higher plants
Fucoxanthol	425, 450, 475		Brown algae, diatoms
Phycobilins			
Phycoerythrins		490, 546, 576	Red algae, some blue-green algae
Phycocyanins		618	Blue-green algae, some in red algae

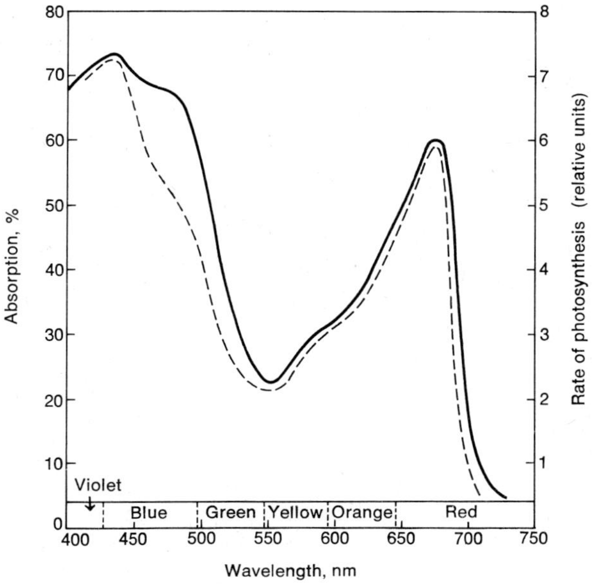

Figure 3-7. Absorption spectrum (solid line) of Elodea leaves and action spectrum of Elodea photosynthesis (broken line). The divergence of the two curves in the blue region results from the fact that much of the absorption is by carotenoids, which are relatively inefficient as photosynthetic pigments. [Data of Hommersand and Haxo. After V. A. Greulach and J. E. Adams, *Plants: An Introduction to Modern Botany,* 2nd. ed., John Wiley & Sons, Inc., New York, 1967]

is proportional to the frequency of the radiation and inversely proportional to its wavelength: $E = h\nu = hc/\lambda$, where ν is the frequency of the radiation, c is the velocity of light, λ is the wavelength in centimeters and h is Planck's constant (6.6×10^{-27} erg sec). Calculating from this equation, radiation of 800 nm wavelength has an energy content of 2.5×10^{-12} erg/quantum whereas radiation of 400 nm wavelength has 5×10^{-12} erg/quantum. For biological use a larger energy unit, kilocalories per einstein (kcal/einstein), is desirable. An **einstein** is a mole equivalent of quanta (6×10^{23} quanta). Radiation of 800 nm wavelength has 36 kcal/einstein whereas that of 400 nm has 72 kcal/einstein. The energy of a quantum is indivisable; a whole quantum is essential for each photochemical event. The minimum quantum energy effective in green plant photosynthesis is about 40 kcal/einstein, but a higher quantum energy is no more effective. Each photochemical event, for example, the excitation of a chlorophyll electron, requires a quantum. Even if an 80 kcal/einstein quantum were absorbed it would drive only one photochemical event, not two.

The excitation of a pigment molecule by the absorption of a quantum results in raising an electron to a higher energy level and consequently in the migration of electrons from one pigment molecule to another and then through a series of electron acceptors, which will be described later. As the electrons flow from one acceptor to another at a lower energy level, the released energy is used in generating ATP and NADPH. If no electron hole is available, the excited electron quickly (in about 1.5×10^{-8} sec) returns to its original energy level. The released energy may be converted in several ways, generally by the emission of light of somewhat longer wavelength (lower energy) than that absorbed. This is **fluorescence,** a phenomenon that can be observed readily in a chlorophyll solution.

A variety of evidence indicates that the pigment molecules of chloroplasts function as photosynthetic units or pigment systems, each containing 200 to 300 molecules of chlorophyll. James Franck proposed that only one chlorophyll a molecule in such a system projected from the lipid layer of a lamella into the aqueous layer, where the electron acceptor molecules are located, and that only this exposed chlorophyll molecule could donate electrons to the acceptors. This chlorophyll molecule corresponds to what has been called **reaction center** chlorophyll. Energy transfer from one pigment molecule to another occurs only within a pigment system.

The concept of such a photosynthetic unit, containing some 200 to 300 molecules of chlorophyll per reaction center, was derived from kinetic considerations by Emerson and Arnold. They found that, under conditions of highest photosynthetic efficiency and light saturation, only one molecule of oxygen was produced per some 2500 chlorophyll molecules. Since about 10 quanta are required for the production of one molecule of oxygen, an estimate of some 200 to 300 chlorophyll molecules per light reaction system is obtained. This concept is also supported by the fact that in chloroplasts there is approximately one molecule of cytochrome f (one of the electron acceptors) and one molecule of P_{700} (a special reaction center chlorophyll) per 300 molecules of chlorophyll. It has been suggested that the small spheres visible in the thylakoid membranes in high resolution electronmicrographs (Chapter 2) may be the physical manifestations of the photosynthetic units.

There are evidently two different pigment systems, each playing a different and essential role in the photochemical reactions of photosynthesis. The first clue to this was provided by Robert Emerson of the University of Illinois in the late 1950s. He was conducting experiments on the rate and efficiency of photosynthesis of algae exposed to monochromatic light of various wavelengths, and he found that those above 685 nm were less efficient in driving photosynthesis than would be expected from the absorption spectra of the chlorophylls. Simultaneous exposure to two beams of light, one above 685 nm and one in the range of 650 to 680 nm, resulted in a rate of photosynthesis greater than the sums of the rates obtained when each beam was supplied separately. For example, in one experiment where the rate of oxygen evolution was 100 units with the shorter red wavelength and 20 units with the longer, it was 160 with the two combined. This enhancement of photosynthesis is known as the **Emerson effect.** This finding led Hill and Bendall [10] to postulate two separate light reactions for photosynthesis, each mediated by a different pigment system. One of these (**pigment system I,** or PSI) is activated by light with wavelengths longer than 685 nm and the other (**pigment system II,** or PSII) by red light of shorter wavelengths (650 nm). The reaction center or electron trap of PSI is P_{700}. The trap of PSII is less certain, but is sometimes given as Chl a II or P_{690}.

NONCYCLIC ELECTRON TRANSPORT AND PHOTOPHOSPHORYLATION. The pathway of electron transport in the two-pigment system model is outlined in Figure 3-8. This system was proposed by Hill and Bendall in 1960 and subsequently modified by them and other investigators including Duysens, French, Kok, and Witt. The approximate oxidation-reduction potential in electron volts (ev) of each electron-acceptor-donor is shown, the most negative potential being at the highest energy level. The process may be considered to begin with the absorption of quanta by the pigment molecules of PSI and the transfer of the energy to the P_{700} reaction center where a charge separation occurs with the resultant production of an oxidant and a reductant. Each event requires the absorption of 2 quanta ($h\nu$), resulting in the transfer of two energized electrons (e^-) from P_{700} to a ferredoxin-reducing substance at the high energy level of -0.6 ev. This reducing substance then donates the electrons to ferredoxin, reducing it. In turn, the ferredoxin donates the electrons to $NADP^+$, reducing it to NADPH.

$$2 \text{ ferredoxin-Fe}^{3+} + 2\,e^- \longrightarrow 2 \text{ ferredoxin-Fe}^{2+}$$
$$2 \text{ ferredoxin-Fe}^{2+} + NADP^+ + 2\,H^+ \longrightarrow 2 \text{ ferredoxin-Fe}^{3+}$$
$$+ \text{ NADPH} + H^+$$

The protons are derived from the water used in photosynthesis. Some of the absorbed light energy has thus been converted into chemical bond energy of NADPH.

Returning now to P_{700} we find that it has two electron holes (2 Chl^+). These are filled by electrons derived from water by way of PSII and its associated enzymes. The absorption of 2 quanta by PSII results in the transfer of two energized electrons from its reaction center (trap, P_{690}, or Chl a II) to an electron acceptor designated as Q, which may be a quinone. The two electrons are

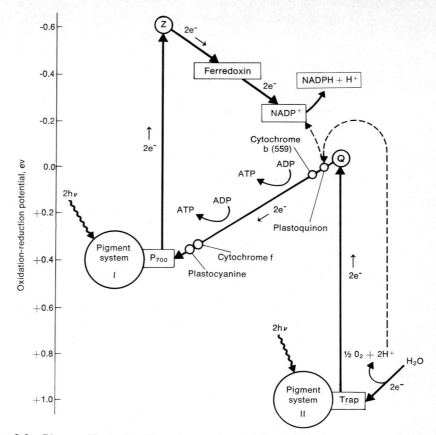

Figure 3-8. Diagram illustrating the pathway of noncyclic electron flow, which involves PSI and PSII. The transfer of H⁺ from water to plastoquinone and then NADP⁺ is shown by the broken lines. In the final analysis the electrons are derived from water and end up in reducing NADP⁺ to NADPH. Generation of two molecules of ATP is shown, but only one may be produced. The processes illustrated must take place twice to produce one molecule of oxygen and the two molecules of NADPH needed for reduction of one molecule of carbon dioxide. See the text for further explanation.

replaced by two electrons from water, which is thus the ultimate electron donor. From Q the electrons are transferred through a series of electron acceptors (plastoquinone (PQ), cytochrome b_{559}, cytochrome f, plastocyanin) at progressively lower energy levels and then to P_{700}. The energy released in these electron transfers is used in synthesizing ATP from ADP and phosphate. Thus, the light energy absorbed has been converted to chemical bond energy of ATP as well as of NADPH. It is not certain whether one or two molecules of ATP are synthesized, but there is increasing evidence that it is two as shown.

It will be noted that the scheme accounts for the production of only one-half a molecule of water, so the entire process must occur twice for the production of a molecule of oxygen. This requires a total of 4 e^- and 4 H⁺ from water, which means that four molecules of water must be oxidized for each molecule of oxygen produced. The details of this reaction are not known, but a manganese-containing enzyme and several electron acceptors appear to be involved in the transfer of electrons from the water to PSII. In skeleton outline the events are probably as follows.

$$4\,H_2O \longrightarrow 4\,H^+ + 4\,OH^-$$
$$4\,OH^- \longrightarrow 4\,(OH) + 4\,e^-$$
$$4\,OH \longrightarrow 2\,H_2O + O_2$$

The net consumption of H_2O is therefore two molecules per molecule of O_2 produced. Both the protons and electrons from the water are used in reducing the plastoquinone: $2\,e^- + 2\,H^+ + PQ \longrightarrow 2\,PQH_2$. The protons are not involved in further reactions except for the production of NADPH.

CYCLIC ELECTRON TRANSPORT AND PHOTOPHOSPHORYLATION. Daniel Arnon [1] had identified noncyclic photophosphorylation and also a second type of photophosphorylation in chloroplasts which he called **cyclic** since the electrons are not derived from water. Instead, the electrons from the reaction center return to it after passing through a series of acceptors. The cyclic pathway is actually a short-circuiting of the usual noncyclic pathway, and involves only PSI (Figure 3-9). The electrons from P_{700} are transferred to the ferredoxin-

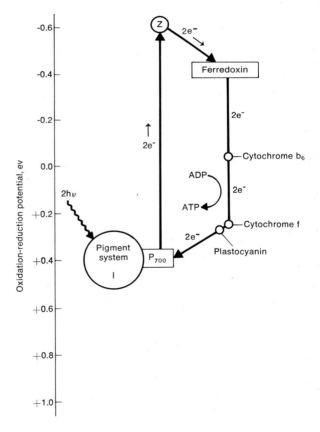

Figure 3-9. Cyclic electron flow and photophorylation, which involves only PSI. The electrons from PSI return to it, thus completing the cycle. Note the participation of cytochrome b_6, which is not involved in noncyclic electron flow, and the fact that no water is used and no NADPH is produced.

reducing substance, and perhaps to ferredoxin as in the noncyclic process, but then they are transferred to cytochrome b_6 (b_{563}), to cytochrome f, plastocyanin, and back to P_{700}. There is enough energy released to drive the synthesis of a molecule of ATP, but it will be noted that no water is used, no oxygen is produced, and no $NADP^+$ is reduced to NADPH.

Cyclic electron transfer and photophosphorylation occur when plants are treated with the photosynthetic inhibitor DCMU [3-(3',4'-dichlorophenyl)-1,1-dimethyl urea], which blocks electron transfer between PSII and PSI. There is doubt as to what extent, if any, cyclic electron transport occurs normally. In any event, it apparently occurs primarily when photosynthesis is proceeding at a low rate. As a short-circuiting of the usual noncyclic process, it is competitive with it.

However, the possibility remains that cyclic photophosphorylation may play an essential role in photosynthesis. In the subsummary equations for photosynthesis given earlier in this chapter it was shown that three molecules of ATP are required for the reduction of one molecule of carbon dioxide. If each noncyclic electron transfer generates two molecules of ATP, the two transfers per molecule of oxygen produced and carbon dioxide consumed would provide four molecules of ATP, which is more than enough. If, however, only one molecule of ATP is generated (two per molecule of oxygen produced) in the noncyclic process, there is an ATP deficit. This could be made up by cyclic photophosphorylation.

QUANTUM EFFICIENCY OF PHOTOSYNTHESIS. For many years investigators of photosynthesis have been interested in the quantum efficiency of the process [18], that is, in the number of quanta of light energy required to reduce one molecule of carbon dioxide or produce one molecule of oxygen. Beginning in 1922, Otto Warburg and his associates conducted many experiments that led them to conclude that the quantum requirement was 4. This seemed reasonable, since four electrons from water must be transferred for reduction of each carbon dioxide molecule and, for almost 20 years, the 4-quantum requirement was accepted. However, from 1938 on numerous investigators, notably Robert Emerson and his coworkers, used improved techniques for determination of the quantum requirement and never obtained a figure less than 8. Usually a higher requirement of 10 to 12 was obtained. However, Warburg continued to conduct experiments which he claimed supported his earlier results and even reported a lower quantum requirement of about 3. To the day of his death he never conceded the validity of the 8-quanta minimum, but now essentially all other investigators do. Aside from the experimental results, the 8-quanta minimum is in accord with the Hill-Bendall two-pigment scheme. Examination of this scheme (Figure 3-8) reveals that, for the production of a molecule of oxygen (and the reduction of a molecule of carbon dioxide), 4 quanta must be absorbed by PSI and 4 more by PSII, for a total of 8.

The Reduction of Carbon

The pathway of carbon dioxide fixation and reduction worked out by Melvin Calvin and his associates [2] at the University of California was for some time generally considered universal among plants, but since about 1965 it has been recognized that another and quite different pathway exists in a considerable number of plants including corn, sugarcane, and Amaranthus. Still other pathways may be discovered in the future among the hundreds of thousands of species of plants whose photosynthetic processes have not yet been investigated. Also, some photosynthetic bacteria fix and reduce carbon dioxide by a pathway that appears to be essentially the reverse of the citric acid cycle of respiration.

Here we shall discuss two pathways of carbon dioxide fixation and reduction: the reductive ribulose diphosphate cycle (Calvin-Benson cycle) and the phosphoenol pyruvate cycle (Hatch-Slack cycle) which occurs in corn and other plants.

THE RIBULOSE DIPHOSPHATE CYCLE. This cycle can be regarded as having three principal components: (1) the carboxylation of ribulose-1,5-diphosphate with the formation of two molecules of phosphoglyceric acid (PGA); (2) the reduction of PGA to triose phosphates by ATP and NADPH from the light reactions; and (3) the regeneration of ribulose-1,5-diphosphate by means of a complicated series of sugar interconversions, involving use of ATP from the light reactions (Figure 3-10). The triose phosphates drained out of the cycle, or the fructose-

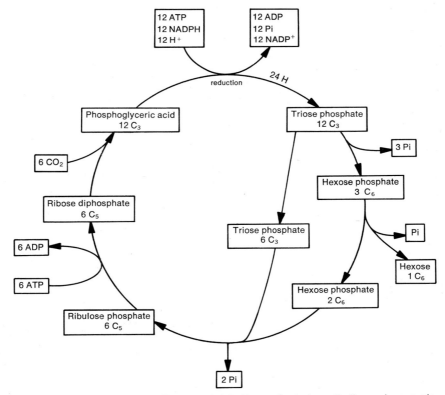

Figure 3-10. The ribulose diphosphate (Calvin-Benson) photosynthetic carbon cycle, simplified by omission of most of the sugar interconversions but providing a carbon balance sheet. Note the use of the ATP, NADPH, and H+ provided by the photochemical reactions of photosynthesis. The six turns of the cycle required for the production of one molecule of hexose are shown.

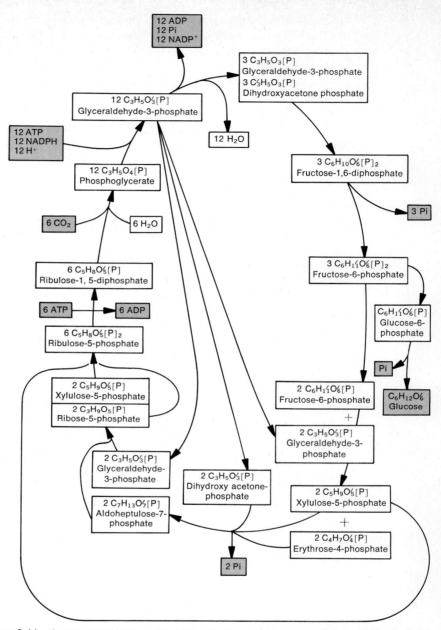

Figure 3-11. A more complete outline of the ribulose diphosphate cycle than Figure 3-10, showing most of the steps involved in the interconversion of the sugars. The eventual result of the sugar interconversions is the regeneration of ribulose-1,5-diphosphate.

1,6-diphosphate usually formed from them, can be regarded as the product of the dark reactions, although to balance the overall summary equation for photosynthesis it is necessary to go on to nonphosphorylated sugars such as fructose or glucose. A variety of other carbohydrates including sucrose and starch may be produced within the chloroplasts, but it seems best not to consider these as primary products of photosynthesis since they can be synthesized from monosaccharides elsewhere in the cell. Actually, carbohydrate photosynthesis is essentially completed with the production of the triose phosphate.

The detailed cycle, as pieced together by Calvin, his coworker Benson, and other investigators, is presented in Figure 3-11 and is largely self-explanatory after careful inspection. That ribulose-1,5-diphosphate (RDP) is actually the carbon dioxide acceptor has been supported by a variety of evidence, for example, RDP accumulates and PGA decreases when the plant is suddenly deprived of carbon dioxide, and also RDP virtually disappears in the dark (Figure 3-12). The enzyme that catalyzes the carboxylation is present in chloroplasts. This enzyme (carboxydismutase) also hydrolyzes the resulting unstable six-carbon acid, oxidizes one carbonyl group to a carboxyl, and reduces another carbonyl group to alcohol, thus giving rise to two molecules of PGA, a three-carbon compound.

Up to this point the products of the light reactions have not been involved directly, but now the energy from ATP and the protons and electrons from the NADPH + H$^+$ are used in reducing the PGA to 3-phosphoglyceraldehyde, a phosphorylated triose. The resulting ADP, inorganic phosphate, and NADP$^+$ are now available for reduction in subsequent light reactions.

Thus, two molecules of triose are produced per molecule of carbon dioxide fixed. Only one out of six triose phosphates is drained out of the cycle as a net product; the other five participate in the complicated series of sugar interconversions that finally result in the production of ribulose-5-phosphate. ATP from the light reactions is used in phosphorylating this to RDP, thus completing the cycle and regenerating the carbon dioxide acceptor that is essential for keeping the cycle turning.

Note that the incorporation of carbon dioxide into carbohydrate is a stepwise process. Only one of the six carbons in the two molecules of triose phosphate produced in one turn of the cycle comes from carbon dioxide, the other five are derived from the RDP. Only after six turns of the cycle is enough carbon

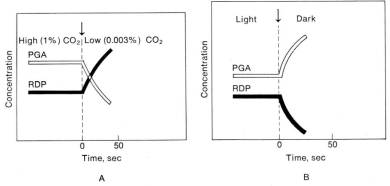

A   B

Figure 3-12. **(A)** The effect of sudden reduction of carbon dioxide concentration (at time 0) on the concentrations of PGA and RDP. **(B)** The effect of turning off the light on the concentrations of PGA and RDP. [Data of M. Calvin and his coworkers. Diagrammatical graphs after E. Rabinowitch and Govindjee, *Photosynthesis*, John Wiley & Sons, Inc., New York, 1969]

dioxide fixed to provide the carbon for the net production of one molecule of hexose. In the course of these six turns 12 molecules of triose phosphate are produced, but 30 of their 36 carbons are used in regenerating a molecule of RDP each turn of the cycle. Of course, in the final analysis, every carbon atom in the organic compounds of a plant (including those in RDP) has been derived from carbon dioxide.

At this point subsummaries of the light and dark reactions involved in the photosynthetic production of a molecule of hexose may be helpful.

$$12\ H_2O + 12\ NADP^+ + 12\ ADP + 12\ H_3PO_4 \longrightarrow$$
$$12\ NADPH + 12\ H^+ + 12\ ATP + 6\ O_2$$

$$6\ ADP + 6\ H_3PO_4 \longrightarrow 6\ ATP$$

$$6\ CO_2 + 12\ NADPH + 12\ H^+ + 18\ ATP \longrightarrow$$
$$12\ NADP^+ + 18\ ADP + 18\ H_3PO_4 + 6\ H_2O + C_6H_{12}O_6$$

Cancelling out each substance that appears on both the right and left sides of the above equations we have left the old summary equation for photosynthesis.

$$6\ CO_2 + 6\ H_2O \longrightarrow C_6H_{12}O_6 + 6\ O_2$$

Thus, although this equation reveals little about the nature of the photosynthetic reactions, it is still valid for the photosynthetic production of hexose.

OTHER PRODUCTS OF PHOTOSYNTHESIS. Sugars are by no means the only organic compounds produced by photosynthesis [3]. The PGA, instead of being converted to triose phosphate, at certain times or in certain plants may be used in making pyruvic acid, other plant acids, amino acids, or fatty acids. Triose phosphate may be converted to glycerol phosphate, and this can then react with fatty acids and form lipids. In certain species of algae lipids, rather than carbohydrates, are the usual products of photosynthesis. All of these processes require ATP and NADPH from the light reactions, but the energy in these substances may also be used in quite different ways, such as the reduction of nitrates to ammonia, the reduction of sulfate to sulfide, and, in some photosynthetic bacteria and blue-green algae that fix nitrogen, the reduction of nitrogen to ammonia. ATP from photosynthesis may also be used in protein synthesis. This diversity of syntheses driven by ATP and NADPH from the light reactions makes it difficult to provide a neat and simple definition of photosynthesis, unless it is restricted simply to the conversion of light energy to chemical bond energy in the light reactions themselves.

THE PHOSPHOENOL PYRUVATE CYCLE. In 1965 Kortschak, Hartt, and Burr [11] working in Hawaii reported that malic and aspartic acids were the principal labeled products in sugarcane after brief periods of photosynthesis with $^{14}CO_2$. In 1966 in Australia Hatch and Slack [9] confirmed this and also found substantial quantities of labeled oxalacetic acid. After longer periods of photosynthesis much of the label was in PGA, hexose phosphates, and sucrose. They proposed that the oxalacetic acid was produced by carboxylation of phosphoenol pyruvic acid.

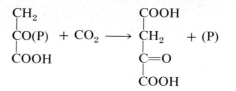

They found this to be the only significant carboxylation, and found the necessary enzyme (phosphoenolpyruvate carboxylase) to be present. The oxalacetic acid could be readily converted to aspartic acid by amination or to malic acid by reduction. However, they suggested that, although RDP was not serving as a carbon dioxide acceptor, it reacted with oxalacetic acid to form 2 PGA and a molecule of pyruvic acid. The PGA could lead to carbohydrates by the usual pathways, whereas the pyruvic acid could be converted to phosphoenol pyruvic acid at the expense of energy from ATP, thus regenerating the carbon dioxide acceptor.

Subsequently this pathway was found to operate in other tropical grasses including corn (*Zea mays*), Sorghum, Paspalum, Axonopus, Digitaria, Chloris, and Eagrostis, as well as in a sedge (Cyperus) and several dicotyledonous plants including pigweed (Amaranthus) and Atriplex. Laetsch [13] has pointed out that these plants all have prominent bundle sheaths containing specialized chloroplasts differing from those in the mesophyll cells. They are also characterized by high rates of photosynthesis at high light intensities and high temperatures and by the ability to reduce the external carbon dioxide concentration to less than 5 parts per million (ppm), in contrast with about 50 ppm for plants having the Calvin-Benson cycle. The sugarcane photosynthetic pathway is highly efficient and productive, particularly under tropical environmental conditions, and it seems likely that many more genera of plants will be found to fix carbon dioxide by way of phosphoenol pyruvic acid.

Efficiency of Photosynthesis in Energy Conversion

The conversion of one kind of energy to another is never 100% efficient. The efficiency is commonly as low as 25% or less, most of the energy being lost as heat, sound, and so on. When a generator converts mechanical energy into electrical energy there is a substantial loss as heat and sound, whereas conversion of electrical energy to light by an incandescent bulb is accompanied by much heat loss. Considered from the most basic standpoint, that is, conversion to chemical bond energy of the light energy absorbed by the photosynthetic pigments, photosynthesis is a relatively efficient energy conversion process [18]. Red light of 680 nm wavelength has about 43 kcal/einstein and thus 344 kcal in the 8 mole quanta required for the reduction of 1 mole of carbon dioxide. For production of 1 mole of hexose sugar this comes to 2064 kcal. The chemical bond energy of 1 mole of hexose is 673 kcal, so the energy efficiency of the conversion is about 32% (673 kcal/2064 kcal). The efficiency of photosynthesis with broad spectrum light is less, since the energy of quanta increases with decreasing wavelength and this extra energy is wasted. However, if the average quantum energy of absorbed sunlight is estimated at roughly 50 kcal/einstein, photosynthetic energy conversion is still about 28% efficient (673 kcal/2400 kcal). The energy loss includes that lost in transfer from one molecule to another in a pigment system and also losses in the subsequent electron transfers.

When, however, the efficiency of photosynthesis is calculated on other bases (such as the percentage of the total solar radiation received by a plant, the percentage of radiation received by an acre of land that is used in photosynthesis, or the percentage of solar radiation received by the earth annually that is used in photosynthesis) the efficiency of photosynthesis is much lower. Some of the light that reaches chloroplasts is reflected or transmitted rather than being absorbed. Also some of the light that reaches a plant misses chloroplasts entirely. This is mostly reflected or transmitted, and most of the fraction that is absorbed is converted into heat. A little of the absorbed light drives photochemical reactions other than photosynthesis, as we shall see later. Of the radiation received by a plant only about 1–3% is used in photosynthesis.

The percentage of the sun's radiation received by an acre of land that is used in photosynthesis varies considerably, of course, with the kinds of plants occupying the area. In 1926 Transeau estimated that for a corn field the figure was 1.6%, but after subtracting the amount of carbohydrate used by the plants in respiration the efficiency was reduced to 1.2% for net gain. For the percentage of the total solar radiation received by the earth that is used in photosynthesis, the estimates range around 0.15%. Although this is a very low percentage, photosynthesis should not be regarded as an ineffective means of trapping solar energy. For one thing, photosynthesis has no close competitor in converting solar energy into a usable state. For another, as we noted earlier, the annual photosynthetic product of the earth is an immense quantity, quite adequate for maintaining the present life of the earth. Furthermore, it should be remembered that the fossil fuels that provide the great bulk of the energy used by man in his industries, transportation, and homes are derivatives of photosynthesis that occurred in the geological past.

Factors That Influence the Rate of Photosynthesis

We now turn to a different aspect of photosynthesis: a consideration of the various factors that may influence the rate at which photosynthesis occurs and of the situations under which each of these may limit the rate. Although research on rates of photosynthesis is less spectacular than that on the nature of the photosynthetic reactions, it is just as important from a general biological standpoint. From the standpoint of plants themselves, the entire biosphere, and also those who cultivate economic plants, it is most important that plants maintain an adequate rate of photosynthesis; in general, anything that brings about an increase in the rate is advantageous.

Apparent and True Photosynthesis

All methods of measuring the rate of photosynthesis, whether based on the amount of carbon dioxide consumed, the amount of oxygen produced, or the

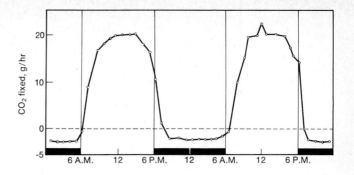

Figure 3-13. Daily periodicity of carbon dioxide exchange of alfalfa plants, showing production of carbon dioxide at night by respiration and the net consumption of carbon dioxide during the days by photosynthesis. [Data of Thomas and Hill. After J. Franck and W. E. Loomis, eds., *Photosynthesis in Plants,* Iowa State University Press, Ames, Iowa, 1949]

amount of organic matter synthesized, result in values that represent the **net** or **apparent** rate of photosynthesis rather than its true rate. The reason is that plants carry on respiration continuously in both the dark and light, and respiration produces carbon dioxide, uses oxygen, and uses sugars and other organic matter (Figure 3-13). The net difference between the rates of the two processes is being measured. To determine the true or total rate of photosynthesis the experimental plants must be placed in the dark before or after the determination of the apparent rate of photosynthesis so the rate of respiration can be measured. This is then added to the apparent rate to obtain the true rate.

Light apparently does not affect the rate of mitochondrial respiration, except as it may increase the temperature of the plant, and this can be taken into account in determining the rate of respiration. However, plants may carry on **photorespiration** [8] in the light, and this can reduce net photosynthesis substantially. Photorespiration is a distinctly different process from ordinary respiration; it occurs in the chloroplasts and probably the peroxisomes rather than in the mitochondria. Photorespiration apparently involves the oxidation of glycolate to glyoxylate and then decarboxylation of the latter, with the release of carbon dioxide. It is most active at high light intensities and high temperatures, increasing about eight times between 20 and 30°C. Moreover photorespiration apparently does not occur in plants with the sugarcane type of photosynthesis, which helps explain the high net productivity of this type of photosynthesis. Photorespiration can be inhibited by treatment of plants with α-hydroxysulphonate, and at high light intensities this results in a high net rate of photosynthesis approaching that of the sugarcane type.

From a standpoint of photosynthetic productivity it is the apparent or net rate that is most important. The rate of photosynthesis divided by the rate of respiration is referred to as the **photosynthetic/respiratory ratio** or **P/R ratio,** which may be calculated on a variety of bases. The most fundamental basis and the one involved in determination of the true rate of photosynthesis involves only the respiration of the photosynthetic tissues during the time photosynthesis is occurring. On this basis the ratio is generally high—on the order of 10 or so—as long as photosynthesis is not being severely limited by a deficiency of some factor. The P/R ratio may also be based on the respiration of the entire plant, including the tissues without chloroplasts, for various periods of time such as a day or a year. Calculated on such bases the P/R ratios are necessarily considerably lower than on the first basis because respiration occurs in all living cells day and night; however, the ratio must be greater than 1 if the plant is to be able to have a surplus of food for assimilation, growth, and accumulation. Finally, the P/R ratio can be based on the total photosynthesis and respiration of all the organisms in an entire ecosystem. This ratio must be no less than 1 if the ecosystem is to be viable and balanced.

Internal Factors Affecting Photosynthesis

The factors that influence the rate of photosynthesis are both internal and external, but it is usually an external factor that limits the rate. The basic internal factors are the chloroplast pigments and the essential enzymes and coenzymes. These are generally capable of carrying on a more rapid rate of photosynthesis than normally occurs under natural environmental conditions, and thus are rarely limiting. However, chlorophyll can be limiting if a plant is chlorotic because of a disease or a deficiency of some essential element such as nitrogen, magnesium, iron, or sulfur. There could also be inadequate quantities of the essential anzymes and coenzymes if there is a deficiency of the elements needed for their synthesis. It is conceivable that a healthy plant under optimal environmental conditions might have its photosynthesis limited simply by the time required for completion of the series of reactions, but this is unlikely because the reactions are very rapid. Accumulation of the end products could theoretically limit the rate, but the sugars either diffuse out of the chloroplasts or are converted to starch so this is unlikely. Oxygen inhibits photosynthesis, and it is possible that at high rates of photosynthesis enough might accumulate to limit the rate; again this is unlikely because of the rapid diffusion of oxygen out of the chloroplasts and the plant.

In vascular plants and bryophytes the anatomical features of the photosynthetic organs may affect the rate of photosynthesis. Such things as the thickness of the tissues, the size and distribution of intercellular spaces, and the size and distribution of the veins in the leaves can influence the amount of light reaching the chloroplasts or the rate at which the reactants can be supplied or the products removed. There is considerable variation in net photosynthetic productivity from species to species. Those species of plants that can thrive in the dim light of a dense forest are, for example, capable of an unusually high net rate of photosynthesis at low light intensities.

The important environmental factors that influence the rate of photosynthesis are carbon dioxide, water, oxygen, light, and temperature. The rate of photosynthesis is generally controlled by one or another of them, rather than by internal factors. Before we discuss these environmental factors some consideration should be given to the principle of limiting factors.

The Principle of Limiting Factors

The concept of limiting factors goes back to about 1840 when Liebig proposed his law of the minimum; this states that the rate of a process influenced by several factors is only as rapid as permitted by the factor available at the relatively lowest level. In 1905 Blackman restated essentially the same concept as the principle of limiting factors. As an example he gave the influence of

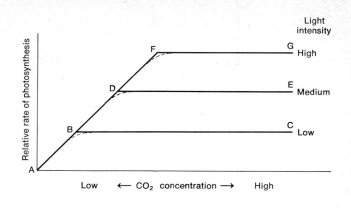

Figure 3-14. Diagram illustrating Blackman's concept of limiting factors. A–F shows the increase in the rate of photosynthesis with an increase in carbon dioxide concentration when carbon dioxide is limiting. BC, DE, and FG show the lack of effect of increasing carbon dioxide concentrations on the rate of photosynthesis when light of three different intensities is limiting. The broken lines show the type of transition actually found in curves plotted from experimental data. [After W. H. Muller, *Botany: A Functional Approach,* The Macmillan Company, New York, 1969]

light and carbon dioxide on the rate of photosynthesis (Figure 3-14). If, at a certain relatively low light intensity, the rate of photosynthesis was being limited by an inadequate concentration of carbon dioxide, the rate would increase with an increase in carbon dioxide, but only to a certain point; beyond this point no further increase could be obtained by further increases in carbon dioxide. At this point carbon dioxide was no longer the limiting factor. If provision of a higher light intensity then made possible a further increase in the rate of photosynthesis, it became evident that the limiting factor had become light intensity. However, if there was once more a leveling off at the higher light intensity, it was evident that some factor other than carbon dioxide was now limiting. Whether or not this was light intensity could be determined by the results when a still higher light intensity was provided.

In actual experimental determinations of this type, there is never a sharp break in the curve as postulated by Blackman but rather a rounded transition (Figure 3-15). This probably results from the fact that in this transition zone photosynthesis is being limited by more than one factor. For example, the more brightly lighted chloroplasts toward the upper surface of a leaf may be limited by carbon dioxide, while the more shaded ones below are being limited by light.

Figure 3-15. Influence of increasing concentrations of carbon dioxide rate of photosynthesis of wheat seedlings at four different light intensities. Note the marked effect of carbon dioxide concentrations above the normal atmospheric level, except at the lowest light intensity. [Data of W. H. Hoover *et al., Smithsonian Inst. Misc. Collect.,* **87:**1 (1933)]

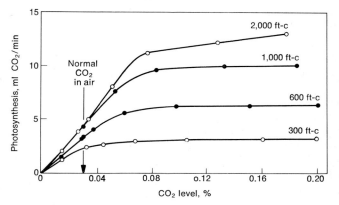

It should also be noted that, with a still further increase in the concentration or intensity of a factor beyond the point at which leveling off has occurred, the factor may become toxic or inhibitory and the rate may then decline, perhaps even to zero. For example, if the temperature is too high enzyme systems may become inactivated or eventually even destroyed. Thus, when any one factor is considered alone it may have an optimal range, with a decrease in photosynthesis as the factor is either increased or decreased beyond this range.

Carbon Dioxide

In nature, carbon dioxide is frequently the limiting factor in photosynthesis. On a sunny summer day the rate of photosynthesis of a well-watered and unshaded plant is quite certain to be limited by carbon dioxide (Figure 3-16). The carbon dioxide simply cannot diffuse into the chloroplasts as fast as it is being used. This is because of the low concentration of carbon dioxide in the atmosphere (0.03%) rather than any hindrance of diffusion by the stomata, through which most of the carbon dioxide enters a land plant. Actually, the stomata provide very efficient pathways for diffusion. Although the stomatal openings generally occupy only 1–3% of the area of a leaf, they may permit 50% or more of the diffusion that would occur if there were no epidermis. Brown and Escomb found in 1900 that carbon dioxide could diffuse into a catalpa leaf at the rate of 7.77 ml/cm^2 of stomatal aperture per hour, a rate 50 times as fast as carbon dioxide diffusion from the air into a 1 N solution of sodium hydroxide.

The air surrounding closely spaced and rapidly photosynthesizing plants may actually contain considerably less than 0.03% carbon dioxide. Studies at Cornell showed that in a corn field there may be less than 0.01% carbon dioxide. Even

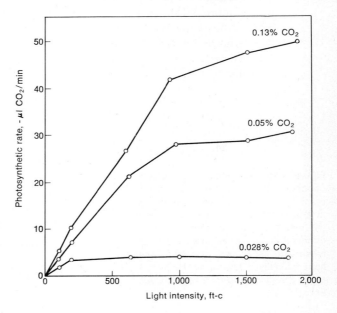

Figure 3-16. Increases in the rate of photosynthesis of wheat seedlings at various light intensities brought about by increasing the carbon dioxide concentration above the normal atmospheric level. [Data of W. H. Hoover et al., Smithsonian Inst. Misc. Collect., **87:**1 (1933)]

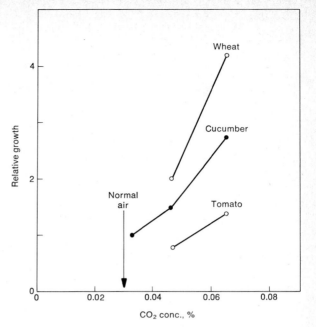

Figure 3-17. Effect of carbon dioxide concentrations greater than the normal level in the atmosphere on the growth of seedlings of tomato, cucumber, and wheat. The increases in growth evidently resulted from increased rates of photosynthesis. [Data of H. G. Lundegardh, *Environment and Plant Development,* Edward Arnold & Co., London, 1931]

the usual atmospheric concentration is generally far below the optimum. Experiments with various species of plants have shown increased rates of photosynthesis up to 0.5% and sometimes even higher. However, over periods of a week or so these higher concentrations may have detrimental effects on the plants. An increase in carbon dioxide much above 1% is generally inhibitory, perhaps principally because of reduction of the pH by the carbonic acid. One effect of such reduced pH is closure of the stomata, resulting in the interesting situation of a carbon dioxide deficiency in the plant in the midst of plenty.*

Provision of additional carbon dioxide to crop plants would undoubtedly increase their photosynthetic productivity and consequently their yields (Figure 3-17), but it is difficult and impractical to make such supplements on any large scale. Carbon dioxide has sometimes been added to the air in greenhouses with good results. The Cornell investigators mentioned previously found that by planting two different kinds of hybrid corn of different heights in alternating sets of rows air turbulence could be created among the plants, thus increasing the carbon dioxide availability. The production of carbon dioxide by the respiration of soil organisms may result in a greater than normal concentration available to low-growing plants. Thus, fertilizers and, in particular, manures and composts may provide a fringe benefit by increasing the number of soil organisms and promoting their growth, thus increasing the carbon dioxide available for photosynthesis.

* See Chapter 9 for a discussion of the mechanism of stomatal opening and closing.

Water

Since less than 1% of the water absorbed by plants is used in photosynthesis it is unlikely that water deficits commonly limit the rate of photosynthesis because of an insufficient quantity of this essential reactant in the process. However, numerous investigators have found that water deficits may markedly limit the rate of photosynthesis. Insufficient water apparently limits the rate of photosynthesis primarily because of a decrease in the degree of hydration of the protoplast and in particular of the chloroplasts. This results in an alteration of colloidal structure and a decrease in enzyme activity, with a consequent reduction in the rate of most metabolic processes. Photosynthesis is even more sensitive to dehydration than other metabolic processes, probably because of the disturbance of the precise molecular orientations within the thylakoids that are so essential for proper operation of the photosynthetic reactions. Respiration may increase substantially when there is great water stress, whereas photosynthesis is drastically reduced (Figure 3-18), resulting in a very low P/R ratio.

Another result of water deficit may be closure of the stomata. This greatly reduces the diffusion of carbon dioxide into the plant so that carbon dioxide could become limiting because of the water deficit.

In nature, water is probably a limiting factor in photosynthesis only when plants are wilted or are approaching wilting. Note that plants may undergo temporary wilting in the afternoons of hot days, even though there is an abundance of soil water, simply because the rate of water absorption cannot keep up with the rapid rate of transpiration. Thus, during hot weather water may be the limiting factor in photosynthesis for part of each day (Figure 3-19).

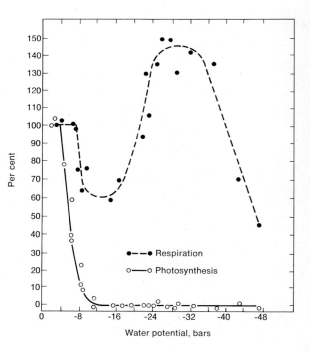

Figure 3-18. The influence of increasing water stress on the rates of photosynthesis and respiration of pine seedlings. Note that the rate of each process is expressed as the percentage of its rate at about −4 bars, when adequate water was available. At that point the actual quantitative rate of photosynthesis was much higher than that of respiration. [Data of H. Brix, *Physiol. Plantarum,* **15**:10 (1962)]

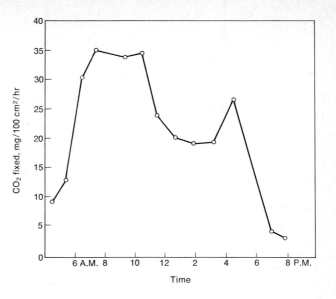

Figure 3-19. Changes in the rate of net photosynthesis of potato plants during a day. The afternoon slump probably resulted from temporary wilting and increased respiration brought about by high temperature. [Data of H. W. Chapman, *Amer. Potato J.*, **28:**602 (1951)]

However, the reduction of photosynthesis by severe wilting may persist over several days even after adequate water is available (Figure 3-20).

Light

Light is, of course, the limiting factor in photosynthesis at night and in the mornings and evenings as the light intensity is increasing or decreasing (Figure 3-13). On heavily overcast days light may also be limiting, as it may be in dense shade. Shade produced by leaves of plants, as under a forest canopy or within the crown of a large tree, is particularly likely to make light limiting because the light transmitted through the leaves is not only reduced in intensity but also in quality; that is, the very wavelengths most effective in photosynthesis

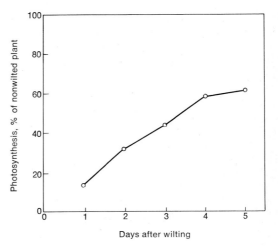

Figure 3-20. Recovery of rate of photosynthesis of sugarcane leaves following a single brief period of severe wilting (day 0). Note that even after 5 days the rate was only about 60% of the rate in the unwilted controls. [Data of F. M. Ashton, *Plant Physiol.*, **31:**266 (1956)]

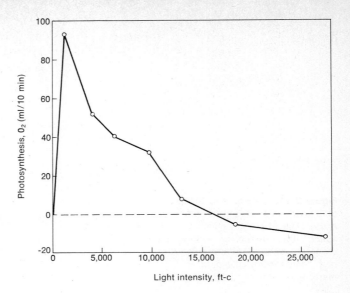

Figure 3-21. Inhibition of photosynthesis in Chlorella by high light intensities. The Chlorella cultures were exposed to each of the light intensities for 30 min before the rate of photosynthesis was measured. [Data of J. Myers and G. O. Burr, *J. Gen. Physiol.*, **24:**45 (1940)]

have been absorbed most completely. There is also a reduction in the photosynthetic quality of sunlight as well as in its intensity as it filters through water because the long red wavelengths are more completely absorbed than are the shorter green and blue ones. Thus light becomes a limiting factor in photosynthesis in the deeper lakes and the oceans. Red algae are able to thrive at greater depths than other kinds because of the accessory red pigment phycoerythrin. This pigment can absorb effectively in the green and blue regions and can then, through resonance, transfer the absorbed energy to chlorophyll. Of course, no photosynthetic plants can survive in the ocean depths where all light has been absorbed and there is continual darkness.

At very high light intensities the rate of photosynthesis, or at least the rate of oxygen evolution, is reduced because of photorespiration and possibly damage of the photosynthetic pigment systems. In Chlorella this occurs increasingly as the light intensity rises above 4000 ft-c; finally, at about 16,000 ft-c, there is oxygen consumption rather than oxygen evolution (Figure 3-21). In most land plants photooxidation begins at higher light intensities than in Chlorella. It should be stressed that this oxygen consumption occurs in the chloroplasts and has nothing to do with oxygen consumption by respiration, which occurs in the mitochondria. Photooxidation is probably the result of absorption of light energy far more rapidly than it can be utilized by the electron acceptor systems; among the effects are destruction of chlorophyll and inactivation of enzymes. Since the carotenes can serve as light screens, they may provide some degree of protection against photooxidation.

Temperature

In nature temperature is less frequently the limiting factor in photosynthesis than any other factor we have considered. In the 15–35°C range it is almost certain that, in nature, some other factor is limiting, and this may even be the case at still lower temperatures. In most tropical species photosynthesis will not occur below about 5°C, and few species can carry on much photosynthesis below 0°C; however, the process has been reported to occur in some lichens at −20°C and in some conifers even at −35°C. High temperatures, even those considerably below the thermal death point of 50–60°C, can cause a reduction in the rate of photosynthesis, probably by promoting photorespiration or by damaging the photosynthetic apparatus (Figure 3-22). However, the amazing thermophilic blue-green algae that live in hot springs thrive at 72°C and undoubtedly carry on photosynthesis at that temperature.

Although low temperature is rarely the limiting factor in photosynthesis in nature, it can be made limiting in experimental setups where high light intensities and abundant carbon dioxide are provided. In such situations the rate may initially increase with temperature up to at least 40°C, but after 10 to 15 min the rates at the higher temperatures start declining and may even drop below the rates at lower temperature. This time factor may result from such factors as thermal damage to the photosynthetic mechanism or perhaps photorespiration.

The rate of respiration increases with temperature more rapidly than the rate of photosynthesis, thus there is a decline in net photosynthesis with increasing temperature. At high temperatures the rate of respiration may even exceed that of photosynthesis (Figure 3-23), resulting in a P/R ratio of less than 1. It seems likely that photorespiration may make a substantial contribution to this.

Temperature also affects the rate of photosynthesis indirectly by influencing the opening and closing of stomata.

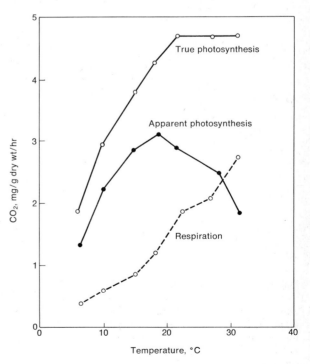

Figure 3-22. Influence of temperature on the rates of photosynthesis and respiration of Sphagnum moss. Note the marked decrease in apparent (net) photosynthesis above 20°C. [Data of M. G. Stålfelt, *Planta, 27*:30 (1937)]

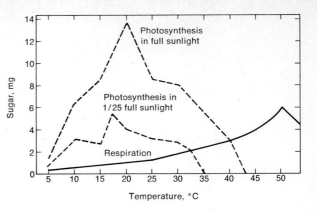

Figure 3-23. Influence of temperature on the rate of respiration of potato leaves and on the rates of photosynthesis in shade and full sunlight. The rates of photosynthesis were measured on the basis of the amount of sugar produced during a 10-min exposure to light at each temperature. [Data of H. G. Lundegårdh, *Environment and Plant Development*, Edward Arnold & Co., London, 1931]

Oxygen

Oxygen has a marked inhibitory effect on photosynthesis, probably for several reasons such as reoxidation of the primary photochemical products, reversible inactivation of some of the Calvin-Benson cycle enzymes, and promotion of photorespiration [8]. Even as little as 1% oxygen reduces the rate of photosynthesis in isolated spinach chloroplasts to about 75% of what it is in a nitrogen atmosphere, whereas 21% oxygen causes a 90% reduction (Figure 3-24). In intact spinach leaves the reduction by 21% oxygen is about 30%. However, the sugarcane type of photosynthesis is not influenced even by nearly 100% oxygen, which seems to explain its high productivity.

Although oxygen greatly reduces photosynthetic productivity in plants having the Calvin-Benson cycle, in nature it has little or no differential influence on the rate of photosynthesis because of the essentially uniform oxygen content of the atmosphere. Further reduction of photosynthesis by oxygen below this universal base level would appear to be possible only when photosynthesis is producing oxygen faster than it can diffuse out of the chloroplasts and the plant.

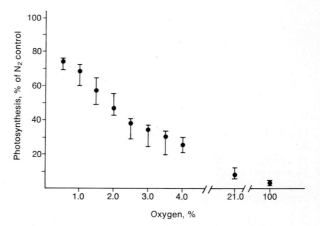

Figure 3-24. Inhibition of photosynthesis by oxygen, as per cent of rate in pure nitrogen (N_2), in spinach chloroplasts. Note the inhibition by even less than 1% oxygen and the severe inhibition by the atmospheric oxygen concentration (21%). [After M. Gibbs, *Amer. Sci.*, **58:**634 (1970)]

References

[1] Arnon, D. I. "The role of light in photosynthesis," *Sci. Amer.,* **203**(5):104–118 (Nov. 1960).

[2] Bassham, J. A. "The path of carbon in photosynthesis," *Sci. Amer.,* **206**(6):89–100 (June 1962).

[3] Bassham, J. A. "The control of photosynthetic carbon metabolism," *Science,* **172**:526–534 (1971).

[4] Calvin, M., and J. A. Bassham. *The Photosynthesis of Carbon Compounds,* W. A. Benjamin, New York, 1962.

[5] Clayton, R. K. *Light and Living Matter: The Physical Part,* vol. 1, *The Biological Part,* vol. 2, McGraw-Hill, New York, 1970 and 1971.

[6] Gabriel, M. L., and S. Fogel. *Great Experiments in Biology,* Prentice-Hall, Englewood Cliffs, N. J., 1955, pp. 152–184.

[7] Gaffron, H. *Photosynthesis,* D. C. Heath, Boston, 1965.

[8] Gibbs, M. "The inhibition of photosynthesis by oxygen," *Amer. Sci.,* **58**:634–640 (1970).

[9] Hatch, M. D., and C. R. Slack. "Photosynthesis by sugar cane leaves," *Biochem. J.,* **101**:103–111 (1966).

[10] Hill, R., and F. Bendall. "Function of the two cytochrome components in Chloroplasts: a working hypothesis," *Nature,* **186**:136–137 (1960).

[11] Kortschak, H. P., C. E. Hartt, and G. O. Burr. "Carbon dioxide fixation in sugar cane leaves," *Plant Physiol.,* **40**:209–213 (1965).

[12] Kriporian, A. D., and F. C. Steward. "Water and solutes in plant nutrition: with special reference to Van Helmont and Nicholas of Cusa." *BioScience,* **18**:286–292 (1968). See also: H. J. Dyer. "A matter of translation," *BioScience,* **18**:759 (1968).

[13] Laetsch, W. M. "Chloroplast specialization in dicotyledons possessing the C_4-dicarboxylic acid pathway of photosynthetic CO_2 fixation," *Amer. J. Bot.,* **55**:875–883 (1968).

[14] Lehninger, A. L. *Biochemistry,* Worth Publishers, New York, 1970.

[15] Levine, R. P. "The mechanism of photosynthesis," *Sci. Amer.,* **221**(6):58–70 (Dec. 1969).

[16] Rabinowitch, E. "Photosynthesis," *Sci. Amer.,* **179**(2):24–35 (Aug. 1948); "Progress in photosynthesis," *Sci. Amer.,* **189**(5):80–85 (Nov. 1953).

[17] Rabinowitch, E., and Govindjee. "The role of chlorophyll in photosynthesis," *Sci. Amer.,* **213**(2):74–83 (July 1965).

[18] Rabinowitch, E., and Govindjee. *Photosynthesis,* John Wiley & Sons, New York, 1969.

[19] Rosenberg, J. L. *Photosynthesis,* Holt, Rinehart, & Winston, New York, 1965.

[20] Stanier, R. Y. "Photosynthetic mechanisms in bacteria and plants: development of a unitary concept," *Bacteriol. Rev.,* **25**:1–17 (1961).

[21] Van Niel, C. B. "The comparative biochemistry of photosynthesis," *Amer. Sci.,* **37**:371–383 (1949).

[22] Wald, G. "The origin of life," *Sci. Amer.,* **191**(2):44–53 (Aug. 1954). For more details see S. W. Fox, "How did life begin?" *Science,* **132**:200–208 (1960), and J. Keosian, *The Origin of Life,* Reinhold, New York, 1964.

[23] Walker, D. A. "Carboxylation in plants," *Endeavour,* **25**:21–26, 1966.

Respiration

4

The maintenance of life requires a continuing expenditure of energy. Some of this energy, particularly in plants, is derived from the kinetic energy of diffusing molecules, but most of it can be traced back to the light energy converted to chemical bond energy by photosynthesis. Utilization of this chemical bond energy by organisms requires oxidation of the organic compounds produced by photosynthesis or of compounds derived from them by subsequent syntheses. These biological oxidations of organic compounds are referred to as **respiration.** All living organisms carry on some type of respiration. The usual type of respiration carried on by most species of plants and animals involves the aerobic oxidation of sugars, and the summary equation for this process is the reverse of the usual summary equation for photosynthesis.

$$C_6H_{12}O_6 + 6\,O_2 \longrightarrow 6\,CO_2 + 6\,H_2O + \text{energy}$$

Like photosynthesis, respiration is a multiple-step process, and the summary equation reveals nothing about the sequence of reactions that occur. Indeed, the equation is a more adequate representation of the combustion of sugar. Both combustion and respiration are oxidations with the same substrates, products, and energy release; however, combustion occurs only at high temperatures with a quick and violent release of energy whereas respiration, catalyzed by many enzymes, occurs at lower temperatures and involves a stepwise energy release. One important fact about respiration is not at all evident from the summary equation: it is essentially a series of reactions that converts the chemical bond energy of the sugar or other substrate to the readily usable chemical bond energy of ATP and reduced hydrogen acceptors such as NADH. These then provide the energy used in doing a great variety of metabolic work, such as the synthesis of fats and proteins, the reduction of substances such as nitrates, the accumulation of ions, cytoplasmic streaming, the movement of flagella, and in animals the contraction of muscles and the generation of nerve impulses.

In addition to using the products of photosynthesis in respiration as an energy source, plants also use them in synthesizing a great variety of other organic compounds. Many of these are **assimilated,** that is, incorporated in the proto-plasts and walls. Others such as starch, fats, alkaloids, and latex **accumulate** in cells, and some of these may be used later in respiration and assimilation. Respiration plays a central role in the syntheses and assimilation, not only in providing the energy needed to drive most of the synthetic reactions but also in providing the substrates for many of them. As we shall see later, this is accomplished by draining intermediates from the respiratory pathways at various points. Respiration is essential, not only for growth, development, and reproduction but also simply for the maintenance of life.

Like other aerobic organisms, photosynthetic plants use oxygen and produce carbon dioxide in respiration. However, this respiratory gas exchange is obvious only in the absence of light or from tissues lacking chloroplasts because, in the presence of light, chlorenchyma cells usually carry on photosynthesis more rapidly than respiration. Thus there is a net consumption of carbon dioxide and a net production of oxygen. Photosynthesis and respiration can occur concurrently in a cell without interference, even though they are generally reverse processes, because they occur in different organelles; photosynthesis takes place in the chloroplasts, and all but the glycolytic reactions of respiration take place in the mitochondria.

Aerobic Respiration

The summary equation for respiration shows the substrate as a hexose, and detailed respiratory pathways generally begin with glucose. However, since cells can readily convert one carbohydrate to another, the pathway might just as well begin with almost any carbohydrate. Fats, proteins, and various organic acids can also be used in respiration. As will be described later these are not necessarily converted to carbohydrates first and can enter the respiratory pathways at appropriate points.

Aerobic respiration can be considered to consist of four identifiable reaction sequences: (1) **Glycolysis** results in the conversion of a molecule of glucose into two molecules of pyruvic acid plus hydrogen that is transferred to two molecules of a hydrogen acceptor (nicotinamide adenine dinucleotide, or NAD$^+$). (2) The **citric acid cycle** (Krebs cycle or tricarboxylic acid cycle): in the course of this cycle pyruvic acid is essentially broken down into carbon dioxide and hydrogen, the latter being donated to hydrogen acceptors. (3) **Oxidative phosphorylation:** in this process electrons from the reduced hydrogen acceptors (and in some steps the protons) are transferred through a series of acceptors at progressively lower energy levels, the released energy being used in synthesizing ATP from ADP and phosphate. (4) **ATP hydrolysis** into ADP and phosphate, with the release of energy usable by the organism in various ways. Although this last process is usually not included as a part of the respiratory pathways, it seems appropriate to do so if respiration is to be considered as an energy releasing process and if we are to emerge with the general summary equation for respiration. The result of the first three reaction sequences is the conversion of chemical bond energy of glucose into chemical bond energy of ATP.

Glycolysis

The initial reactions of glycolysis (Figure 4-1) that lead from glucose to fructose-1,6-diphosphate result in phosphorylation of the substrate rather than oxidation, and these reactions may be regarded as preliminaries that vary with the carbohydrate used.

The first substrate oxidation occurs when the 3-phosphoglyceraldehyde is

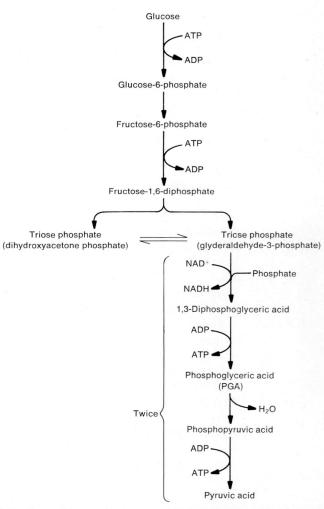

Figure 4-1. The glycolytic pathway. Glucose is shown as the substrate but other sugars can also be used. See text for discussion.

converted to 1,3-diphosphoglyceric acid by reaction with a molecule of inorganic phosphate (Pi) and by removal of two hydrogen atoms. The two electrons of these hydrogen atoms are used in reducing a molecule of NAD^+ to NADH. One of the protons is incorporated in the NADH and the other is released as a hydrogen ion (H^+). In the next step the 1,3-diphosphoglyceric acid is converted to 3-phosphoglyceric acid, the phosphate being transferred to ADP and producing a molecule of ATP. Later on when phosphoenol pyruvic acid is converted to pyruvic acid a second molecule of ATP is synthesized. Since each molecule of glucose gives rise to two triose molecules, four ATP and two NADH are synthesized at the expense of the energy provided by the glycolytic oxidations. However, two ATP were used in the initial phosphorylations, so the net production of ATP is only two molecules per molecule of glucose used. Thus, glycolysis can be summarized as follows.

$$\left.\begin{array}{l} C_6H_{12}O_6 + 2\ NAD^+ \\ \text{Glucose} \\ 2\ Pi\quad +\ 2\ ADP \end{array}\right\} \longrightarrow \left\{\begin{array}{l} 2\ CH_3 \cdot CO \cdot COOH + 2\ NADH + 2\ H^+ \\ \text{Pyruvic acid} \\ \quad\quad 2\ ATP \end{array}\right.$$

From an energy standpoint, the net result of glycolysis is the conversion of some of the chemical bond energy of ATP and NADH. However, about 93% of the energy from glucose is now in the two molecules of pyruvic acid.

Citric Acid Cycle

The pyruvic acid produced by glycolysis does not enter the citric acid cycle directly, but is first converted to **acetyl-coenzyme A** by decarboxylation and combination with coenzyme A (CoA). In this reaction two hydrogens are removed (one from acetaldehyde, CH_3CHO, and one from the —SH group of CoA) and these are used in reducing NAD^+ to NADH and in the production of H^+ (Figure 4-2). The acetyl-CoA enters the citric acid cycle (Figure 4-3) by reacting with oxalacetic acid (four-carbon), producing citric acid (six-carbon) and releasing CoA, which can then be used again. The remainder of the cycle involves a series of decarboxylations, dehydrogenations, hydrations, and dehydrations that convert one acid to another with the eventual regeneration of oxalacetic acid, which is thus available for another turn of the cycle. NAD^+ is reduced to NADH when α-ketoglutaric acid is oxidized to succinyl-CoA, and also when malic acid is oxidized to oxalacetic acid. When isocitric acid is oxidized to oxalosuccinic acid $NADP^+$ rather than NAD^+ serves as the hydrogen acceptor, but the NADPH in turn reduces NAD^+ to NADH. When succinic acid is oxidized to fumaric acid the released hydrogen is transferred to still another hydrogen acceptor **(flavine adenine dinucleotide, FAD),** reducing the FAD to $FADH_2$.

The net result of the formation of acetyl-CoA from pyruvic acid and the citric acid cycle is the complete oxidative decomposition of pyruvic acid into carbon dioxide and hydrogen, the hydrogen being used in the reduction of hydrogen acceptors and the production of H^+. Energy from the chemical bonds of pyruvic acid has been transferred to the reduced hydrogen acceptors and to the molecule of ATP that is generated when succinyl-CoA is oxidized to succinic acid.

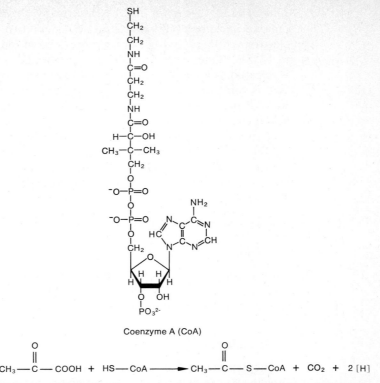

Coenzyme A (CoA)

$$CH_3-\overset{O}{\overset{\|}{C}}-COOH + HS-CoA \longrightarrow CH_3-\overset{O}{\overset{\|}{C}}-S-CoA + CO_2 + 2\,[H]$$

Figure 4-2. The structural formula of coenzyme A (CoA) and the reaction of CoA with pyruvic acid. CoA is shown as HS-CoA since the reaction takes place at the —SH group.

A total of five molecules of hydrogen acceptors are reduced. Along with the H^+ produced, this constitutes a total of 10 hydrogens. The pyruvic acid can obviously provide only four of these; the other six are derived from the three molecules of water that constitute the net water consumption of the cycle. If for simplification we represent the hydrogen acceptors as A, the reduced hydrogen acceptors and H^+ as AH_2, and if the two molecules of pyruvic acid derived from a molecule of glucose are shown, the reactions from pyruvic acid through the citric acid cycle can be summarized as follows.

$$\left.\begin{array}{l} 2\,CH_3 \cdot CO \cdot COOH + 6\,H_2O + 10\,A \\ 2\,Pi + 2\,ADP \end{array}\right\} \longrightarrow \left\{\begin{array}{l} 6\,CO_2 + 10\,AH_2 \\ 2\,ATP \end{array}\right.$$

The atoms that made up the glucose used in respiration are now all in either carbon dioxide or the 12 molecules of reduced hydrogen acceptors, two of which were produced during glycolysis. The chemical bond energy of the glucose, less that lost as heat, is now in the 12 molecules of reduced hydrogen acceptors and in four molecules of ATP (two of which were produced during glycolysis).

Oxidative Phosphorylation

The next sequence of reactions in aerobic respiration involves the transfer of electrons (and in some steps protons) through a series of **cytochromes** [7]

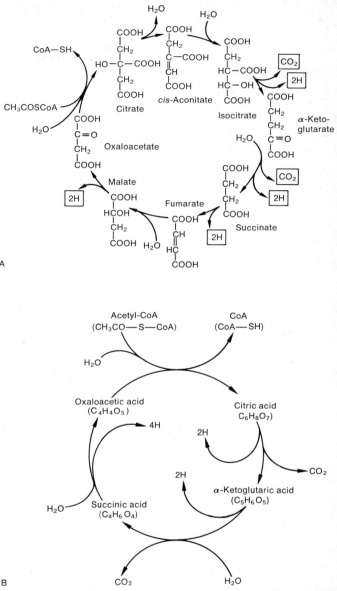

Figure 4-3. The citric acid cycle (Krebs or tricarboxylic acid cycle). **(A)** Detailed cycle giving structural formulas of most of the intermediates. Not shown is the reaction of α-ketoglutarate with CoA-SH with the formation of succinyl-CoA, which is then converted to succinate + CoA. In the latter step a molecule of ATP is synthesized from ADP + Pi. NAD^+ serves as the hydrogen acceptor except in the step from succinate to fumarate, where FAD is reduced to $FADH_2$. **(B)** Simplified cycle omitting several intermediates but showing net input and output.

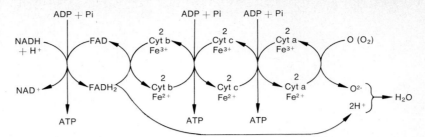

Figure 4-4. Diagram illustrating oxidative phosphorylation. As the electrons from NAHD are transferred through the series of electron acceptors at progressively lower energy levels the released energy is used in synthesizing ATP. Cytochrome a is also referred to as cytochrome oxidase. Another electron acceptor (coenzyme Q) probably functions in the series, between FAD and cytochrome b, whereas cytochrome c_1 may be intermediate between cytochrome b and cytochrome c.

and other electron acceptors, each acceptor being at a progressively lower energy level (Figure 4-4). The electrons and protons from NADH + H⁺ are transferred to FAD, the first acceptor in the series, reducing it to $FADH_2$ and oxidizing the NADH back to NAD⁺. In turn, the $FADH_2$ donates its electrons to **coenzyme Q** (CoQ) and only the electrons are transferred from CoQ through the cytochrome series. The protons do not participate further until the electrons reach oxygen, the final electron acceptor, at which time the oxygen, the electrons, and the protons combine to form water.

The important point about the electron transfer is that it is coupled with the phosphorylation of ADP to ATP. As the electrons from NADH are transferred down the energy stairway, enough energy is released to drive the synthesis of three molecules of ATP but, because the $FADH_2$ from the citric acid cycle enters the series at the second step, it provides only enough energy for generation of two molecules of ATP. Thus oxidative phosphorylation results in the production of a maximum of 34 molecules of ATP (30/10 NADPH + 4/2 $FADH_2$) per molecule of glucose used. The terminal oxidations can be summarized as follows.

$$\left.\begin{array}{l} 12\ AH_2 + \ 6\ O_2 \\ 34\ Pi \quad + 34\ ADP \end{array}\right\} \longrightarrow \left\{\begin{array}{l} 12\ A \quad + 12\ H_2O \\ 34\ ATP \end{array}\right.$$

Adding the four molecules of ATP produced during glycolysis and the citric acid cycle, a total maximum of 38 molecules of ATP can be synthesized at the expense of the energy originally derived from glucose. Thus, the net result of respiration may be considered as conversion of the chemical bond energy of glucose into chemical bond energy of ATP, which is readily usable in doing various kinds of metabolic work.

This energy conversion is by no means 100% efficient. A mole of glucose yields 673 kcal of free energy when oxidized. Estimates of the free energy released when the terminal phosphate bond of ATP is hydrolyzed vary from 7 to 12 kcal/mole. Using the low value, 38 moles of ATP could provide 266 kcal of usable energy. On this basis respiration is about 40% efficient as an energy conversion process, a relatively high degree of efficiency as compared with energy conversions in general. The remainder of the energy from glucose is converted into heat at various points in the respiratory pathways. The heat is useful in maintaining a stable body temperature in homothermic animals but is wasted energy in other organisms.

ATP Energy Release

The terminal oxidations are sometimes considered as being the final steps of aerobic respiration, but it is perhaps best to include in respiration one more step: the release of energy from ATP when it is hydrolyzed back to ADP.

$$38 \text{ ATP} + 38 \text{ H}_2\text{O} \longrightarrow 38 \text{ ADP} + 38 \text{ H}_3\text{PO}_4 + 266 \text{ kcal energy}$$

In any event, it is necessary to include this step if one is to come out with the standard summary equation for respiration. If the summary equations given for the four stages of respiration are added together and all substances that are both used and produced are cancelled out, the old summary equation is left. It should be noted that the energy transferred from glucose to ATP is really not useful until it is released.

Plants use energy from ATP and NADH in many ways, perhaps most extensively as an energy source for the synthesis of a wide variety of organic compounds, many of which are essential to life. We have already seen some such reactions and others will be considered later. For example, the synthesis of fats from carbohydrates requires much energy from ATP and NADH. Energy from ATP is also used in accumulation of solutes against a concentration gradient and in cell division, assimilation, and other aspects of growth. Energy is required for the maintenance of the differential permeability of cell membranes and for the maintenance of the structural and functional integrity of cell organelles in general. The flagellar movements of sperm and of certain algae and fungi occur at the expense of ATP, which brings about sequential contraction of the ring of fibers within the flagella. Phosphorescent organisms, which include some species of fungi and bacteria as well as animals such as fireflies and some deep-sea fish, convert ATP energy into light.

Respiration also provides energy that can be used in biochemical reductions in another way. Instead of participating in oxidative phosphorylation, NADH and other reduced hydrogen acceptors may be drained out of the respiratory pathways and used as electron and proton donors in various processes, such as the reduction of nitrates ($-\text{NO}_3$) to ammonia (NH_4-) and the synthesis of fatty acids. A number of examples will be considered in subsequent chapters.

Hexose Monophosphate Shunt

Although the glycolysis–citric acid cycle pathway of aerobic respiration appears to be the usual one in plants as well as in animals, there is at least one other important pathway: the hexose monophosphate shunt (also called the pentose phosphate pathway or direct oxidation). In this glucose is first phosphorylated to glucose-6-phosphate which is then oxidized to 6-phosphogluconic acid, the energy being used in reducing a molecule of NADP^+ to NADPH. Next the 6-phosphogluconic acid is reduced to 6-phospho-3-keto-gluconic acid, with the formation of a second molecule of NADPH. The

6-phospho-3-ketogluconic acid is very unstable and is quickly decarboxylated, the products being ribulose-5-phosphate and carbon dioxide.

This sequence of reactions results in only partial (1/6) oxidation of a glucose molecule and must occur six times for the complete oxidation of a molecule of glucose (Figure 4-5). The ribulose-5-phosphate is the substrate for a complex series of sugar interconversions that eventually results in the regeneration of glucose-6-phosphate, five molecules being produced for six turns of the cycle. Only one molecule of glucose-6-phosphate is introduced into the cycle for each six turns. Complete oxidation of a molecule of glucose results in production of 12 molecules of NADPH, which may donate its electrons and proton to NADH. In oxidative phosphorylation this can generate a maximum of 36 molecules of ATP (a net of 35 minus the ATP used in phosphorylating the glucose).

However, the hexose monophosphate shunt may be more important as a synthetic pathway than as an energy-providing process. The ribulose-5-phosphate can be converted into either ribulose-1,5-diphosphate or ribose-5-phosphate, the latter being important in the synthesis of nucleotides and other essential compounds. Subsequent reactions can result in the production of the monophosphate esters of a wide variety of monosaccharides, including xylulose, sedoheptulose, erythrose, glyceraldehyde, dihydroxyacetone, fructose, and eventually glucose, making the reactions cyclic. As we shall see in the next chapter, the usual respiratory pathway also plays a synthetic as well as an energy-releasing role.

Comparison of Respiration and Photosynthesis

It may be interesting at this point to consider to what degree respiration and photosynthesis are the reverse of one another in detail as well as in their overall summary reactions. Several such inverse similarities are evident at once. Water is produced and oxygen used at the end of respiration, whereas water is broken down and oxygen produced at the beginning of photosynthesis. Generation of ATP by electron transport through a system involving cytochromes occurs at the beginning of photosynthesis and at the end of respiration. Hydrogen acceptors are reduced toward the beginning of photosyntheses and at the end of respiration. However, a good many details are not the reverse of one another. For example, the electron transport systems are not the same, although in both processes electron flow is stepwise down a similar energy stairway. Light energy is used in photosynthesis, but it is not produced by respiration except in the few phosphorescent organisms. The energy used in generating ATP in photosynthesis comes from light, and in respiration it comes from the chemical bond energy of the substrate. In photosynthesis the hydrogen acceptor is $NADP^+$, whereas in respiration it is principally NAD^+.

At the opposite ends of the two processes a reverse situation exists in that sugars or other organic compounds are the product of photosynthesis and the substrate of respiration. If glucose is considered as an eventual product of photosynthesis, the initial stages of glycolysis as far as PGA are essentially the reverse of the final stages of photosynthesis from PGA on. Here the reverse similarity of the RDP cycle and the usual glycolysis–citric acid cycle of respira-

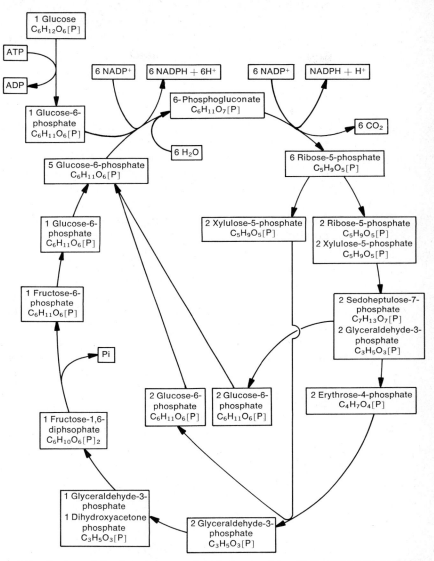

Figure 4-5. Outline of the hexose monophosphate shunt (pentose phosphate pathway) showing the six turns of the cycle required for the complete oxidation of one molecule of glucose to carbon dioxide and hydrogen. The NADH produced can be used in various reductions or can transfer its electrons and protons to NAD^+. The NADH can then participate in oxidative phosphorylation, with the production of a maximum of 36 molecules of ATP. Subtracting the molecule of ATP used in converting glucose to glucose-6-phosphate, the net production is 35. Note the regeneration of five molecules of glucose-6-phosphate for every six turns of the cycle. The various sugars may be drained out of the cycle and used in syntheses of different kinds.

tion ends. However, it should be noted that the hexose monophosphate shunt is, to a large degree, the reverse of the sugar interconversions in the RDP cycle and also that the sugarcane type of carbon dioxide fixation is the reverse of several steps in the respiratory pathway. Furthermore, Ochoa [11] and others have evidence for the participation of other respiratory intermediates, such as isocitric and α-ketoglutaric acid, in photosynthesis.

Anaerobic Respiration

Various species of bacteria and fungi carry on anaerobic respiration, that is, oxidation of foods without the use of oxygen. Higher plants may also carry on anaerobic respiration when there is a lack of oxygen. There are a good many different kinds of anaerobic respiration, but they all have in common the fact that oxygen is not the final e^- and H^+ acceptor and that the substrate is only partially oxidized so that the end products include a variety of organic compounds such as alcohols, lactic acid, or citric acid. Anaerobic respiration is sometimes referred to as **fermentation,** but the two terms are not precise synonyms because some fermentations, like that of acetic acid, are aerobic oxidations. The fermentations carried on by bacteria and fungi are widely utilized for the commercial production of the end products.

We shall select alcoholic fermentation for consideration as an example of anaerobic respiration. Alcoholic fermentation is one of the most widespread types and the most important from a commercial standpoint. It is the principal respiratory pathway of yeasts [12]; these are of great economic importance both because of the ethanol they produce and because the carbon dioxide they produce is used in the leavening of breads. Yeasts are also important as a source of vitamins. Some varieties of yeast have desirable flavors and are cultured in mass as a food supply in some countries where there is a shortage of food.

Yeasts can also carry on aerobic respiration, but even in well-aerated cultures this accounts for no more than one-third of the total respiration and usually much less. However, aerobic respiration substantially increases the rate of cell division.

Alcoholic fermentation also occurs in several other species of microorganisms and can be carried on by most, if not all, tissues of higher plants when there is a deficiency of oxygen.

The Reactions of Alcoholic Fermentation

The summary equation for alcoholic fermentation is as follows.

$$\underset{\text{Glucose}}{C_6H_{12}O_6} \longrightarrow \underset{\text{Ethanol}}{2\ C_2H_5OH} + 2\ CO_2$$

The detailed steps are the same as those of glycolysis all the way to pyruvic acid, which is then decarboxylated to acetaldehyde. However, instead of reacting with CoA and entering the citric acid cycle, the acetaldehyde is reduced to ethanol by accepting hydrogen from the $NADH + H^+$ produced during glycolysis. These reactions can be represented by the following subsummary equations.

$$C_6H_{12}O_6 + 2\ NAD^+ \left.\right\} \longrightarrow \left\{\right. 2\ CH_3 \cdot CO \cdot COOH + 2\ NADH + 2\ H^+$$
$$2\ Pi\ \ + 2\ ADP \qquad\qquad 2\ ATP$$
$$2\ CH_3 \cdot CO \cdot COOH \longrightarrow 2\ CH_3 \cdot CHO + 2\ CO_2$$
$$2\ CH_3CHO + 2\ NADH + 2\ H^+ \longrightarrow 2\ C_2H_5OH + 2\ NAD^+$$

The readily available energy provided by alcoholic fermentation consists only of the net of two molecules of ATP produced during glycolysis, a total of about 14 kcal for the 2 moles of ATP. Most of the bond energy originally present in the glucose is now in the alcohol, which apparently cannot be used (under anaerobic conditions) by yeast as an energy source and continues to accumulate. Thus, anaerobic respiration is a very inefficient means of generating ATP as compared with aerobic respiration (two molecules compared with 38 of ATP per molecule of glucose), but it does have the advantage of not requiring oxygen. The yeast cells are killed by the alcohol they produce when it reaches a concentration of 12–14%. This is the reason that naturally fermented beverages never contain more than this percentage of alcohol.

Anaerobic Respiration in Vascular Plants

As has been mentioned higher plants carry on anaerobic respiration when there is a lack of oxygen. The reactions are apparently the same as in yeast. Under natural conditions anaerobic respiration occurs in a limited number of situations. Roots deprived of oxygen because the soil is waterlogged and plants submerged by flood waters commonly carry on anaerobic respiration. Most vascular plant tissues cannot survive long on anaerobic respiration alone, and some die after only a day or two. This may result either from a lack of ATP or from the accumulation of toxic concentrations of alcohol. Of course, some species of plants such as rice, willows, and cattails thrive even though their roots are continuously submerged as do submerged aquatics. This is probably due largely to their ability to carry on adequate rates of aerobic respiration with the oxygen dissolved in the water rather than to a capacity for indefinite survival on anaerobic respiration.

It is possible that toward the interior of bulky living tissues, such as the thick stems and leaves of succulents and the larger fleshy fruits, there may be an oxygen deficit (because of use of much of the oxygen by more external cells) and, consequently, some anaerobic respiration, although the experimental evidence on this is somewhat conflicting. Grapes and some other fruits have skins that are not very permeable to oxygen, and may carry on anaerobic respiration. There is definite evidence that in germinating seeds of many species anaerobic respiration may constitute a considerable fraction of the total, particularly in the early stages when the seed coats are only slightly permeable to oxygen. This might be anticipated as regards rice, which normally germinates under water. With as much as 8% oxygen available, the rate of anaerobic respiration equals the rate of aerobic respiration. However, even more aerobic seeds such as those of wheat, corn, peas, and sunflower may carry on considerable anaerobic respiration during the early stages of germination, although corn seedlings die within a day or so if the atmosphere is devoid of oxygen.

The ability of higher plants to survive for at least some hours or days with anaerobic respiration is in contrast with the situation in animals where anaerobic conditions generally result in death in a matter of minutes. During periods of intense muscular activity in animals oxygen may not be delivered to the cells as rapidly as it is used, and anaerobic respiration may occur for a few minutes with the accumulation of lactic acid, but the resulting oxygen debt must be repaid quite promptly.

Other Oxidation Systems in Plants

Plants have a considerable number of oxidation systems other than the respiratory pathways we have just described. One of these is photorespiration, which was discussed in Chapter 3. In the early years of this century there was considerable interest in various oxidation-reduction enzymes of plants other than the dehydrogenases, decarboxylases, cytochromes, and so on, that are involved in the usual respiratory pathways. These include polyphenol oxidases, peroxidases, catalase, and ascorbic acid oxidase. The precise roles and importance of these enzymes in the overall metabolism of plants is still not well understood, but their functions may be more central than is now generally considered to be the case.

Polyphenol oxidase (also called tyrosinase or catechol oxidase) catalyzes the oxidation of the amino acid tyrosine and a wide variety of other phenolic compounds (Figure 4-6) through various intermediates to black substances such as melanin and polymerized hydroquinones. Some of the intermediates are red. The reddish, brown, or black color that appears at the cut surfaces of plant organs such as potato tubers and apples is a result of these oxidations.

Ascorbic acid oxidase is also of common occurrence in plants. It catalyzes the oxidation of ascorbic acid (vitamin C) to dehydroascorbic acid (Figure 4-6). Since the reaction is reversable, it may function in electron transport systems. Ascorbic acid can also be oxidized by the quinones produced by the oxidation of phenolic compounds, thus preventing the continuation of the reactions to the dark pigments. Fruit salads containing citrus fruits, which have a high ascorbic acid content, do not darken as rapidly as those without them.

Another widespread plant oxidase is **indole-3-acetic acid oxidase,** which oxidizes indole-3-acetic acid (IAA), an important plant hormone. As will be seen in Chapter 12, this enzyme plays a role in the regulation of IAA concentration. Other plant oxidases may also function primarily in ways other than energy release.

Peroxidases, which are present in both plants and animals, are dehydrogenases that catalyze the oxidation of phenolic compounds. The reaction is unique in that hydrogen peroxide serves as the hydrogen acceptor. If the substrate is designated as XH_2, the reaction is as follows

$$XH_2 + H_2O_2 \longrightarrow X + 2 H_2O$$

Catalase, an enzyme essentially universal among organisms, catalyzes the decomposition of hydrogen peroxide.

$$2 H_2O_2 \longrightarrow 2 H_2O + O_2$$

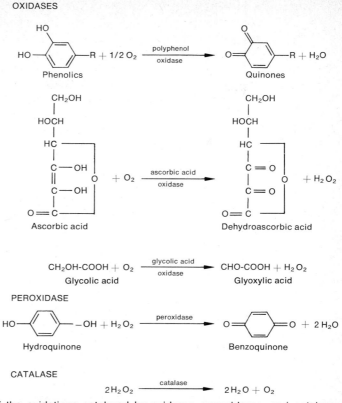

OXIDASES

PEROXIDASE

CATALASE

$$2H_2O_2 \xrightarrow{\text{catalase}} 2H_2O + O_2$$

Figure 4-6. Some of the oxidations catalyzed by oxidases, peroxidases, and catalase.

The reaction also occurs spontaneously, but at a much lower rate. The metabolic role or roles of catalase are not known. It may simply prevent the accumulation of toxic quantities of hydrogen peroxide, which is a product of a number of metabolic reactions. Catalase activity correlates well with the general level of metabolic activity of a tissue. Because catalase activity can be measured easily by the rate of oxygen evolution, it has been used as a means of estimating the relative metabolic level of various tissues.

The beta (β) oxidation of fatty acids, an important respiratory pathway that provides large amounts of usable energy, will be considered in Chapter 5.

The Respiratory Quotient (RQ)

The respiratory quotient (RQ) (or respiratory ratio) is the ratio of the amount of carbon dioxide produced by respiration to the amount of oxygen used, that is, $RQ = CO_2/O_2$. The importance of this ratio is that it can provide clues as to the kind of substrate being used in respiration and as to what degree the respiration is aerobic or anaerobic. Obviously, the RQ must be determined when there is no photosynthesis in the tissue.

As can be seen from the reactions of aerobic respiration, RQ = 1 when any carbohydrate is completely oxidized to carbon dioxide and water. When only anaerobic respiration is occurring, the RQ is infinity because oxygen consumption is zero. Simultaneous aerobic and anaerobic respiration result in some value greater than 1 for the RQ, the higher the value the greater the proportion of anaerobic respiration. RQs of more than 1 also result when the substrate of aerobic respiration consists of organic acids, since these are more highly oxidized than carbohydrates. For example, the RQ for oxalic acid is 4 and for malic acid 1.33, as indicated by the following summary equations.

$$2\ COOH \cdot COOH +\ O_2 \longrightarrow 4\ CO_2 + 2\ H_2O$$
$$CH_2 \cdot CHOH \cdot (COOH)_2 + 3\ O_2 \longrightarrow 4\ CO_2 + 3\ H_2O$$

When fats or proteins are being used in respiration, the ratio is less than 1 because these substances are more highly reduced than carbohydrates. For proteins the RQ is usually between 0.8 and 0.9. For fats the RQ decreases with an increase in the length of the fatty acid chains, the RQ for the fat tripalmitin being about 0.7.

$$C_{51}H_{98}O_6 + 72.5\ O_2 \longrightarrow 51\ CO_2 + 49\ H_2O$$

As will be seen in the next chapter, considerable oxygen consumption without any carbon dioxide evolution occurs during the β oxidation of fatty acids; this process produces acetyl-CoA which can then go into the citric acid cycle.

Various complications make it necessary to use a RQ with caution in drawing conclusions about the substrate for respiration or the type of respiration occurring. It is difficult to tell whether a RQ above 1 is resulting from a combination of anaerobic and aerobic respiration or from use of organic acids in aerobic respiration. Also, several different substrates may be used simultaneously in respiration. Some processes such as the oxidation of carbohydrates to organic acids use oxygen without any carbon dioxide production, whereas nonphotosynthetic carboxylations that result in organic acid synthesis use carbon dioxide but do not evolve oxygen.

Rates of Respiration in Plants

There is an immense variation in the rates at which plants carry on respiration, the rate being influenced by the type and age of the tissue and by the species of plant as well as by environmental factors. In general, meristematic tissues have high rates of respiration as do leaves. Germinating seeds have a large proportion of meristematic tissue and high rates of respiration. Stems, roots, and immature fruits generally have lower respiration rates. For example, one measurement of spinach leaf respiration showed an oxygen consumption of 515 mm^3/(g fresh wt)(hr), whereas for carrot leaves the figure was 440. In the same units, the rate was 233 for the cambium of an ash tree, 154 for the phloem, and only 47 for the sapwood; 25 for a carrot root, 30 for the flesh and 95 for the skin of an apple, and 715 for the embryos of germinating barley seeds as compared with only 76 for the endosperms. Such figures should be considered as indicative only because under other environmental conditions the same tissues could be expected to have different rates. Some of the differences between tissues are related to the amount of protoplasm per gram of fresh weight. Thus, when the rates for various tissues of ash trees were calculated per milligram of nitrogen in each by Goodwin and Goddard, the rates were less diverse than when calculated on a simple fresh weight basis. For example, the rates on a nitrogen basis were 120 for the cambium, 112 for the phloem, and 130 for the sapwood xylem. The nitrogen content provides an estimate of the protein content and thus, presumably, of the amount of protoplasm.

The rate of respiration per unit fresh weight of a plant tissue or organ decreases with its aging; senescent organs generally have much lower rates than when they were younger and more active metabolically. However, when the rate of respiration is calculated per unit of nitrogen, it may actually increase with age.

Factors Influencing the Rate of Respiration

The rate of respiration in plants may be influenced by both internal and external factors. Among the environmental factors temperature, oxygen, and water are the most important. Carbon dioxide concentrations greatly in excess of those normally present in the atmosphere will cause substantial reductions in the rate of respiration (Figure 4-7), perhaps principally by causing stomatal closure and thus a decrease in available oxygen. However, such high concentrations of carbon dioxide are rarely if ever encountered in nature. Marked deficiencies of essential elements, particularly nitrogen, sulfur, phosphorus, iron, and copper or other trace elements, may bring about a reduction in respiration, probably by preventing synthesis of adequate quantities of the respiratory enzymes and coenzymes.

Wounding a plant tissue will result in a temporary increase in its rate of respiration. Hopkins [5] found that cut potato tubers had increases in sugar

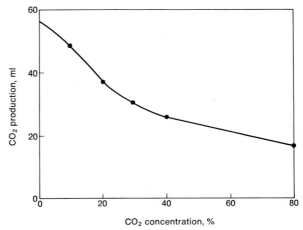

Figure 4-7. Influence of carbon dioxide concentration on the rate of respiration of germinating white mustard seeds. Note that the carbon dioxide concentrations used are much greater than those encountered in nature. [Data of F. Kidd, *Proc. Roy. Soc. London,* **B89:**136 (1915)]

(presumably from hydrolysis of starch) of up to 68% and suggested that this increase in substrate was responsible for the increase in respiration. It is also possible that oxygen is more readily available at the cut surfaces. Wounding generally initiates meristematic activity and formation of a wound callus. These metabolically active cells could be expected to have a higher rate of respiration, but the increase in respiration generally has disappeared by the time these cells are fully active. Wounding is not an important factor as regards natural rates of respiration but, in experimental determinations of respiration rates where pieces of plant tissue are used, the wounding may result in higher rates than existed in the intact tissues.

Audus [1] found that merely gentle bending or rubbing of leaf blades could increase their rate of respiration by as much as 100%, but he was unable to provide an explanation. The increase often persisted for several hours, but response to the stimulus decreased when it was repeated over a period of time. Similar increases in respiration of potato tubers have been found after they were handled. It is likely that such mechanical stimulation, as well as wounding, may have resulted in abnormally high rates of respiration in various experimental determinations. In nature, the bending and twisting of leaves and stems by the wind may well bring about increases in the rate of respiration.

Poisons, including carbon monoxide, cyanides, and fluorides, block different reactions in the respiratory pathways, and their use has been most valuable in determining the sequence of reactions in respiration. Since such substances are common components of polluted air, it is likely that they may reduce the rate of respiration in urban and industrial regions.

Internal Factors

Since green plants produce their own food by photosynthesis and subsequent processes, food should be considered as an internal factor. Although an inadequate quantity of substrate can theoretically limit the rate of respiration, this is rarely if ever the limiting factor in nature. In a healthy plant the supply of food is always adequate. Food is likely to become limiting only if a plant is kept from carrying on photosynthesis over an extended period of time, as by keeping it in the dark. Inadequate hydration of the protoplast may be limiting at times. A deficit of enzymes because of severe mineral deficiencies could also be limiting.

However, the principal internal factor that is most likely to limit the rate of respiration is a feedback mechanism involving the rate at which the ATP produced by respiration is used. If ATP is being produced faster than it is being used, the supply of ADP becomes reduced or even exhausted, thus slowing down or stopping further ATP production. In turn, this inhibition can result in an accumulation of intermediates, including reduced hydrogen acceptors, and so a decreased rate of respiration. On the other hand, increasingly rapid use of the ATP provides more and more ADP that can be reduced to still more ATP. This may provide the explanation for the well known fact that there is a positive correlation between respiratory rates and the general level of metabolic activity of a tissue.

One specific example of the above process may be the **salt respiration**

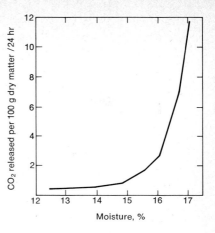

Figure 4-8. Influence of the water content of wheat grains on their rate of respiration. [Data of C. H. Bailey and A. M. Gurjar, *J. Agr. Res.*, **12**:685 (1918)]

described by Lundegårdh and Burström [10]. They found that when plants were transferred from water to a solution of salts there was a definite increase in the rate of respiration, probably because of increased use of ATP in accumulation of the ions against a concentration gradient.

Water

Water is the limiting factor in respiration in mature and dry seeds and spores in which the rate of respiration is very low. As a dry seed absorbs water its rate of respiration slowly increases up to a certain percentage of water content, and then it increases rapidly with only a slight further increase in water. For wheat grains this rapid increase occurs between 16 and 17% water content (Figure 4-8). This is presumably the point at which the cells are becoming adequately hydrated for active respiration. Similar increases in respiration occur as water is being absorbed by plants such as lichens and the resurrection plant (a clubmoss), which can be desiccated to an air dry condition without being killed.

Water may also be the limiting factor in respiration in wilted plants, particularly after a prolonged period of wilting. This is probably largely due to inadequate hydration of the cells, although closure of the stomata brought about by wilting could create an oxygen deficit. However, as plants approach wilting there is sometimes an increase in the rate of respiration (Figure 3-18).

Oxygen

As long as the air surrounding a plant contains about the usual 20% of oxygen and the stomata are open, there is usually an adequate supply of oxygen. The rate of respiration is then limited by other factors, as indicated by the fact that further increases in the concentration of oxygen do not result in an increase in the rate of respiration. However, there is a decline in the rate of respiration as the oxygen concentration of the air is progressively reduced below 20% (Figure 4-9). Below 5% there is a rapid decline and, of course, aerobic respira-

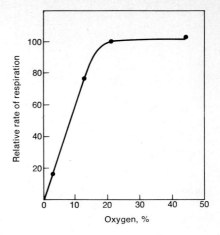

Figure 4-9. Influence of oxygen concentration on the rate of respiration. [Data of F. C. Steward, W. E. Berry, and T. C. Broyer, *Ann. Bot.*, **50**:354 (1936)]

tion ceases completely at 0% oxygen. As has been noted earlier, germinating seeds may carry on more anaerobic than aerobic respiration at low levels of oxygen, but above 10% oxygen the respiration is all aerobic.

Under natural conditions oxygen is likely to be the limiting factor in respiration principally for roots, tubers, and other underground parts of plants and particularly when the soil is waterlogged. Because of the low solubility of oxygen in water anaerobic respiration may occur.

Temperature

In contrast with the situation in photosynthesis, temperature is usually the limiting factor in respiration, as indicated by the fact that an increase in temperature almost always brings about an increase in the rate of respiration.

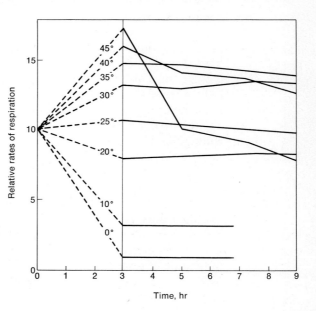

Figure 4-10. Influence of temperature on the rate of respiration of *Pisum sativum* (pea) seedlings. All seedlings were initially kept at 25°C. Measurement of the rates of respiration began 3 hr after lots of seedlings had been transferred to each of the indicated temperatures. Note the decrease in rate of respiration with time at the higher temperatures, particularly 40 and 45°C. [Data of D. S. Fernandes, *Rec. Trav. Bot. Néerlandis*, **20**:107 (1923)]

However, the optimal temperature for respiration is considerably lower than the temperatures at which disruption of enzyme systems begins. For pea seedlings the optimum is about 30°C, but there is considerable variation in the optima of different species. As might be expected, tropical species generally have higher optima than temperate zone species.

In respiration, as in photosynthesis, there is a time factor in the influence of temperature (Figure 4-10). Up to the optimum the rate increases with temperature and the rate is maintained over a long period of time. At temperatures above the optimum, such as 40 and 45°C in Figure 4-10, there may be an initially higher rate but this rapidly declines. At temperatures such as 50 and 55°C, which cause damage to enzyme systems, even the initial rate of respiration is much lower than at the optimum and the decline in rate is very rapid. It is possible that the less prompt and less rapid declines at 40 and 45°C are the result of a slower inactivation of enzymes.

At 0°C and below the rates of respiration of most plants are extremely slow, although measurable respiration has been found in some plants at as low as −20°C.

References

[1] Audus, L. J. "Mechanical stimulation and respiration in the green leaf," *New Phytol.,* **34**:557 (1936); **38**:284 (1939); **39**:65 (1940); **40**:86 (1941).

[2] Beevers, H. *Respiratory Metabolism in Plants,* Row-Peterson, Evanston, Ill., 1961.

[3] Green, D. E. "The mitochondron," *Sci. Amer.* **210**(1):63–73 (Jan. 1964).

[4] Green, D. E., and Y. Hatefi. "The mitochondrion and biochemical machines," *Science,* **133**:13–19 (1961).

[5] Hopkins, E. F. "Variation in sugar content of potato tubers caused by wounding and its possible relation to respiration," *Botan. Gaz.,* **84**:75 (1927).

[6] James, W. O. *Plant Respiration.* Clarendon Press, Oxford, 1953.

[7] James, W. O., and R. M. Leech. "The plant cytochromes," *Endeavour,* **19**:108–114, 1960.

[8] Lehninger, A. L. "Energy transformations in the cell," *Sci. Amer.* **202**(5):102–114 (May 1960); "How cells transform energy," *Sci. Amer.,* **205**(3):62–73 (Sept. 1961).

[9] Lehninger, A. L. *Energy of the Living Cell: Molecular Basis of Energy Transformations in the Cell,* Benjamin, New York, 1965.

[10] Lundegårdh, H., and H. Burström. "Untersuchungen über die Salzaufnamhe der Pflanzen," *Biochem. Zeit.,* **261**:235 (1933).

[11] Ochoa, S., and W. Vishniac. "Carboxylation reactions and photosynthesis," *Science* **115**:297 (1952).

[12] Rose, A. H. "Yeasts," *Sci. Amer.,* **202**(2):136–142 (Feb. 1960).

[13] Stiles, W., and W. Leach. *Respiration in Plants,* 3rd ed., Methuen, London, 1952.

Some Other Biochemical Activities of Plants

5

In addition to photosynthesis and respiration hundreds of other biochemical reactions occur in plant cells. Most of these are either synthetic processes, in which simpler compounds are built up into more complex ones; interconversions, in which a substance such as an amino acid or a simple sugar is converted into another substance of the same class; or decompositions, in which a more complex substance is broken down into simpler compounds by processes such as hydrolysis (digestion) and phosphorolysis.

Plants synthesize a vast array of organic compounds, many of them very complex. In the final analysis, all of these compounds are derived from the products of photosynthesis. The substances produced in greatest abundance are various carbohydrates, lipids, and proteins, which can be referred to collectively as foods. They are important both as respiratory substrates and as substances from which the protoplasts (including the various organelles) and walls of the cells are constructed by processes referred to as assimilation. Proteins are of particular importance since the enzymes that catalyze the various biochemical reactions are all proteins. Between 90 and 95% of the dry weight of plant tissues is composed of carbohydrates, lipids, and proteins.

However, many of the compounds synthesized in lesser quantities are equally essential in the life of the plant. These include the nucleic acids (DNA and RNA) that carry and transcribe the hereditary potentialities, the various plant hormones and vitamins, the coenzymes (which in many cases are derived from vitamins), and the chlorophylls and carotenoids that are so essential in the photosynthetic mechanism. Some of the other substances synthesized by plants are important primarily as cell wall constituents, for example, cellulose and other polysaccharides such as the hemicelluloses, pectins, lignin, and waxes.

In addition to compounds such as those mentioned above, which are essential in the metabolic activities and structure of plants, many other kinds of organic compounds are synthesized by plants. These include many different alkaloids, such as nicotine, quinine, strychnine, morphine, and atropin; such things as rubber latex, turpentine, resins, and the anthocyanin pigments that impart red to blue colors to petals and other organs; and the volatile oils (essential oils), such as those of peppermint, rose, lemon, and lavender, which are responsible for the pleasant odors of various plants. That these substances are, in general, not essential in plant metabolism is indicated by the fact that they are not produced by all species of plants and, in some cases, are products of only a single species, genus, or family. It is possible that they are simply metabolic byproducts that play no important role in the plants that produce them. However, there is evidence that some of the alkaloids may function as either antibiotics or germination inhibitors and the essential oils and anthocyanins may possibly attract pollinating insects. Future investigations may reveal that at least some of these various substances are more important in the metabolism of plants than is now suspected. In any event, many of them are important items of commerce. Although the alkaloids seem to have little effect on the plants that produce them, most of them have marked physiological effects on animals and many are used as drugs.

The synthetic capacities of plants are much more extensive than those of animals which, in general, do not produce such a wide variety of substances that are, at best, of peripheral metabolic significance. Furthermore, only plants can synthesize certain substances that are of great metabolic importance in both groups of organisms. Plants synthesize all the vitamins, whereas animals have only limited capacities for producing a few of them from complex precursors derived from plants. Plants can synthesize various amino acids from sugars and nitrates and, from these, can produce all the other amino acids essential in the synthesis of proteins, whereas animals have only a limited capacity for converting one amino acid into others. In addition, plants synthesize dozens of amino acids that are never used in protein synthesis. Thus, animals are dependent upon plants not only for the food and oxygen produced by photosynthesis but also for such essential substances as amino acids and vitamins.

It is beyond the scope of this book to consider all of the many syntheses and other metabolic processes carried on by plants. To do so would require many volumes the size of this one. Even plant biochemistry textbooks [3, 14] can give reasonably complete consideration only to the most important processes and a few examples of the others. In this chapter it will be necessary to restrict the discussion to some of the most important metabolic processes involving carbohydrates, lipids, and proteins and other nitrogenous compounds.

Carbohydrate Metabolism

The carbohydrates [3] play a basic and central role in plant metabolism. Sugars are the principal product of photosynthesis and the principal substrate in respiration. They can readily be converted into glycerol and fatty acids, which can then react and form fats, or into amino acids, from which proteins are synthesized. Indeed, practically all of the numerous synthetic pathways of plants, if traced backward far enough, arrive at some sugar. The sugars that play these important metabolic roles are **monosaccharides,** or simple sugars.

They cannot be hydrolyzed into other sugars with smaller molecules, although they can be converted to them by other processes such as those involved in photosynthesis and respiration. The monosaccharides can also be converted into more complex carbohydrates by the union of two or more molecules. **Disaccharides** are synthesized from two molecules of monosaccharides, **trisaccharides** from three, and **tetrasaccharides** from four. The **polysaccharides,** such as starch and cellulose, are built up from hundreds or thousands of monosaccharide molecules in most cases, although the dextrins, which have as few as six monosaccharide residues, are frequently considered to be polysaccharides.

Monosaccharides

There are many different kinds of monosaccharides, but only a dozen or two of these are known to play important roles in plant metabolism. These include various trioses, tetroses, pentoses, hexoses, and heptoses (which, respectively, contain three, four, five, six, and seven carbon atoms). The octoses, nonoses, and decoses (eight, nine, and ten carbons) are much less important and will not be considered here. Technically, sugars are aldehydic or ketonic derivatives of aliphatic polyhydroxy alcohols, and both aldehyde and ketone sugars are found in each of the groups mentioned above.

TRIOSES. Because of their limited number of carbons there are only two structural isomers among the trioses (Figure 5-1): an aldehyde sugar (glyceraldehyde) and a ketone sugar (dihydroxyacetone). Neither of the trioses is found in the free state in plants in any appreciable concentration, but their phosphate esters (3-phosphoglyceraldehyde and dihydroxyacetone phosphate) are important intermediates in both photosynthesis and respiration. These phosphate derivatives are readily convertable from one to the other and participate in a number of other processes such as the reduction of 3-phosphoglyceraldehyde to glycerol, $C_3H_5(OH)_3$.

All of the sugars, as well as the trioses, participate in biochemical processes primarily as their phosphate esters; these, in contrast with the sugars themselves, are very reactive biochemically. The principal reaction in which a sugar itself participates is its phosphorylation, and even this does not occur commonly with the trioses because their phosphate esters are usually produced from, or converted into, either fructose-1,6-diphosphate or phosphoglyceric acid.

TETROSES. The tetroses (Figure 5-1) are also intermediates in various metabolic pathways. For example, erythrose-4-phosphate is an intermediate in both the ribulose diphosphate cycle of photosynthesis and the respiratory hexose monophosphate shunt. Tetroses are never present in plants at any appreciable concentration. Unlike the pentoses and hexoses, they are not used extensively in the synthesis of polysaccharides.

PENTOSES. The pentoses (Figure 5-2) do not accumulate in plants in quantity, but they are very important as intermediates in various metabolic pathways and in the synthesis of several cell wall polysaccharides and of various sub-

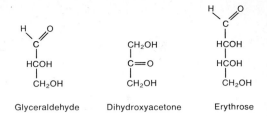

Figure 5-1. Structural formulas of the two trioses and erythrose, a tetrose.

Glyceraldehyde   Dihydroxyacetone   Erythrose

stances essential in many metabolic processes. It will be recalled that the phosphate esters of xylulose, ribose, and ribulose are intermediates in both the photosynthetic sugar interconversions and the hexose monophosphate shunt, and that ribulose-1,5-diphosphate is a carbon dioxide acceptor in photosynthesis.

Both ribose and deoxyribose (as their 1-pyrophosphate-5-phosphate esters) can react with adenine, guanine, or other purines, or with thymine, cytosine, uracil, or other pyrimidines to form nucleotides. The purines and pyrimidines, both of which are heterocyclic nitrogenous bases, and the nucleotides will be considered later in this chapter.

Deoxyribose differs from ribose in that it has an —H on carbon 2 rather than an —OH. Thus, deoxyribose is $C_5H_{10}O_4$ whereas ribose is $C_5H_{10}O_5$. Other sugars may also have deoxy derivatives.

The nucleotides are constituents of a number of compounds of great meta-

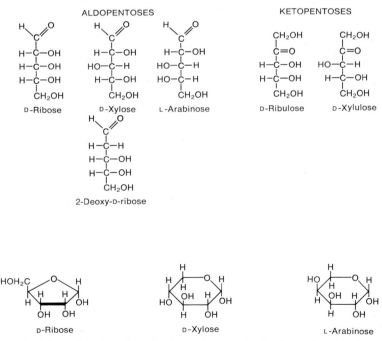

Figure 5-2. Structural formulas of pentoses that play important roles in plant metabolism. Several ring structures are given below.

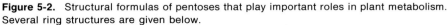

bolic importance. Deoxyribonucleic acid (DNA), which carries the genetic code as will be described later, is a high polymer of four different nucleotides: adenosine, guanosine, thymidine, and cytidine. The several kinds of deoxyribonucleic acids, which are involved in the transfer of the genetic information of DNA to the sequence of amino acids in proteins (and will also be considered later), are also polymers synthesized from four principal nucleotides: deoxyadenosine, deoxyguanosine, deoxyuridine, and deoxycytosine.

Nucleotides are also components of a number of important coenzymes. For example, adenosine is used in the synthesis of CoA (Figure 4-2), NAD, and NADP (Figure 3-4). Pentoses are also used in the synthesis of various glycosides, compounds in which sugars are linked to a variety of organic radicals.

The nucleotides (monophosphates) may acquire a second phosphate and thus become diphosphates such as adenosine diphosphate (ADP). Addition of a third phosphate to a diphosphate results in triphosphates such as ATP (Figure 3-4). The importance of ATP as a ready energy source has already been stressed. Other triphosphates including uridine triphosphate (UTP) and guanosine triphosphate (GTP) also serve as energy or phosphate donors in some metabolic processes.

Pentoses are also used in the synthesis of various polysaccharides, known collectively as pentosans. The arabans are pentosans synthesized from arabinose and are generally mucilaginous or gummy. They may be closely associated with the pectic compounds of the walls. The gums that exude from cherry trees are arabans, and commercial mucilages are arabans from other sources. Xylans are synthesized from another pentose, xylose, and are rather abundant in cell walls, particularly in wood, seed coats, and straw. Up to a quarter of the cell walls in the xylem may be xylans.

HEXOSES.   The hexoses are by far the most abundant monosaccharides in plants and are the principal sugars used in the synthesis of disaccharides, other more complex sugars, and also the various polysaccharides. Although some two dozen hexose isomers with the empirical formula $C_6H_{12}O_6$ are possible, only four of these are of particular importance in plants: glucose, fructose, mannose, and galactose (Figure 5-3). Actually only glucose and fructose occur in any quantity as free sugars; the mannose and galactose are used primarily in the synthesis of more complex carbohydrates.

Of these four sugars fructose is a ketose; the other three are aldoses that differ from one another structurally in the arrangement of the —H and —OH groups on the carbons. Each of the hexoses (as well as each of the other monosaccharides) has D and L stereoisomers that are mirror images of one another, but it is interesting to note that only the D isomers occur naturally. Although the structural formulas of monosaccharides are often written as straight chains, the molecules, at least from the pentoses up, are actually heterocyclic rings with the aldehyde or ketone =O forming an oxygen bridge. In glucose and other hexoses the oxygen bridge may be between carbons 1 and 5, thus forming a six-atom **pyranose** ring, or between carbons 1 and 4, which results in a five-atom **furanose** ring. The furanose rings are less stable and more reactive than the pyranose rings. Because of the ring structure, carbon 1 becomes assymetrical, and permits an additional type of isomer

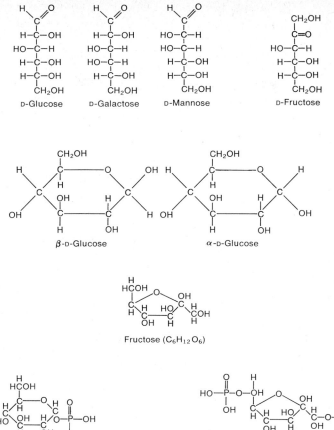

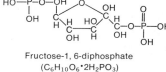

Fructose ($C_6H_{12}O_6$)

Glucose-1-phosphate
($C_6H_{11}O_6 \cdot H_2PO_3$)

Fructose-1, 6-diphosphate
($C_6H_{10}O_6 \cdot 2H_2PO_3$)

Figure 5-3. Structural formulas of four hexoses that are abundant in plants, with several ring structures below. Two of the phosphate esters of hexoses are also shown.

depending on the position of its —H and —OH groups. These are referred to as the alpha (α) and beta (β) isomers (Figure 5-3). The importance of this isomerism will become evident when we consider the different disaccharides and polysaccharides derived from α- and β-D-glucose.

Glucose is generally considered to be the primary substrate in respiration, and it is used more extensively in the synthesis of complex sugars and polysaccharides than any other monosaccharide; however, in some ways it probably plays a less important role in plant metabolism than was once believed. For example, glucose is no longer considered the primary product of photosynthesis nor the principal carbohydrate translocated through the phloem. Although glucose-6-phosphate is an intermediate in glycolysis and the hexose monophosphate shunt and glucose-1-phosphate is used in the synthesis of more complex carbohydrates, the phosphate esters of fructose also play extensive roles as intermediates. It will be recalled that both fructose-6-phosphate and fructose-1,6-diphosphate are involved in glycolysis, the carbon cycle of photosynthesis, and the hexose monophosphate shunt.

OTHER MONOSACCHARIDES. The monosaccharides with more than six carbon atoms are of limited occurrence in plants. However, it will be recalled that the seven-carbon sedoheptulose-7-phosphate is an intermediate in both the hexose monophosphate shunt and the photosynthetic carbon cycle and that sedoheptulose-1,7-diphosphate is also involved in the latter.

Disaccharides

Disaccharide sugars are produced by the condensation of two molecules of monosaccharides. Although disaccharides theoretically could be expected to be produced from any of the monosaccharides, only the hexoses are used in the synthesis of the disaccharides commonly found in plants. Furthermore, one of the monosaccharides used is always glucose, and only a few of the possible hexose disaccharides are important in plants.

By far the most abundant of the disaccharides is sucrose (Figure 5-4), which is synthesized from a molecule of α-D-glucose and a molecule of β-D-fructose. Sucrose frequently accumulates in plants in considerable quantities. It, as well as starch, may be synthesized quite promptly from the sugars produced by photosynthesis. Some plants, notably members of the lily family, accumulate sucrose instead of starch. Sucrose is present in the phloem in high concentrations and is believed to be the principal carbohydrate translocated through

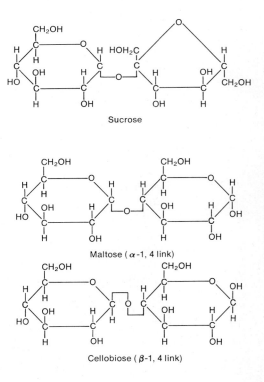

Figure 5-4. Structural formulas of three important plant disaccharides. Maltose is synthesized from α-D-glucose and cellobiose from β-D-glucose. Note the difference between the α- and β-1,4 links.

the phloem. Ordinary table sugar, whether white or brown, is sucrose, and it is also the principal sugar in maple syrup and sorghum syrup. Sugarcane stems and sugarbeet roots are the most common commercial sources of sucrose because of the high concentration (up to 20% of the fresh weight) present.

There are two other disaccharides that are generally present in plants, but neither accumulates as abundantly as sucrose. One is maltose and the other cellobiose. Both are composed of two glucose residues, but in maltose it is α-D-glucose whereas in cellobiose it is β-D-glucose. Maltose is the product of starch hydrolysis by amylases and is most abundant in germinating seeds and other tissues where extensive starch hydrolysis is in progress. In turn the maltose is hydrolyzed to glucose by maltase, and the glucose is then used in respiration or other metabolic processes. The accumulation of maltose is therefore neither very permanent nor very extensive. Cellobiose is the product of cellulose hydrolysis but, because cellulose is not commonly hydrolyzed by most plants, there is little cellobiose present. However, some fungi, bacteria, and protozoa produce cellulase and cellobiase and are able to hydrolyze cellulose to cellobiose and then to glucose and thus utilize cellulose as a food. The other disaccharides are not as important in plant metabolism. It may be noted that lactose, the milk sugar of mammals which is synthesized from glucose and galactose, is not common in plants. Although plants produce galactose, they apparently lack the enzyme necessary for synthesis of lactose.

We shall use sucrose as an example of the synthesis and breakdown of disaccharides. Several different pathways of sucrose synthesis are known, perhaps the most common of these in higher plants being the one catalyzed by **sucrose synthetase.** In this process glucose first reacts with uridine triphosphate (UTP) (a compound homologous with ATP) with the formation of uridine diphosphate glucose (UDP-glucose). The glucose part of this molecule is then transferred to fructose, producing sucrose and UDP.

$$UTP + glucose \longrightarrow UDP\text{-glucose} + H_3PO_4$$

$$UDP\text{-glucose} + fructose \longrightarrow sucrose + UDP$$

The UDP-glucose may also transfer glucose to fructose-6-phosphate in a process activated by a different enzyme (sucrose phosphate synthetase) with the production of sucrose phosphate and UDP. The sucrose phosphate is then hydrolyzed to sucrose by the action of a phosphatase. A third synthetic pathway appears to be more common in microorganisms than higher plants. This reaction is catalyzed by **sucrose phosphorylase** and involves the synthesis of sucrose from glucose-1-phosphate and fructose.

$$\underset{\text{Glucose-1-phosphate}}{C_6H_{11}O_6 \cdot H_2PO_3} + \underset{\text{Fructose}}{C_6H_{12}O_6} \longrightarrow \underset{\text{Sucrose}}{C_{12}H_{22}O_{11}} + \underset{\text{Phosphate}}{H_3PO_4}$$

Since all three of the synthetic pathways are reversible, theoretically they can also bring about the degradation of sucrose into glucose and fructose. However, it is probable that the usual pathway of sucrose degradation is by hydrolysis, a reaction catalyzed by **sucrase** (also called invertase).

$$\underset{\text{Sucrose}}{C_{12}H_{22}O_{11}} + \underset{\text{Water}}{H_2O} \longrightarrow \underset{\text{Glucose}}{C_6H_{12}O_6} + \underset{\text{Fructose}}{C_6H_{12}O_6}$$

Figure 5-5. The monosaccharide residues of several plant trisaccharides, tetrasaccharides, and pentasaccharides, with their disaccharide precursors. The fructose-glucose-galactose oligosaccharides are important kinds of carbohydrates translocated through the phloem as is sucrose.

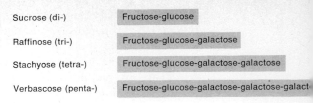

Sucrose (di-)	Fructose-glucose
Raffinose (tri-)	Fructose-glucose-galactose
Stachyose (tetra-)	Fructose-glucose-galactose-galactose
Verbascose (penta-)	Fructose-glucose-galactose-galactose-galact

The hydrolysis of foods into the simpler compounds from which they w originally synthesized is commonly referred to as **digestion,** the various drolyses being identical in plants and animals. However, in animals the drolyses occur primarily in the digestive tract, whereas in plants they intracellular and occur in living cells throughout the plants.

More Complex Sugars

Sugars synthesized from more than two molecules of monosaccharides neither very numerous nor very abundant in plants. They do not particij in plant metabolism themselves, but they can be hydrolyzed by plants to t usable monosaccharide components. They appear to be particularly abunc in the phloem, especially of various species of trees, and are important tr location carbohydrates along with sucrose.

The principal trisaccharides ($C_{18}H_{32}O_{16}$) of plants are gentianose melezitose, which are both built up from one molecule of fructose and of glucose, and raffinose, which is synthesized from one molecule eacl fructose, glucose, and galactose. Addition of still another galactose resul the formation of the tetrasaccharide stachyose, and the pentasaccharide bascose is formed when a third galactose is added (fructose-glucose-galac galactose-galactose). The fact that the principal tetrasaccharides and pe saccharides are sugars derived from fructose, glucose, and galactose (Fi 5-5) is an interesting example of synthetic specialization in one out of n possible ways.

Polysaccharides

Although it might be expected that polysaccharides (Figure 5-6) coul synthesized from any of the monosaccharides, plants specialize in synthes them from pentoses (the **pentosans**) and hexoses (**hexosans**), particularl latter. There are a few polysaccharides made from both pentoses and hex We have already considered some of the pentosans in the section on pent so the present discussion will be devoted to the hexosans.

The principal hexosans of plants are starch and cellulose, the former t the principal carbohydrate that accumulates in most species of plants an latter the basic component of plant cell walls. Starch is composed of α-D-gl residues and cellulose of β-D-glucose residues. Some species of plants, pa larly many of the composites, accumulate inulin rather than starch. Inu a polysaccharide composed principally of fructose residues, with a few gl residues, and is crystalline rather than granular like starch. In addition to cellulose, certain other polysaccharides serve as cell wall constituents. These include the xylans and arabans, two types of pentosans mentioned earlier, and several hexosans, including mannans (from mannose) and galactans (from galactose). The mannans are most abundant in the cell walls of gymnosperm tracheids and of the endosperms of some seeds. The galactans are usually associated with the pectic compounds of cell walls. These and other noncellulose cell wall polysaccharides are sometimes referred to collectively as **hemicelluloses;** these differ from cellulose in not being essential for the structural integrity of the walls and in being quite readily hydrolyzable into their component monosaccharides, which can then be utilized metabolically.

The pectic compounds of cell walls, which are particularly important as constituents of the middle lamellae and primary cell walls, are strictly speaking not polysaccharides because they are polymers of galacturonic acid rather than

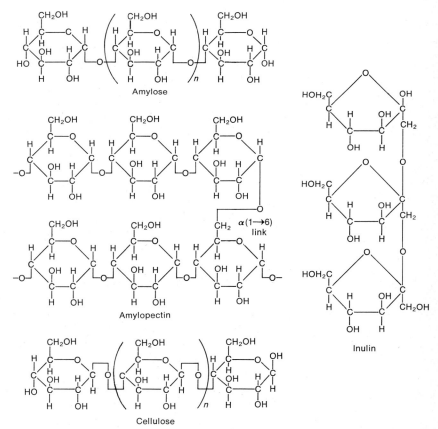

Figure 5-6. The repeating units of several plant polysaccharides. Note the α-(1 → 6) linkage in amylopectin, which results in a branched starch molecule, in contrast with the unbranched amylose type of starch. Note also the α linkages of starch versus the β linkages of cellulose (See Figure 5-4). There are about 35 fructose residues in an inulin molecule, of which three are shown. Each inulin molecule also contains two glucose residues: one at one end and the other near the middle.

sugars. However, because sugar acids, such as galacturonic acid, are oxidation products of the monosaccharides, and also as a matter of convenience, we shall follow the usual practice of considering the pectic compounds along with the polysaccharides.

In general, the polysaccharides differ from the sugars not only in being much larger molecules but also in lacking a sweet taste and in being insoluble in water. Although some polysaccharides such as starch do form stable dispersions in water, these are actually colloidal sols rather than true solutions because a single polysaccharide molecule is within the size range of a colloidal micelle. Dextrins, which are intermediates in the hydrolysis and synthesis of starch, have six or more glucose residues per molecule. They are sweet and soluble even though they are often classified as polysaccharides. There are also molecules composed of three, four, or five glucose residues, but these should be considered as tri-, tetra-, and pentasaccharide sugars. Corn syrup is a mixture of dextrins and sugars produced commercially by the hydrolysis of corn starch.

We will now consider in somewhat greater detail the principal polysaccharides synthesized by plants: starch, cellulose, inulin, and the pectic compounds.

STARCH. In plants starch synthesis occurs only within chloroplasts and leucoplasts, where the enzymes necessary for its synthesis are localized. The starch accumulates in the form of grains that range in size from 1 to 150 μ, the grains in chloroplasts usually being much smaller than those in the leucoplasts. The grains in chloroplasts are also more transient because the starch generally is decomposed within a day or so, whereas the leucoplast grains may remain for months before being hydrolyzed and used. Each leucoplast produces only one starch grain; thus, under the microscope the grains may appear to be free in the cytoplasm, but in electron micrographs a double bounding membrane is visible. Starch grains vary in structure as well as in size, and are so characteristic of the species that an expert can determine the source of isolated grains by microscopic examination. Starch grains are often laminated, but the usual lamellae are not formed under conditions of continuous illumination of the plant, even though the starch is produced in an underground structure such as a root or tuber.

The starch making up a grain is usually of two different kinds: **amylose,** with a molecule that is a long, unbranched helix, and **amylopectin,** with a molecule that is a highly branched helix. Usually some 70 to 80% of the starch in a grain is amylopectin, but in peas it constitutes as little as 30% of the starch whereas the waxy starch of some varieties of corn is 100% amylopectin. Amylose disperses in water much more readily than amylopectin and is sold commercially as "soluble starch." There is considerable variation with species in the number of glucose residues per molecule of either amylose or amylopectin; the range is from several hundred to 1000 or more, so it is really more proper to consider the starches as a class of compounds rather than as a single compound.

There are several different pathways of starch synthesis, and apparently all of them may occur within a single species. Most of the synthetic enzymes catalyze the formation of the α-(1→4)-glycosidic linkages that are present in both amylose and amylopectin, but one (the **Q-enzyme**) catalyzes the formation of the α-(1→6) linkages that result in formation of the branches of amylopectin.

Perhaps the best known pathway of starch synthesis is the one catalyzed by **starch phosphorylase** in which glucose-1-phosphate is linked (1→4) molecule by molecule until the starch molecule is complete. However, the synthesis cannot be initiated unless there is a primer (or acceptor) molecule present. This primer can be anything from a glucose trisaccharide (maltotriose) to a dextrin with 20 or so glucose residues. The reaction may be represented as follows.

$$n\ C_6H_{11}O_6 \cdot H_2PO_3 + \text{primer} \longrightarrow (C_6H_{10}O_5)_n + n\ H_3PO_4$$
$$\text{Glucose-l-phosphate} \qquad\qquad \text{Amylose} \qquad \text{Phosphate}$$

The reaction is reversible, and provides a means of breaking down starch to glucose-1-phosphate by phosphorolysis. One of the principal factors determining the direction of the reaction is pH. The conversion of glucose-1-phosphate to starch has an optimum pH of around 5, whereas the optimum in the reverse direction is about 7. The conversion of starch to glucose-1-phosphate is also promoted by a high concentration of phosphate. As we shall see later on, this reversible reaction is involved in the opening and closing of stomata.

Another pathway of starch synthesis is catalyzed by the enzyme **UDP-glucose-transglycosylase,** the substrate being UDP-glucose rather than glucose-1-phosphate. Again a primer is needed, and it must contain at least one α-(1→4)-glycosidic link. The primer can be anything from maltose on up to starch itself, the product in the latter case being a higher molecular weight starch. The reaction may be written as

$$\text{UDP-glucose} + \text{primer} \longrightarrow \text{starch} + \text{UDP}$$

A similar reaction involving ADP-glucose instead of UDP-glucose has been found in corn and rice, and this may prove to be an important pathway of starch synthesis.

Still another means of starch synthesis is catalyzed by what has been named the **D-enzyme** by Peat and his coworkers, who first discovered it in potatoes in 1953. In this case the substrate can be any of the complex sugars with two or more α-(1→4)-glucose linkages, that is, from maltotriose on. The acceptor can be the same sugars or the larger dextrin molecules or even starch. An interesting feature of this reaction is that a molecule of glucose is always formed. For example, if maltotriose is both the substrate and acceptor the products would be maltopentose and glucose.

$$C_{18}H_{32}O_{16} + C_{18}H_{32}O_{16} \longrightarrow C_{30}H_{52}O_{26} + C_6H_{12}O_6$$
$$\text{Maltotriose} \qquad \text{Maltotriose} \qquad \text{Maltopentose} \qquad \text{Glucose}$$

The above reaction has resulted in the synthesis of a pentasaccharide from two molecules of a trisaccharide, but the product of one reaction can serve as the primer for a subsequent similar reaction, thus giving rise to dextrins and eventually to starch. Like the other pathways of starch synthesis this one is reversible.

The three pathways of starch synthesis described above all result in α-(1→4) linkages and give rise to amylose or the α-(1→4) linkages of amylopectin, but the synthesis of amylopectin also requires an enzyme that can catalyze the formation of the α-(1→6) linkages that result in the branching of the amylopectin molecule. Such an enzyme was found in potatoes by Baum and Gilbert in 1953 and was named the Q-enzyme. Both the substrate and the primer must contain at least four glucose residues with α-(1$\rightarrow$4) linkages (that is, at least the tetrasaccharide, maltotetrose), but dextrins with about 40 glucose residues seem to be the most common substrate and the primer may be anything up to a growing amylopectin molecule.

Although starch can be broken down by the reversal of the various synthetic pathways, plants apparently degrade starch primarily by hydrolysis (digestion). The hydrolytic reactions are not reversible, and do not serve as another means of starch synthesis. Most, if not all, species of plants produce two different enzymes that hydrolyze starch: **α-amylase** and **β-amylase,** the latter being particularly abundant in germinating seeds. The two amylases hydrolyze starch in different ways.

β-Amylase hydrolyzes starch by splitting off one maltose molecule after another from the end; if the starch is an amylose with an even number of glucose residues, it is eventually completely hydrolyzed to maltose. If it has an odd number of glucose residues, the final molecule is a maltotriose. β-Amylase can hydrolyze the α-(1$\rightarrow$4) linkages of amylopectin, but only to within two glucose residues of a branch. A smaller amylopectin molecule or, at best, branched dextrins still remain. The maltose produced by the hydrolysis of starch can be hydrolyzed on down to glucose by the enzyme maltase.

$$(C_6H_{10}O_5)_{2n} + n\, H_2O \longrightarrow n\, C_{12}H_{22}O_{11}$$
$$\underset{\text{Amylose}}{} \qquad \underset{\text{Water}}{} \qquad \underset{\text{Maltose}}{}$$

$$C_{12}H_{22}O_{11} + H_2O \longrightarrow 2\, C_6H_{12}O_6$$
$$\underset{\text{Maltose}}{} \qquad \underset{\text{Water}}{} \qquad \underset{\text{Glucose}}{}$$

In contrast with β-amylase, α-amylase can hydrolyze α-(1$\rightarrow$4) linkages toward either end of an amylose molecule or at various points along the length of the molecule, and the products of the hydrolysis are dextrins rather than maltose. Initially at least the dextrins may be of rather high molecular weight with numerous glucose residues, but eventually much of the starch is hydrolyzed to dextrins with either six or twelve glucose residues, that is, with one or two complete turns of the starch helix. The various dextrins may be subsequently hydrolyzed to maltose by β-amylase. The α-amylase can hydrolyze amylopectin only to within three glucose residues of a branch, but since it can attack the middle of an α-(1$\rightarrow$4) chain it has the potentiality of reducing amylopectin to smaller dextrins than can β-amylase.

Complete hydrolysis of the branch points of amylopectin cannot occur without another enzyme that can break the α-(1$\rightarrow$6) linkages. Such an enzyme was isolated by Hobson and his coworkers in 1951 and was named the **R-enzyme.**

The synthesis of starch and its degradation to sugars by various pathways is influenced by a number of factors of both the internal and external environment, and these factors may determine whether starch synthesis or degradation is predominant at any particular time. The influence of pH on starch phosphorylase activity has already been mentioned, but it should be noted that each of the enzymes involved in starch metabolism has its own optimal pH; therefore, the overall influence of pH on starch metabolism may be quite

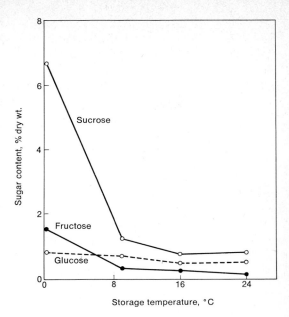

Figure 5-7. Influence of temperature on starch-sugar interconversions in potato tubers. Analyses were made after the tubers had been at the indicated temperatures 2 weeks. Note the marked conversion of starch to sugars below 10°C. [Data of B. Arreguin-Lozano and J. Bonner, *Plant Phsiol.*, **24**:720 (1949)]

complex. As might be expected, a high concentration of sugars favors starch synthesis. Thus, during the day when photosynthesis is occurring starch synthesis is active, whereas at night when the sugar concentration is lower the starch is broken down into sugars which are then translocated out of the leaves.

Low temperatures generally favor starch degradation over starch synthesis. Thus, the leaves of evergreens generally have a much higher sugar content in the winter than the summer. When the temperature drops below 10°C, potato tubers begin converting starch to sugars (Figure 5-7), largely through the action of starch phosphorylase, and this results in an undesirable sweet taste when the potatoes are eaten. However, potatoes also convert starch to sugar when the temperature is above 35°C. This may result from a water deficit, which is known to favor conversion of starch to sugar in wilted plants. There are marked species differences as regards the effect of temperature on starch-sugar interconversions. For example, in bananas starch hydrolysis is rapid at ordinary room temperatures but essentially stops below 10°C.

CELLULOSE. The macromolecules of cellulose are composed of from 3000 to 10,000 β-D-glucose residues. The β-(1→4) linkages have their oxygen bridges on alternating sides, in contrast with the α-(1→4) linkages of starch where all the oxygen bridges are on one side (Figure 5-6). In cellulose molecules all the —OH groups are at the sides, and are equally distributed on the two sides because of the alternating β linkages. The —OH groups are thus available for hydrogen bonding with the ring oxygens of adjacent cellulose molecules. In the helical starch molecules all the —OH groups are on the inside and are not available for hydrogen bonding with other starch molecules. Thus, the marked differences between the grains of starch and the cellulose microfibrils result basically from the fact that starch has α linkages whereas cellulose has

β linkages. Despite its chemical similarity to cellulose, starch simply does not have the structural requirements essential for a fibrous cell wall component.

Considerably less is known about the details of cellulose metabolism than about starch metabolism. However, there is evidence that the principal pathway of cellulose synthesis is from guanosine diphospho-glucose (GDPG), the glucose being the β-D isomer. As in starch synthesis, a primer seems to be necessary and is generally a celludextrin. It is possible that cellulose may also be synthesized by pathways comparable with the other kinds of starch synthesis, but a completely separate set of enzymes is required to make or break the β-(1→4) linkages. The enzymes involved in starch metabolism are completely ineffective in cellulose metabolism.

Cellulose can be hydrolyzed by the enzyme **cellulase,** the product being the disaccharide cellobiose. The cellobiose, in turn, can be hydrolyzed to β-D-glucose by **cellobiase,** so the situation is homologous with the hydrolysis of starch to maltose and the maltose to α-D-glucose. Unlike the amylases, the cellulases are not common among green plants, although they are presumably involved in isolated situations such as dissolving away the end walls of vessel elements as vessels are being formed. However, cellulase is produced by a good many species of fungi and bacteria and by some protozoa and these organisms are thus able to utilize cellulose as a food, in contrast with the vast majority of other organisms. These cellulose-hydrolyzing organisms play an extremely important ecological role. They bring about the decay and prevent the continuing accumulation of cellulose of dead plants. Sometimes, however, their activities are considerably less welcome, as when they operate on lumber that has been made into fences, houses, and other structures.

INULIN. Next to starch, inulin is the most important accumulation polysaccharide of plants, with the possible exception of the cell wall hemicelluloses. Unlike the other hexosans we have been considering, inulin is composed principally of fructose residues rather than those of glucose. Inulin, rather than starch, constitutes the accumulation carbohydrate of many species of composites including dahlia, dandelion, chickory, and goldenrod. The Jerusalem artichoke and some other species accumulate starch in their shoots and inulin in their tubers and roots. Inulin is dispersed in the cell sap as a colloidal sol.

The inulin molecule is considerably smaller than starch molecules, consisting of about 35 β-D-fructose residues and two α-D-glucose residues, one of them at the middle of the molecule and the other at one end. The fructose residues are joined by β-(2→1) linkages (Figure 5-6). Inulin is synthesized from sucrose, one molecule of sucrose receiving a fructose from another so as to produce a glucose-fructose-fructose trisaccharide and a molecule of glucose. Repetition of this reaction results in an increasing number of fructose residues until an inulin molecule finally results. This method of synthesis explains the presence of the terminal glucose residue. The presence of a glucose residue in the middle of the molecule suggests that the final inulin molecule may be formed by the union of two smaller molecules synthesized in the manner described. Inulin can be hydrolyzed by **inulase,** the products being fructose plus two molecules of sucrose. The latter can, of course, be hydrolyzed by sucrase to glucose and fructose.

Figure 5-8. Pectic acid is composed of α-D-galacturonic acid residues, *n* in the formula being about 100. Pectin is the methyl ester of pectic acid, and may have some 200 galacturonic acid residues. Protopectin has a much larger molecule than either pectic acid or pectin and an intermediate degree of methylation. Calcium pectate, the calcium salt of pectic acid, is abundant in the middle lamella.

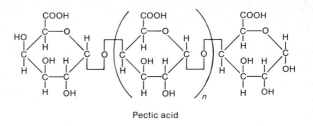

Pectic acid

PECTIC COMPOUNDS. The pectic compounds are important cell wall constituents. There are three principal kinds: pectic acid, pectin, and protopectin (Figure 5-8). All three are polymers of galacturonic acid, which has the same structure as galactose except that carbon 6 is in a carboxyl group rather than a —CH_2OH group. The galacturonic acid residues are linked through α-(1→4) bonds, UDP-galacturonic acid being used in the synthesis. The UDP-galacturonic acid may be derived from UDP-galactose, but it seems more likely that it is derived from UDP-glucose by way of UDP-glucuronic acid.

Pectic acid seems to be the most basic of the three pectic compounds. It is a straight-chain molecule composed of about 100 galacturonic acid residues and is probably a precursor of the other pectic compounds. Pectic acid is soluble and may occur in the protoplasts of cells.

Pectin differs from pectic acid in two ways: its molecules are longer and most of the —COOH groups have been esterified with methyl groups to —$COOCH_3$. Pectin is an important constituent of primary cell walls and may make up a high percentage of the dry weight of various tissues, particularly of fleshy fruits and fleshy roots such as those of carrots. Apple fruits may contain as much as 15% pectin and lemon skins as much as 35% on a dry-weight basis. Most of the pectin sold commercially is derived from the rinds of citrus fruits left over from the production of juice or canned segments or from carrot roots. Pectin, unlike either pectic acid or protopectin, will form firm gels and is the gelling agent in fruit jellies. Ripe fruits are often unsuitable for making jellies because much of the pectin is converted to pectic acid during ripening, but addition of commercially available pectin makes gelation possible.

Protopectin has even larger molecules than pectin. It is the principal constituent of the middle lamella and serves as a cement that holds the cells together. It exists at least partially as the calcium or magnesium salt, and the divalent calcium or magnesium ions serve to link together adjacent protopectin molecules. As fruits ripen the protopectin of the middle lamella is hydrolyzed, and the fruits become softer and easier to chew than the green fruits where the cells are held together much more firmly. In overripe fruits such as apples the decomposition of the middle lamella may proceed to the point where the fruits become mealy.

Hydrolysis of the α-(1→4) linkages of the pectic compounds is catalyzed by **pectinase** (also called pectin polygalacturonase), whereas hydrolysis of the methyl ester bonds is catalyzed by **pectin methyl esterase.** Complete hydrolysis results in galacturonic acid, which can then presumably be converted to sugars that can be utilized in various metabolic pathways.

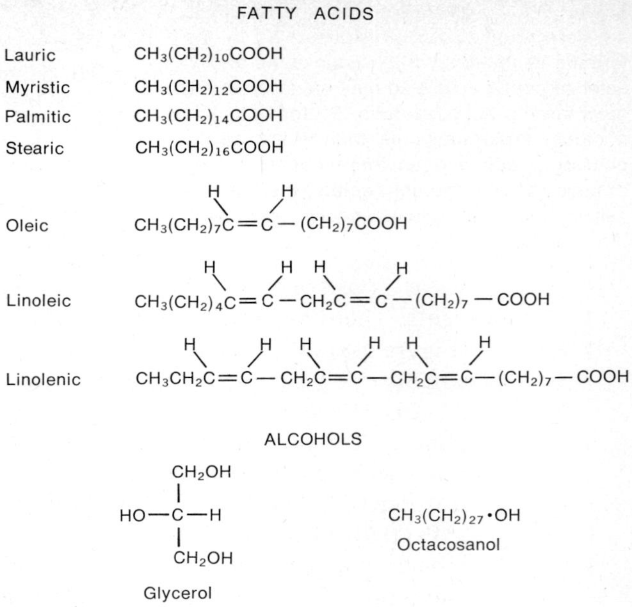

Lauric	$CH_3(CH_2)_{10}COOH$
Myristic	$CH_3(CH_2)_{12}COOH$
Palmitic	$CH_3(CH_2)_{14}COOH$
Stearic	$CH_3(CH_2)_{16}COOH$

Oleic $\quad CH_3(CH_2)_7C=C-(CH_2)_7COOH$

Linoleic $\quad CH_3(CH_2)_4C=C-CH_2C=C-(CH_2)_7-COOH$

Linolenic $\quad CH_3CH_2C=C-CH_2C=C-CH_2C=C-(CH_2)_7-COOH$

ALCOHOLS

CH_2OH
|
HO—C—H $CH_3(CH_2)_{27}\cdot OH$
|
CH_2OH   Octacosanol

Glycerol

Figure 5-9. Some of the fatty acids and alcohols used in the synthesis of fats (triglycerides), phospholipids, and waxes. The alcohol used in the synthesis of fats and phospholipids is always glycerol.

Lipid Metabolism

We now turn to another class of compounds that are important in the metabolism of plants: the lipids. These are esters of the fatty acids and alcohols (Figure 5-9). The most abundant lipids are the true fats **(triglycerides),** which are synthesized from glycerol, $C_3H_5(OH)_3$, a trihydroxy alcohol, and three molecules of fatty acid. A second, and perhaps more essential, class of lipids is the **phospholipids,** which as we have noted in Chapter 2 are important constituents of the various cell membranes. The phospholipids are synthesized from glycerol, two molecules of fatty acids, and one molecule of phosphoric acid that may have linked to it either choline or ethanolamine (Figure 5-10). The third important class of lipids is composed of **waxes,** which are esters of fatty acids and long-chain monohydroxy alcohols. The waxes are important primarily as constituents of cutin and suberin where they serve as waterproofing materials. Thus, the waxes and phospholipids serve as structural components of cells, whereas the fats are an accumulation type of food that can readily be hydrolyzed and used in respiration or other metabolic processes. Phospholipids can also accumulate.

The lipids are much more highly reduced substances than the carbohydrates. They require a substantial expenditure of energy (obtained from respiration) for their synthesis from carbohydrates.

The Fats

Most, if not all, plant cells have the capacity for synthesizing fats and at least some fat is generally present in plant cells, but by far the highest concentrations of fat are found in the embryos or endosperms of seeds [24]. Most plant tissues have fat contents of less than 5% of their dry weight, but up to 50% of the dry weight of seeds may be fats. The seeds of almost 90% of the species that have been analyzed accumulate more fat than either starch or proteins, although substantial quantities of these may also be present. A few fruits such as those of the avocado also have high fat contents. Since the fats, like the other lipids, are insoluble in water, they occur in the cytoplasm as an emulsion and fat droplets are visible under a microscope.

Fats that are liquid at room temperature are referred to as oils, and most plant fats, like those of fish but unlike those of mammals, are oils. Oils contain a higher proportion of unsaturated fatty acids than do solid fats. Many plant oils are important items of commerce, including corn, cottonseed, soybean, peanut, coconut, and olive oils. These may either be sold as oils or converted to solid fats such as oleomargarine or vegetable shortening by catalytic saturation of the unsaturated fats with hydrogen.

FAT SYNTHESIS. The synthesis of fats [9] from carbohydrates involves three more or less distinct sequences of reactions: the synthesis of glycerol, the synthesis of various fatty acids, and the esterification of glycerol with fatty acids. In glycerol synthesis the sugar enters the pathway of glycolysis (Chapter 4) and the reactions proceed as usual up to the formation of dihydroxyacetone phosphate. This is then drained out of the glycolytic reactions by reacting with

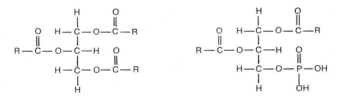

Triglycerides (fats) Phosphatidic acid

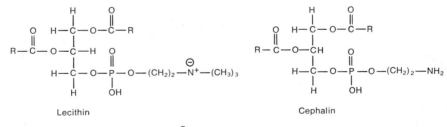

Lecithin Cephalin

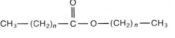

Waxes

Figure 5-10. Generalized formulas of triglycerides (fats), phospholipids, and waxes. R represents the hydrocarbon chain of the fatty acid residues. In general, more than one kind of fatty acid is used in the synthesis of a specific fat or phospholipid. In waxes the number of CH_2 groups (n) in both the fatty acid and alcohol residues is large, generally on the order of 20.

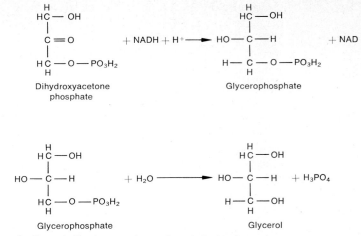

Figure 5-11. The reduction of dihydroxyacetone phosphate to glycerol.

NADH, the reduced product being glycerophosphate, which is then dephosphorylated to glycerol, as shown in Figure 5-11. This reduction synthesis gives glycerol more chemical bond energy than the sugar from which it was synthesized. It will be noted that one molecule of hexose gives rise to two molecules of glycerol.

The synthesis of fatty acids is more complex than the synthesis of glycerol, but the sugar used in fatty acid synthesis also enters the glycolytic pathway. In this case the glycolytic reactions proceed all the way to pyruvic acid, which then gives rise to acetyl-CoA. At this point fatty acid synthesis diverges from respiration. Instead of entering the citric acid cycle, two molecules of acetyl-CoA react to form a molecule of acetoacetyl-CoA and a molecule of CoA. The acetoacetyl-CoA is converted to butyric acid in a series of reactions shown in Figure 5-12. In the course of these reactions reduction by addition of hydrogen occurs at two points: NADH serves as the hydrogen donor when acetoacetyl-CoA is converted to β-hydroxybutyryl-CoA and FADH$_2$ is the hydrogen donor when the unsaturated crotonyl-CoA is converted to butyryl-CoA. The second hydrogenation may be omitted when unsaturated fatty acids are being synthesized. Although reduction is by hydrogenation, another result is a loss of oxygen, which becomes a component of water. The longer the fatty acid molecule, the smaller the percentage of oxygen in it, and the greater its degree of reduction.

The above series of reactions may be repeated with one molecule of acetyl-CoA and one molecule of butyryl-CoA as the substrate, and the product in this case is a six-carbon fatty acid (caproic). Repetitions of this procedure can thus give rise to fatty acids of increasingly greater chain length and molecular weight, but always with an even number of carbon atoms.

It may be noted that, although caproic acid ($C_6H_{12}O_2$) contains the same number of carbon and hydrogen atoms as a molecule of hexose ($C_6H_{12}O_6$), it contains only one-third as much oxygen. On the other hand, two molecules of glycerol ($C_3H_8O_3$) contain the same number of carbon and oxygen atoms as the molecule of hexose from which they were derived but have four more hydrogen atoms. Thus, both of the substances from which fats are produced are more highly reduced than carbohydrate and have a higher energy content. A gram of fat contains just about twice as many calories as a gram of carbohydrate. This high energy content of fat makes it an excellent accumulation food.

In general, the fatty acids used in fat synthesis have high molecular weights. Six-carbon fatty acids are just about the smallest used, and most of the fatty acids have from 12 to 24 carbons in their chains. The C_{16} and C_{18} chain lengths predominate. Also, as has been noted earlier, the fatty acids used in making plant oils are predominantly unsaturated.

The final step in fat synthesis is the esterification of glycerol with three molecules of fatty acids. Although all three of the fatty acids reacting with a molecule of glycerol can be of the same kind, it is much more likely that all three will be different fatty acids or that at least two different fatty acids will be used. Thus, an immense number of different fats can be synthesized.

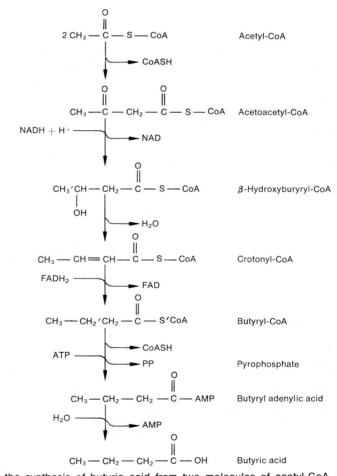

Figure 5-12. Steps in the synthesis of butyric acid from two molecules of acetyl-CoA.

Furthermore, any particular plant oil is usually a mixture of several different fats. Each species of plant is, however, quite consistent in synthesizing certain fats in certain proportions; thus the oil from any particular species has its own particular properties, as varied as those of corn oil, olive oil, tung oil, and linseed oil. It may be noted that the last two oils are very highly unsaturated and oxidize to a hard, tough substance when exposed to the air, and these are useful in making paints.

FAT BREAKDOWN.   The first step in the decomposition of a fat is its hydrolysis to glycerol and fatty acids, a reaction catalyzed by the enzyme lipase. The subsequent oxidation of the glycerol and fatty acids occurs as an approximate reversal of the pathways by which they were synthesized (Figure 5-13). The oxidations that occur in the course of these reactions give rise to reduced hydrogen acceptors that can then be used in making ATP, and these oxidations should be considered as a part of respiration when fat is the substrate.

The NADH produced when glycerol is dehydrogenated and converted to dihydroxyacetone phosphate will, when oxidized during oxidative phosphorylation, result in the production of three molecules of ATP; and the dihydroxyacetone phosphate can enter the glycolytic pathway, giving rise to the usual amount of ATP. Thus, oxidation of glycerol results in the production of three more molecules of ATP than oxidation of a triose. As fatty acids pass through the oxidation pathway, a molecule of $FADH_2$ (which is good for two molecules of ATP) and a molecule of $NADH + H^+$ (which is good for three molecules of ATP) are produced. Because the β carbon of the fatty acid chain is oxidized in these reactions, the process is called **beta oxidation.** The five molecules of ATP, less the one molecule used in the first step, result in the net production of a maximum of four ATP each time an acetyl-CoA is cut off the end of a fatty acid molecule. Since the acetyl-CoA can enter the citric acid cycle, each beta oxidation provides four more molecules of ATP than does the oxidation of a molecule of triose. ATP production occurs each time an acetyl-CoA is cut off, so the amount of ATP produced by beta oxidation increases with the length of the fatty acid molecule.

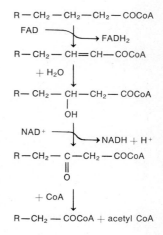

Figure 5-13.   Steps in the oxidation of fatty acids. R represents the remainder of the hydrocarbon chain of a fatty acid. Note that the length of the fatty acid chain has been reduced by two carbons and that a molecule of acetyl-CoA has been produced. The oxidations can be repeated until all of the fatty acid has been converted to acetyl-CoA. Each time the reactions occur a molecule of $FADH_2$ and a molecule of NADH are produced, and in oxidative phosphorylation they can provide five molecules of ATP.

The large amount of ATP produced by the oxidation of fats than by the oxidation of carbohydrates does not represent a net energy gain by the plant, since the same amount of energy is used in fat synthesis as is provided by the oxidation of fatty acids and glycerol.

The dihydroxyacetone phosphate resulting from glycerol oxidation and the acetyl-CoA resulting from fatty acid oxidation do not necessarily go on through the respiratory pathways. Instead, they may be converted to sugar by going through the reverse of the glycolytic pathway; they may be used in the synthesis of other kinds of fats; or they may enter a variety of other metabolic pathways. Also, the NADH and $FADH_2$ produced in the course of fat oxidation can serve as hydrogen donors in processes other than oxidative phosphorylation.

Phospholipids

The phospholipids (Figure 5-10) differ from fats (triglycerides) in that either phosphoric acid or a substituted phosphoric acid group reacts with one —OH of the glycerol rather than a fatty acid. If it is phosphoric acid the resulting phospholipids are called **phosphatidic acids. Lecithins** are phospholipids with a phosphate-choline group, and **cephalins** have a phosphate-ethanolamine group. Each of the three classes of phospholipids includes a variety of specific compounds differing from one another in the kinds of fatty acid residues in the molecule. Phosphatidic acids are the most abundant phospholipids in leaves whereas lecithins and cephalins are more abundant in seeds and seedlings, but any cell apparently contains some of all three classes.

Since phospholipids are essential constituents of cell membranes (Chapter 2), every living cell is evidently capable of synthesizing them. A property of phospholipids that makes them suitable membrane components and contributes toward the stabilization of the membrane structure and the differential permeability of membranes is that the phosphate portion of the molecule is water soluble whereas the remainder of the molecule is fat soluble. As we shall see in Chapter 7, one of the proposed mechanisms for ion transport across membranes (enabling cells to accumulate ions against a concentration gradient) involves interconversions between phosphatidic acid and lecithin in a possible carrier system.

Waxes

Waxes are also esters of fatty acids and alcohols, but unlike fats and phospholipids glycerol is not the alcohol component. Instead, the alcohol is one of a number of different long-chain monohydroxy alcohols with 24 to 36 carbon atoms. In a very few waxes a dihydroxy alcohol is involved. Frequently the fatty acid and the alcohol used in synthesizing a wax have the same number of carbon atoms in their chains, and in any event both are generally large molecules. It will be noted that waxes are very highly reduced compounds that are almost hydrocarbons, with only two atoms of oxygen at the point where the esterification occurred (Figure 5-10).

Some plant waxes are of commercial importance. For example, bayberry wax is used in making candles and several different waxes are used in making high quality shoe polishes, furniture polishes, and automobile waxes. However, most commercial waxes are hydrocarbons derived from petroleum.

Plant waxes are found principally as constituents of the cutin that covers the epidermal cells of leaves, herbaceous stems, and fruits and the suberin with which the walls of cork cells are impregnated. Cutin is a mixture of waxes with fatty acids (that are usually oxidized) and soaps (salts of fatty acids). Suberin also contains some fats, and the waxes and fatty acids are different from those of cutin. Both cutin and suberin are highly impermeable to water and play an important role in preventing excessive water loss from plants.

Nitrogen Metabolism

An essential aspect of plant metabolism is the use of nitrates and other inorganic nitrogen compounds in the synthesis of the various important nitrogen-containing organic compounds such as the amino acids, from which proteins are synthesized, and the purines and pyrimidines [23]. The last two are constituents of such important compounds as the nucleic acids, ATP, and various coenzymes. Other essential nitrogenous compounds are the tetrapyrroles, which provide the basic molecular structure of chlorophylls, cytochromes, phycocyanin, phycoerythrin, and phytochrome. A number of vitamins and plant hormones also contain nitrogen, as do all alkaloids.

The Nitrogen Cycle

Plants absorb nitrates and other nitrogen compounds from the soil, along with the various essential mineral elements, but the ultimate source of most of the nitrogen is the free nitrogen (N_2) of the atmosphere rather than the rocks of the Earth's crust. A fertile soil contains about 1000 lb of available nitrogen per acre, but the atmosphere above the acre contains about 100 million lb of nitrogen. The pathways whereby the free nitrogen is converted to nitrogen compounds, and whereby these are then converted from one nitrogen compound to another by various organisms constitute the nitrogen cycle (Figure 5-14).

NITROGEN FIXATION. The fixation of atmospheric nitrogen requires much energy, which may be provided either by electricity or by nitrogen-fixing organisms. Electrical nitrogen fixation occurs during thunderstorms and results in the formation of various oxides of nitrogen, which are carried to the soil by rains and are eventually present mostly as nitrates. Electrical fixation adds about 5 lb/acre/year of nitrogen to the soil, but this is minor compared with biological fixation. Nonsymbiotic nitrogen-fixing organisms add about 25 lb, whereas legume crops, such as clover and alfalfa, add from 100 to 250 lb by symbiotic nitrogen fixation.

Biological nitrogen fixation is a complicated process [5, 10] involving the stepwise reduction of free nitrogen to ammonia through a series of intermediates. The known nitrogen-fixing organisms include members of at least nine genera of blue-green algae and an even larger number of genera of bacteria.

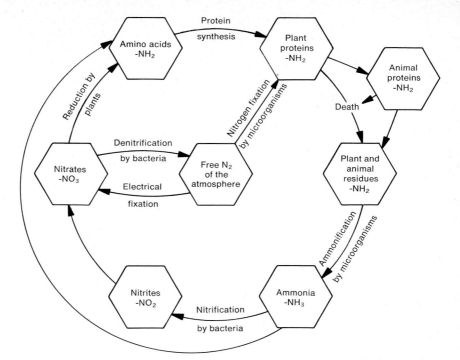

Figure 5-14. Principal steps in the nitrogen cycle. Note the various oxidations and reductions of nitrogen in the course of the cycle.

The nitrogen-fixing bacteria of some genera are free-living or nonsymbiotic. Those of other genera are symbiotic with vascular plants and occur in nodules (Figure 5-15) formed by the host plant in response to the bacterial invasion. Among the genera of free-living nitrogen-fixing bacteria are Azotobacter, Clostridium, and Rhodospirillum. Clostridium is the only anaerobic nitrogen fixer. Rhodospirillum and several other genera are photosynthetic. Of the symbiotic nitrogen-fixing bacteria the most important are the numerous species of Rhizobium, each of which is symbiotic with one or a few closely related species of leguminous plants. Other genera of bacteria are symbiotic with vascular plants other than legumes. These vascular plants include Alnus (alder), Casuarina, Ceanothus, Eleagnus, and Myrica (myrtle). Psychotria, a tropical woody plant, is exceptional in that the nodules induced by the symbiotic bacteria are in its leaves rather than in the roots as is usual. The symbiotic bacteria can live outside the host plant, but can fix nitrogen only when in the nodules of the plant (Figure 5-16).

The Rhizobium bacteria generally enter the legume through root hairs and move to the inner cells of the root cortex through an infection thread, which is formed by the infolding and intracellular extension of the plasmalemma of the invaded cell (Figure 5-17). The bacteria then enter cortical cells, causing them to divide and so give rise to the large nodules within which the nitrogen fixation occurs. The cells participating in nodule formation are generally tetraploid. Dissected nodules are pinkish because of the presence of a hemo-

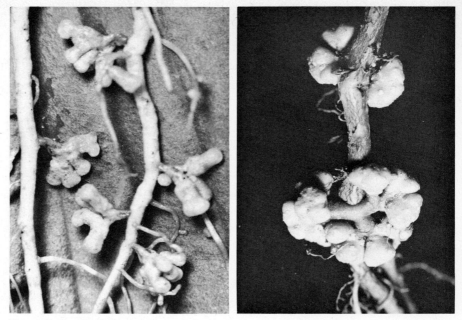

Figure 5-15. Nodules on roots of clover (left) and alder (right). [Courtesy of Carolina Biological Supply Company]

globin (leghemoglobin) that cannot be synthesized by either the bacteria or the legume alone.

The reduction of free nitrogen to ammonia (Figure 5-18) is catalyzed by a complex enzyme system designated as nitrogenase (N_2ase). A supply of ATP, electron donors (principally reduced ferredoxin or flavodoxin), and hydrogen

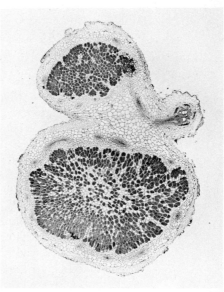

Figure 5-16. Photomicrograph of a section through a nodule. The nitrogen-fixing bacteria are in the dark cells. [Courtesy of Ripon Microslides]

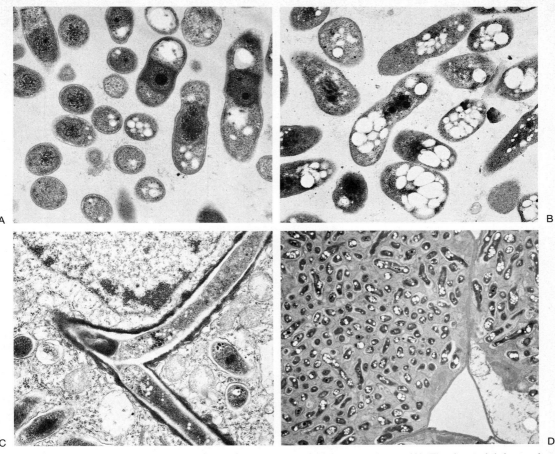

Figure 5-17. Electronmicrographs of *Rhizobium japonicum*. **(A)** The bacterial form of Rhizobium, $\times 33,000$. **(B)** The bacteroidal form of Rhizobium, $\times 33,000$. **(C)** Infection thread carrying Rhizobium into soybean root cells, $\times 28,500$. **(D)** Cells of Rhizobium inside root nodule cells of soybean, $\times 10,400$. [Courtesy of R. R. Hebert, R. D. Holsten, and R. W. F. Hardy, Central Research Department, E. I. du Pont de Nemours & Company]

donors such as NADH and NADPH is essential. In blue-green algae and photosynthetic bacteria these may be provided directly by photosynthesis, whereas in the other nitrogen-fixing organisms they are provided by respiration. Divalent metal cations are also essential. Magnesium ions (Mg^{2+}) are most effective, but manganese (Mn^{2+}), ferrous (Fe^{2+}), cobalt (Co^{2+}), and nickel (Ni^{2+}) are also essential, whereas zinc (Zn^{2+}) and cupric (Cu^{2+}) are inhibitory. Molybdenum is an essential constituent of nitrogenase. Diimide ($HN{=}NH$) and hydrazine ($H_2N{-}NH_2$) have been identified as intermediates in the reduction sequence, although there may be others. These compounds, like the nitrogen being reduced, are enzyme-bound. Further details may be found in the review by Hardy and Burns [10]. All nitrogen fixers except Azotobacter also evolve hydrogen, some of the available electrons being used to reduce H^+.

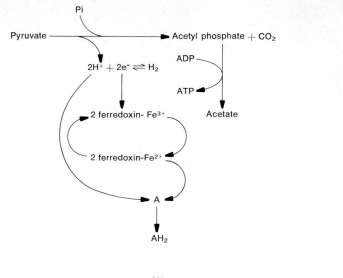

$$N \equiv N + \boxed{E} \longrightarrow \boxed{N \equiv N - E} \xrightarrow[A]{AH_2} \boxed{HN = NH - E} \xrightarrow[A]{AH_2} \boxed{H_2N - NH_2 - E} \xrightarrow[A]{AH_2} 2\,NH_3 + \boxed{E}$$

$$(N_2 + 3AH_2 \longrightarrow 2NH_3 + 3A)$$

Figure 5-18. Pathway of nitrogen fixation in *Clostridium pasteurianum,* as proposed by R. H. Burris. **Above:** Electron transport and production of reduced hydrogen acceptor (AH_2) and ATP, with pyruvic acid as the substrate. The role of the ATP is not certain. The production of hydrogen gas is a short-circuiting of the process. **Below:** The reduction of nitrogen to ammonia, showing stepwise addition of hydrogen. The nitrogen and intermediates are complexed with the nitrogenase enzymes (E). Note that three molecules of pyruvic acid would have to be oxidized to provide the electrons and protons for the reduction of nitrogen to ammonia. [Pathways modified from R. H. Burris, *Ann. Rev. Plant Physiol.,* **17:**155-184 (1966)]

Nitrogen fixation does not necessarily follow exactly the same pathways in all organisms. For example, in legumes Co^{2+} is essential, apparently because it is a constituent of vitamin B_{12}, which may play a role in the synthesis of leghemoglobin. Co^{2+} is not essential for legumes without nodules. The role of leghemoglobin is not clear, but it may transport either nitrogen or oxygen or prevent inhibition of fixation by oxygen.

NITRATE REDUCTION. Plants other than those in which symbiotic nitrogen fixation is occurring are dependent on nitrogen compounds absorbed from the soil or body of water in which they are growing. These nitrogen compounds may be derived either from free-living nitrogen-fixing organisms or from the biological recycling of nitrogen that will be described later. Although plants can absorb and use a variety of nitrogen compounds including ammonium compounds, nitrites, and organic nitrogen compounds such as urea, most of the soil nitrogen is present as nitrates. The plants must reduce these nitrates to ammonia before they can synthesize amino acids or other organic nitrogen compounds.

Nitrate reduction to ammonia [1] probably occurs in four steps, with nitrite, hyponitrite, and hydroxylamine as possible intermediates (Figure 5-19). This reduction requires much energy, which is supplied by proton and electron donors such as NADH, NADPH, and reduced ferredoxin among others. These electron and proton donors may be provided by either respiration or photosynthesis. Both nitrate reduction and amino acid synthesis often occur in roots, and in this case the nitrogen compounds translocated to the shoot are usually amino acids.

Neither ammonia nor the intermediates in nitrate reduction normally accumulate in plants in any appreciable quantity. The ammonium compounds are used rapidly in synthesizing amino acids and other compounds. In any quantity ammonia is toxic to plants, at least partly because it inhibits ATP production in both respiration and photosynthesis. Despite the large amount of energy used in nitrate reduction, plants usually grow better when supplied with nitrates than ammonium compounds. With an abundant ammonia supply so much of the available carbohydrate is used in synthesizing amino acids and then proteins that there is not enough carbohydrate for cell wall formation or starch accumulation. The result is likely to be lush vegetative growth with tissues that are too soft and succulent, poorly developed vascular tissues, and only a few small fruits. When nitrates are supplied the rate of nitrogen reduction provides a control. Unlike ammonia and the intermediates, accumulated nitrates are not toxic to plants. However, overfertilized plants with a high nitrate content may be injurious to man or other animals that eat them, because they may reduce some of the nitrate to nitrite, which apparently interferes with the transport of oxygen by hemoglobin.

All green plants and some species of bacteria and fungi can reduce nitrates to ammonia, but animals and a good many species of bacteria and fungi do not have the necessary enzymes. Some bacteria and fungi that cannot reduce nitrates can synthesize amino acids if supplied with ammonium compounds, but amino acid synthesis in animals is limited to certain transaminations so that animals are quite dependent on plants for their basic supply of amino acids.

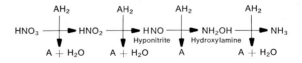

$$(HNO_3 + 4AH_2 \longrightarrow NH_3 + 3H_2O + 4A)$$

Figure 5-19. Proposed steps in the reduction of nitrate to ammonia. The reduced hydrogen acceptors (AH_2) may be NADH or NADPH, provided by respiration or photosynthesis. Reduced ferredoxin is involved as an electron donor, at least in leaves. Nitrate reductase, which catalyzes the reduction of $-NO_3$ to $-NO_2$, has FAD as a prosthetic group. The electrons from $FADH_2$ are transferred to molybdenum and then to the $-NO_3$. The intermediates are evidently enzyme-bound.

NITROGEN CONVERSIONS BY MICROORGANISMS. When plants or animals die their amino acids, proteins, and other organic nitrogen compounds are broken down by the bacteria and fungi that cause decay, and ammonia is produced. Decomposition of nitrogenous excretions of animals, including urea and uric acid, also results in production of ammonia. The ammonium compounds do not remain in the soil long because bacteria of two genera (Nitrosomonas and Nitrosococcus) oxidize the ammonia to nitrite. The energy released by this oxidation is used by these bacteria in the chemosynthetic production of carbohydrates from carbon dioxide and water. The nitrites produced are quite promptly oxidized to nitrates by bacteria of the genus Nitrobacter, the released energy again being used in chemosynthesis. Thus, the nitrogen compounds present in the soil are predominantly nitrates, which plants then absorb and reduce.

Several different species of soil bacteria can reduce nitrates to nitrogen, thus removing it from the cycle, at least until it may be fixed later on. Fortunately, this denitrification can occur only under anaerobic conditions, and can be prevented by adequate drainage and aeration of the soil. Well-aerated soil also promotes the growth and activity of the decay organisms, the nitrifying bacteria, and most species of nitrogen-fixing organisms.

Amino Acid Synthesis

Amino acids are synthesized initially by the reaction of ammonia with α-keto acids (Figure 5-20), primarily α-ketoglutaric acid. The resulting amino acid is glutamic acid. This process is catalyzed by glutamic dehydrogenase and requires hydrogen from NADH.

The other amino acids are synthesized by transaminations, reactions that involve the transfer of the $-NH_2$ group from an amino acid to an α-keto acid, thus producing a different amino acid. The donor amino acid is thus reconverted to the keto acid from which it was originally synthesized. One transamination is outlined in Figure 5-20. Other transaminations proceed in a similar manner, but apparently a different transaminase is required for each pair of amino and keto acids. The transaminases have a coenzyme, pyridoxal phosphate, that accepts the $-NH_2$ group from the amino acid and then transfers it to the keto acid.

The α-amino acids synthesized by plants and used in protein synthesis are listed in Figure 5-21. In α-amino acids the amino group is attached to the α carbon (the one next to the carboxyl carbon), so the general formula for α-amino acids may be given as

$$R-CH-COOH$$
$$|$$
$$NH_2$$

The R ranges in complexity from a single hydrogen atom in glycine to complicated radicals that include aromatic rings as in phenylalanine, tyrosine, and tryptophan. Synthesis of the more complicated amino acids obviously involves various preliminary reactions for production of their R groups, but these will not be considered here.

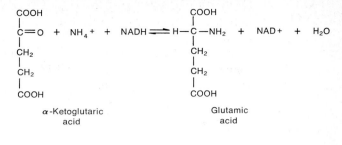

AMINATION

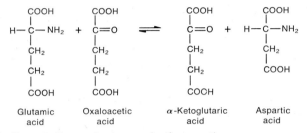

TRANSAMINATION

Figure 5-20. An amination reaction and a transamination reaction.

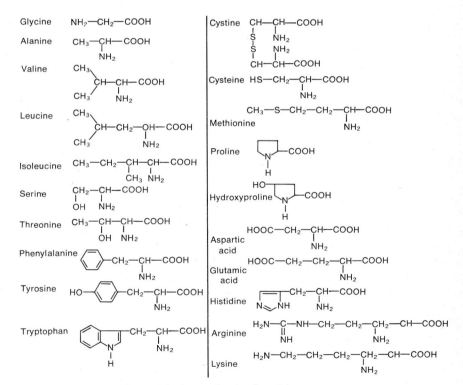

Figure 5-21. Amino acids used in the synthesis of proteins.

Several points regarding the amino acids are worthy of note. Proline and hydroxyproline have their nitrogen atom in a heterocyclic ring so that they are really imino (—NH) acids rather than amino acids, even though they are constituents of proteins and are commonly called amino acids. Some amino acids (tryptophan, histidine, arginine, lysine) contain nitrogen groups in addition to their α-amino group that make them basic in pH, whereas aspartic and glutamic acids both contain two —COOH groups and are in the acidic range. Cysteine, cystine, and methionine are the only amino acids that contain sulfur and are important in protein synthesis. The —SH (sulfhydryl) groups of cysteine and cystine can form —S—S— bonds that are an essential feature of protein structure and function in the enzymatic activity of proteins. Proteins generally contain a small percentage of sulfur, and the sulfur compounds produced when proteins are burned are largely responsible for the unpleasant odor. The cystine molecule is composed of two molecules of cysteine linked by —S—S—, so sometimes only cysteine, rather than both, is included among the protein-forming amino acids.

Glutamine and asparagine are amides of glutamic acid and aspartic acid respectively, and are synthesized by replacement of the —OH group of the second —COOH by an amino group (Figure 5-21). These amides are often quite abundant in plants and, in addition to serving as components of proteins, they function as reservoirs of amino nitrogen and also serve in its translocation. It seems probable that they prevent accumulation of toxic quantities of ammonia when the ammonia/carbohydrate ratio is high. The death of plants by starvation (when photosynthesis occurs at too low a rate) is probably brought about by ammonia toxicity before all the available food for respiration has been exhausted. Under these conditions proteins are hydrolyzed and used in respiration. The amino acids are deaminated before entering respiratory pathways and, when all the available aspartic and glutamic acids have been converted to their amides by the ammonia released, further deamination of amino acids leads to accumulation of toxic quantities of ammonia.

In addition to the α-amino acids used in protein synthesis plants produce a great variety of other amino acids [8]. These include various α-, β-, and γ-amino acids, some of them modifications of protein amino acids and some quite different. Altogether, over 215 different nonprotein amino acids have been found in plants, some known in only a single species so far, and it is quite possible that the list will eventually include hundreds more. Plants also synthesize a great variety of amino acids. The metabolic roles of the nonprotein amino acids are not well understood. Some are probably intermediates in synthetic pathways, whereas others are amides that seem to function in amino accumulation and transport as do asparagine and glutamine. Some may not have any metabolic role of importance at all. It is likely that in the evolution of plants a considerable number of mutations have resulted in enzymes that catalyze syntheses of metabolically useless substances, which, however, are not harmful and so do not bring about elimination of the mutant genes by natural selection.

Purines and Pyrimidines

Another very important group of nitrogenous organic compounds consists of the heterocyclic nitrogenous bases, in particular the purines and pyrimidines

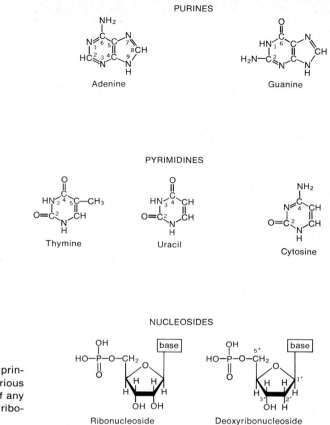

Figure 5-22. Structural formulas of the principal purines and pyrimidines. The various nucleotides are formed by attachment of any one of these nitrogenous bases to a ribo- or deoxyribonucleoside as shown.

[3]. These are used in the synthesis of a great variety of essential compounds. For example, kinetin, a plant hormone that will be considered in Chapter 13, is a purine derivative. As has been noted in the preceding discussion of the pentoses, the nucleotides, which are synthesized from purines or pyrimidines and ribose phosphate or deoxyribose phosphate, are used in the synthesis of many metabolically essential compounds such as various vitamins, coenzymes, ATP and ADP, and DNA and RNA.

Plants synthesize a considerable number of different purines and pyrimidines, and the structural formulas of some of these are given in Figure 5-22. Of these, the most important purines are adenine and guanine and the most important pyrimidines are thymine, cytosine, and uracil. The pathways whereby purines and pyrimidines are synthesized have been worked out in considerable detail, and are described in biochemistry textbooks [3].

Nucleic Acids

Deoxyribonucleic acid (DNA) and ribonucleic acid (RNA) are compounds of major biological importance. DNA carries the genetic code and RNA transcribes the DNA code and, in turn, controls the sequence in which amino acids are linked together in protein synthesis. The nucleic acids are usually linked to proteins, forming the compounds known as nucleoproteins.

DNA is found principally in the chromosomes, but some is also found in other organelles such as chloroplasts and mitochondria. It was first shown by Watson, Crick, and Wilkins [22] that the DNA molecule is a long double helix, each helix being composed of numerous deoxyribose phosphate residues linked to one another. The two helices are linked together by hydrogen bonds between a nucleotide purine of one helix and the pyrimidine opposite it on the other helix (Figure 5-23). The difference in the sizes and shapes of purines and pyrimidines makes it impossible for two purines (or two pyrimidines) to form hydrogen bonds with one another in the DNA molecule. Furthermore, a purine with an —NH_2 on carbon 6 can link only with a pyrimidine with an —OH on carbon 6, and vice versa. This means that in a DNA molecule adenine (A) is always crosslinked to thymine (T), and guanine (G) to cytosine (C), a specificity that is very important in the duplication of genetic information.

Each DNA molecule participates in two important reactions: self-replication and the synthesis of RNA that carries a code complimentary to the DNA code. These processes require the participation of specific enzymes. A more detailed discussion of them can be found in other publications [for example, reference 22].

DNA replication occurs during the interphase of cell division so that each of the two cells resulting from the mitotic division has the same complete set of DNA molecules as the parent cell. DNA replication requires breaking the hydrogen bonds that hold the two strands together so that each helix can then assemble a new companion helix with the same nucleotide sequence as its old one. For example, if the nucleotides are represented by their initial letters and the first six in one DNA helix happen to be ATGCCG, then the first six in the new companion helix must be TACGGC, just the same as in the old companion helix. Meanwhile, the latter assembles a new companion helix with

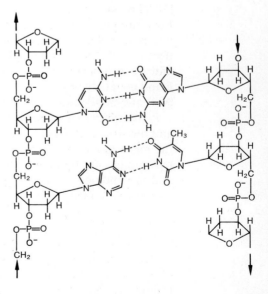

Figure 5-23. Structure of a short section of a DNA molecule. Note the hydrogen bonds between cytosine and guanine (above) and adenine and thymine (below).

the sequence ATGCCG. The unzippering and assembly of new companion helices continues to the end of each DNA molecule with the resulting production of two molecules, each identical with the original one. Each of the two molecules has one helix from the parent molecule and one newly synthesized helix. In the rare instances where a mistake in synthesis of a new helix occurs, the result is a mutation.

RNA molecules differ from DNA molecules in several respects: the sugar used in making RNA nucleotides is ribose rather than deoxyribose and uracil rather than thymine is used along with the adenine, cytosine, and guanine. There are several kinds of RNA, the most important ones apparently being transfer RNA (tRNA, also called soluble RNA, sRNA) and messenger RNA (mRNA). The tRNA molecules are much smaller than those of mRNA and are composed of a single strand of nucleotides doubled back on itself so that part of the molecule is a double helix. Each of the several kinds of tRNA can link with only one of the different amino acids that are used in synthesizing proteins; further, that particular tRNA can fit only the mRNA coded for that particular amino acid. When we consider protein synthesis we shall have more to say about the role of tRNA.

The mRNA molecules are long single strands of nucleotides arranged in such a sequence that they copy the DNA code for the synthesis of a particular protein or, perhaps, for a polypeptide chain that constitutes a portion of a protein molecule. As has been noted, the sequence of nucleotides in mRNA is determined by the sequence of nucleotides in the DNA against which it is synthesized. Presumably the DNA molecule must be unzipped at the time it is participating in mRNA synthesis. If a portion of the DNA code is ATGCGC, the sequence of nucleotides in the mRNA produced will be UACGCG because uracil (like thymine) can link only with adenine and cytosine can link only with guanine.

One of the important concepts that has emerged from the extensive research on nucleic acids and proteins since 1955 is that the DNA and mRNA codes for an amino acid consist of a sequence of three nucleotides. Once it was evident that the code was determined by the sequence of the nucleotides in the nucleic acid molecules, it was obvious that on theoretical grounds a sequence of at least three nucleotides was needed to code for the 20 different amino acids used in protein synthesis. Individually, the four nucleotides could code for only four amino acids and even in pairs of two the nucleotides could code for only 16 (4^2) in every possible combination. A group of three (a triplet) in every possible combination could code for 64 (4^3), which was more than needed, whereas in fours the number of possible combinations would be 256 (4^4), which was far too many. Subsequent research indicated that the code really did consist of nucelotide triplets in all 64 possible combinations and that there were several different triplet codes for most of the amino acids, although any particular triplet always coded for only one of the amino acids. Also, there were apparently two mRNA triplets that did not code for any amino acid (UAA and UAG) but probably served as a signal for termination of a polypeptide chain. The amino acid coded for by each triplet has been determined at least tentatively [11, 22]. Thus, in the example we have been using, UAC codes for tyrosine and GCG for alanine.

Proteins

The proteins are compounds of extremely great biological importance. They are the principal organic structural components of the protoplast and, even more important, all of the enzymes that catalyze the various biochemical processes are proteins. We have seen that proteins are essential constituents of cell membranes, and they are also important constituents of organelles such as the nucleus, ribosomes, chloroplasts, and of the so-called undifferentiated cytoplasm. They are even present in primary cell walls, particularly those in the process of formation. The distinction between structural proteins and enzyme proteins may not be a valid one. At least some of the membrane proteins are enzymes, and it is quite possible that most if not all of the so-called structural proteins are enzymes. However, plants accumulate proteins, particularly in seeds, and these accumulated proteins cannot be considered as either structural or enzymatic but apparently serve as reservoirs of amino acids. It may be noted that plants accumulate carbohydrates, fats, and proteins, whereas animals accumulate only fats (except for temporary accumulation of glycogen).

Proteins are the principal organic constituent of protoplasm, although not necessarily of an entire plant tissue that also includes the cell walls and accumulated carbohydrates and fats. Of the dry weight of protoplasm 5% or more consists of proteins. Seeds with accumulated proteins may have a very high protein content, ranging up to 40% or more of their dry weight. Wheat grains contain 8% or more of protein, which is found in both the embryo and endosperm and in particular in the aleurone layer of the latter, where the proteins may be in crystals. Between one-third and one-half of the proteins of leaves are found in the chloroplasts, the rest being mostly in the other cytoplasmic structures with a small percentage of the total in the nuclei (principally as nucleoproteins).

The molecules of the various kinds of proteins vary greatly in size, ranging from molecular weights of around 13,000 to several million. Even the smallest protein molecules are composed of around 100 amino acid residues, and most of the common plant proteins have from 300 to 3000 amino acid residues per molecule. The various proteins differ from one another, not only in molecular size but also in the sequence of amino acids in their molecules, the kinds and proportions of the amino acids, and their molecular configurations [13, 16, 20]. For all practical purposes, the number of different proteins possible is unlimited, even if one considers only the various sequences in which 100 to several thousand amino acids of 20 different kinds can be linked together. A change in even a single amino acid in the chain results in a different protein and can result in a change or loss of the catalytic capabilities of an enzyme. There are well over a million different species of organisms, and each species has dozens or perhaps hundreds of proteins found in no other species. No other class of chemical compound, except for the nucleic acids, comes close to the proteins in the number of different specific compounds it includes.

PROTEIN SYNTHESIS.   Protein synthesis [22, 25] involves a basic problem not encountered in the synthesis of other macromolecules such as starch or cellulose. The latter are composed of one glucose residue after another, whereas a protein consists of up to 20 different amino acids that must be linked in one very precise order which is different for each of the many kinds of proteins. How an organism could consistently synthesize its own particular kinds of proteins remained a mystery until research on DNA and RNA revealed their roles in protein synthesis.

Synthesis of proteins begins with preliminary reactions of the amino acids that must occur before they can be linked together into proteins. The first step is activation of the amino acids by reaction with an activating enzyme and ATP, resulting in an amino acid–enzyme–adenosine monophosphate complex (aa-E-AMP) plus inorganic phosphate. Next the aa-E-AMP reacts with a tRNA specific for that particular amino acid producing an aa-tRNA complex and freeing the activating enzyme and AMP. Subsequent steps take place on a ribosome-mRNA complex. The aa-tRNA diffuses to a mRNA molecule, which must be in contact with one or more ribosomes if it is to function in protein synthesis (Figure 5-24).

A mRNA molecule is very long compared with the diameter of a ribosome and is attached to a series of ribosomes, forming what is called a polyribosome [18]. A ribosome has two structural subunits, each composed of RNA and protein (Chapter 2). A tRNA molecule with its attached amino acid can attach to a mRNA molecule only at a point where the latter is in contact with a ribosome. Each kind of tRNA carries a triplet code that can attach only to the complementary mRNA code. For example a tRNA with the code UAC can attach to the mRNA only where it has the triplet AUG. Since this kind of tRNA can link only with the amino acid methionine, only methionine can be incorporated in the growing polypeptide chain at this point. If the next mRNA triplet is UUU, only phenylalanine can be incorporated in the chain. As the amino acids become linked by peptide bonds, their tRNA molecules are freed. Each tRNA molecule can then react with and transport another molecule of its specific amino acid. As each ribosome moves along the mRNA molecule, each triplet code is recognized by the proper tRNA in sequence; thus the amino acids are added one by one to the growing protein molecule in the sequence specified by the mRNA and in turn by the DNA. In this way the specific protein coded for by the DNA is synthesized.

The peptide linkage that holds one amino acid to adjacent ones in the protein chain is between the —COOH group of one and the —NH_2 group of the other, a molecule of water being formed each time a peptide linkage is made. Linkage of only two amino acids results in a dipeptide; of three, a tripeptide; and of 4 to 50 or so, a polypeptide. The smallest proteins probably consist of a single polypeptide, but larger protein molecules are composed of varying numbers of polypeptide molecules that are probably linked together after their synthesis.

CONJUGATED PROTEINS.   Many proteins are attached to nonproteins, and these are referred to as conjugated proteins. The nonprotein component is referred to as the prosthetic group. Among the prosthetic groups are many different coenzymes, pigments, nucleic acids, carbohydrates, phospholipids, and various metallic ions. Conjugated proteins play important roles both enzymatically and structurally.

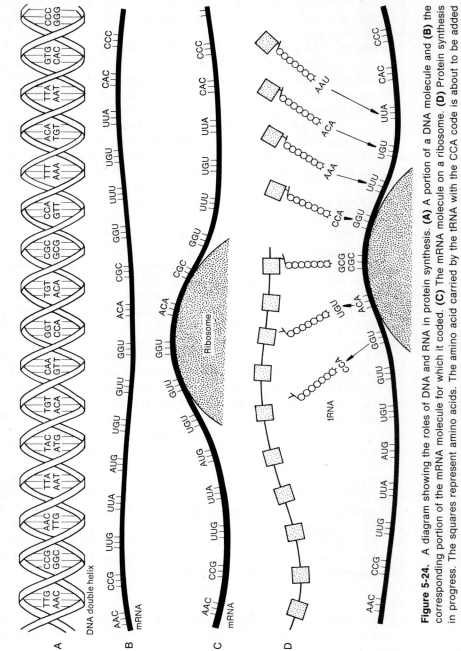

Figure 5-24. A diagram showing the roles of DNA and RNA in protein synthesis. **(A)** A portion of a DNA molecule and **(B)** the corresponding portion of the mRNA molecule for which it coded. **(C)** The mRNA molecule on a ribosome. **(D)** Protein synthesis in progress. The squares represent amino acids. The amino acid carried by the tRNA with the CCA code is about to be added to the growing polypeptide chain by a peptide linkage. [After W. D. McElroy, *Cell Physiology and Biochemistry, 3rd. ed. Prentice-Hall, Inc., 1971*]

PROTEIN DEGRADATION. In plants the enzymatic decomposition of proteins into their constituent amino acids and protein synthesis occur simultaneously and probably continuously. One means of protein degradation is hydrolysis [15]. Another means that has been proposed [23] involves the amino acid activating enzyme in a reversal of the reactions it catalyzes during protein synthesis. Thus an aa-E-AMP complex would be phosphorylated, resulting in the production of considerable ATP. Some investigators think that this may be a more common means of protein degradation than hydrolysis, but it is probable that hydrolysis is more usual. The amount of each type of reaction may vary from time to time and from organ to organ.

In any event, a considerable number of different protein-hydrolyzing enzymes have been isolated from plants. Some of these enzymes hydrolyze proteins only to polypeptides, whereas others hydrolyze proteins or polypeptides all the way to their constituent amino acids. At least one protein-hydrolyzing enzyme, **papain,** is of commercial importance as a meat tenderizer. Papain is present in the latex of the tropical papaw (*Carica papaya*) in considerable concentration and can be extracted and crystallized without great difficulty. Another plant protease, **bromelin,** is found in quite high concentration in pineapple fruits. Fresh pineapple can not be used in gelatin desserts because the bromelin hydrolyzes the gelatin protein into an amino acid broth.

As previously mentioned both protein synthesis and protein degradation generally occur simultaneously; this results in a considerable turnover between amino acids and proteins. In young actively growing plant tissues protein synthesis occurs more rapidly than protein degradation, resulting in a net increase in proteins. However, in senescent tissues, during starvation, and in dehydrated tissues, as well as in certain other situations, protein degradation exceeds protein synthesis. Mere detachment of the leaves of many species of plants, even though the leaves are immediately placed in water, results in a quite rapid loss of protein. It is interesting to note that this apparently does not occur in species such as African violets and begonia whose leaves are capable of regenerating buds and roots and so can be used in vegetative propagation.

Other Important Nitrogenous Compounds

As we have already noted, a considerable variety of nitrogenous compounds other than the amino acids, proteins, purines, pyrimidines, and nucleic acids are synthesized by plants and play important roles in metabolism. These include various coenzymes, vitamins, hormones, and pigments. Much has been learned about the pathways by which these various compounds are synthesized, but any detailed consideration of these syntheses is beyond the scope of this book.

Many coenzymes are nitrogenous compounds, and several of them such as NAD, NADP, flavin mononucleotide (FMN), and CoA include adenine among their constituent groups. The B vitamins are also nitrogenous compounds and seem to be important primarily as constituents of coenzymes. Thiamin (vitamin B_1) is converted to thiamin pyrophosphate (TPP), which functions in the conversion of pyruvic acid to acetyl-CoA and other processes. Riboflavin (vitamin B_2) is a constituent of FMN and FAD. Nicotinic acid is a constituent of NAD and NADP. Pyridoxyl phosphate, a derivative of pyridoxine (vitamin B_6), serves as a coenzyme in amino acid metabolism. Pantothenic acid is a constituent of CoA.

Several of the plant hormones are nitrogenous compounds. Indole-3-acetic acid, the principal auxin, contains the same indole ring as tryptophan, the amino acid from which it is synthesized. Another hormone, kinetin, is 6-furfurylamino purine, an adenine derivative.

A most important class of nitrogenous compounds consists of the pyrrole derivatives [12]. If four pyrrole molecules (Figure 5-25) are linked together into a ring the product is **porphyrin,** which may be considered as the fundamental tetrapyrrole. Porphyrin can be modified in various ways so as to produce a variety of compounds of great biological importance: the chlorophylls and the prosthetic groups of the cytochromes, hemoglobin, and hemocyanin. (It should be noted that the actual synthesis of pyrrole and the tetrapyrroles is considerably more complex than the above statements suggest.) The chlorophylls contain an atom of magnesium in the center of the tetrapyrrole ring; the heme of hemoglobin and the cytochromes all contain an atom of iron; and the heme of hemocyanin (found in the blood of some anthropods and mollusks) contains copper. In addition, the various tetrapyrroles differ from one another in the various side groups attached to the ring. Notable among these is the long phytol tail of the chlorophylls (Figure 3-5).

The phycobilin pigments of the blue-green and red algae (phycocyanin, which is blue, and phycoerythrin, which is red) are also pyrrole derivatives, but in these the pyrrole residues are linked in a chain rather than a ring (Figure 3-5). Similar straight-chain pyrrole derivatives are the bile pigments of animals and the prosthetic group of phytochrome, a pale blue pigment that is of wide distribution in the plant kingdom and serves as the photoreceptor pigment in photoperiodism and many other developmental responses of plants that are influenced by light.

All of the various pyrrole derivatives we have mentioned are colored because they absorb the various wavelengths of visible light differentially, and thus they can be designated as pigments. The capacity of the chlorophylls, phycocyanin, phycoerythrin, and phytochrome for absorbing light is an important part of their functioning. However, as far as is known, the absorption of light by the cytochromes, hemoglobin, hemocyanin, or the bile pigments is of no direct functional significance.

Still another class of nitrogenous compounds synthesized by some species of plants are the **alkaloids** [19] which, like the purines and pyrimidines, are heterocylic nitrogenous bases. Animals are unable to synthesize alkaloids as are many species of plants. Species that can synthesize alkaloids are scattered throughout the plant kingdom but are mostly concentrated in a few families, including the potato, poppy, legume, buttercup, madder, and hemp families. Alkaloids are also synthesized by some fungi including the hallucinogenic mushrooms, the poisonous mushrooms, and the fungi causing the ergot disease of cereal grains. The latter produce a variety of alkaloids including LSD, and can cause serious illness or death if inadvertently included among the grain ground into flour. Penicillin can be classed as an alkaloid. The structural

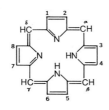

Pyrrole

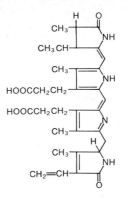

Phycoerythrobilin

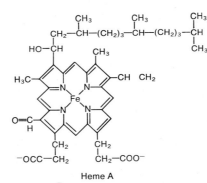

Porphin

Protoporphyrin IX

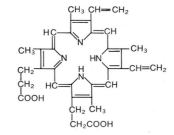

Heme A

Figure 5-25. A pyrrole molecule and several tetrapyrroles. Phycoerythrobilin is a straight-chain tetrapyrrole that is the chromatophore of phycocyanin. Porphin is a cyclic tetrapyrrole from which various tetrapyrroles such as chlorophylls, cytochromes, and the heme of hemoglobin are derived, protoporphyrin IX being an intermediate. Heme A is the prosthetic group of class A cytochromes. See Figure 3-5 for a chlorophyll molecule.

formulas of several plant alkaloids are given in Figure 5-26. It will be noted that their structures are rather diverse and that the alkaloids do not constitute a well-defined chemical class as do the purines or pyrimidines. As a matter of fact, there can be considerable confusion. Note, for example that caffeine is actually a purine chemically even though it is commonly classed as an alkaloid.

Unlike the nitrogenous compounds we have discussed previously, the alka-

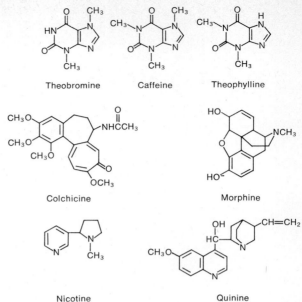

Figure 5-26. Structural formulas of seven of the numerous alkaloids synthesized by plants.

Theobromine

Caffeine

Theophylline

Colchicine

Morphine

Nicotine

Quinine

loids apparently do not play metabolic roles of any great importance in plants. One thing that supports this view is that many species of plants, including entire families in many cases, do not synthesize alkaloids. However, it is possible that species that do produce them convert alkaloids into metabolically important substances. Hordenine, an alkaloid found in young barley plants, is apparently used in the synthesis of lignin. It has been suggested that nicotine serves as a donor of methyl groups in the synthesis of other compounds. There is evidence that some alkaloids may serve as germination inhibitors or as antibiotics. It has been suggested that because alkaloids can chelate with metal ions they could function in the translocation of mineral elements. (Chelating agents have available electrons, and can bind cations of metals such as iron and zinc which otherwise might be converted to insoluble salts.)

Despite their apparent lack of importance in plant metabolism, most alkaloids have marked influences on animals and many are valuable drugs. For example, quinine is used in treating malaria, cocaine and morphine are pain killers, colchicine is used in treating gout, and atropine is a sedative and muscle relaxant. Misuse of alkaloids such as morphine and cocaine results in narcotic addiction. Most alkaloids are poisonous when injested in any quantity and some, like strychnine and coniine (the hemlock alkaloid that killed Socrates), are violent poisons. At the other extreme the caffeine of coffee, tea, and cola is consumed daily by many people without apparent serious results.

It appears that synthesis of alkaloids [7] generally begins with appropriate amino acids, although most of the synthetic pathways are still not completely worked out. Alkaloid synthesis sometimes occurs only in certain organs of a plant, and the alkaloids may remain localized there or may be translocated to other parts of the plant. Potato plants accumulate alkaloids in their leaves and fruits, but not in their tubers. Ray Dawson found that tobacco plants synthesize nicotine only in their roots, but over 85% of it is translocated to the leaves. Tobacco shoots grafted to tomato roots contained no nicotine in their leaves, whereas tomato shoots grafted to tobacco roots did.

Some Other Classes of Compounds Synthesized by Plants

Plants synthesize many other kinds of organic compounds than those already considered. Some of these play important metabolic or structural roles, whereas others have no known function in the plants that produce them although a number of these are important items of commerce. Of the many classes of compounds synthesized by plants we shall select six for brief consideration here: the terpenes, anthocyanins, tannins, sterols, and lignins.

Terpenes

The terpenes may be considered as compounds composed of chains or rings of isoprene (Figure 5-27) residues, ranging in number from two to many.

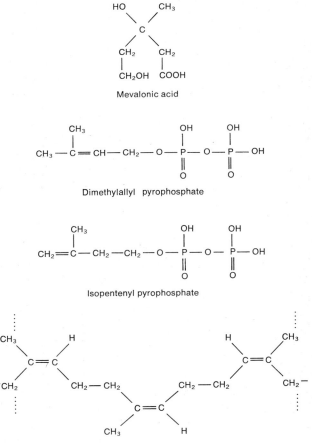

Figure 5-27. Mevalonic acid, which is synthesized from acetyl-CoA, is the precursor of isoprene, from which the terpenes are built up. The pyrophosphate derivatives of isoprenes, shown in the center, are the active molecules used in terpene synthesis. From two to many isoprene units make up a terpene, for example turpentine (2), oils and resins (3,4,6), carotenoids (8), rubber and guttapercha (many). At the bottom three of the isoprene units from a long rubber molecule are shown.

Isoprene is synthesized from acetic acid by a rather complicated pathway. One of the intermediates is mevalonic acid [21], a substance that has a number of important functions in plant metabolism. Although the terpenes constitute a distinct class of compounds chemically, the physical properties of the various terpenes may be quite different. The terpenes composed of two to four isoprene residues are mostly volatile oily liquids, although some are components of resins. Pinene, the principal constituent of turpentine, and camphor are both simple terpenes composed of two isoprene residues. The so-called essential oils of plants, including lemon, peppermint, rose, and lavender oils, are mostly terpenes. These volatile oils usually have pleasant odors, and a number of them are extracted and used commercially. Although the terpenes are basically hydrocarbons they may be oxidized or reduced to alcohols, aldehydes, ketones, or acids, and the presence of such groups makes possible a wide array of essential oils with different characteristic odors and other properties.

Only a few of the many species of plants synthesize essential oils. The terpenes composed of six isoprene units are all resins. Synthesis of resins is limited mostly to pines and other conifers. However, the terpenes composed of eight isoprene units such as the carotenoids (Figure 3-5) are synthesized universally by green plants and also by some species of fungi. It will be recalled that carotenoids (carotins and xanthophylls) are found in both chromoplasts and chloroplasts, and in the latter they play a role in photosynthesis. The carotenoids in chromoplasts have no known function, but they impart yellow, orange, or even red (as in tomatoes) colors to tissues in which they occur.

The polyterpenes, which are composed of from 500 to 5000 or so isoprene residues, include rubber and gutta percha. Rubber is synthesized by some 2000 different species of plants, but most of the commercial rubber derived from plants comes from the bark of the rubber tree, *Hevea brasiliensis*. In plants rubber is found in latex, which is in laticifers (Chapter 2). The rubber is dispersed in the water of the latex as particles of microscopic size (0.01 to 50 μ) along with other substances such as proteins, amino acids, sugars, lipids, and smaller-molecule terpenes. The latex of some species of plants does not contain rubber.

Anthocyanins

The anthocyanins [6] are water-soluble compounds that are commonly found in the cell sap of the vacuoles. Most anthocyanins are colored and are responsible for the red, pink, violet, magenta, and blue colors of flowers. The anthocyanins are pH indicators, so a single anthocyanin may vary in color with the pH of the cell sap; litmus is an example. Some anthocyanins (the leucoanthocyanins) are colorless. Although the colored anthocyanins are found extensively in the petals or sepals of flowers they also occur in other plant structures. Anthocyanins are found in blueberries and other fruits, leaves such as those of coleus, begonia and red cabbage, and even in the stems of some species. Some kinds of anthocyanin require light for their synthesis whereas others do not. Plants that usually do not produce anthocyanins may do so under conditions of nitrogen deficiency, presumably because of a high accumulation of carbohydrate that cannot be used in amino acid synthesis.

Chemically, the anthocyanins are glycosides composed of a sugar linked to a nonsugar known as an anthocyanidin. The anthocyanidins are composed of two benzene rings linked together by a heterocyclic ring containing three carbons and an oxygen atom (Figure 5-28). There are three principal classes of anthocyanidins, and these differ from one another in the number of —OH groups on the second benzene ring. Additional kinds result when one or two of the —OH groups are methylated to —O—CH_3. One anthocyanin may also differ from another because of the number and kind of sugar molecules attached.

Another class of water-soluble pigments, closely related to the anthocyanidins, are the flavonols which have a yellow color of various shades. The flavonols are responsible for most of the clear or pale yellow flower colors,

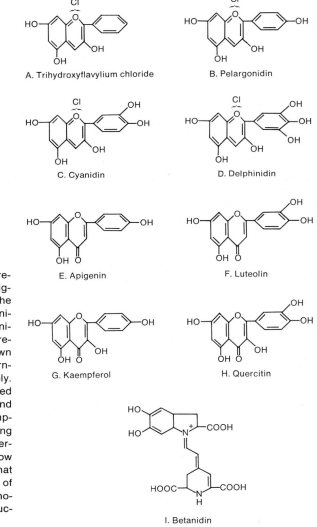

Figure 5-28. Structural formulas of representatives of several classes of plant pigments. **(A)** Trihydroxyflavylium chloride is the parent nucleus of the principal anthocyanidins. **(B–D)** Three of the many anthocyanidins, which form glycosides with sugars, resulting in anthocyanins. The three shown are found in red geraniums, violet cornflowers, and blue larkspurs, respectively. The colors of anthocyanins are influenced by pH. **(E–F)** Two flavones. Apigenin is found in yellow dahlia. **(G–H)** Two flavenols. Kaempferol is found in various plants, including black locust and yellow rose petals. Quercitin is found in oak bark, corn, and yellow apples. **(I)** Betanidin forms glycosides that are called betacyanins. The pigment of redbeet roots is a betacyanin not an anthocyanin. Note the markedly different structure of the betanidin molecule.

A. Trihydroxyflavylium chloride

B. Pelargonidin

C. Cyanidin

D. Delphinidin

E. Apigenin

F. Luteolin

G. Kaempferol

H. Quercitin

I. Betanidin

but the deeper yellows or oranges are usually due to carotenoids. The principal structural difference between anthocyanidin and flavonol molecules is that in the latter the first carbon of the heterocyclic ring is oxidized to a ketone group. There is a range of different flavonols homologous with the different anthocyanidins. The two groups of pigments are referred to collectively as the flavonoids.

The color of redbeet roots results from the presence of a betacyanin. Betacyanins are known to occur only in eight families of the order Centrospermae, and plants that synthesize them do not synthesize anthocyanins. Betacyanins can be hydrolyzed into a sugar and a pigment that, unlike the flavonoids, contains nitrogen (betanidin, Figure 5-28).

Except for the fact that some leucoanthocyanins may function as growth substances, the flavonoids and betacyanins are not known to play any essential metabolic role. Although colored petals may attract pollinating insects, it is improbable that the pigmentation of stems, leaves, roots, or wind-pollinated flowers or cones has any such role.

Tannins

The tannins are complex aromatic compounds with numerous phenolic —OH groups and are quite diverse in chemical structure. They have a bitter astringent taste, precipitate proteins, and form dark-colored complexes with heavy metals such as iron. Tannins are found in the wood, bark, leaves, and fruits of many species but are more abundant in some than others. Dried tea leaves contain about 15% tannins. Some unripe fruits, such as persimmons and plums, contain high concentrations of tannins that make them astringent, but the tannins are decomposed as the fruits ripen. The bark of oaks, chestnuts, hemlock, sumacs, and the quebracho tree of South America are rich in tannins and have all been used as sources of tannins for tanning leather. Their wood also contains considerable tannin, and quebracho wood is now the principal commercial source. The bark of some oaks contains up to 40% tannins, and oak galls may have as much as 80% on a dry-weight basis. Tannins are most abundant in cell walls, although they are sometimes present in the cell sap of vacuoles.

Tannins are, then, a very abundant constituent of certain plant tissues, but the pathways by which they are synthesized are not well understood. Some tannins apparently cannot be hydrolyzed or otherwise degraded by the plants that produce them, whereas others such as those in persimmon fruits can be. It is quite possible that their constituents can be used as metabolic substrates after degradation. It is also possible that tannins may play a useful role as cell wall constituents, but as far as is known now the tannins are probably just another group of nonessential compounds synthesized by plants.

Sterols

The sterols are complex alcohols with a tetracyclic ring structure. Most, if not all, species of plants synthesize sterols, but their functions are not yet known. Fungi, as well as green plants, synthesize sterols and yeast is the most abundant commercial source of sterols. The sterol that is apparently synthesized by more species of plants than any other is ergosterol, which can be converted to vitamin D by irradiation. This conversion can be carried on by animals exposed to light, so plants may be considered the ultimate source of this vitamin as well as the others.

Although sterols may be unimportant in plant metabolism, they play important roles in animal metabolism. The animal sterols include cholestrol and a variety of hormones such as cortisone, androsterone, and progesterone. As far as is known plants are unable to synthesize these or other animal sterols, but various plant sterols can be used as substrates for the synthesis of animal hormones by pharmaceutical companies. Most of the commercially available animal steroid hormones are now derived from plant sterols.

Lignins

The lignins [4, 17] are abundant constituents of the cell walls of tracheids, vessel elements, and fibers and other sclerenchyma cells, and are universally synthesized by vascular plants. Lignins may make up as much as half of the materials in these cell walls. They are deposited between the cellulose micelles of the walls and, although they do not increase the tensile strength of the walls, they do increase the compressional strength; thus lignins probably play an important part in the mechanical support of wood.

Plants are apparently unable to degrade the lignin they produce, and few organisms have enzymes that can break down lignin. However, several species of bacteria and fungi can degrade lignin and use the products as food, thus contributing to the eventual complete decay of wood. Decayed wood that is gray has been attacked first by lignin-utilizing microorganisms, the remaining cellulose and other wall constituents having a gray color. Decaying wood that is a reddish-brown color because of the remaining lignin has, on the other hand, been attacked by bacteria and fungi that decompose the cellulose. In general, lignin is decayed more slowly than cellulose so that lignin is an abundant constituent of the humus of soils.

Although the molecular structure of lignin is still not too well known, lignin molecules are large and complex. They are composed of numerous benzene rings variously linked together and substituted. The lignins can be considered to be polyflavones. Lignins are apparently synthesized from at least three different kinds of substituted benzenes that are similar in structure to the amino acids phenylalanine and tyrosine and may be produced by similar synthetic reactions. One of these units is vanillin, the component of artificial vanilla flavoring.

Many millions of tons of lignin are removed from wood each year in the process of making paper. Sodium bisulfite is used to dissolve the lignin and is responsible for the disagreeable odor of paper factories. The resulting lignin sulfites frequently are discharged into streams as waste, polluting them to the degree that most of the stream organisms are killed. Burning the lignin is a more desirable alternative. A small amount of lignin is used in making vanillin and certain types of plastics, and lignin is a potential source of numerous aromatic compounds now derived largely from petroleum.

References

[1] Beevers, L., and R. H. Hageman. "Nitrate reduction in higher plants," *Ann. Rev. Plant Physiol.,* **20:**495–522 (1970).

[2] Bonner, James. "Protein synthesis and the control of plant processes," *Amer. J. Bot.,* **46:**58–62 (1959).

[3] Bonner, James, and J. E. Varner (eds.). *Plant Biochemistry,* Academic Press, New York, 1965.

[4] Brown, S. A. "Chemistry of lignification," *Science,* **134:**305–313 (1961).

[5] Burris, R. H. "Biological nitrogen fixation," *Ann. Rev. Plant Physiol.,* **17:**155–184 (1966).

[6] Clevenger, Sarah. "Flower pigments," *Sci. Amer.,* **210**(6):85–92 (June 1964).

[7] Dawson, R. F. "Biosynthesis of the Nicotinia alkaloids," *Amer. Sci.,* **48:**321–340 (1960).

[8] Fowden, L. "The non-protein amino acids of plants," *Endeavour,* **21:**35–42 (1962).

[9] Green, D. E. "The synthesis of fat," *Sci. Amer.,* **202**(2):46–52 (Feb. 1960).

[10] Hardy, R. W. F., and R. C. Burns. "Biological nitrogen fixation," *Ann. Rev. Biochem.,* **37:**331–358 (1968).

[11] Jukes, T. H. "The genetic code," *Amer. Sci.* **51:**227–245 (1963); "The genetic code II," *Amer. Sci.,* **53:**477–484 (1965).

[12] Kamen, M. D. "A universal molecule of living matter," *Sci. Amer.,* **199**(2):77–82 (Aug. 1958).

[13] Kendrow, J. C. "The three-dimensional structure of a protein molecule," *Sci. Amer.,* **205**(6):96–110 (Dec. 1961).

[14] Kretovich, V. L. *Principles of Plant Biochemistry,* Permagon Press, Oxford, 1966.

[15] Neurath, H. "Protein-digesting enzymes," *Sci. Amer.,* **211**(6):68–79 (Dec. 1964).

[16] Phillips, D. C. "The three-dimensional structure of an enzyme molecule," *Sci. Amer.,* **215**(5):78–90 (Nov. 1966).

[17] Nord, F. F., and W. J. Schubert. "Lignin," *Sci. Amer.,* **199**(4):104–113 (Oct. 1958).

[18] Rich, A. "Polyribosomes," *Sci. Amer.,* **209**(6):44–53 (Dec. 1963).

[19] Robinson, T. "Alkaloids," *Sci. Amer.,* **201**(1):113–121 (July 1951).

[20] Stein, W. H., and S. Moore. "The chemical structure of proteins," *Sci. Amer.,* **204**(2):81–92 (Feb. 1961).

[21] Wagner, A. F., and K. Folkers. "The organic and biological chemistry of mevalonic acid," *Endeavour,* **20:**177–185 (1961).

[22] Watson, J. D. *Molecular Biology of the Gene,* 2nd ed., W. A. Benjamin, New York, 1970.

[23] Webster, G. C. *Nitrogen Metabolism in Plants,* Row-Peterson, New York, 1959.

[24] Wolff, I. A. "Seed lipids," *Science,* **154:**1140–1149 (1966).

[25] Yanofski, C. "Gene structure and protein structure," *Sci. Amer.,* **216**(5):80–94 (May 1967).

Mineral Nutrition and Metabolism

6

We now turn our attention from the numerous organic compounds synthesized and used by plants to the simpler inorganic compounds that play essential roles in plant metabolism. These include water, several gases (oxygen, carbon dioxide, and nitrogen), acids, bases, and salts. In this chapter we will be concerned primarily with the essential mineral salts and the inorganic acids and bases derived from them, particularly from the standpoint of their ions.

At the present time 16 of the more than 100 chemical elements are definitely known to be essential for plant metabolism, and a few others may be. The three elements that are most abundant in plants, constituting over 98% of the fresh weight of a plant, are carbon, hydrogen, and oxygen. As we have seen, these elements are the principal constituents of organic compounds and are derived from the carbon dioxide and water used in photosynthesis. The remaining essential elements are absorbed as ions from the soil or water, and (except for most of the nitrogen) are derived originally from the rocks of the earth. The importance of these other elements in plant metabolism is far out of proportion to their scarcity [10]. The three required in the greatest quantity are nitrogen, phosphorus, and potassium, but calcium, sulfur, and magnesium are also required in relatively large amounts. These six are known as **major elements.** The remaining essential elements—iron, copper, zinc, boron, manganese, molybdenum, chlorine, and possibly sodium and cobalt—are used by plants only in minute quantities and are known as **trace elements.**

Development of Concepts About Mineral Nutrition

The development of our present-day understanding of the mineral nutrition and metabolism of plants took place more or less concurrently with the development of concepts about photosynthesis, and both were dependent on the development of chemistry. In 1656 Glauber, a German investigator, concluded that potassium nitrate was the "essential principle of vegetation" since he found that it promoted plant growth when applied to the soil. In 1699 John Woodward [20] concluded that "some earthy substance" as well as water was essential for plant nutrition on the basis of rather crude experiments in which he cultured spearmint plants in rain water, river water, and conduit water. Van Helmont might have reached similar conclusions if he had not considered the slight loss in soil weight in his experiment as an experimental error. Little further progress was made during the subsequent hundred years in understanding the mineral nutrition of plants.

Finally, in 1804 Theodore de Sassure, who had contributed much to an understanding of photosynthesis, provided a sound experimental foundation for our present knowledge of mineral nutrition. He found that various mineral elements are essential for plant growth, that these are absorbed by plants in proportions different from that found in the soil, and that some elements were absorbed even though they appeared to be of no use to plants. He also demonstrated that nitrogen was derived from the soil rather than directly from the air.

By 1861 all the major elements that we now know to be essential had been identified. This identification was made possible by raising experimental plants in waxed pots containing well-washed sand and supplying solutions of salts in distilled water containing all but one of the elements believed to be essential. In 1860 Julius von Sachs improved the technique by eliminating the sand and suspending plants with their roots in the salt solutions, but neither he nor others were able to add to the already established list of essential elements because of contaminants in the salts. It was not until the early 1920s that contaminants could be eliminated from the salts to a degree that permitted determination of the various trace elements essential for plants.

There is no possibility of finding additional essential major elements, but more essential trace elements may still be discovered. Some elements are essential in trace quantities for certain species of plants, and may be found to be more generally essential than now believed. In more recent years Arnon and his associates have demonstrated sodium to be essential for blue-green algae. Evans has found cobalt to be essential for the growth of legumes in association with symbiotic nitrogen-fixing bacteria (Figure 6-1). Cobalt is also essential for some microorganisms, where it is required for synthesis of vitamin B_{12}. Silicon is essential for diatoms because it is used in making their siliceous cell walls. It may also be essential for the various species of Equisetum, whose stems contain abundant silicon. The shoots of grasses and some other plants also contain much silicon. Although most of these plants can apparently complete their life cycles successfully without silicon, it probably contributes to such factors as stem rigidity and may be of survival value in nature. Several rare elements, such as scandium and vanadium, have been reported essential for certain fungi. Iodine is evidently not essential for plants, but the iodine absorbed by plants is useful in human and animal nutrition since it is a constituent of their thyroxin.

Along with the basic research on the mineral nutrition of plants there has been extensive applied research on the use of fertilizers to promote good growth

Figure 6-1. Left to right: Soybean plants supplied with 1.0 or 0.1 parts per billion (ppb) of cobalt flourished, whereas those with no cobalt were stunted and chlorotic. Substitution of certain other elements for cobalt did not correct the deficiency symptoms. [Courtesy of H. J. Evans. From S. Ahmed and H. J. Evans, *Soil Sci.,* **90:**205 (1960) with the permission of the publisher. Copyright © 1960 by William & Wilkins Co.]

of agricultural and horticultural crop plants. The pioneer in this field was the German agricultural chemist Justus von Liebig who, as early as 1840, stressed the importance of replenishing the mineral elements of farm soils by the use of fertilizers. One reason that a major part of agricultural research has dealt with soil fertility is that it is one of the most easily controlled environmental factors as well as frequently being the limiting factor in plant growth.

Each year farmers, horticulturists, and home gardners spend many millions of dollars on fertilizers of various types, principally the so-called chemical fertilizers [13]. Some of these are designed to supply only a single essential element such as nitrogen, phosphorus, or calcium, but fertilizers that provide nitrogen, phosphorus, and potassium together are most widely used. Although these are called complete fertilizers, they are complete only in the sense that they supply the elements most likely to be deficient in soils. They may, of course, also contain some of the other essential elements as impurities. A really complete fertilizer containing all the essential elements is sometimes useful because even the trace elements [1, 16] may at times be deficient. Some elements like iron are abundant in soils but may be in the form of insoluble compounds that are not available to plants.

Also widely used are the organic fertilizers such as manures, green manures (legumes or other plants plowed under), and composts. There is no evidence that these are superior to chemical fertilizers as sources of mineral elements, as claimed by certain faddists [17], but they may contribute toward improved soil structure.

The cultivation of plants without soil by the use of solutions containing all the essential mineral elements, a procedure known as **hydroponics,** is a rather common practice, particularly in greenhouses [9]. The plants are placed in some substance such as sand, gravel, vermiculite or perlite, or they may be supported mechanically with their roots immersed in the solution.

Roles of the Mineral Elements

Although fertilizers are commonly referred to as plant foods, the mineral elements are really not true foods that can be used in respiration or assimilation like the carbohydrates, fats, and proteins. Fertilizers should rather be considered as raw materials. Mineral elements are used most extensively in the synthesis of a variety of important organic compounds of which they are essential constituents, but they also play a variety of roles as ions of inorganic compounds [5].

Mineral Elements as Constituents of Organic Compounds

We have already considered in some detail the various important organic compounds of which nitrogen is a constituent such as the amino acids, proteins, purines, pyrimidines, chlorophyll, and many coenzymes. Phosphorus is another element that is a constituent of numerous important organic compounds including the nucleic acids, phospholipids, ADP, ATP, NAD, NADP, and the phosphate esters of the sugars. Calcium is essential for the synthesis of the calcium pectate of the middle lamellas, and there is evidence that the calcium salt of lecithin is essential as a membrane constituent. Calcium is necessary for normal mitosis, probably as a component of the chromosomes or mitotic spindles. Magnesium is an essential constituent of chlorophyll and is also found in the middle lamella as magnesium pectate. This element is a constituent of ribosomes and seems to function by binding together the subunits that make up a ribosome. Sulfur is a constituent of the amino acids cystine, cystiene, and methionine and of proteins. The —SH groups of proteins are important in their enzymatic activities and also in crossbonding of the protein molecules. Sulfur is also a constituent of the B vitamins thiamin and biotin and of CoA, the —SH groups again playing essential roles. Iron is an essential constituent of the cytochromes, peroxidases, catalases, and metallo-flavoproteins, serving as an electron acceptor and donor. Copper is a component of cytochrome oxidase, ascorbic acid oxidase, polyphenol oxidase, and plastocyanin. Like iron, it is alternately reduced and oxidized as it accepts and donates electrons. Zinc is a constituent of a number of enzymes including alcohol dehydrogenase, glutamic dehydrogenase, lactic dehydrogenase, carbonic anhydrase, and alkaline phosphatase. Molybdenum is a constituent of nitrate reductase and other enzymes. Cobalt is an essential constituent of vitamin B_{12} coenzymes.

Several diverse roles of boron in plants have been proposed. In 1953 Gauch and Dugger [11] suggested that boron is required for sugar translocation. However, although borates form complexes with some sugars, none of these has been isolated from plants. Lee and Aronoff [12] reported that borates complex with 6-phosphogluconic acid, thus inhibiting the action of its dehydrogenase and preventing the eventual synthesis of excessive quantities of phenolic acids, which accumulate in boron-deficient plants and cause necrosis and death. Rajaratnam and his coworkers [14] reported that boron is essential for flavonoid synthesis.

Mineral Elements as Enzyme Activators

A number of the essential elements function as activators of enzymes. Although these probably form temporary complexes with the enzymes that they activate, their catalytic role is different enough from that of most elements that are structural components of enzymes to justify their separation into another category. Unlike elements such as iron, copper, and molybdenum, these do not participate in oxidation reactions. Perhaps they are involved in the maintenance of the proper structural conformation of the enzyme proteins. In some cases enzyme activators may also act as enzyme inhibitors, particularly when these are too concentrated or are in imbalance with other elements.

Calcium is required as a cofactor or enzyme activator by enzymes involved in a number of reactions, including some enzymes involved in the hydrolysis of ATP and phospholipids. Magnesium is required nonspecifically by a large number of enzymes involved in phosphate transfer. Manganese is specifically required for the activity of some dehydrogenases, decarboxylases, kinases, oxidases, and peroxidases. It is also essential for the photosynthetic production of oxygen, as is chlorine. In some cases where a divalent cation is required as an enzyme activator several different cations can function, but frequently a specific divalent cation is required.

Evans and Sorger [8] have tabulated a list of 46 enzymes or enzyme systems that require univalent cations as activators, and there are no doubt still others. The reactions catalyzed include a great variety such as glycolysis, starch synthesis, oxidative phosphorylation, oxygen uptake, dehydrogenations, amino acid synthesis, protein synthesis, and a variety of other syntheses.

The potassium ion (K^+) is effective as an enzyme activator in every case, and in a good many of these other univalent cations cannot substitute for it. However, in a majority of the cases other univalent cations such as Na^+, Rb^+, Cs^+, Li^+, or NH_4^+ may be able to substitute for K^+. Of these alkaline metals only potassium is essential for plants in general. Sodium has been reported to be essential for some halophytes such as *Atriplex vesicaria* and *Halogeton glomeratus* as well as some blue-green algae. Furthermore, potassium is usually by far the most abundant of these elements in plants and it is undoubtedly the predominant univalent cation that serves as an enzyme activator. Since potassium is the only one of the major elements that is not a stable component of the various compounds synthesized by plants, there was for a long time a question as to why it is required in such large quantities. Evans and Sorger [8] suggest that the extensive requirement for potassium as an enzyme activator can probably account for essentially the entire quantity of potassium required.

Other Roles of Mineral Elements

Although mineral elements are important in plants primarily as constituents of organic compounds and as enzyme activators, the ions of the mineral salts have a number of other influences on plant metabolism. Some of these are nonspecific and may involve the ions of any element absorbed by plants, whether essential or not. It should be noted that plants absorb any kind of ion present in the soil solution regardless of whether or not they are either essential or toxic, so plants generally contain a good many elements that play no essential role in their metabolism.

One nonspecific influence of the mineral salt ions is a contribution toward the osmotic potential of the cells, although generally sugars and other organic solutes are the more important osmotically active solutes. Since phosphate (PO_4^{3-}) and carbonate (CO_3^{2-}) ions form weak acids they contribute toward the buffering of plants cells, although the weak organic acids and their salts are more important buffer systems. Of course, the cations of all these buffer systems except for H^+ are the various mineral cations such as Ca^{2+}, Mg^{2+}, K^+, and Na^+.

Ions also influence the permeability of the various cell membranes. In general, univalent ions such as K^+ and Na^+ increase membrane permeability whereas divalent ions such as Ca^{2+} and Mg^{2+} or trivalent ions reduce permeability. A suitable ratio between these groups of ions is therefore important for normal differential membrane permeability. Similar antagonistic interactions of ions influence muscle contraction and nerve impulse conduction in animals.

A number of different mineral ions can be toxic to plants, sometimes highly toxic [3, 18]. The toxic ions are aluminum, arsenic, and most of the heavy metals including silver, lead, and mercury. Also toxic when present in more than the usual trace quantities are boron, copper, magnesium, manganese, molybdenum, and zinc. The toxic effects are chemical in nature, involving principally interference with the normal functioning of other mineral elements in electron transfer and other catalytic roles.

Very high concentrations of ions of almost any kind can bring about such serious consequences as the denaturation of proteins and the flocculation of other protoplasmic colloids, resulting in death of the cell. Such high concentrations of ions are rare in cells but may occur during freezing as water is converted to ice and ceases functioning as a solvent. Neither freezing nor the plasmolysis of plant cells by excessively high concentrations of ions in the surrounding soil solution or water are commonly considered as toxicity.

Mineral Deficiency Symptoms

Plants that have access to inadequate quantities of one or more of the essential mineral elements do not grow and develop normally and exhibit characteristic mineral deficiency symptoms [15, 19]. These include (1) **chlorosis,** a pale green or yellow color resulting from reduced chlorophyll synthesis; (2) **necrosis,** localized death of tissues such as buds, leaf tips or margins, or scattered spots on leaves; (3) **anthocyanin formation** in stems or other structures where it does not normally occur; (4) **stunted growth;** (5) **unusually slender and woody stems** of herbaceous species; (6) **poor reproductive development,** including at times complete failure of fruit or seed production.

There is a characteristic complex of deficiency symptoms for each of the

Figure 6-2. Tobacco plants in sand culture showing deficiency symptoms resulting from a lack of seven of the essential elements. The control plant (Ck.) was supplied with a complete mineral nutrient solution. Each of the other plants received all but one of the essential elements, as indicated on the labels. [Courtesy of W. Rei Robbins]

essential elements. The characteristic symptoms show up best in experiments or demonstrations in which plants have been cultured in solutions of salts, each solution lacking only one of the essential elements (Figure 6-2). It is much more difficult to use the symptoms for diagnosis of specific deficiencies of plants under cultivation or in nature because the symptoms may not be as severe as when there is a complete absence of an element, several elements may be deficient concurrently, and some virus diseases produce symptoms similar to those brought about by mineral deficiencies. Furthermore, the deficiency symptoms for any particular element are often somewhat different in one species than another. Even with these complications a specialist in the mineral nutrition of plants can often identify the deficient elements from the symptoms, particularly in certain crop plants, although chemical analysis of the plants or the soil generally can give more precise information.

The deficiency symptoms are, of course, products of metabolic disturbances brought about by inadequate supplies of an element. In some cases the metabolic causes of a symptom are quite obvious; but in others they may be obscure and quite complex. We will now consider the principal symptoms brought about by a deficiency of each of the elements and, when known, their metabolic causes.

Nitrogen

One of the first symptoms of nitrogen deficiency is uniform chlorosis of the older leaves, which may later die and turn brown. The younger leaves may show little or no chlorosis because they are capable of mobilizing nitrogen which is retranslocated to them from the older leaves. However, when there is severe nitrogen deficiency all the leaves become chlorotic. The chlorosis probably results not only because of inadequate nitrogen for chlorophyll synthesis but also because of reduced chloroplast protein production.

Another characteristic nitrogen-deficiency symptom is reduced growth, brought about primarily by lack of sufficient proteins and nucleic acids for continued production of new cells. However, it will be recalled that nitrogen is also a constituent of many other metabolically essential compounds, such as the various coenzymes, and it is likely that in severe nitrogen deficiency these other nitrogenous compounds are also in short supply. Although the chlorosis generally reduces the rate of photosynthesis, lack of carbohydrate is usually not the limiting factor in the growth of nitrogen-deficient plants and carbohydrates may actually accumulate in unusually high concentrations. Not surprisingly, nitrogen deficiency limits growth more severely than a deficiency of any other element. Plants completely deprived of nitrogen are frequently unable to grow larger than the seedling stage (Figure 6-2).

Another nitrogen-deficiency symptom in a number of species including corn and tomato is the abnormal production of anthocyanins in the stems and leaves, resulting in red coloration. This seems to be brought about by the accumulation of unusually high concentrations of sugars that cannot be used in the synthesis of amino acids or other nitrogenous compounds and are then used in anthocyanin synthesis.

The stems of nitrogen-deficient plants not only have greatly reduced terminal growth but also tend to be unusually slender and woody, the latter probably resulting from a disproportionate amount of secondary cell wall thickening since carbohydrates are abundant. Unless nitrogen is totally lacking, root growth of nitrogen-deficient plants is considerably better than shoot growth because the roots have prior access to the limited supply of nitrates (or other nitrogen compounds).

Phosphorus

The most characteristic symptoms of phosphorus deficiency are the strange dark grayish-green color of the leaves; necrotic spots on the leaves, fruits, or other structures; and sometimes malformed leaves. Other symptoms include stunted growth, anthocyanin formation, and death of the older leaves, all similar to nitrogen-deficiency symptoms but generally less severe. Like nitrogen, phosphorus is readily mobilized to the younger leaves and meristems from the older leaves, so the latter show deficiency symptoms earlier. The phosphorus-deficient plants also accumulate carbohydrates. Their stems are likely to be weak and spindly but not woody, and the vascular tissues are poorly developed. The stunted growth of phosphorus-deficient plants probably results to a great extent from limitations on nucleic acid synthesis, but reduced quantities of ATP, NAD, NADP, and the various other important compounds containing phosphorus undoubtedly contributed toward a general decrease in, and disruption of, various metabolic pathways. There is relatively greater inhibition of root growth in phosphorus deficiency than in nitrogen deficiency.

Potassium

The first indication of potassium deficiency is mottled chlorosis followed by necrosis of the leaf margins and tips. Curling and crinkling of the leaves may occur. The internodes of the stems are usually abnormally short (Figure 6-2), and the stems are weak and may have brown streaks on them. Potassium, like nitrogen and phosphorus, is highly mobile within the plant, so the symptoms appear first in the older leaves. At first, potassium-deficient plants accumulate carbohydrates and amino acids because protein synthesis is not occurring normally, but later on carbohydrates decrease, probably because of a decrease in photosynthesis and an increase in respiration.

Sulfur

Many of the symptoms of sulfur deficiency are similar to those of nitrogen deficiency although generally less severe: uniform chlorosis, anthocyanin formation, and stunted growth. This could be anticipated since both elements are constituents of proteins, and in both cases there is accumulation of carbohydrates and soluble nitrogen compounds as well as a decrease in proteins. This is apparently brought about by increased protein hydrolysis as well as decreased protein synthesis. However, in contrast with nitrogen deficiency the chlorosis appears first in the younger leaves, and the chlorotic leaves are pale green, never yellow or white.

Calcium

Since calcium is quite immobile in plants (except when in the xylem) calcium-deficiency symptoms appear first in the buds and younger leaves. The most characteristic symptom of calcium deficiency is necrosis of the tips and margins of the youngest leaves and then of the buds. This is usually preceded by a pale green chlorosis and downward hooking of the leaf tips as well as by general deformation of the young leaves. Another characteristic symptom is an unusually short, stubby, highly branched root system, the roots having a brownish color. Even before death of the terminal meristems there is an interference with meristematic activity, resulting in stunted growth, and cell enlargement and vacuolation occur much closer to the apex than usual.

Iron

The principal symptom of iron deficiency is a characteristic chlorosis (Figure 6-3), although there is usually also some growth inhibition. Since iron is another immobile element, the chlorosis occurs first in the younger leaves. A characteristic feature of iron chlorosis is that it is interveinal, and the principal veins remain green except in cases of extreme and prolonged iron deficiency.

Magnesium

As might be expected, the principal symptom of magnesium deficiency is chlorosis, although there is also general growth inhibition. As in the case of

Figure 6-3. Leaf from a sunflower plant deficient only in iron (left) compared with a leaf from a plant supplied with all the essential elements (center). The leaf at the right was supplied with iron, but an excess of manganese made the iron unavailable. [Courtesy of W. Rei Robbins]

iron deficiency, the chlorosis is initially interveinal but, because magnesium is quite mobile, the chlorosis appears first in the older leaves. In severe magnesium deficiency all the leaves may become yellow or white, the latter color resulting from loss of the carotenoids as well as the chlorophylls (Figure 6-2). Also, the leaves of magnesium-deficient plants tend to abscise prematurely.

Copper

The first symptom of copper deficiency is generally necrosis of the tips of young leaves, with the necrosis then extending downward along the leaf margins. Severe copper deficiency may result in leaves that look wilted and premature leaf abscission.

Zinc

The symptoms of zinc deficiency [4] are quite characteristic and surprisingly striking for a trace element. Internode growth is greatly inhibited, resulting in a rosette condition, and the leaves are greatly distorted and unusually small. It is possible that these symptoms result largely because of an inadequate supply of IAA, since zinc is necessary for its synthesis. There is also interveinal chlorosis of the older leaves followed by white necrotic spots, and these may be the only symptoms when the zinc deficiency is not severe. Reproductive development is often poor.

Boron

The most distinctive symptom of boron deficiency is black necrosis of the terminal buds and the nearby young leaves, particularly at the bases of their blades. As a consequence of the loss of apical dominance there is extensive branching, but the terminal buds of the branches soon become necrotic also. The leaves may also become thick, brittle, and distorted. Reproductive development is poor, and fleshy structures such as fruits, fleshy roots (Figure 6-4), tubers, and cauliflower inflorescences may have extensive necrotic areas or abnormalities such as internal cork formation in apples.

Manganese

The principal symptom of manganese deficiency is interveinal chlorosis, usually but not always on the younger leaves, in combination with numerous small necrotic spots. Etlidge [6] has shown that in tomato leaves, at least, the chlorosis seems to be a result of chloroplast damage which involves the loss of both chlorophyll and starch grains, causes vacuolation, and finally complete disintegration.

Molybdenum

The first symptoms of molybdenum deficiency are generally interveinal chlorosis of the older leaves and marginal necrosis of the leaves. In some species such as citrus fruits the chlorosis and necrosis may be limited to rather large

Figure 6-4. The redbeet at the left received all essential elements, but the other three were deficient in boron. Note the large black necrotic areas in the boron-deficient roots. [Courtesy of R. C. Burrell]

spots. Flower formation may not occur, and if it does the flowers generally abscise early. Since molybdenum is essential for nitrate reduction, its absence may bring about nitrogen-deficiency symptoms unless ammonium or organic nitrogen compounds are available [7].

Molybdenum is required in smaller quantities than any other definitely established trace element, 1 part per 100 million of culture solution being enough to prevent molybdenum-deficiency symptoms. As little as 1 oz/acre of molybdenum trioxide (MoO_3) will correct molybdenum deficiency of soils. Any very great concentration of molybdenum is highly toxic to plants.

References

[1] Anderson, A. J., and E. J. Underwood. "Trace element deserts," *Sci. Amer.,* **200**(1):97–106 (Jan. 1959).

[2] Broyer, T. C., A. B. Carlton, C. M. Johnson, and P. R. Stout. "Chlorine—a micronutrient element for higher plants," *Plant Physiol.,* **29**:526 (1954).

[3] Burt, J. C. "Desert in the Appalachians," *Nature Mag.,* **49**:486–488 (1958).

[4] Camp, A. F. "Zinc as a nutrient in plant growth," *Soil Sci.,* **60**:156 (1945).

[5] Devlin, R. M. *Plant Physiology,* 2nd ed., Reinhold, New York, 1969.

[6] Etlidge, E. T. "Effects of manganese deficiency on the histology of *Lycopersicon esculentum," Plant Physiol.,* **16**:189 (1941).

[7] Evans, H. J. "Role of molybdenum in plant nutrition," *Soil Sci.,* **81**:199–208 (1956).

[8] Evans, H. J., and G. J. Sorger. "Role of mineral elements with emphasis on the univalent cations," *Ann. Rev. Plant Physiol.,* **17**:47–76 (1966).

[9] Gericke, W. F. *The Complete Guide to Soilless Gardening,* Prentice-Hall, Englewood Cliffs, N. J., 1940.

[10] Gilbert, F. *Mineral Nutrition of Plants and Animals,* University of Oklahoma Press, Norman, Okla. 1948.

[11] Gauch, H. G., and W. M. Dugger. "The role of boron in the translocation of sucrose," *Plant Physiol.,* **28**:457 (1953); "The physiological role of boron in higher plants: a review and interpretation," *Univ. Md. Agr. Exp. Sta. Tech. Bull.* A-80.

[12] Lee, S., and S. Aronoff. "Boron in plants: a biochemical role," *Science,* **158**:798–799 (1967).

[13] Pratt, C. J. "Chemical fertilizers," *Sci. Amer.,* **212**(6):62–71 (June 1965).

[14] Rajaratnam, J. A., *et al.* "Boron: possible role in plant metabolism," *Science,* **172**:1142 (1971).

[15] Sprague, H. B. *Hunger Signs in Crops,* David McKay Co., New York, 1964.

[16] Stiles, W. *Trace Elements in Plants,* Cambridge University Press, London, 1961.

[17] Throckmorton, R. I. "Organic farming—bunk," *Reader's Dig.,* **61**(4):45–48 (Oct. 1952).

[18] Trelease, S. F. "Bad earth," *Sci. Mon.,* **54**:12–28 (1942).

[19] Wallace, T. "Mineral deficiencies in plants," *Endeavour,* **5**:58–64 (1946).

[20] Woodward, J. "Some thoughts and experiments on vegetation," *Phil. Trans. Royal Soc. London,* **21**:382 (1699).

Cellular Traffic

7

There is an essentially continual movement of substances into and out of cells and from one part of a cell to another. To a considerable extent this results from the use and production of substances in the various metabolic processes that we have been considering. The exchanges between autotrophic cells and their external environment involve principally a few simple inorganic compounds: water, carbon dioxide, oxygen, and the ions of various salts. However, other substances including some organic compounds and toxic gases may also enter the cells, and sugars and some other organic compounds may be lost from cells. Nitrogen fixation results in the movement of atmospheric nitrogen into cells. Heterotrophic bacteria and fungi must obtain from their environment organic substances such as sugars and amino acids and in some cases certain vitamins.

In multicellular plants there is also extensive movement of substances from one cell to another, including such organic compounds as sugars, amino acids, vitamins, and hormones as well as the various inorganic substances. This internal transport plays an essential role in the maintenance and coordination of the metabolic processes of the various organs and in the coordination of the growth and development of the plant.

In this chapter we shall consider the movement of substances into and out of cells. The long-distance transport of substances through the xylem and phloem of vascular plants will be considered in subsequent chapters.

Membrane Permeability

Any discussion of the movement of substances into and out of cells, or even from one cell organelle to another, requires consideration of the fact that the substances must be able to pass through the differentially permeable cell membranes, except when taken up by pinocytosis (Chapter 2). All cytoplasmic membranes are differentially permeable, permitting some substances to pass through them more easily than others and completely blocking the passage of still other substances. Substances entering or leaving plant cells must also pass through the cell walls, but the walls of most cells are permeable to the particles of all substances and do not constitute a barrier. Exceptions are the impermeable suberized walls of cork cells and the cutinized walls of epidermal cells, which are relatively impermeable to water and other substances.

Permeability to Various Substances

Cell membranes are highly permeable to water and gases and are also generally quite permeable to most other substances with molecular weights of less than about 60. Permeability to nonionic water-soluble substances is generally inversely proportional to the size of their molecules. Membranes are commonly more permeable to undissociated molecules than to their ions, perhaps partly because of the charges on the ions and partly because of the hydration of the ions. If the size of an ion including the water of hydration is considered, permeability is generally related inversely to the number of charges on the ions. Membranes are more permeable to weak acids than strong ones because weak acids have a lower percentage of ionization. An increase in the pH increases the percentage of dissociation of carbonic acid (H_2CO_3) and proportionately reduces permeability to it. In general, permeability to ions is low but, as we shall see later, ions are commonly carried across membranes by active transport systems requiring respiratory energy.

There is a marked difference in the permeability of cell membranes to polar (water soluble) and nonpolar (fat soluble) compounds. The molecules of polar substances, which include nonionizable substances such as sugars as well as electrolytes, have differential electrical charges, whereas nonpolar molecules are uncharged. Hydrocarbons are nonpolar. Long-chain fatty acids and alcohols are predominantly nonpolar, although the ends with the —COOH or —OH groups are polar. Permeability to nonpolar substances, in contrast with permeability to water-soluble polar substances, increases with molecular weight and size, that is, with increasing nonpolarity and fat solubility. Cell membranes are highly permeable to petroleum hydrocarbons. Because of their toxicity, it is fortunate that these rarely come in contact with cell membranes in nature.

Basis of Differential Permeability

Despite years of extensive research, the basis for the differential permeability of cell membranes is still not well understood. A good bit of the difficulty results from the fact that the molecular organization of cell membranes is still not clear (Chapter 2), although it is evident that these membranes are composed predominantly of proteins and lipids [3].

Both of the principal hypotheses regarding the differential permeability of cell membranes were first proposed years ago. In 1867 Traube proposed the molecular sieve hypothesis, based on permeability to polar substances. In 1898 Overton proposed the lipid solubility hypothesis, based on his work on permeability to fatty acids with a range of molecular weights. Both of these mechanisms of differential permeability clearly operate in nonbiological membranes. Differentially permeable membranes of collodion, parchment paper, copper ferrocyanide, and various plastics are molecular sieves, having pores that exclude molecules over a certain size. Layers of fat or other lipids are permeable to substances in proportion to their degree of lipid solubility. It appears likely that both mechanisms also operate in cell membranes, but until membrane structure is better understood the way in which they operate remains uncertain [4]. The greatest difficulty is with the molecular sieve hypothesis. Cell membranes were once presumed to have pores of molecular size through which polar substances could pass, but some hypotheses of membrane structure seem to make this improbable, and so far electron micrographs have failed to reveal structurally defined pores except in the nuclear membranes [13].

Factors Influencing Membrane Permeability

The differential permeability of cell membranes changes from time to time and can be greatly modified by various experimental treatments. Membrane permeability increases during senescence, and when a cell dies its membranes become completely permeable. Respiration is evidently essential for the maintenance of differential permeability, and reduction of the rate of respiration by low oxygen, low temperature, or respiratory inhibitors increases membrane permeability. Ca^{2-} has been reported as necessary for the maintenance of normal membrane structure and permeability. Permeability is influenced by the balance of univalent and divalent ions, by pH, by dehydration, and probably by regulatory molecules like phytochrome. It should be stressed that the differential permeability of cell membranes is far more complex than that of nonbiological membranes.

The Significance of Differential Permeability

The fact that cell membranes are differentially permeable is of immense biological importance. Differential permeability permits essential exchanges of substances between a cell and its environment and between one cell organelle and another and, at the same time, prevents the dissipation and loss of many substances of metabolic importance such as proteins and coenzymes. Life would obviously be impossible if the cell membranes were completely impermeable, but it would also be impossible if the membranes were completely permeable. The appearance of differentially permeable membranes early in the origin of life was a most critical event in making possible the structural and functional integrity of metabolizing cells.

Plasmodesmata

In connection with the movement of substances from one plant cell to another it should be noted that the plasmodesmata connecting adjacent cells through their walls provide channels through which substances can pass without crossing a membrane. However, the extent and importance of the role of plasmodesmata in intercellular transport is not known.

Gas Exchanges of Plant Cells

As has been noted, cell membranes are very permeable to gases and offer essentially no impediment to their movement in and out of cells. Although carbon dioxide and oxygen are the principal gases exchanged between a cell and its environment, any gas that is adjacent to a cell can enter it, even toxic gases such as sulfur dioxide. The movement of gases into, within, and out of cells is a matter of simple diffusion.

Diffusion of Gases

Diffusion is the net movement of the molecules of a substance from one region to another under their own kinetic energy. This net movement occurs only when the molecules of the substance have a greater total free energy or molecular activity per unit volume in one part of a system than another. As far as gases are concerned, this molecular activity is determined by two principal factors: concentration and temperature. If the temperature is uniform throughout the system being considered, diffusion is from a region of higher to lower concentration of that particular substance, the concentration of any other gas present being immaterial. Since gases diffusing in or out of cells are dissolved in water, the solubility of a gas in water influences its concentration. When a cell is carrying on photosynthesis carbon dioxide diffuses in as it is being consumed, whereas oxygen diffuses out at the same time and place because photosynthetic oxygen production is increasing the internal oxygen concentration above the external concentration.

Diffusion is simply a matter of probability: if oxygen is more concentrated inside a cell than outside, more of its molecules can be expected to move out than in during any unit of time. This will happen even though the oxygen molecules are moving in all directions, into as well as out of the cell. If, in an isothermal system, the concentration of a certain gas is uniform throughout, there are just as many moleucles moving in one direction as another and there is no *net* movement of molecules; the system is then in dynamic equilibrium. It should be noted that diffusion is not synonymous with the kinetic movement of molecules, even though it is a result of such movement. Nitrogen molecules are continuously entering and leaving cells but, because most cells neither use nor produce nitrogen, there is no concentration gradient and so no diffusion of nitrogen. Of course, nitrogen does diffuse into root nodules where nitrogen fixing organisms are converting nitrogen into other substances.

However, diffusion may occur when the concentration of a substance throughout a system is uniform provided that the temperature is higher in one part of it than another. Since an increase in temperature increases the energy and rate of movement of molecules, more molecules move from the warmer region per unit time than enter it even though the molecules are no more concentrated in the warmer region. If the temperature difference is maintained, a new dynamic equilibrium will soon be established with the same total molecular activity throughout but with a greater concentration of the substance in the cooler region because total molecular activity is a product of the number of molecules per unit volume and their rate of movement. This situation prevails in nature where leaves and other organs exposed to direct sunlight may be 10°C or so warmer than the surrounding air. Thus, there is a brief diffusion of nitrogen from a leaf when it is first exposed to the sun, and a brief diffusion of nitrogen in when the leaf is shaded again.

Factors Influencing the Rate of Diffusion

Many years ago Thomas Graham found that the rate of diffusion (and the rate of molecular movement) of a gas was inversely proportional to the square root of its relative density (molecular weight), the relative density of hydrogen being arbitrarily set as 1. Thus, the rate of diffusion of hydrogen is $1/\sqrt{1} = 1$, whereas the rate of diffusion of oxygen is $1/\sqrt{16} = 1/4$. In other words, hydrogen diffuses four times faster than oxygen, and about five times faster than carbon dioxide $(1/\sqrt{22})$. Even though photosynthesis uses just as many molecules of carbon dioxide per unit time as molecules of oxygen produced, the carbon dioxide diffuses into the cell more rapidly than the oxygen diffuses out. Although the molecules of carbon dioxide are heavier, they are much more soluble in water.

The rate of diffusion in simple systems is determined by the length and steepness of the activity gradient, the temperature, and the density of the medium. The rate of diffusion varies directly with the steepness of the gradient, which is generally determined by the difference in concentration of the substance in the two parts of the system. Thus, if diffusion is leading toward a dynamic equilibrium, the rate of diffusion becomes slower and slower as the concentration differences become less; finally at equilibrium the rate is zero. The rate of diffusion is inversely proportional to the length of the gradient. Thus if the two parts of the system are connected with one another by a tube, the rate of diffusion will be twice as fast through a tube 10 cm long as through a tube 20 cm long.

A second factor that influences the rate of diffusion is temperature, since an increase in temperature increases the rate of molecular movement. With a 10°C increase in temperature there is an increase of from 1.2 to 1.4 times in the rate of diffusion, a Q_{10}* characteristic of physical processes.

Finally, the rate of diffusion is inversely proportional to the density of the medium through which it is occurring; therefore the concentration of other gases present will affect the rate of diffusion of a particular gas even though that concentration has no influence whatsoever on the direction of diffusion of the particular gas in question. Also, gases diffuse more slowly through a liquid such as water than through the air or other gases. This reduction of the rate of diffusion by the density of the medium is a result of an increasing number of molecular collisions.

Movement of Ions Into and Out of Cells

The movement of salts into and out of cells is considerably more complicated than the movement of gases. Since it is likely that salts enter and leave cells predominantly in the form of their ions rather than their undissociated mole-

---

*Q_{10} is the temperature coefficient of a process; that is, the increase in the rate of a process with a 10°C increase in temperature.

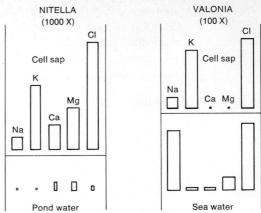

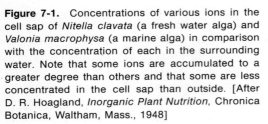

Figure 7-1. Concentrations of various ions in the cell sap of *Nitella clavata* (a fresh water alga) and *Valonia macrophysa* (a marine alga) in comparison with the concentration of each in the surrounding water. Note that some ions are accumulated to a greater degree than others and that some are less concentrated in the cell sap than outside. [After D. R. Hoagland, *Inorganic Plant Nutrition,* Chronica Botanica, Waltham, Mass., 1948]

cules, we shall speak in terms of ion transport. The ions and molecules of salts, like all other substances, diffuse, but simple diffusion apparently plays only a minor role in the movement of ions across cell membranes. There are many data showing that ions of various kinds enter cells even when their concentration in the vacuoles or other parts of cells is much higher than in the surrounding water or soil solution (Figure 7-1). This cannot possibly be accounted for by simple diffusion. Furthermore, some kinds of ions accumulate in the vacuoles to a much greater degree than others, whereas in a few cases, notably that of sodium, the concentration may be considerably lower in the cells than in the surrounding solution. That the membranes of living cells play an essential role in the movement of ions and their accumulation is indicated by the fact that, when cells are killed, the various ions quickly attain a dynamic diffusion equilibrium with the surrounding solution.

There has been much research on ion transport and a number of different mechanisms have been identified. Some of these mechanisms are purely physical processes and are referred to collectively as **passive absorption,** but the mechanism that is apparently most important in the accumulation of ions against a diffusion gradient requires metabolic energy from respiration and is referred to as **active transport.** We shall now consider briefly several proposed mechanisms of ion transport across cell membranes.

Diffusion

Although diffusion is not an important factor in the absorption of ions by cells, it appears that free diffusion occurs between certain regions of cells and the surrounding medium and that, in these regions, there tends to be a diffusion equilibrium of any particular ion with its concentration in the external medium. These regions of cells are referred to as outer space or apparent free space [7] and definitely include the cell walls and possibly the undifferentiated cytoplasm, but the vacuoles, the mitochondria, and other organelles surrounded by membranes are not included. Since the accumulation of ions against a diffusion gradient occurs primarily within the vacuoles, it has been proposed by some workers that active absorption at the expense of metabolic energy occurs primarily in the vacuolar membrane rather than in the plasma membrane.

When a plant tissue is placed in water part of its ions diffuse into the water until an equilibrium for each kind of ion is attained, but the ions accumulated in the vacuoles and organelles remain there. (Ions and other solutes may leach out of plants during a heavy rain.) If a tissue that has been in water until diffusion equilibria are established is placed in a solution of salts of known composition, there is a rapid initial inward diffusion of the ions followed by a slower metabolic accumulation of ions. By poisoning the metabolic transport system with respiratory inhibitors, an estimate of the nonmetabolic ion uptake by the tissue can be secured. Quantitative determinations of the nonmetabolic ion uptake or release, along with estimates of the volumes of cell walls, vacuoles, and protoplasts of the tissue, are used in calculating the volume of free space. However, the accuracy of the volume estimates is not high enough to determine with certainty whether or not the undifferentiated cytoplasm is a part of the free space [8].

Ion Exchange

Ion exchange is a purely physical process that can account for a greater uptake of any particular kind of ion than would be possible by simple diffusion. Ions may be adsorbed on cell walls and membranes, and thus cannot diffuse freely. However, an adsorbed ion can be replaced by another ion with the same charge. For example, a Ca^{2+} can exchange with two absorbed H^+, Na^+, or K^+, or with one Mg^{2+}; a sulfate ion (SO_4^{2-}) could exchange with two OH^-. However, it is likely that ions taken up by exchange do not play any very important role in the general economy of the cell, at least as long as they remain adsorbed.

A similar ion exchange takes place between soil particles and the soil solution. Soil particles such as clay micelles have multiple negative charges and so adsorb numerous cations, and these may exchange with ions in the soil solution. The forces holding the ions on the surface of the soil particles vary in strength with the kind of ion, H^+ being held most tenaciously, followed in order by Ca^{2+}, Mg^{2+}, K^+, NH^+, and Na^+. Thus, Na^+ is readily replaced by any of the other ions, whereas Ca^{2+} is readily replaced only by H^+. The adsorption of cations on soil particles is of considerable importance because it prevents the rapid leaching of cations from the soil. However, anions other than phosphate are not held by the soil and do leach out readily. Plants themselves can bring about cation exchange, since the carbon dioxide from the respiration of root cells reacts with water to form carbonic acid, thus providing H^+ that can release other cations and so make them available for absorption.

Ordinarily, the entry and exit of ions is electrically balanced; that is, the positive charges of the cations are equal to the negative charges of the anions entering during the same time period, although ion exchange makes possible the entry of any specific kind of ion out of proportion to its concentration in the external solution. However, if the cell contains either cations or anions to which the membrane is impermeable, the entering ions may be electrically unbalanced. Thus, if a cell contains more anions than cations that can not pass through the membrane, there will be an excess of cations absorbed against a concentration gradient until the internal solution is electrically balanced. Conversely, if a cell contains an excess of cations, more anions than cations will be absorbed. Organic acids may be involved in such ion exchange. These ion exchanges are represented mathematically by the Donnan equilibrium, further discussion of which can be found elsewhere [for example, page 46 of reference 11].

Mass Flow

As we shall see in Chapter 9, there is a mass flow of water upward through the xylem. Any ions or other solutes present in the water are, of course, carried up along with the water. This mass flow of the solution is generally much more rapid than diffusion of either the water or solutes could possibly be. There is also often a mass flow of water and its solutes through the free space of plant tissues. However, the Casparian strip of the cells of the root endodermis prevents the flow of water and its solutes through the free space of the walls of these endodermal cells, so they must pass through the cytoplasm of the endodermal cells.

Active Transport and Accumulation of Ions

The vacuoles of plant cells and various membrane-bounded organelles such as mitochondria usually contain a far higher concentration of ions of various kinds than does the surrounding solution, yet each kind of ion can continue to enter the vacuoles or organelles against its concentration gradient [6]. The movement of ions against such steep diffusion gradients requires the expenditure of considerable externally supplied energy, just as energy is required to move an object up a hill although it can roll downhill at the expense of its own potential energy. The energy used in ion accumulation is metabolic energy supplied by respiration, and perhaps in some cases by photosynthesis. Many experiments have shown that changes in respiration rates brought about by such factors such as temperature, oxygen concentration, or respiratory poisons result in parallel changes in the rate of ion accumulation by root cells (Figure 7-2) or cells of other organs.

Vacuolar membranes (tonoplasts) or other membranes across which ion transport and accumulation occurs are only slightly permeable to many kinds of ions, and once such ions are inside they cannot leave the vacuole or organelle to any extent by simple diffusion or ion exchange. This raises the question of the importance of ions that have been accumulated in vacuoles in the mineral metabolism of a plant. Perhaps those ions used metabolically have mostly not been accumulated. However, as the cells become senescent their membranes may become more permeable, thus releasing ions that can be translocated to younger and more metabolically active cells.

The first suggestion as to how ions could be actively transported across membranes and so accumulated in high concentrations came from van den Honert in 1937. He proposed a carrier system whereby certain molecules of the membrane would pick up ions at the outer surface of the membrane, ferry

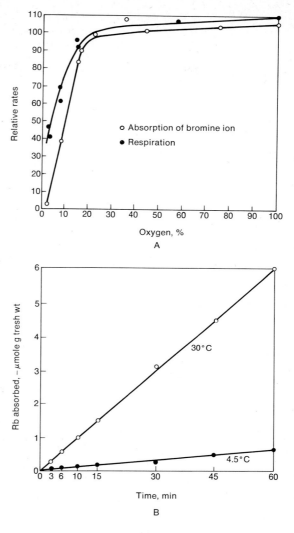

Figure 7-2. **(A)** Effect of oxygen concentration (of oxygen-nitrogen mixtures bubbled through the solutions) on the rates of respiration and absorption of bromine ions in slices of potato tissue. [Data of F. C. Steward, *Protoplasma,* **15:**29 (1933)] **(B)** Influence of temperature on the absorption of rubidium ions (Rb⁺) by excised barley roots. [Data of E. Epstein, D. W. Rains, and W. E. Schmid, *Science* **136:**1051–1052 (1962)] Reproduced by permission. Copyright 1962 by the American Association for the Advancement of Science.

them across the membrane, and release them at the inner surface. Much subsequent experimental evidence supports this carrier concept. The nature of the carriers and the specific ways in which they function are still the subject of intensive investigation.

The rate of active absorption of specific ions increases with their external concentration only up to the point where the process becomes saturated. This result would be expected if a limited number of carrier molecules exist which become saturated with ions in much the same way as substrates may saturate enzymes. The carrier theory is also in accord with the fact that different kinds of ions are actively absorbed at different rates and accumulated in different ratios than those existing in the surrounding solution. This assumes that there are different carrier systems or different binding sites on the carriers specific for different kinds of ions. This assumption is supported by evidence from studies such as those of Epstein and Hagen [5], who found that certain ions such as K^+, Cs^+, and Rb^+ interfere with each other's absorption and are therefore presumably carried on the same binding site, whereas Na^+ and Li^+ do not interfere with the absorption of these three ions and are evidently transported by a different carrier or binding site. SO_4^{2-}, NO_3^-, and PO_4^{3-} are not competitive with one another, but selenate ions (SeO_4^{2-}) interfere with sulfate absorption.

There have been several proposals regarding the nature of the carrier molecules. One theory is that they are enzymes called permeases. Any acceptable theory as to the nature and operation of the carrier system must account for the metabolic energy requirement as well as identify specific carriers that are actually components of membranes. Lundegårdh [10] proposed a transport mechanism involving transfer of anions through a cytochrome system, much as electrons are transferred. This theory has been criticized because cation movement would be nonmetabolic and because energy from ATP is not implicated. It has been shown that DNP (2,4-dinitrophenol) which inhibits ATP production also inhibits ion accumulation.

One of the most plausible theories of ion transport, which overcomes the above objections and seems to be in accord with the known facts, is that of Bennet-Clark [2]. He proposed a cyclic interconversion of choline, acetylcholine, lecithin, and phosphatidic acid that provides a carrier system for both anions and cations (Figure 7-3). The reactions require ATP and the phospholipids involved are known membrane constituents. It is probable that several different carrier systems exist. The system proposed by Bennet-Clark may or may not be one of them, but even so it provides a reasonable suggestion as to how carrier systems could operate.

Although most kinds of ions are accumulated in cells far in excess of their external concentrations, Na^+ is commonly less concentrated in cells than in

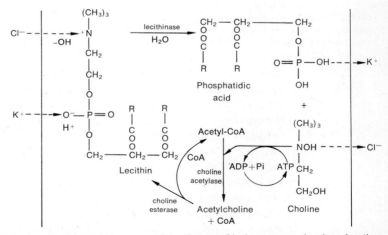

Figure 7-3. The phosphatide cycle proposed by Bennet-Clark as a mechanism for the active transport of ions across a membrane (shaded area). [After T. A. Bennet-Clark, in R. L. Wain and F. Wightman, eds., *Chemistry and Mode of Action of Plant Growth Substances*, Butterworths, London, 1956, pp. 284–291]

the surrounding solution, particularly in marine organisms (Figure 7-1). This involves a carrier system, known as the sodium pump [1], which actively transports Na^+ out of cells rather than into them. The outward transport of Na^+ is thought to be coupled with one kind of inward transport of K^+. The maintenance of relatively low levels of Na^+ in the protoplasts is important from the standpoint of ion antagonism and balance and is probably particularly critical in marine organisms. In plant cells a suitable Na^+ level in the protoplasts could also be achieved by pumping Na^+ into the vacuoles.

Research on the active transport of ions has been restricted principally to the ions of inorganic salts, but it is likely that organic ions such as the anions of organic acids may be transported by carrier systems. There is evidence that amino acids and sugars are often carried from cell to cell by active transport. Sugars are probably transported through membranes in the form of their phosphate esters or perhaps as borate-sugar complexes, although it has been suggested that free sugars may be transported actively. Other organic compounds may also be carried across membranes by active transport. For example, the accumulation of anthocyanins in vacuoles evidently requires an active transport mechanism.

Diffusion of Solutes Into and Out of Cells

The rate at which solutes enter or leave cells or move from one part of a cell to another by simple diffusion is influenced both by their rate of diffusion and the permeability of membranes to them. In general, the rate of diffusion of solute particles is influenced by the same factors as the rate of diffusion of gases. Because their molecular weight is generally considerably higher than that of gases and because the density of the medium through which diffusion occurs is much greater, the rate of diffusion of solute particles is thousands of times less than for undissolved gases. Diffusion is thus an ineffective means of solute transport through any appreciable distance. However, it is effective through short distances, as from one cell to another. An important factor in the establishment of diffusion gradients in plants is the continual production and use of various substances. This results in maintenance of relatively steep gradients and makes the attainment of diffusion equilibra rare.

Movement of Water Into and Out of Plant Cells

The movement of water across cell membranes requires separate consideration since it involves certain complications not applicable to the movement of other substances and quite different from the complications involved in active transport. There have been numerous investigations designed to determine whether water is ever actively absorbed at the expense of metabolic energy, but so far there is no clear evidence of any such water transport and it seems probable that water movement into and out of plant cells is entirely a matter of osmosis.

Osmosis is commonly defined as the diffusion of a solvent through a differentially permeable membrane, the solvent in biological systems commonly being water. There is now considerable doubt as to whether the osmotic movement of water is entirely, or even primarily, a special case of diffusion, since it is frequently much more rapid than would be possible by simple diffusion. It is believed that much of the water movement across differentially permeable membranes may actually be mass flow. However, for the sake of simplicity and because the net results are the same as if only diffusion were involved, we shall continue to refer to the movement of water across differentially permeable membranes as diffusion.

Like the diffusion of gases or other substances the diffusion of water is from a region of its greater free energy or total molecular activity to a region of its lesser free energy or molecular activity, or what may be referred to as a higher to lower **water potential.** In the discussions that follow we shall assume for the sake of simplicity that the systems are isothermal so that the temperature influence can be ignored.

The basic thermodynamic units for water potential are dynes per square centimeter (dyne/cm^2) or ergs per cubic centimeter (erg/cm^3), but in biological applications pressure units are usually used. Although any pressure unit can be used, including millimeters of mercury (mm Hg) or pounds per square inch, bars or atmospheres (atm) are usually employed in osmotic applications because they are of a suitable magnitude. Since the **bar** is a metric unit it will be used here; 1 bar equals 0.987 atm, 750.12 mm Hg, 10^6 dyne/cm^2, or 10^6 erg/cm^3.

At any particular temperature water potential is determined by three principal factors: **osmotic potential, pressure potential,** and **matric potential** [9]. The symbols used for these are respectively Ψ_π, Ψ_p, Ψ_m, and water potential is Ψ. Osmotic potential results from the presence of solute particles in the water and reduces water potential proportional to the concentration of solutes. Pressure potential increases water potential by the amount of any imposed pressure (in plant cells the wall pressure). Matric potential is the decrease in water potential resulting from the imbibition or adsorption of water by such things as cellulose or soil particles. It is not directly involved in osmosis. The water potential in osmotic systems can be represented by the simple equation

$$\Psi = \Psi_p + \Psi_\pi$$

Various other terms have been used for water potential, osmotic potential, and pressure potential [11, p. 62], most commonly diffusion pressure deficit (DPD) for water potential, osmotic pressure (OP) for osmotic potential, and turgor pressure or wall pressure (TP or WP) for pressure potential.

Influence of Solutes on Water Potential

One interesting thing about the influence of solute particles in reducing water potential is that any one kind of particle is just as effective as any other kind regardless of its size, mass, or electric charge. An ion is just as effective as an intact molecule. Thus, a 1 M solution of a salt such as sodium chloride (NaCl), which is 100% dissociated, would theoretically reduce water potential twice as much as a 1 M solution of a nonionized substance such as sucrose. A sucrose molecule is no more effective than a glucose molecule, even though it is about twice as large and heavy. A large colloidial micelle has just the same effect as K^+.

Solute particles have the same nonspecific effect on reduction of the freezing point of water, reduction of vapor pressure, and increase of the boiling point of water. Thus, measurement of any one of these can be used as a means of determining the potential of water in any solution. It is not necessary to know what kinds of solutes are present, and indeed in plant saps there is an immense variety. Freezing-point depression has been most widely used for determining the water potential of plant saps or extracts, although in recent years improved instruments have made possible rapid and accurate measurements of vapor pressure.

The reduction of water potential by the presence of solutes has sometimes been explained on the basis of a reduction in the concentration of the water, but this is inadequate. The concentration of water is reduced only when the solution has a greater volume than the water used in making it. For example, if a liter of water is used in making a 1 molal sucrose solution, the solution volume is 1120.68 ml, so the concentration of water has decreased. However, some solutes such as sodium chloride bring about only a small increase in volume, and so only a slight reduction in water concentration, whereas still others such as magnesium sulfate ($MgSO_4$) have a solution volume less than the volume of the water used so that there has actually been an increase in water concentration. In every case, however, there has been a reduction in water potential.

Influence of Imposed Pressure on Water Potential

If temperature and the presence of solutes were the only factors influencing water potential there would be little point in making a special case of osmosis, the diffusion of water through a differentially permeable membrane. However, pressure potential also plays an extremely important role in the osmotic movement of water. If water were placed in a water-tight cylinder and subjected to a pressure of 5 bars by the piston of a hydraulic press, its potential (free energy or molecular activity) would be increased by 5 bars. If a solution with an Ψ_π of -5 bars were similarly subjected to the 5 bars pressure, it would have the same potential as pure water despite the presence of the solutes.

The important point to be made here is that osmosis results in the development of pressure accompanied by an equal back pressure and that such pressures would not develop in the absence of a differentially permeable membrane. In a simple demonstration of osmosis, consisting of a solution in a tube separated from water by a membrane permeable to water but not permeable to the solute (Figure 7-4), water will diffuse into the solution as long as the water potential in the solution is less than in the pure water. However, it should be noted that a solute concentration equilibrium will *never* be attained because the pure water will always lack solutes regardless of how much water diffuses into the solution. A diffusion equilibrium *will* eventually be attained (provided the tube is long enough) because as the height of the solution increases so does the hydrostatic pressure on it, and this imposed

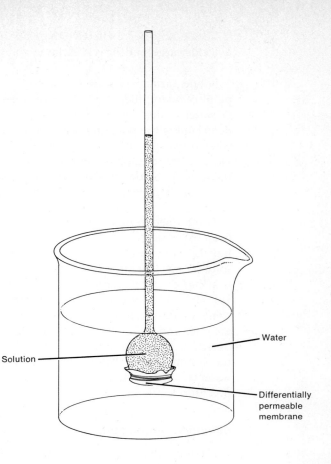

Figure 7-4. In a thistle-tube osmotic system a dynamic equilibrium is attained when the hydrostatic pressure of the water in the tube increases the water potential in the solution enough to counteract the reduction of water potential by the presence of solute particles.

Solution

Water

Differentially permeable membrane

pressure eventually becomes great enough to offset completely the reduction in water potential by the solutes; therefore the diffusion of water ceases.

In an osmotic system such as a cell, or an osmotic cell model made of a differentially permeable membrane (Figure 7-5), where the solution is completely enclosed within a differentially permeable membrane, imposed pressure also plays an important role in influencing the water potential and the attainment of a diffusion equilibrium. If a cell model is filled with a solution with a Ψ_π of -8 bars and placed in water, only a relatively small amount of water will diffuse into the model before an equilibrium is attained. Diffusion will stop even though the solute concentration inside is still far greater, provided that the membrane can stretch only slightly. The cell model will become **turgid** (inflated) as water diffuses into it, and a pressure against the membrane develops. This pressure is known as **turgor pressure** and is counterbalanced by an equal but opposite back pressure of the membrane on the enclosed solution, referred to as **wall pressure.** This imposed pressure increases the water potential in the solution and continues to increase in magnitude as more water diffuses in. When the wall pressure (Ψ_p) reaches 8 bars (in the example being considered) it completely counteracts the reduction in water potential due to the presence of solutes so that the water in the contained solution has a potential equal to that of pure water. As a result, diffusion ceases and a dynamic equilibrium exists even though there is by no means an osmotic potential equilibrium.

It should be noted that at equilibrium the actual wall pressure will be somewhat less than 8 bars because the osmotic potential of the solution was decreased somewhat by the water that diffused into it. The precise osmotic potential of the internal solution at equilibrium, and therefore the precise wall pressure, can be calculated if the initial volume of the solution and the volume of water that diffused in are known. Since the amount of water entering a plant cell before an equilibrium is established is often very small, this complication will be ignored in most of our subsequent considerations.

If the cell containing a solution with a Ψ_π of -8 were placed in a solution with an Ψ_π of -3 rather than in pure water, the water would still diffuse in since its potential is higher outside than inside. In this case however, the turgor pressure at equilibrium would be only 5 bars because the internal and external water potentials will be equal (both -3 bars) when the imposed pressure reaches this magnitude. Note that the external and internal osmotic potentials are not the same at equilibrium, Ψ_π being -8 inside and -3 outside (ignoring volume changes).

If, however, the cell is placed in a solution with an Ψ_π of -12 bars, water will diffuse out rather than in. Since no turgor pressure or wall pressure can develop, the only differential factor influencing internal water potential in this case is the osmotic potential and water will therefore diffuse out until the internal Ψ_π equals the external Ψ_π. If the external volume is large in com-

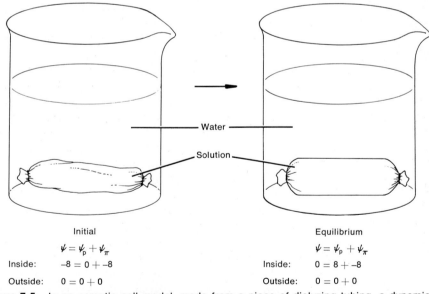

	Initial		Equilibrium	
	$\psi = \psi_p + \psi_\pi$		$\psi = \psi_p + \psi_\pi$	
Inside:	$-8 = 0 + -8$	Inside:	$0 = 8 + -8$
Outside:	$0 = 0 + 0$	Outside:	$0 = 0 + 0$

Figure 7-5. In an osmotic cell model, made from a piece of dialyzing tubing, a dynamic equilibrium is attained when turgor pressure generates enough pressure potential to counteract the reduction in water potential brought about by the presence of solute particles. At the left is the nonturgid cell model just after being placed in the water.

parison with the internal volume of solution, the Ψ_π throughout the system at equilibrium would be only slightly more than -12 bars. In such a case we could ignore the outside increase in volume but not the inside decrease in volume and water concentration.

Osmosis in Plant Cells

The preceding considerations of osmosis may seem to be rather complex, even when volume changes are ignored, but osmosis in plant cells is considerably more complex. The osmotic potential in a cell is in an almost continuous state of flux because solutes are always being produced and used in metabolic processes and are entering and leaving cells continuously. The osmotic potential of the external solution (never pure water under natural conditions) is also likely to be variable, although less variable than the internal potential. There are also very marked changes in the turgor pressure of cells from time to time. The result of these fluctuations in osmotic potential and pressure potential is an essentially continuous diffusion of water either into or out of cells. In cells there is rarely if ever an equilibrium such as develops in a purely physical system, at least for any appreciable period of time.

There are increases in cell volume with increases in turgor pressure, so the volume factor cannot be ignored in precise measurements but, since cell walls have limited extensibility, volume increase is limited. Cell membranes are quite extensible and, without the restraining effect of wall pressure exerted by cell walls, a cell placed in a solution with a high water potential would swell greatly and eventually burst. In rare situations, such as exposure of pollen grains to rain water, even cells surrounded by their walls may burst because the turgor pressure becomes so high that the tensile strength of the wall is exceeded and it ruptures. Cell walls are generally strong enough to resist turgor pressures of 30 bars or more, which are not uncommon in plant cells. The restraining influence of cell walls is one of their most important roles. It should be noted, however, that the imposed pressure (Ψ_p) on the water in a cell is contributed to by the pressure of adjacent cells as well as by the wall pressure of the cell itself.

Despite the fact that there is a continual flux in the relative amount of water in cells, variation in the Ψ_π in a cell is generally within surprisingly narrow limits. These limits vary with the species, the organ in which the cell is located, and the external water concentration in the plant's natural habitat. In general, the Ψ_π of the root cells of herbaceous mesophytes is around -5 bars, whereas in the leaf cells it is likely to be in the neighborhood of -10 to -18 bars, depending on the species. The leaf cells of trees generally have higher solute concentrations; the water potential usually ranges between values of -20 and -35 bars. The Ψ_π of the soil solution is ordinarily only -1 or -2 bars in humid regions but may be much higher in arid regions.

Plants native to the oceans, salt marshes, and saline soils have unusually high solute concentrations in their cells. As a matter of fact these plants could not survive in these environments if they did not, because of the high solute content of the external water. Their Ψ_π values generally range between -50 and -100 bars, but a record low of -200.9 bars was found in the saltbrush

(*Atriplex confertifolia*), a species that grows in saline soils of the arid West. The high solute content of these halophytic plants is partly a result of the high concentration of salts in the surrounding water, but species differences in the capacity for absorbing and accumulating salts are also involved because mesophytic plants are unable to adapt to similar saline conditions. However, there can be an increase in the solute concentration of the root cells of mesophytic plants with more moderate increases in the salt content of the soil solution.

In addition to solute concentration and imposed pressure there is another factor that has an important influence on the water potential in cells: the adsorption or imbibition of water molecules by the micelles of hydrophilic colloids or by ions. These binding forces may be very strong and may bring about a very great reduction in the kinetic energy of the bound water molecules and thus in their water potential. These binding forces are referred to as matric forces, and their contribution to reduction of water potential is represented as Ψ_m. Although these matric forces are present in the protoplasts of all cells, they play a relatively more important role in influencing the water potential of meristematic cells than of the highly vacuolated parenchyma cells, and they are the most important factor in determination of the water potential of cell walls, where considerable water is imbibed by the cellulose and other wall constituents.

The water potential in a cell (Ψ_c) is, then, determined by four sets of factors and can be represented by the following equation.

$$\Psi_c = \Psi_p + \Psi_\pi + \Psi_m + \Psi_t$$

In the sequence listed the factors are wall pressure or other imposed pressure (p), reduction in water potential by solutes (π), matric forces (m), and temperature (t). Ψ_π and Ψ_m are always negative quantities because they can only reduce water potential below that of pure water. Ψ_p is usually a positive quantity because it increases water potential, but under certain conditions negative turgor and wall pressures develop. These will be considered later. An increase in temperature results in an increase in water potential. Ψ_c is a negative quantity except in fully turgid cells immersed in pure water, a situation that does not exist in nature. It should be noted that Ψ_c represents the amount by which the water potential in a cell is less than the potential of pure water.

For use in the quantitative examples of osmosis in parenchyma cells and other cells with large vacuoles that we shall consider later on, the above equation can be simplified by elimination of two of the factors without serious loss of accuracy. Temperature can be deleted if, as is usually the case, the cell and the external solution (or an adjacent cell) are at the same temperature. The matric factor can also be ignored, since we will be concerned primarily with the gradient between the external solution and the cell sap of the vacuole. Also, Ψ_π and Ψ_m tend to come into equilibrium with one another. Thus, the equation becomes

$$\Psi_c = \Psi_p + \Psi_\pi$$

The water potential of the external solution can be influenced by the same factors as the water potential in a cell, but usually Ψ_π is the only one that needs to be considered because the external solution is generally not subjected to any imposed pressure (normal atmospheric pressure is not a differential factor). However in the case of soils, when the entire water content of the soil is considered rather than just the salt concentration of the soil solution, the matric forces that bind water to the soil particles and hold it in capillary spaces cannot be ignored. The water potential of soils will be considered in Chapter 9 in connection with the absorption of water by roots.

Some Quantitative Examples of Osmosis in Plant Cells

A few quantitative examples should help clarify the osmotic movement of water into plant cells. In these examples we will ignore such complications as the continual flux in solute content of the cells, matric forces, and in general volume changes so as to simplify the situation as much as possible. Furthermore, we will begin with the simplest possible system: a unicellular alga immersed in a solution that is large in volume compared with the cell.

If the algal cell has a Ψ_π of -20 bars and no turgor pressure and is placed in a solution with a Ψ_π of -12 bars, water will diffuse into the cell because the water potential is greater outside than inside. Diffusion will continue until the turgor pressure and wall pressure reach 8 bars, thus raising the internal water potential to the same value as external potential. The diffusion of water now ceases and a dynamic equilibrium exists.

$$\Psi_c \quad = \quad \Psi_p \quad + \quad \Psi_\pi$$
$$-12 \text{ bars} = 8 \text{ bars} + -20 \text{ bars}$$

If, after attainment of equilibrium, this cell is transferred to a solution with a Ψ_π of -10 bars, water will again diffuse into the cell because the external water potential is greater than the internal. A new equilibrium will be reached when the Ψ_c becomes -10 bars, the same as the potential of water in the surrounding solution, as a result of the increase of wall pressure to 10 bars.

$$-10 \text{ bars} = 10 \text{ bars} + -20 \text{ bars}$$

Suppose this cell is now transferred to a solution with a Ψ_π of -16 bars. Water will diffuse out of the cell, because the water potential is higher inside than outside, and it will continue to diffuse out until the water potential is the same inside as outside, that is, -16 bars. Thus, at equilibrium the pressure potential will have dropped to 4 bars.

$$-16 \text{ bars} = 4 \text{ bars} + -20 \text{ bars}$$

If we place the cell we have been considering in a solution with a Ψ_π of -22 bars, water will diffuse out of the cell until the water potential inside is -22 bars; as a result of this outward diffusion of water all turgor pressure will be lost and the situation at equilibrium will be

$$-22 \text{ bars} = 0 \text{ bars} + -22 \text{ bars}$$

Water diffused out of the cell until its internal potential became as low as its outside potential. As a result of this the protoplast contracts and shrinks away from the cell wall and the wall pressure becomes zero, even in the initial stages when the separation of the membrane and wall has occurred only at certain places. This process is called **plasmolysis.**

In plasmolysis negative turgor or wall pressure does not develop because there is no inward pull on the wall. If the cell under consideration were placed in a solution of still lower water concentration such as -26 bars, additional water would diffuse out of the vacuole until the internal water potential became -26 bars; the protoplast would shrink still more and probably become a small sphere entirely separated from the wall, but the wall pressure would still be zero. Cells can recover from moderate plasmolysis without apparent injury if placed in water or a solution with a higher water potential than the original Ψ_π of the cell. However, plasmolyzed cells often die as a result of adverse effects of dehydration on metabolic processes and protoplasmic organization.

In none of the above examples has the cell under consideration attained its full turgidity. This can be achieved only by placing it in distilled water. If this is done the water potential inside the cell becomes 0 at equilibrium, and Ψ_p and Ψ_π are of equal magnitude.

$$0 \text{ bars} = 20 \text{ bars} + -20 \text{ bars}$$

It should be noted that the Ψ_π of a cell sets a limit to the maximum turgor pressure that can develop in a cell. If the cell had an Ψ_π of -40 bars then its turgor pressure when immersed in pure water would be 40 bars.

We will now use several examples from multicellular plants to illustrate points that cannot be made with a unicellular alga. The first of these examples involves the diffusion of water from a cell to an adjacent cell of similar size. To simplify the example we will assume that both cells have the same volume. We shall refer to the cells as A and B (Figure 7-6) and assume that they have the following osmotic quantities initially.

$$\text{Cell A: } -13 \text{ bars} = 12 \text{ bars} + -25 \text{ bars}$$
$$\text{Cell B: } -17 \text{ bars} = 4 \text{ bars} + -21 \text{ bars}$$

Water will diffuse from cell A to cell B because the water potential is higher in cell A, even though the osmotic potential is higher in cell B. As is always the case, a dynamic equilibrium will be attained when the water potential is the same throughout the system (in this case the two cells). However, since no relatively large volume of external solution is involved and since the cells

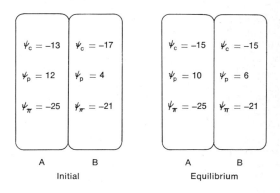

Figure 7-6. When water diffuses from a cell to an adjacent cell of the same volume, the equilibrium cell water potential will be the average of the initial cell water potentials of the two cells.

have equal volumes, the equilibrium water potential will be the mean of the two initial water potentials. Thus, at equilibrium the osmotic quantities of the two cells will be as follows

$$\text{Cell A: } -15 \text{ bars} = 10 \text{ bars} + -25 \text{ bars}$$
$$\text{Cell B: } -15 \text{ bars} = 6 \text{ bars} + -21 \text{ bars}$$

Note that the equilibrium has been reached because of the decrease in pressure potential in cell A and its increase in cell B, a result of the diffusion of water from A to B.

Our final example will deal with a cell, such as a mesophyll cell of a leaf, that is losing water by evaporation into the air of the intercellular spaces. If the rate of transpiration is high, such a cell may become greatly desiccated, but the cell does not plasmolyze as it would if it were losing water into a solution. Instead, as water continues to diffuse out of the cell, the cell membrane adheres to the cell wall and pulls it inward. Thus a negative wall pressure develops and the water in the cell is in a state of tension, or negative pressure potential as in the following example.

$$-25 \text{ bars} = -3 \text{ bars} + -22 \text{ bars}$$

It will be noted that Ψ_c is less than Ψ_π, a situation that can exist only when the turgor pressure and wall pressure are negative. Negative turgor pressures are quite common in the cells of leaves in the afternoons of hot summer days, when the rate of transpiration exceeds the rate of water absorption, or under drought conditions. There is an appreciable decrease in cell volume, resulting in wilting of the leaves. If the low water potential is prolonged, the cells may

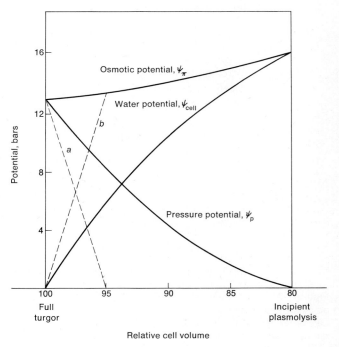

Figure 7-7. Diagram showing the changes in the osmotic values of a cell as it loses water and goes from full turgidity to incipient plasmolysis (or the reverse). The solid lines are for a highly extensible cell and the dotted lines for a slightly extensible cell (a, pressure potential; b, cell water potential; osmotic potential curve same for both). [After P. J. Kramer, *Plant and Soil Water Relations: a Modern Synthesis,* McGraw-Hill Book Co., New York, 1969]

be killed by desiccation; however, cells can generally survive through even several days of negative pressure potentials with no apparent permanent injury, although during such a period most of the metabolic processes occur at a very low rate.

There should probably be another reminder that in the above examples we have been dealing with simplified situations and that actually in plant cells there ara various complications such as changes in solute concentration, rare attainment of dynamic equilibria, and changes in cell volume that cannot be ignored in precise quantitative considerations. The graphic representation in Figure 7-7 of the changes in the osmotic quantities of cells as water diffuses into them and they become more turgid includes changes in cell volume.

Imbibition

The imbibition of water by cellulose and other cell wall constituents and by the hydrophilic colloids of the protoplast deserves further consideration. In imbibition the molecules of a liquid or gas (almost always water or water vapor in biological systems) diffuse into a solid substance (or semisolid substances such as gelatin) causing it to swell. There must be strong intermolecular binding forces between the molecules of the imbibant and the substance being imbibed. Such binding forces exist between cellulose molecules and water molecules, but not between cellulose molecules and ether molecules, so cellulose can imbibe water but not ether. However, rubber can imbibe ether but not water. The matric forces that bind imbibed water molecules may be very great, the water potential of the imbibed water often being as low as −1000 bars, much less than in any solution naturally occurring in plant cells. The first water imbibed by a dry imbibant is held by greater matric forces than that imbibed later.

The great reduction in the potential of the imbibed water is a result of the greatly decreased kinetic energy or rate of movement of the water molecules, and the lost energy appears as heat. The release of heat always occurs during imbibition. The swelling of the imbibant is a result of the interposition of the water molecules between the molecules of the imbibant. Although the imbibant swells, sometimes to a very great extent (Figure 7-8), the total volume of the system (imbibant + substance imbibed) always decreases (Figure 7-9). This is a result of the fact that the bound molecules are highly compacted, that is, they are closer to one another than in the free liquid. Another point that should be noted is that imbibition, unlike osmosis, does not require the presence of a differentially permeable membrane. For example, wood readily imbibes water; the swelling and sticking of doors and windows in periods of high humidity often results from the imbibition of water vapor from the air.

If an imbibant is confined in some way it will develop very large pressure potentials. For example, if a glass jar is filled with pea seeds and then with water and the lid is screwed on, the swelling seeds will exert a pressure of several hundred bars against the wall of the jar. The glass generally does not have enough tensile strength to exert an equivalent back pressure, and the jar explodes. If the jar were made of strong metal the following situation might exist at equilibrium.

$$\Psi_i \;=\; \Psi_p \;+\; \Psi_m$$
$$0 \text{ bars} = 800 \text{ bars} + -800 \text{ bars}$$

The water potential of the imbibed water (Ψ_i) would be 0 bars since it would be in equilibrium with pure water.

If a salt solution with a Ψ_π of -90 bars had been used instead of water, the Ψ_i at equilibrium would have been the same and the pressure developed would be 710 bars.

$$-90 \text{ bars} = 710 \text{ bars} + -800 \text{ bars}$$

If the imbibant were placed in water but not confined, no pressure would develop and the water potential of the imbibed water would equal the matric forces holding the water.

$$-800 \text{ bars} = 0 \text{ bars} + -800 \text{ bars}$$

The effect of adding solutes to the surrounding water is to reduce the swelling of an unconfined imbibant below the maximum obtained when it is in pure water because less water is imbibed at the time equilibrium is attained. For example, salt-free kelp stipe swells much more when placed in distilled water than when placed in sea water.

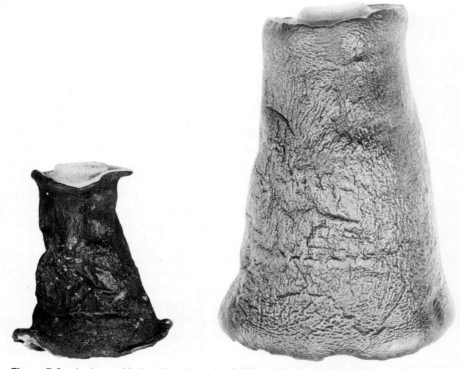

Figure 7-8. A piece of kelp stipe when dry (left) and the same piece after imbibing water and attaining a dynamic equilibrium (right). Its weight increased from 6.0 g to 54.9 g (915%). [Courtesy of E. V. Miller]

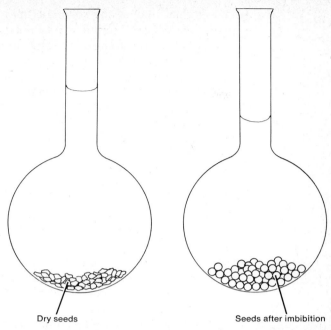

Figure 7-9. Even though there is a marked increase in the volume of the imbibant (here pea seeds) as it imbibes water, the total volume of the system decreases.

Dry seeds   Seeds after imbibition

Most of the initial water absorption by dry seeds is by imbibition rather than by osmosis. The walls of plant cells are always saturated with water, much of it being held by the high matric forces of imbibition. The water potential in the walls tends toward an equilibrium with that in the protoplast and vacuoles, so evaporation of water from walls results in the diffusion of water from the cell into the walls. Because of the magnitude of the matric forces in the walls, they still contain considerable imbibed water when the cell itself is quite thoroughly desiccated. Only complete desiccation of a plant tissue, as by drying in an oven at relatively high temperatures, will remove all the imbibed water from the cell walls.

Significance of Osmosis and Imbibition in Plants

Plants can obtain water from their environment only when the water potential in their cells is lower than that of the soil water or the body of water in which they may be living. The relatively low water potentials of protoplasts and the cell sap of the vacuoles results from the relatively high solute concentration, derived partly from accumulated ions of salts and partly from metabolic products such as sugars. Only plants with unusually low osmotic potentials are able to survive and thrive when the solute concentration in the surrounding water is high, as in the oceans, salt marshes, and alkaline soils. Water potential gradients are also important in other situations, such as the germination of pollen grains. If germination is to occur, the water potential in the pollen must be lower than that of the stigmatic fluid so that water will diffuse in and generate the essential turgor pressure. However, too high an external water potential, as in rain drops, may result in such high turgor pressure that the cell wall is ruptured. Movement of water from cell to cell requires a gradient of water potentials and, as we shall see later, even the mass flow of water through the xylem results from water potential gradients. Transpiration occurs because the water potential in cells is much higher than that of the atmospheric water vapor, and the difference in water potentials is the principal factor causing the continuing movement of water from the soil into and through the plant. Despite the very low matric potential of cell wall components, the water potential in walls tends to come to an equilibrium with that in the proto-plasts. These aspects of plant water relations will be considered more fully in Chapter 9.

The turgor pressure of plant cells developed by osmosis, and dependent on the presence of differentially permeable membranes, plays an extremely important role in plants. Turgid cells are essential for the support of nonwoody tissues, as evidenced by the wilting that occurs when cells lose their turgidity. Cell turgidity and hydration are essential for the maintenance of adequate rates of photosynthesis and other metabolic processes as well as for growth. A number of plant movements such as the opening and closing of stomata, the "sleep movements" of leaves, and the rapid leaf movements of sensitive plants and the Venus's fly trap involve changes in cell turgidity.

Cell walls have an important function in the water relations of plants. Since the walls of mature cells have great tensile strength and are only slightly extensible, they make possible the development high pressure potentials in cells and so play a major role in determining the water potential of cells. The walls thus prevent excessive diffusion of water into cells that would otherwise result in the ballooning and probable disruption of the cells. The contractile vacuoles of protozoa achieve the same result in quite a different way.

References

[1] Baker, P. F. "The sodium pump," *Endeavour,* **25:**166–172 (1966).
[2] Bennett-Clark, T. A. "Salt accumulation and the mode of action of auxin: a preliminary hypothesis," in R. L. Wain and F. Wightman (eds.), *Chemistry and Mode of Action of Plant Growth Substances,* Butterworths, London, 1956, pp. 284–291.
[3] Branton, D. "Membrane structure," *Ann. Rev. Plant Physiol.,* **20:**209–233 (1969).
[4] Collander, R. "Permeability of plant cells," *Ann. Rev. Plant Physiol.,* **8:**335–348 (1957).
[5] Epstein, E., and C. E. Hagen. "A kinetic study of the absorption of alkali cations by barley roots," *Plant Physiol.,* **27:**457 (1952).
[6] Hendricks, S. B. "Salt transport across cell membranes," *Am. Sci.,* **52:**306–333 (1964).
[7] Kramer, P. J. "Outer space in plants," *Science,* **125:**633–635 (1957).

[8] Kramer, P. J. *Plant and Soil Water Relationships: A Modern Synthesis,* McGraw-Hill, New York, 1969.

[9] Kramer, P. J., E. B. Knipling, and L. N. Miller. "Terminology of cell-water relations," *Science,* **153:**889–890 (1966).

[10] Lundegårdh, H. "Anion respiration: the experimental basis of a theory of absorption, transport and exudation of electrolytes by living cells and tissues," *Symp. Soc. Exp. Biol.,* **8:**262 (1954).

[11] Salisbury, F. B., and C. Ross. *Plant Physiology,* Wadsworth Publishing Co., Belmont, Calif., 1969.

[12] Slatyer, R. O. *Plant-Water Relationships,* Academic Press, New York, 1967.

[13] Stadelmann, E. J. "Permeability of the plant cell," *Ann. Rev. Plant. Physiol.,* **20:**585–602 (1969).

[14] Wheeler, H., and P. Hanchey. "Pinocytosis and membrane dilation in uranyl-treated plant roots, *Science,* **171:**68–71 (1971).

The Vascular System 8

In the preceding chapters the considerations of plant structure and function have dealt essentially with plant cells, even though reference has been made from time to time to tissues, organs, or plants as a whole. We have discussed the structure of plant cells, the metabolic processes they carry on, and the movement of substances in and out of cells. Yet it should be evident that in any multicellular organism there are levels of organization beyond the cell, and that somehow or another the individual cells making up the organism are coordinated so that a suitable structural organization develops and a smoothly functioning organism results. In the remaining chapters of this volume attention will be centered on those aspects of plant structure and function that are meaningful only at supracellular levels of organization: the orderly transport of substances from one part of a plant to another, often over considerable distances; coordination of the metabolic activities of plants by nutrients and hormones; coordination of growth and development; and the reproduction of the organism. However, it should be noted that even these levels of structure and function still involve cells.

Although higher plants are not as specialized as higher animals, either structurally or functionally, and do not possess a nervous system, which in animals plays such an important role, there is intercommunication and coordination among the various cells, tissues, and organs of multicellular plants. This is brought about primarily by chemical means: the transport of substances such as foods, hormones, water, and salts from one part of the plant to another. In the nonvascular plants—algae, fungi, and bryophytes—such transport is essentially from one cell to another by the processes discussed in the last chapter; however, diffusion and the other processes involved in cellular transport are far too slow for the effective movement of water and solutes through distances of more than a decimeter or so. There must be some system of ducts through which water and solutions can flow from one part to another in a plant of any appreciable size, and in vascular plants the xylem and phloem of the vascular system play this role.

It is not just a matter of chance that all the land plants of any great size are vascular plants. Without ducts through which water can flow, the leaves of a plant even a few feet high, not to mention trees that may attain a height of almost 150 meters (m), could not possibly receive water as rapidly as it is lost by transpiration. Similarly, food could not be delivered from the leaves to the roots as rapidly as it is used. Of course, some of the marine plants, particularly brown algae such as kelps, attain a considerable size and may be as much as 60 m long. However, no cell in the thallus is more than a few centimeters from the surrounding sea water, and the entire thallus is generally photosynthetic so that there is no problem of long distance water or food transport. Each portion of the thallus is much more autonomous than the roots, stems, and leaves of vascular plants. Nevertheless, it is interesting to note that some of the larger species of brown algae have specialized cells similar to the sieve tubes of vascular plants that may function in solute transport [3].

This chapter is devoted to a consideration of the structural organization of the vascular tissues. The transport of water and solutes through the vascular system will be discussed in the next two chapters. Since our primary concern is relating the structure of the vascular tissues to their transport functions, we will not consider the diverse vascular systems of the lower vascular plants and their phylogenetic significance. This is a fascinating matter in its own right, and is considered in other textbooks [2, 3].

Although the vascular tissues function primarily as pathways of water and solute transport, the xylem of woody plants also plays an important role in providing mechanical support of the stems. From an economic standpoint, xylem is very important as the source of lumber and other wood products and as the source of most paper. The phloem fibers of a number of species are used in making fabrics and ropes.

The Vascular System of Stems

The structure and arrangement of the vascular system in the stems of various major groups of vascular plants is diverse enough to make it desirable to select a few representative examples. Herbaceous monocotyledons, herbaceous dicotyledons, woody dicotyledons, and conifers will serve this purpose. Various other groups of vascular plants, such as the few species of woody monocotyledons, the cycads, the ferns, the horsetails, and the clubmosses, have vascular systems with characteristic differences [4], but consideration of these is beyond the scope of this book and not essential for relating the structure of the vascular system to its functions.

Herbaceous Monocotyledons

As examples of herbaceous monocotyledon stems we shall select two grasses, wheat and corn, which are typical of grass stems but not necessarily of other herbaceous monocotyledons. The vascular system of wheat stems is similar to

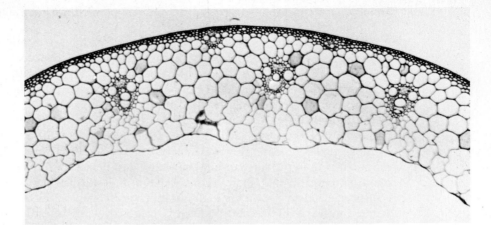

Figure 8-1. Cross section of a segment of a wheat stem cut through an internode, showing the vascular bundles. Note the large central hollow space of the stem (below). [Courtesy of the Carolina Biological Supply Company]

those of many other grasses, including rye, oats, rice, and barley. In cross sections of the stems, the vascular bundles are found in two circles (Figure 8-1), the inner circle having larger bundles than the outer one. The outer circle of bundles is generally embedded in the continuous cylinder of sclerenchyma located inside the epidermis, whereas the inner circle of bundles is surrounded by parenchyma cells. At a node the vascular bundles are relatively large and close together and interconnect. Some of these bundles enter the sheath of the leaf attached at the node whereas others continue on up the stem. In wheat and many other grasses, but not all, the stem is hollow in the internodes, but not at the nodes. What we refer to as circles of bundles in cross sections of a stem are, of course, actually cylinders of bundles when the stem is considered from a three-dimensional standpoint. The outer part of each bundle is the phloem, with its sieve-tube elements and companion cells, and the inner part is the xylem, with its vessels and parenchyma (Figure 8-1). Each bundle is surrounded by a sheath of sclerenchyma cells. As in other monocotyledons, there is no cambium between the xylem and phloem, so there can be no production of secondary xylem and phloem.

The vascular system of corn stems is typical of some other grasses that have thick stems, such as sugarcane and sorghum. The stems of most of these species do not have hollow internodes, and the vascular bundles are embedded in the ground parenchyma tissue and are distributed throughout most of the area of the stem (Figure 8-2). The ground parenchyma of a corn stem in which the vascular bundles are embedded is sometimes referred to as the pith, but the term is not applicable since the bundles extend all the way to the center of the stem. In the internodes the bundles are essentially separate and parallel, but at the nodes they anastomose and are connected by transverse bundles. The cross-sectional view of an individual vascular bundle of corn (Figure 8-3) shows that it has the same general arrangement of tissues as the vascular bundle of wheat. Note the clearly delimited phloem with its sieve tubes and companion

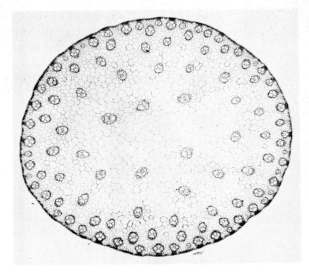

Figure 8-2. Cross section of a young corn stem cut through an internode. Note the solid stem and the vascular bundles scattered through the ground parenchyma. [Courtesy of Ripon Microslides]

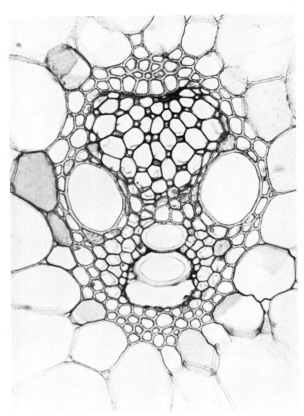

Figure 8-3. Cross section of a corn vascular bundle, showing its cellular organization. Note the sieve tubes and companion cells of the phloem and the large vessels in the xylem. The space (lacuna) under the bottom vessel was occupied by a vessel that disintegrated. [Courtesy of Ripon Microslides]

cells and the arrangement of the large vessels of the xylem. Except in some of the smaller bundles near the outside of the stem, this same arrangement of vessels occurs in all vascular bundles of corn, giving them something of the appearance of a mask in cross section. As in wheat, leaf sheaths encircle corn stems but these are not shown in the cross-sectional views.

Herbaceous Dicotyledons

In most species of herbaceous dicotyledons there is only a single cylinder of bundles (Figure 8-4), although in the region of a node bundles may be seen outside the circle in cross sections. These are the **leaf traces** that extend from the vascular cylinder into the petiole of a leaf. The single cylinder of bundles in herbaceous dicotyledon stems clearly delimits the external cortex from the internal pith (Figure 8-5). In general, the pith has a considerably greater diameter than the cortex and is composed of parenchyma cells, whereas the cortex is composed primarily of chlorenchyma cells and sometimes includes sclerenchyma or collenchyma. The bundles of the vascular cylinder often anastomose in the internodes as well as at the nodes, resulting in a vascular cylinder that is something of a network.

There is considerable variation in the detailed structure of the vascular system from species to species of herbaceous dicotyledons. One of the most important differences is in regard to the presence or absence of a cambium

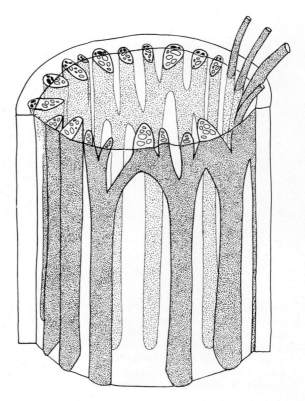

Figure 8-4. Diagrammatic three-dimensional drawing of the vascular cylinder of a herbaceous dicotyledon stem showing anastomosis of the bundles and (upper right) leaf traces leaving the vascular cylinder. In the front segment of the stem the epidermis, cortex, and phloem are shown as being removed.

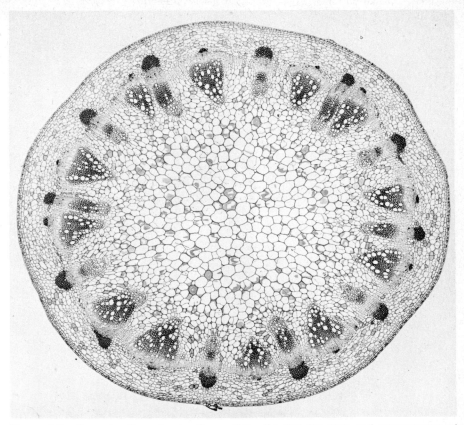

Figure 8-5. Cross section of a sunflower stem showing the type and arrangement of vascular bundles characteristic of many herbaceous dicotyledons. Large vessels and the smaller xylem fibers are evident in the wedge-shaped xylem of the bundles. The phloem shows as a lighter area outside the xylem in each bundle, but this magnification is not high enough to show the individual sieve tubes, companion cells, or other cells clearly. The dark caps of the bundles outside the phloem are composed of cortical fibers. Note the pith rays between the bundles. [Courtesy of the Carolina Biological Supply Company]

between the xylem and phloem. Some species such as buttercups (Figure 8-6) and garden beans lack a cambium, as do the herbaceous monocotyledons, and cannot produce secondary xylem or phloem. Others such as geranium, tobacco, and sunflower have a cambium that produces secondary xylem and phloem, sometimes in substantial quantity. Such stems increase in diameter as the season progresses and may become quite woody. In some species such as geranium an interfascicular cambium develops between the vascular bundles, resulting in the production of a continuous cylinder of secondary xylem and phloem. Other species differences include variations in the size, number, and spacing of the vascular bundles, and the presence or absence of caps of sclerenchyma tissue on the outer side of the bundles. The phloem is always external to the xylem, but some species such as cucumber and tomato have phloem internal to the xylem as well as external to it. Such vascular bundles are referred

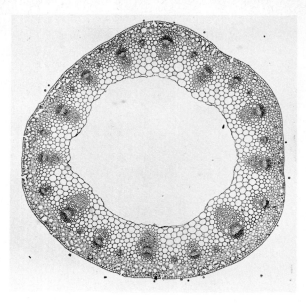

Figure 8-6. Cross section of a stem of Ranunculus (buttercup), another herbaceous dicotyledon. Note the alternation of large and small bundles and the hollow center. [Courtesy of Triarch Products]

to as **bicollateral,** in contrast with the more usual **collateral** type that has phloem only toward the outside of the xylem.

Woody Dicotyledons

The woody dicotyledons include many different species of trees and shrubs as well as some vines, all of them perennials. Despite the marked difference in size and appearance between herbaceous stems and the older woody stems, for example, between a sunflower stem and the trunk of an oak tree, there is really no sharp line of demarcation between woody dicotyledon stems and the stems of those herbaceous dicotyledons with cambium. We have noted that some herbaceous dicotyledons produce considerable secondary xylem and phloem and become semiwoody. Some species such as the castor bean, which are considered to be herbaceous annuals in temperate regions, are actually woody perennials in tropical regions where they can survive through the winter. In some cases a single genus may include both woody and herbaceous species, although the latter have a cambium and produce at least some secondary xylem and phloem. The basic difference is really just a matter of how long the plant lives and continues to produce secondary xylem and phloem. The continuing growth of woody stems results in increasing stem diameter, the splitting of the tissues external to the cambium, and the development of corky bark.

The tissues of young woody stems (Figure 8-7) are essentially the same as those of herbaceous dicotyledon stems, although the vascular bundles are generally closer together with narrower rays between them. Also, woody stems usually produce somewhat more secondary xylem and phloem the first year than herbaceous stems with secondary growth.

SECONDARY GROWTH. The cambium consists of a cylinder of elongated meristematic cells and ray initials that is only one cell thick, although when its cells are dividing the undifferentiated cells just produced may make the cambium appear to be several cells thick. The cells cut off toward the inside differentiate into the tracheids, vessel elements, fibers, or parenchyma of the xylem, whereas those cut off toward the outside differentiate into the sieve-tube elements, companion cells, fibers, and parenchyma of the phloem. More cells are cut off toward the inside than the outside, so the annual production of secondary xylem is greater than that of secondary phloem. While the fascicular cambium is producing xylem and phloem, the interfascicular cambium between the bundles is producing parenchyma cells, thus extending the rays. As the xylem continues to grow in diameter, the radial rays become farther and farther apart. Some of the cambial cells between them then become ray initials, thus forming new rays. As the stem grows in diameter the addition of new rays each year results in maintenance of about the same volume of vascular tissue between the rays.

In temperate regions there is a marked seasonal periodicity of both secondary growth from the cambium and primary growth from the apical meristem of the buds. In general, growth is very rapid for a few weeks after it is initiated in the spring and then slows down, although fluctuations in the growth rate during the growing season are not uncommon (Figure 8-8). The apical growth generally ceases during the summer, and the secondary growth slows down

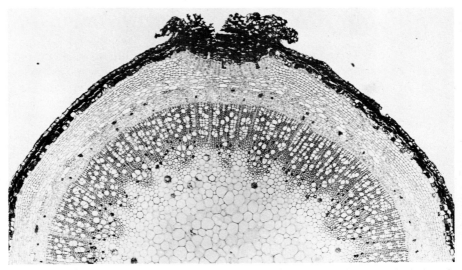

Figure 8-7. Cross section of a one-year-old stem of *Sambucus canadensis* (elderberry). There has already been considerable production of secondary xylem and phloem from the cambium, and cork formation has begun in the outer part of the cortex. Note the lenticel at the top. The pith rays in the secondary xylem and phloem are narrow compared with the wide pith rays in herbaceous dicotyledon stems. However, note that the bundles of primary xylem are widely separated by pith rays, as in herbaceous stems. [Courtesy of Carolina Biological Supply Company]

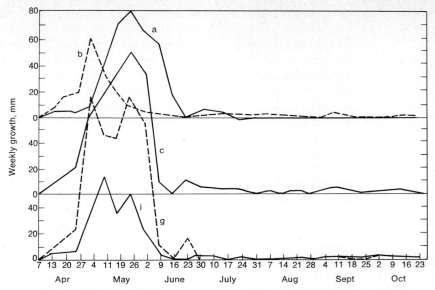

Figure 8-8. Types of seasonal growth periodicity (in height) of red, white, jack, and Scotch Pine trees 3–8 years old in Indiana. Type c was the most common and type g was rare. Note that growth was restricted largely to spring and early summer. [Data of R. Friesner, *Butler Univ. Bot. Stud.,* **5:**145 (1942)]

considerably and finally stops entirely in the autumn when the plants become dormant.

ANNUAL RINGS. The vessels and fibers produced during rapid spring growth are large in diameter and have relatively thin walls, but as the season progresses the cells become smaller and smaller and have thicker walls. The result of these differences in diameter is the formation of clearly defined annual rings (Figure 8-9). As the term implies, one ring is generally produced each year and the ages of temperate zone trees can be determined by counting their annual rings. However, under certain conditions, such as defoliation of a tree by a late freeze or by a massive insect infestation, growth may be interrupted and two rings may be formed in a year. Such rings are unusually narrow and can generally be detected. It should be noted that if the age of a tree is to be determined precisely the rings must be counted at the base of the trunk, at a point no higher than the height of the tree when it was 1 year old. If there are 50 annual rings at this level there will be only 49 in the region just above that corresponds with the growth in height of the tree during its second year, and eventually there will be only one annual ring in the portion of the stem produced during the current year. Thus, the secondary xylem of a woody stem actually consists of a series of inverted concentric cones, the oldest and most central one being the shortest and the youngest and outermost one being the longest (Figure 8-10).

The amount of secondary xylem produced varies from year to year because of the influence of environmental factors such as water availability, tempera-

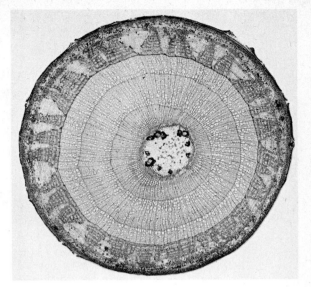

Figure 8-9. Cross section of a Tilia (linden or basswood) stem in its third year of growth. Note the distinct annual rings of xylem and the splitting of the phloem into segments, a result of the increased diameter of the stem. Ray cells have divided and filled in the gaps. [Courtesy of Carolina Biological Supply Company]

ture, and light intensity. Thus, variations in width of the annual rings can provide considerable information on the climatic conditions during the life of the tree and on the local environment of the tree. For example, a sudden marked increase in the width of the annual rings indicates a release from shading by larger trees. Wider rings on one side of the trunk than the other indicates that there was unequal shading of the tree.

CELLULAR COMPOSITION OF SECONDARY XYLEM. The arrangement of the cells in the xylem can be visualized best if sections taken in three different planes

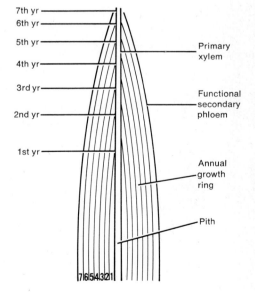

Figure 8-10. Diagrammatic representation of a median longitudinal section of a 7-year-old woody stem, showing the position of the annual rings of xylem in relation to one another and the other vascular tissues. The labels at the left show the length of the stem at the end of each growing season. For a more normal proportion the vertical axis should be elongated. [After V. A. Greulach and J. E. Adams, *Plants: An Introduction to Modern Botany,* John Wiley & Sons, Inc., New York, 1967]

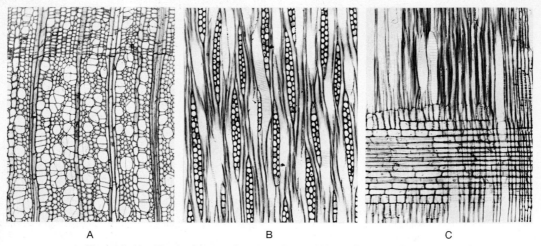

A   B   C

Figure 8-11. Photomicrographs of sections of Magnolia wood (secondary xylem) in three planes. **(A)** Cross section. **(B)** Tangential longitudinal section. **(C)** Radial longitudinal section. [Courtesy of Ripon Microslides]

are examined under a microscope (Figures 8-11 and 8-12). The planes are a cross section and two longitudinal sections, one in a radial plane and the other in a tangential plane. Since the sections shown are largely self-explanatory they will not be described in detail, but a few points must be noted. The longitudinal sections illustrate the fact that the rays do not extend far in a vertical direction and are actually ribbons of parenchyma cells, generally lens-shaped in tangential section. Thus, the vascular cylinder is anastomosing, with interconnecting

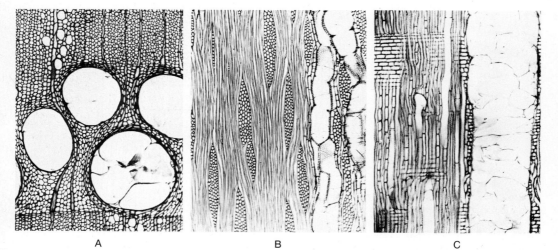

A   B   C

Figure 8-12. Photomicrographs of sections of oak wood in three planes. **(A)** Cross section. **(B)** Tangential longitudinal section. **(C)** Radial longitudinal section. Note tyloses in one vessel in the cross section and in the vessel in the radial section. [Courtesy of Ripon Microslides]

bundles rather than distinctly separated parallel bundles. Note the large diameter of the vessels compared with the other cells. The vessels in secondary xylem are pitted, reticulate, or scalariform (Figure 2-16).

SAPWOOD AND HEARTWOOD. The transport of water is limited to the younger annual cylinders of xylem, which are referred to as the sapwood. The older cylinders of xylem, which no longer conduct water, are referred to as the heartwood. The oldest layer of sapwood generally becomes nonfunctional each year, so the number of annual rings in the sapwood tends to remain more or less constant. The loss of capacity for water conduction by the heartwood is brought about by, or at least accompanied by, a number of changes in the xylem. Sometimes the xylem parenchyma cells grow into the vessels through their pits, thus blocking the vessels (Figure 8-12). These invaginations are called **tyloses** [4]. The walls of the vessels and other cells generally become impregnated with various organic substances such as tannins, oils, gums, and resins. The parenchyma cells lose most of their starch and other accumulated food, and finally the parenchyma cells of the xylem and the rays may die. The primary function of the heartwood is mechanical support, and as a matter of fact the changes that occur may contribute toward increased compressional strength and increased resistance to decay of the heartwood. However, the heartwood does at times decay, resulting in a hollow trunk. The transport of water is not appreciably hampered by a hollow trunk, but the mechanical strength of the trunk is greatly reduced.

SECONDARY PHLOEM. Since the cambium produces new cylinders of both xylem and phloem each year, the youngest layers are the nearest to the cambium in both cases. It should be noted that the oldest phloem is the outermost layer whereas the oldest xylem is the innermost one. The annual additions to the phloem do not form distinct rings, as in the case of the xylem. Only the last few annual layers of the phloem (or often only the last one) are functional in transport.

EFFECTS OF SECONDARY GROWTH. As the diameter of the xylem increases year after year, the tissues outside the cambium are subjected to considerable stress which results in the splitting of the older tissues (Figure 8-9). In the cambium itself there are occasional cell divisions in a tangential direction; therefore the cylinder of cambium remains intact as it continues to lay down xylem and increases in circumference. However, there is insufficient tangential cell division in the tissues outside the cambium to keep pace with the increase in diameter. The stress developed is relieved only when the periderm and phloem split. Even by the end of the first year of growth of a woody stem there is generally some splitting of the epidermis and cortex, and they are usually quite completely disrupted and sloughed off after only a few years of growth.

Wherever a split occurs, or even before, a patch of parenchyma cells resumes cell division and becomes converted into cork cambium **(phellogen).** Division of the cells of the cork cambium results in the production of layers of cork cells toward the outside and a smaller number of parenchyma cells (the phelloderm) toward the inside. Unlike the vascular cambium, the cork cambium does not form a continuous cylinder nor does it continue its meristematic activity indefinitely. As the tissues continue to split, new areas of cork cambium continue to develop and often overlap the older ones, somewhat in the fashion of shingles, so that eventually there is a more or less continuous layer of cork. Because the suberized walls of the cork cells are essentially impermeable to water, they not only prevent excessive water loss from the stem (replacing the epidermis) but also prevent access of water and solutes to the cells external to the cork. This causes these external cells to die. New cork cambia continue to develop inside the older ones and soon reach the older layers of the phloem. By the time a woody stem is only a few years old the epidermis and cortex have generally been entirely cut off by splitting and the development of cork, and the outermost tissues of the stem consist of a mixture of cork and the older, nonfunctional layers of phloem.

In woody stems the tissues external to the cambium are commonly referred to as **bark.** The inner bark consists of the younger layers of secondary phloem that have living sieve tubes, companion cells, and parenchyma cells and that function in the translocation of solutes; the outer bark consists of the older dead and nonfunctional phloem intermixed with cork. As growth in diameter continues, the splitting of the outer bark continues and generally results in the deep furrows characteristic of the bark of most species, although in some species such as sycamore and birches the bark splits off in sheets. In any event, the outer layers of bark continue to slough off. Although an intact woody stem contains all the pith and xylem it has ever produced, any older stem contains only the phloem and cork produced during the more recent years.

Conifer Stems

All conifers and other gymnosperms are woody plants. Since the arrangement of the vascular tissues and bark and the pattern of secondary growth are essentially the same in conifers such as pine (Figure 8-13) as they are in woody dicotyledons, it will not be necessary to discuss these here. The important difference between woody dicotyledon and conifer stems that does require consideration is in the cellular composition of the xylem and phloem.

Photomicrographs of sections of pine wood grouped in Figure 8-14 clearly show the cellular makeup of typical conifer xylem. It is much more homogeneous than the xylem of angiosperms, consisting almost entirely of tracheids and the parenchyma cells of the narrow rays, although there are scattered resin ducts. These resin ducts are not cells, but are cylindrical intercellular cavities lined with the parenchyma cells that synthesize and secrete resin. As in woody dicotyledons, the cells (here the tracheids) produced in the spring are larger and have thinner walls than those produced during the summer and there are distinct annual rings. The tracheids are single cells, unlike the long vessels of angiosperms that are composed of a series of vessel elements, and they serve both the transport functions of vessels and the supporting function of fibers.

It is generally believed that in the evolution of the angiosperms both the vessel elements and fibers were derived from tracheids. The pits of tracheids in pines are generally of the bordered type and are clearly visible in the radial and tangential sections in Figure 8-14. The doughnut-shaped appearance of

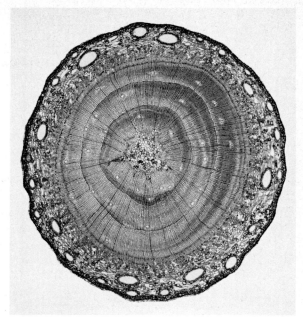

Figure 8-13. Cross section of a pine stem in the spring of its fourth year of growth, showing distinct annual rings. Note that the xylem is composed entirely of tracheids, except for the scattered resin ducts. There are larger resin ducts in the bark. The cells of the phloem are not clear at this magnification. [Courtesy of Carolina Biological Supply Company]

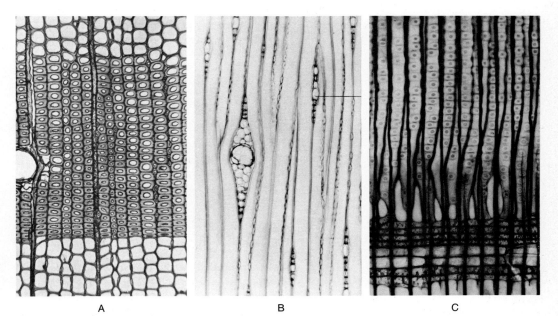

A   B   C

Figure 8-14. Photomicrographs of sections of pine wood. **(A)** Cross section. **(B)** Tangential longitudinal section. **(C)** Radial longitudinal section. [Courtesy of the U. S. Forest Service Products Laboratory]

the bordered pits in the radial section results from flaring of the cell wall to the pit (Figure 2-14) and the disk-shaped torus, which is a thickening of the central area of the pit membrane that extends across the pit. The membrane is composed of the middle lamella and the primary walls of the adjacent tracheids. Like simple pits, the bordered pits probably facilitate the movement of water from cell to cell, although the thickened secondary walls are themselves quite permeable. It has been suggested that the torus can function as a valve because it can fill the pit on one side or another, but such a role is only conjectural. However, in conifer heartwood the tori are usually permanently pressed against the border on one side. Bordered pits are also found in some vessels and tracheids of angiosperms.

The cellular composition of the phloem in conifers also differs from that of angiosperms. The conducting cells are the individual sieve cells (sieve elements) described in Chapter 2 rather than sieve tubes composed of an integrated series of sieve-tube elements, and also companion cells are lacking. However, some of the phloem parenchyma cells are adjacent to the sieve cells and may play some of the roles of companion cells.

The Vascular System of Roots

Herbaceous Roots

The cellular composition of the xylem and phloem of roots is the same as in stems, but the arrangement of the vascular tissues is different. The vascular cylinders of roots are more clearly delimited from the cortex than those of stems because they are surrounded by a pericycle and an endodermis (Figures 8-15 and 8-16). The pericycle is composed of parenchyma cells, some of which

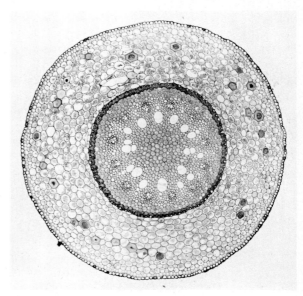

Figure 8-15. Cross section of a Smilax root, an example of a root with pith. Note the alternating bundles of xylem and phloem and the distinct endodermis. [Courtesy of Ripon Microslides]

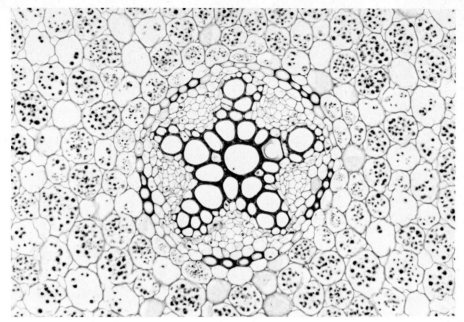

Figure 8-16. Cross section of the stele of a Ranunculus (buttercup) root, along with some of the adjacent cortical parenchyma. This is an example of a root lacking pith and having a solid central core of xylem. For a cross section of an entire root see Figure 2-1. [Courtesy of Carolina Biological Supply Company]

may become meristematic and thus give rise to (a) the primordia and apical meristems of branch roots, (b) part of the vascular cambium, (c) phellogen or cork cambium in roots with secondary growth. In some species the pericycle is only one cell thick in a radial direction, whereas in others it is several cells thick. The endodermis is a cylinder of cells only one cell thick. The endodermal cells have Casparian strips in their walls (Figure 2-9).

The primary xylem and phloem are in separate bundles that alternate with one another rather than being in collateral bundles as in stems. The number of xylem and phloem strands is generally characteristic for a species, although age or cultural conditions may alter the number. In dicotyledons the number of xylem (or phloem) strands is small, usually from two (**diarch**) to five (**pentarch**). In monocotyledons the number of strands is usually much larger (**polyarch**).

The first xylem differentiated is on the outer side of the provascular strand and differentiation proceeds toward the center of the root. In many species of dicotyledons xylem is differentiated all the way to the center, resulting in a solid, fluted column of xylem and a lack of pith (Figure 8-16). Monocotyledons generally have roots with pith (Figures 8-15 and 8-17), and the pith usually is of considerable diameter.

It should be obvious that the arrangement of the vascular tissues in roots is quite different from that in stems. At the point where roots and stems meet, there must be a transition region where the xylem and phloem are converted

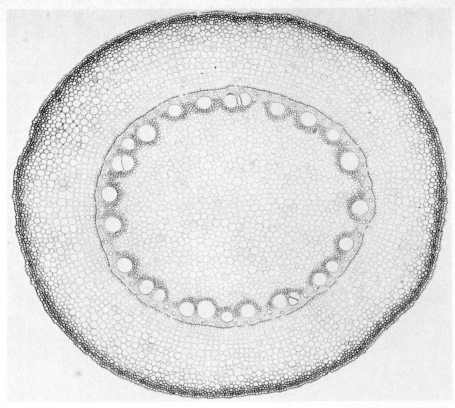

Figure 8-17. Cross section of a corn prop root, an example of a root with pith. Note the numerous bundles of xylem with their large vessels and smaller fibers and the intervening bundles of phloem. [Courtesy of Carolina Biological Supply Company]

from their arrangement in roots to their arrangement in stems. This transition zone is located in what was the hypocotyl of the seedling, that is, the region between the cotyledons and the radicle or young root. In this interchange region the pathways of the xylem and phloem strands are quite complex and different in various species. Consideration of the transition region in detail is beyond the scope of this discussion but may be found elsewhere [4].

Woody Roots

The roots as well as the stems of woody plants increase in diameter as a result of secondary growth from the cambium, and the roots eventually attain diameters comparable with those of the stems. Like the older woody stems, roots consist mostly of xylem (wood) and have a bark composed of phloem and cork outside the cambium.

The primary xylem and phloem of most woody dicotyledon roots is generally arranged much as it is in the buttercup root, and only the initial stages of secondary growth from the cambium differ from those in the stem. The vascular cambium at first differentiates only between the phloem strands and the adjacent xylem, not over the ends of the projecting xylem. As secondary xylem is laid down by these isolated arcs of cambium, the region between the xylem projections is filled with xylem so that the cylinder of xylem is converted from a fluted form to a circular one in cross-sectional view. The primary phloem is pushed outward by the production of the secondary xylem and phloem (Figure 8-18). Once the region originally occupied by the primary phloem has been filled with secondary xylem, pericycle cells next to the xylem projections develop into cambium and provide a complete cylinder of cambium as in stems. The cambium continues to produce secondary xylem and phloem, and a complete cylinder of each is laid down. From this point on secondary growth

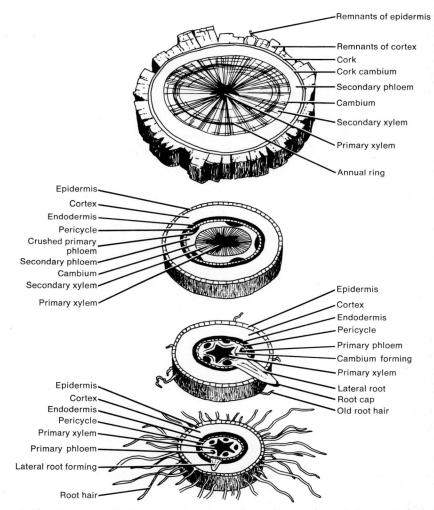

Figure 8-18. Diagrammatic drawings of cross sections of a woody root at progressively higher (older) levels. The section through the root hair zone has only primary tissues. The other sections show successive stages in the development of secondary xylem and phloem, and (top) the beginning of formation of the corky bark. [After W. H. Muller, *Botany: A Functional Approach,* The Macmillan Co., New York, 1969]

in roots is the same as in stems, and the arrangement of xylem and phloem is now the same as in woody stems, that is, a cylinder of phloem surrounds the xylem rather than the alternating arrangement of the primary xylem and phloem. Vascular rays develop in the secondary xylem and phloem. Those opposite the projections of the primary xylem are frequently much wider than the others.

As the vascular cylinder grows in diameter the epidermis and cortex split, but usually the cork cambium first develops in the pericycle rather than the cortex. A complete layer of cork that cuts off the epidermis, cortex, and endodermis may develop before any of the pericycle or phloem has been lost; eventually, however, the pericycle and older layers of phloem split, cork cambia arise in the phloem, and the pericycle and older phloem are isolated by the cork and slough off. The bark is now like that of the stem: an outer bark of mixed cork and nonfunctional older phloem and an inner bark of phloem that functions in the translocation of solutes.

Although there are annual growth increments of xylem in woody roots as in stems, the annual cylinders are generally less distinct. Other differences between woody roots and stems is that in roots the bark is generally thicker in proportion to the wood than in stems and has a smaller number of fibers, the rays occupy a greater percentage of the volume, and the vessels are generally larger in diameter and more uniform in size.

The Vascular System of Leaves

The vascular system is continuous, not only through the roots and on up through the stems but also from the stems to and through the leaves. The vascular cylinder of stems is dissected by leaf gaps and branch gaps, which occur above the point where vascular tissue extends outward from the vascular cylinder into the leaves or branches (Figure 8-19). Since lateral buds that may develop into branches are located in the axils of leaves (the point just above the junction of a leaf and the stem), the branch gaps are found just above the leaf gaps.

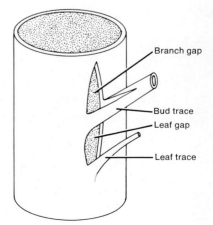

Branch gap

Bud trace

Leaf gap

Leaf trace

Figure 8-19. Diagrammatic representation of the vascular cylinder of a stem at a node, showing the emergence of a leaf trace and a bud trace of the axillary bud and the resulting gaps in the cylinder. The leaf trace enters the petiole and becomes the main vein of the leaf blade. When the bud develops into a branch the bud trace becomes the vascular cylinder of the branch stem. [After Eames and MacDaniels, *Introduction to Plant Anatomy*, McGraw-Hill Book Company, New York, 1925]

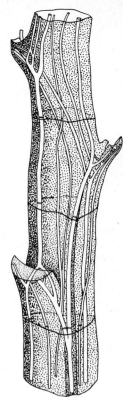

Figure 8-20. Drawing of the vascular system of a potato stem, showing leaf traces. Not all of the vascular bundles are shown. [After E. F. Artschwager, *J. Agric. Res.*, **14**:221 (1918)]

For convenience the vascular system of leaves may be considered in two parts: the vascular strands that extend from the stem into and through the petiole (if there is one) and the veins of the leaf blade.

The Vascular Tissue of Petioles

There is great variation in the arrangement of vascular tissues in petioles from species to species, so we can describe only a few typical examples. The number of vascular strands (leaf traces) that extend from the vascular cylinder into a single leaf varies from one to many (Figure 8-20). In some species several leaf traces may merge into one before they enter the petiole, whereas in others each trace enters the petiole independently (Figure 8-21). Although the number, size, and arrangement of the vascular bundles in the petioles of leaves varies greatly from species to species, it is highly uniform in individuals of any particular species. When deciduous leaves abscise, bundle scars are found within the leaf scar on the stem. The number and arrangement of these bundle scars are so consistent within a species that they can be used in identifying trees or shrubs in winter. For the purposes of the present discussion species variations in the structure and arrangement of the vascular bundles of petioles are of minor importance, but it is important to note their adequacy as channels for the flow of water and solutes into and out of leaves.

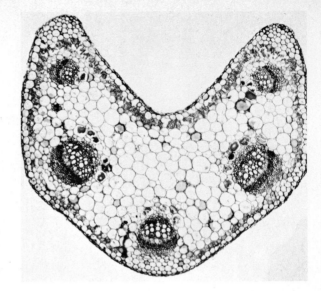

Figure 8-21. Cross section of a petiole of a peanut leaf, showing its five vascular bundles which emerged from the vascular cylinder of the stem as leaf traces. [Courtesy of John A. Yarbrough]

One structural feature that is quite consistent from species to species is the location of the phloem on the under side of the bundles and the xylem on the upper side. This is, of course, due to the fact that the phloem is outside the xylem in the stem. Where the bundles are arranged in a circle or semicircle, the xylem is toward the inside. However, species that have bicollateral bundles in their stems, with phloem inside as well as outside of the xylem, may have a similar bicollateral arrangement in the bundles of the petiole.

The Vascular Tissues of Leaf Blades

The vascular bundles of the petioles continue on into the midrib of the blade or, in the case of the pinnately compound leaves, into the rachis and petiolules. In some leaves, particularly palmately compound or lobed ones, several main branches of the bundles may diverge from the end of the petiole. The main bundles, or **veins** as they are referred to in leaves, branch into smaller ones and these into still smaller ones that permeate the entire leaf blade. It has been estimated that on the average there are around 100 cm of veins per square centimeter of blade and that the interveinal spaces are usually no more than 130 μ in diameter. Thus, no cell of the mesophyll is more than a few cells away from a vein, providing for effective movement of water and solutes to and from the cells.

The arrangement of the veins varies considerably with the species, but there are two major types: parallel veins and netted veins. Parallel venation is characteristic of the monocotyledons and netted venation of the dicotyledons, (Figure 8-22) and this difference is one of the important general distinctions between the two groups of plants. However, a few species of monocotyledons have netted veined leaves and a few species of dicotyledons have parallel veins.

Grass leaves will serve as an example of the parallel-veined type. A number

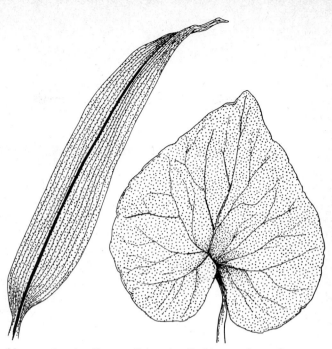

Figure 8-22. Drawings of leaves showing the parallel and netted types of venation.

of main veins, varying with the species and the width of the blade but often on the order of a dozen or so, enter the leaf from the stem and extend lengthwise of the blade approximately parallel with one another, although they converge toward the tip of the leaf. The vein in the midrib is generally larger than the others, which may all be of about the same diameter or may be of two sizes alternating with one another (Figure 8-23). The main veins are crossconnected with one another by smaller veins which are usually arranged like the rungs of a ladder. Some of the main veins extend all the way to the tip of the leaf whereas others terminate back from the tip or merge into neighboring main veins.

In netted-veined leaves the pattern of venation is considerably more complex (Figure 8-24). Leaves are thicker where there are large veins than elsewhere (Figure 8-25), so the course of the principal veins can usually be determined simply by observation of the leaf surface. Even some of the larger subsidiary

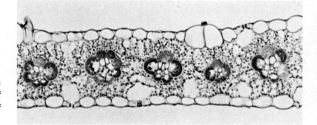

Figure 8-23. Photomicrograph of a portion of a corn leaf, showing cross sections of several of the parallel veins. [Courtesy of Ripon Microslides]

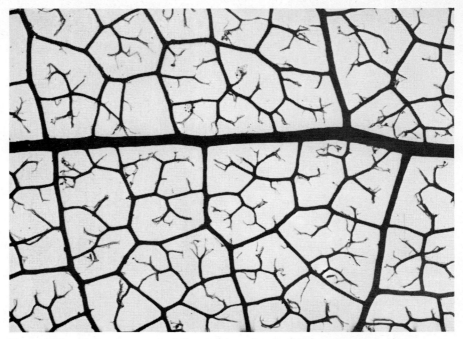

Figure 8-24. Photomicrograph of the skeleton of a dicotyledon leaf (prepared by removing the mesophyll) showing the complex pattern of branching of the netted veins. [Copyright by General Biological Supply House, Inc., Chicago]

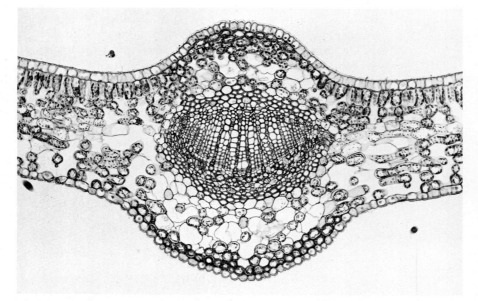

Figure 8-25. Photomicrograph of a portion of a tea leaf, showing the cross section of one of the large veins and the bulging of the leaf above and below the vein. In the vein note the xylem (above), the phloem (below), and the bundle sheath. See also Figure 2-1. (Courtesy of Carolina Biological Supply Company)

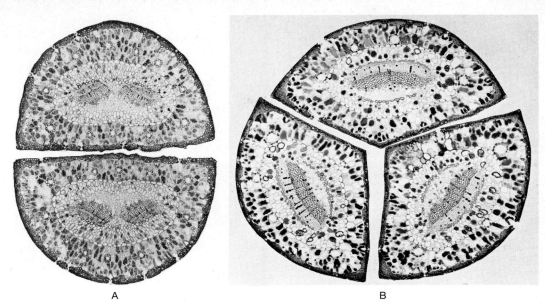

A   B

Figure 8-26. Photomicrographs of cross sections of pine leaves (needles), showing the central veins. **(A)** Leaves from a species with two leaves per fascicle (leaf cluster). [Copyright by General Biological Supply House, Inc., Chicago] **(B)** Leaves from a species with three leaves per fascicle. [Courtesy of Carolina Biological Supply Company]

branch veins may be observed in this way, particularly when the leaf is held to the light, although the smaller veins are microscopic. The veins are highly anastomosing and actually form a network. The area bounded by a net of medium-sized veins is generally permeated by a highly branched and somewhat irregular system of the smallest veins which are connected to the bounding veins at one or more points (Figure 8-24). Many of the smallest veins are dead end, but in some species they may also interconnect and form a secondary net.

The veins are enclosed by bundle sheaths, composed of compactly arranged parenchyma cells. Some of the larger veins of dicotyledon leaves have additional layers of parenchyma cells called bundle-sheath extensions. The ends of the smallest veins may lack xylem and phloem and consist only of the bundle sheaths. The largest veins contain a substantial amount of both xylem and phloem (Figure 8-25), and there is, of course, a decrease in the amount of both as the size of a vein decreases. The arrangement of the veins in the leaves of gymnosperms is usually quite different from that in angiosperms. For example, in pine needles (leaves) the veins (usually one or two) are central (Figure 8-26).

The Vascular System of Reproductive Organs

The arrangement of the vascular tissues that supply the various parts of flowers (sepals, petals, stamens, and pistils) and the fruits that develop from flowers varies greatly with the species and is often quite complex (Figure 8-27).

Figure 8-27. Semidiagrammatic drawing of the vascular tissues supplying a *Ranunculus repens* (buttercup) flower. The vascular tissues form a cylinder, but here they are shown cut and flattened out. The letters represent the flower parts supplied by various vascular bundles: C, carpel; P, petal; S, stamen; Se, sepal. [After S. P. Tepfer, *Univ. Calif. Publ. Bot.,* **25:**513 (1953)]

A consideration of vascular system of these reproductive organs is beyond the scope of the present discussion, but it should be stressed that, like all other organs of vascular plants, the reproductive organs are adequately supplied with vascular tissue. In general, the vascular tissues supplying the sepals and pistils are better developed than those supplying the petals and stamens, and in the latter there may be few or no branches of the main bundle.

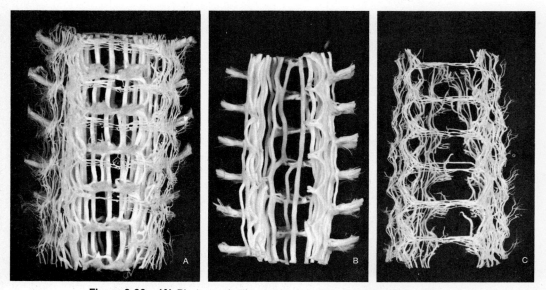

Figure 8-28. **(A)** Photograph of vascular bundles dissected from a 2-in. portion of an ear of Golden Bantam corn. **(B)** and **(C)** are further dissections of **(A)**. [From R. A. Laubengayer, *Amer. J. Bot.,* **36:**236 (1949)]

Figure 8-28 shows dissections of the vascular system of a corn cob made by R. A. Laubengayer. Although a corn cob is actually a modified stem bearing numerous pistillate flowers (the immature grains of corn with their silks), the dissections illustrate well the complicated arrangement of the vascular tissues that supply reproductive structures.

In closing this chapter it is well to stress again the fact that no living cell of any vascular plant is any great distance from some portion of the vascular system. In many cases it is only a matter of microns or millimeters, and in the most extreme cases not more than a centimeter or so. Thus, all the cells are within a short distance of flowing water and solutions and the much slower transport by diffusion is necessary only through limited distances.

References

[1] Bailey, I. W. *Contributions to Plant Anatomy,* Ronald Press, New York, 1954.
[2] Bierhorst, D. W. *Morphology of Vascular Plants,* The Macmillan Company, New York, 1971.
[3] Bold, H. C. *Morphology of Plants,* 2nd ed., Harper & Row, New York, 1967.
[4] Esau, K. *Plant Anatomy,* 2nd ed., John Wiley & Sons, New York, 1965.
[5] Esau, K. *Vascular Differentiation in Plants,* Holt, Rinehart & Winston, New York, 1965.

The Absorption, Translocation, and Loss of Water

9

In Chapter 7 we considered the movement of water into and out of plant cells and in Chapter 8 the vascular system of plants. We are now prepared to consider the water relations of plants as a whole: the absorption of water from the soil; its movement across the tissues of roots into the xylem; its upward flow through the xylem of the roots, stems, and leaves; its movement through the tissues of the stem and leaves; and finally its loss into the atmosphere by transpiration. The water of the soil, the plant, and the atmosphere constitutes a continuum sometimes designated by the acronym SPAC. In the soil the water continuum is essentially liquid, whereas in the intercellular spaces and the atmosphere it is vapor. Within a plant the continuum of liquid water permeates the cell walls, the protoplasts, the vacuoles, and the lumens of the vessels or tracheids of the xylem. Throughout the SPAC there is a gradient of free energy of water or water potential. This is highest in the soil, decreases as the water moves through the plant, and is lowest in the atmosphere. It is this energy gradient that is responsible for the movement of the water from the soil into the plant, through the plant, and out into the atmosphere.

It should be noted that there is a SPAC for all terrestrial plants with aerial organs and roots, or comparable organs embedded in the soil, whether the plants are vascular or nonvascular. Although the system is modified in the case of epiphytes and submerged aquatics, there is still a water continuum between any plant and its environment. In this chapter we will ignore such special cases and center our attention on the absorption, translocation, and loss of water by vascular plants with roots in the soil and aerial shoots.

The rate of water movement into, through, and out of a plant is determined not only by the water potential gradients but also by the resistance to water movement through the soil, the various tissues of the plant, and the atmosphere. Resistance, being the reciprocal of conductivity, is expressed in seconds per centimeter (sec/cm) or comparable units. In 1948 van den Honert pointed out the analogy of water movement through the SPAC to the flow of electricity through a circuit and devised equations comparable with Ohm's law (current = potential/resistance):

$$\text{Rate of water movement} = \frac{\Psi_1 - \Psi_2}{r_1 + r_2}$$

Ψ_2 and Ψ_2 are the water potentials in any two parts of the system such as the xylem and the leaf or the leaf and the air, and their difference is the water potential gradient. The resistances of these (xylem, leaf or leaf, air) are represented by r_1 and r_2. In some parts of the system, as from root surface to root xylem, only one resistance (here root resistance) is involved. If water potential is expressed as ergs per cubic centimeter (ergs/cm^3) or dynes per square centimeter (dynes/cm^2) the calculations result in suitable units for expression of rate of water movement. The roles of resistance and water potential in the SPAC can be represented diagramatically by the use of the standard symbols for electrical circuits (Figure 9-1). Further analogy to an electrical system may be made by considering water storage as a capacitor.

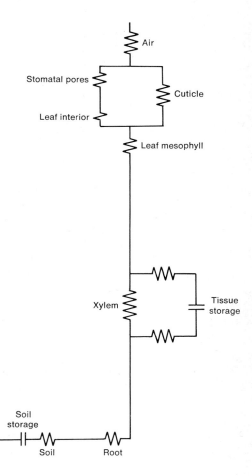

Figure 9-1. Diagram comparing the flow of water through the soil-plant-atmosphere continuum (SPAC) with the flow of an electrical current along a potential gradient and through a series of resistances and capacitances. [After P. J. Kramer, *Plant and Soil Water Relations*, McGraw-Hill Book Company, New York, 1969]

Figure 9-2. Guttation by a strawberry leaf.
[Courtesy of J. Arthur Herrick]

The Loss of Water from Plants

Despite the immense importance of water in the life of plants over 99% of the water absorbed by plants is lost into the atmosphere. Less than 1% is actually used by the plant in biochemical reactions such as photosynthesis and hydrolysis, in the hydration of protoplasm and cell walls, in the maintenance of cell turgor, or in other ways. Almost all of this water loss is by the process of **transpiration,** which involves the evaporation of water from cells into intercellular spaces and then the diffusion of the water vapor out of the plant. A minute fraction of the water loss is by **guttation,** which involves exudation under pressure of drops of liquid water from ends of veins that terminate in the leaf margin (the **hydathodes**) (Figure 9-2). Guttation occurs only under conditions of high soil moisture and high atmospheric humidity. Under similar environmental conditions liquid water may be lost from cut or broken stems **(bleeding).** The secretion of nectar might also be construed as a loss of liquid water from plants, but all these losses of liquid water are minor compared with the loss of water vapor by transpiration.

Transpiration occurs through the stomata, the lenticels, and the cuticle, but the major portion of transpiration is stomatal, only a small fraction being lenticular or cuticular. Stomata are usually thought of as being located in the epidermis of leaves, but they are also found in the epidermis of other aerial organs, including herbaceous stems and the various flower parts and fruits, even though the major portion of stomatal transpiration occurs from leaves in most species of plants. Because of the importance of stomatal structure and function in transpiration we shall consider stomata along with other aspects of leaf structure pertinent to transpiration before proceeding with a discussion of transpiration as a process.

Leaf Structure in Relation to Transpiration

Leaves are typically organs with broad, thin **blades (lamellae)** covered on both sides by a cutinized epidermis containing stomata (on one side or both) and with a mesophyll composed of loosely arranged chlorenchyma cells permeated by veins (Figures 2-1 and 8-25). The selective pressures that have resulted in the evolution of an organ of this type have presumably been related to its effectiveness as a site for photosynthesis. Typical leaves provide a large area for light absorption with a minimum of tissue, and their thinness permits light to penetrate through the mesophyll. The large intercellular spaces and the stomata provide for rapid diffusion of gases into and out of the leaf, and the extensive system of veins provides for rapid entry of water and salts and the rapid removal of the photosynthetic product. The fact that such leaf structure also provides for rapid transpiration is incidental and unavoidable.

There are, of course, many species differences in leaf structure even among the numerous species that have typically broad, thin blades. These differences include the size and shapes of the blades, the thickness of the blades, the number and distribution of the stomata, the structure and thickness of the cuticle, and the structure of the mesophyll (Figure 9-3). Also, the leaves of some species deviate greatly from the typical broad, thin blade, for example the needle-shaped leaves of many conifers (Figure 8-26) and the thick, fleshy leaves of such genera as Dianthus (Figure 9-3), Sedum, and Kalanchoë. Even more extreme deviations occur in plants such as cacti and Equisetums that have highly specialized leaves (spines and scales respectively), which are nonphotosynthetic, and green stems that carry on photosynthesis.

The total area of the walls of mesophyll cells exposed to the intercellular spaces is large, ranging from 5 to 30 times the area of the external leaf surface. Even the cells of the palisade layers have most of their wall area exposed to intercellular spaces. In stained leaf sections the palisade cells may appear to be in close contact with one another throughout their length, but fresh sections reveal the fact that these cells are generally in contact only at their upper ends and that throughout most of their length they are separated by intercellular

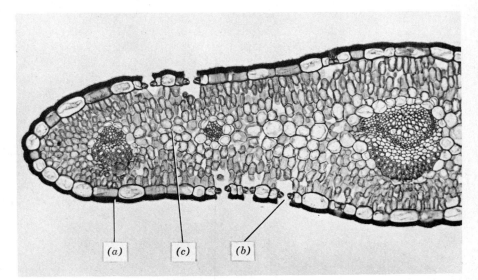

Figure 9-3. Cross section of a Dianthus leaf showing three xeromorphic features: (a) Thick cuticle, (b) sunken stomata, and (c) compact mesophyll. [Copyright by General Biological Supply House, Inc., Chicago]

spaces. The surfaces of the mesophyll cell walls are lightly cutinized and so hydrophobic. This may mean that evaporation of water from the cell walls may occur at a different rate than from a free water surface, but the extent and significance of such a difference has not been determined.

Structure and Distribution of Stomata

A stoma is composed of two guard cells with a stomatal pore between them (Figure 9-4), and in some species the guard cells are surrounded by subsidiary cells that are different in size and shape from the other epidermal cells (Figure 2-6). There is considerable variation in the structure and shape of guard cells from species to species as well as in the structural features that cause the stomata to be open when the cells are turgid and closed when they are flaccid, but we will center our attention on the bean-shaped guard cells that are characteristic of many dicotyledon species. In these guard cells the structural feature that makes possible the opening and closing of the stomata is the thickening of the cell walls that are adjacent to the stomatal pore in contrast with the thinner walls of the remainder of the cells. As the guard cells become more turgid, the thinner parts of their walls stretch more than the thicker parts and cause the thickened parts to cup inward which brings about stomatal opening. When the guard cells lose turgor, their thickened walls come to rest against each other, closing the stoma.

There is a considerable range in the size of stomata from species to species. For example, when fully open, the stomatal pore of common string bean plants is about $7\,\mu$ long and $3\,\mu$ wide, whereas in oats the pores are about $38 \times 8\,\mu$. Although microscopic, the stomatal pores are of course immense in comparison with the sizes of the gas molecules that diffuse through them. More than 6000 water molecules side by side could pass through the 3-μ width of a bean stomatal pore.

There are also marked species differences in the number and distribution of stomata (Table 9-1). Although many plants, particularly trees, have stomata only on the lower epidermis of their leaves, many other species have stomata on both the upper and lower epidermis. Where the latter is the case, there are often more stomata per unit area on the lower epidermis than on the upper epidermis; however, in some species there are approximately equal numbers,

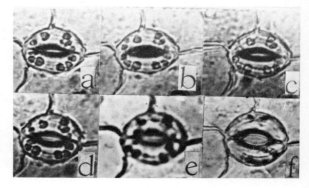

Figure 9-4. Photomicrographs of guard cells showing various degrees of stomatal opening (a–f). [From P. Alvim, *Amer. J. Bot.,* **36:**781 (1949)]

Table 9-1 Distribution of stomata in the leaves of several species, average number per cm^2

Species	Upper epidermis	Lower epidermis	Species	Upper epidermis	Lower epidermis
Alfalfa	16,900	13,800	Lilac	0	33,000
Apple	0	29,400	Maize	5,200	6,800
Bean	4,000	28,100	Oat	2,500	2,300
Begonia	0	4,000	Pea	10,100	21,600
Black oak	0	58,000	Scarlet oak	0	103,800
Cabbage	14,100	22,600	Sunflower	8,500	15,000
Castorbean	6,400	17,600	Tomato	1,200	13,000
Coleus	0	14,100	Wandering Jew	0	1,400
Geranium	1,900	5,900	Wheat	3,300	1,400
Jimson weed	11,400	18,900	Willow oak	0	72,300

whereas in others (notably some grasses) most stomata are on the upper epidermis. In plants such as waterlilies with floating leaves, stomata are present only on the upper epidermis.

It is evident from Table 9-1 that there is great variation from species to species in the number of stomata per unit area. The 1400 per cm^2 for Wandering Jew is about as low a number as found in any species with stomata on the lower epidermis only, whereas the 103,800 for scarlet oak is about as high as found in any species, although several other species of oaks have comparable numbers. In general, there is an inverse relation between the size of stomata and the number per unit area, but this is by no means precise or universal.

Stomatal Opening and Closing

It has long been known that stomata generally close at night, open in the morning with the advent of light, and close again either when it becomes dark or, if a water deficit develops, by the middle of the afternoon or sometimes even earlier. That stomata are open when the guard cells are turgid and closed when they are not has also been known for many years. However, the catenary sequence that links light with the turgidity of the guard cells has been the subject of extensive investigation over the years. Only in relatively recent years has an acceptable explanation been provided, and even this is subject to further modification. The sequence of events appears to be as follows:

1. With the advent of light photosynthesis begins in the guard cells.

2. This reduces the concentration of carbon dioxide and thus of carbonic acid, thereby bringing about an increase in pH from about 5 up to 7.

3. The reduced acidity promotes the reaction of starch with phosphate in the guard cells, with the production of glucose-1-phosphate (G-1-P). (At pH 7 this reversible reaction catalyzed by starch phosphorylase goes from starch to G-1-P; at pH 5 it goes from G-1-P to starch.) The G-1-P is then converted to G-6-P, which in turn is converted to glucose and phosphate.

4. The additional solutes reduce the water potential of the guard cells, causing water to enter them by osmosis from adjacent epidermal cells.

5. As a result the turgor pressure of the guard cells increases, causing the stoma to open.

The sugar produced by photosynthesis may also contribute to the increased solute content of the guard cells, but this contribution is minor compared with the sugar derived by the phosphorolysis of starch. Note that conversion of the G-1-P to glucose and phosphate is an essential feature because the G-1-P does not increase the solute content above what it was before the phosphate was used in the breakdown of starch.

The closing of stomata when photosynthesis ceases in the evening is brought about by the reverse of each of the steps outlined above. However, energy from respiration is used in bringing about stomatal closure since ATP is needed for the production of the G-1-P used in synthesizing the starch.

That the pH of the guard cells is an important factor in their opening and closing is indicated by the fact that stomata can be caused to close in the light by exposing them to the vapors of volatile acids, such as hydrochloric or acetic, or by increasing the carbon dioxide of the atmosphere to unusually high concentrations. Similarly, vapors of ammonium hydroxide will cause closed guard cells to open.

Although the above theory of the relationship between light and the opening of stomata has been generally accepted by most plant physiologists, it does not provide the only possible explanation. For example, in 1968 Fischer [4] provided evidence in support of an entirely different theory. He suggested that light stimulated the active accumulation of potassium ions by guard cells in sufficient quantity to provide the necessary reduction in the water potential in the guard cells. Also, Zelitch [25] has proposed a hypothesis that involves glycolate produced during photosynthesis and states that, in any case, it seems most likely that photosynthetic products of some kind, rather than the decrease in carbon dioxide and the consequent increase in pH, are involved in the opening of the stomata in light.

The closing of stomata when there is a water deficit in the plant would seem to involve a much simpler explanation, since the lack of water would cause a reduction in the turgor pressure of the guard cells. However, there is some evidence that the water deficit may also cause a decrease in the pH of the guard cells, thus bringing into play the starch phosphorylase mechanism described above.

Temperature may also influence the opening and closing of stomata. At temperatures of around 0°C or less the stomata of most species remain closed even during the day, whereas at temperatures of around 40°C stomata may open at night. Between 0 and 25–30°C the degree of stomatal opening in at least some species has been found to be proportional to the temperature. Also, Scarth has reported that reduction of the oxygen concentration of the air can result in stomatal opening in the dark. Although the mechanisms whereby such factors as temperature and oxygen influence stomatal opening or closing are not known with certainty, it is possible that they operate by affecting the starch phosphorylase system.

Diffusion Through Stomata

The diffusion of gases through small pores such as those of stomata differs in a number of important ways from the diffusion of gases from a free surface, and these differences have marked influences on the diffusive capacities of stomata and the rate of transpiration [19]. Furthermore, because diffusion through any particular stoma may be influenced by diffusion through adjacent stomata it is necessary to take account of the fact that the epidermis is a multiperforate septum. Our knowledge of diffusion of gases through small pores and multiperforate septa has been gained by investigations of diffusion both through stomata and through purely physical systems involving perforated septa made of thin metal sheets or other suitable substances.

One of the most striking facts about diffusion of water vapor through stomata is that the rate may equal 50% or more of the evaporation from a free water surface equal in area to the leaf surfaces, even though the stomatal openings occupy only 1 to 3% of the total area of the epidermis. This results to a considerable degree from the fact that the rate of diffusion of gases through small pores into still air is more nearly proportional to the perimeters of the pores than to their areas (Table 9-2). The diffusing gas (water vapor in this

Table 9-2 Relationship between the areas and perimeters of small pores and the rate of diffusion of water vapor through them

Diameter of pores, mm	Loss of water, g	Relative amounts of water lost	Relative areas of pores	Relative perimeter of pores
2.64	2.66	1.00	1.00	1.00
1.60	1.58	0.59	0.37	0.61
0.95	0.93	0.35	0.13	0.36
0.81	0.76	0.29	0.09	0.31
0.56	0.48	0.18	0.05	0.21
0.35	0.36	0.14	0.01	0.13

SOURCE: Sayre, J. D. *Ohio J. Sci.*, **26**:233 (1926).

case) accumulates in a hemispherical **diffusion shell** over the pore (Figure 9-5). Near the edge the concentration gradient in this diffusion shell is steeper than toward the center, so the diffusion is more rapid. The smaller the pore, the greater the proportion of the molecules diffusing from near the perimeter.

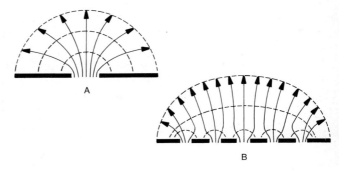

Figure 9-5. Diagrams of diffusion shells resulting from diffusion of water vapor through small pores into still air. **(A)** Diffusion through a single pore. **(B)** Diffusion through closely spaced pores, showing mutual interference. The broken lines enclose areas of equal molecular concentration, the three concentrations shown being the same in A as in B. [After G. G. J. Bange, *Acta. Bot. Neer.*, **2**:255 (1953)]

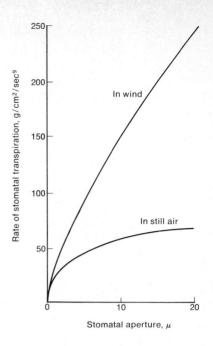

Figure 9-6. Influence of the width of the stomatal opening on the rate of stomatal transpiration of Zebrina in still air and in wind. [Data of G. G. J. Bange, *Acta. Bot. Neer.,* **2:**255 (1953)]

Although the combined areas of the stomatal openings are only a small fraction of the leaf area, their combined perimeters constitute an appreciable fraction of the leaf perimeter.

As stomata are closing the area of the stomatal pores decreases much more than the perimeter. Consequently, in still air the rate of transpiration does not decrease greatly until the stomata are nearly closed (Figure 9-6). In this case transpiration is being limited essentially by the resistance of the moist air on the leaf surface, and stomatal resistance is of little significance. However, in moving, dry air the diffusion shells are removed, the resistance of the boundary layer of air is low, and stomatal resistance becomes the limiting factor. In this situation the rate of transpiration decreases in proportion to the decrease in area of the stomatal openings.

Since the epidermis is a multiperforate septum, diffusion through the stomata collectively must be considered. There is a decrease in the rate of diffusion per pore with a decrease in the distance between pores. This results from the progressively greater overlapping of diffusion shells, which increases the concentration of water vapor (or other gas) in the overlapping regions and reduces the rate of diffusion. Although there is an increase in the rate of diffusion per pore with an increase in distance between pores up to the point where the diffusion shells no longer overlap, there is a decrease in diffusion per septum as the distance between pores increases further because of a decrease in the number of pores per unit area.

Even if the individual diffusion shells do not overlap, the water vapor diffusing from them tends to build up a larger superimposed diffusion shell covering the leaf as a whole, although the concentration of water vapor here is lower than in the individual diffusion shells. This layer of water vapor over the leaf is sometimes considered to consist of two sublayers: a lower laminar layer and an upper turbulent layer. The thickness of the layer of air next to the leaf, which has a higher humidity than the surrounding atmosphere, has been calculated to be about 10 mm in still air, around 3 mm in a slow wind, and only about 0.4 mm in a strong wind.

Factors Influencing the Rate of Transpiration

The rate of water vapor loss by transpiration is highly variable. It may fluctuate rapidly as well as greatly from moment to moment as well as diurnally and seasonally. The rate of transpiration is influenced by two main sets of factors: the environmental factors and the factors that determine the rate of water movement through the plant, that is, the water potential gradients and the resistances of the various tissues. The latter determine what may be called the diffusive capacity of a plant.

The principal environmental factors that influence transpiration are temperature, atmospheric humidity, availability of soil water, and light. Light plays a somewhat indirect role by influencing both temperature and stomatal opening. The other factors influence transpiration through their effect on the steepness of the vapor pressure gradient between the intercellular spaces of the plant and the surrounding air. From the standpoint of transpiration, the water potential gradients and resistances of a plant may be considered as determining its diffusive capacity.

PLANT FACTORS. The most important factor affecting the diffusive capacity of a plant is the opening and closing of stomata (Figure 9-7). When the stomata are closed resistance to transpiration is high and diffusive capacity is very low, being limited to such diffusion as can occur through the high resistance cuticle

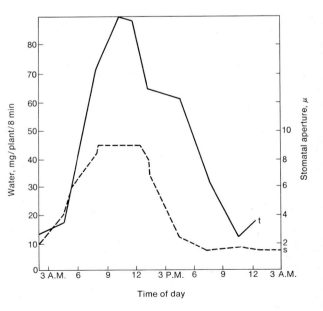

Figure 9-7. The relationship between the rate of transpiration and the width of the stomatal aperture (broken line) in *Verbena ciliata.* The greater increase in transpiration during the day is a result of increased vapor pressure gradients. Note the partial closing of the stomata beginning about 2 P.M. and its influence on the rate of transpiration. [After F. E. Lloyd, *The Physiology of Stomata,* Carnegie Institution of Washington, 1908]

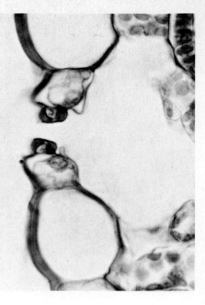

Figure 9-8. Photomicrograph of a section through a stoma of a corn leaf. Note the stomatal aperture, the guard cells and subsidiary cells, and the large intercellular space below the stoma. [Courtesy of the Carolina Biological Supply Company]

and, in plants with bark, through the lenticels. The sizes and numbers of stomata have less influence on diffusive capacity than might be anticipated. Species with large stomata generally have a relatively small number per unit area, and vice versa. Also, as stomata become more numerous there is greater overlapping of their diffusion shells. As long as the stomata are open (Figure 9-8) plants of most species have a high diffusive capacity, and transpiration is then limited by the air resistance and the steepness of the vapor pressure gradient. However, some xerophytes have less than 100 stomata per cm^2 and in such extreme cases the rate of transpiration may be limited simply by the small number of stomata. In some species, particularly xerophytic ones, the stomata are located some distance below the epidermal surface with a tube leading from them to the surface (Figures 9-3 and 9-9). Because the rate of diffusion is inversely proportional to the length of the diffusion pathway, **sunken stomata** have a lower diffusive capacity than those located flush with the other epidermal cells.

Since the cuticle has a high resistance to water movement, cuticular transpira-

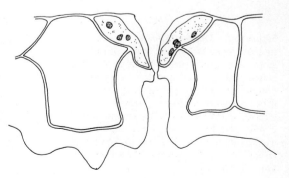

Figure 9-9. Drawing of a section through a sunken stoma from the lower epidermis of a xeromorphic species. Note the tube between the guard cells and the leaf surface and also the thick cuticle. See also Figure 9-3.

tion is negligible when the stomata are open because the diffusing water molecules follow the path of least resistance. When the stomata are closed the rate of cuticular transpiration is influenced by the compactness and thickness of the cuticle. Cuticular resistances range from around 10 sec/cm in mesophytes to highs of 100 sec/cm or more in xerophytes, and generally range between 10 and 40 sec/cm for most crop plants. The diffusive capacity of plants with a high cuticular resistance and closed stomata is extremely low. The presence of a cuticle is, indeed, essential for the survival of land plants.

At one time it was thought that leaf hairs (trichomes) might have an appreciable influence on the rate of transpiration. Abundant dead hairs could be expected to help hold the boundary layer of moist air, and so reduce the rate of transpiration in moving air. Living hairs could be expected to increase the evaporating surface greatly. However, in 1920 Sayre found that carefully shaving the abundant hairs from mullein leaves had little effect on the rate of transpiration. This and other experiments have led to the conclusion that hairs have little if any effect on diffusive capacity.

In some species the leaves either fold up or roll into a tube when there is a water deficit (Figure 9-10), thus reducing the diffusive capacity. In particular, the leaves of various species of grasses roll upward into a tube when there is a water deficit. It has been suggested that this results from the loss of turgor by longitudinal rows of large, thin-walled cells (**bulliform cells**) located in the upper epidermis (Figure 9-11), although some investigators have questioned this. The greater number of stomata per unit area in the upper epidermis of

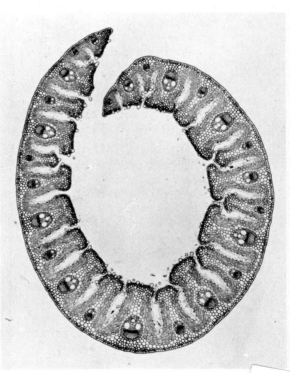

Figure 9-10. Photomicrograph of a cross section of a leaf of *Ammophila arenaria* (beach grass) that has rolled into an almost-closed tube as a result of a water deficit. Note the convoluted upper surface of the leaf. [Courtesy of Triarch Products]

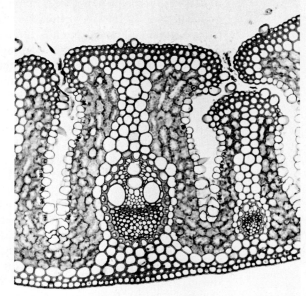

Figure 9-11. More highly magnified portion of the leaf shown in Figure 9-10. Note the stomata in the upper epidermis, the bulliform cells, and the lack of stomata in the lower epidermis. Note also the relatively large amount of sclerenchyma and small amount of chlorenchyma in the mesophyll. [Courtesy of Triarch Products]

many grasses than in the lower epidermis may possibly have resulted from selective pressures related to the rolling of the leaves.

So far we have considered diffusive capacity only from the standpoint of its influence on the rate of transpiration per unit area, but it should be noted that it is really the total diffusive capacity of a plant that is important under natural conditions. This involves not only the rate per unit area but also the total epidermal area of the shoot, including stems and flower parts as well as leaves. To take extreme examples, a plant with numerous large leaves obviously has a greater total diffusive capacity than a plant with a few small leaves even though its diffusive capacity per unit area might be less. However, total diffusive capacity generally does not increase as much as total epidermal area. For one thing, the rate of transpiration per unit area may be influenced by leaf size itself, and large leaves often have a lower rate than smaller ones. Also, the root/shoot ratio is an important factor (Figure 9-12). For example, if two plants of a certain species have root systems of about the same size but one has fewer leaves than the other, the rate of transpiration is likely to be higher in the plant with fewer leaves because the roots can keep the leaves more adequately supplied with water. Although removal of part of the leaves from a plant results in a decrease in total transpiration, this decrease is not proportional to the area of the leaves removed because the rate of transpiration of the remaining leaves increases. It has been found that removal of half the leaves from a tree can result in an increase of 20 to 90% in the transpiration rate of the remaining leaves. This is partly because of the increased root/leaf ratio and partly because of other factors such as reduced shading of the remaining leaves and increased air flow through the crown.

The root/shoot ratio may influence not only the diffusive capacity of the plant (because a high ratio may delay or even prevent a water deficit and consequent stomatal closure) but also to a slight degree the vapor pressure gradient. If the root/shoot ratio is high it is more likely that the leaves will be well supplied with water, thus resulting in a higher vapor pressure in the intercellular spaces. Distribution, as well as size, of the root system may also influence the vapor pressure gradient. During a drought a plant with a deep root system may be well supplied with water whereas a nearby plant of another species with shallow roots may be wilted.

Although various structural features of plants, along with the opening and closing of stomata, determine the diffusive capacity and in some cases also influence the vapor pressure gradient, these structural features are not necessarily as adaptive as they were believed to be in the early years of this century. Some of the xeromorphic structures have been found to have little or no influence in reducing diffusive capacity, and those structures, such as sunken stomata, that do reduce diffusive capacity may not reduce it to the extent that diffusive capacity rather than the vapor pressure gradient is limiting the rate of transpiration. Even in species with the lowest diffusive capacities the rate of transpiration is commonly limited by the vapor pressure gradient. Some xeromorphic leaves have structural features that actually promote high rates of transpiration. For example, Turrell [20] found that xeromorphic leaves generally have more cell wall surface exposed to the intercellular spaces than mesomorphic leaves.

Plants native to deserts have high transpiration rates in many cases and are able to survive for other reasons than a low diffusive capacity. Some have an unusually high root/shoot ratio, and this may become even higher as a result of leaf abscission during droughts. The succulents such as cacti accumulate considerable quantities of water in their tissues during the rainy seasons, and many desert annuals complete their entire life cycles during the rainy seasons and are actually not xerophytes at all. Other species such as the creosote bush

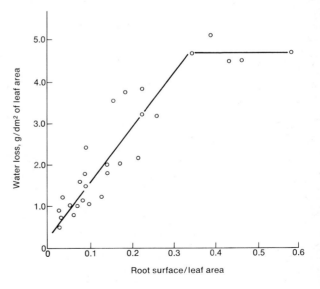

Figure 9-12. Influence of the ratio of root surface to leaf area of rooted lemon cuttings on the rate of transpiration. [Data of J. Bialoglowski, *Proc. Amer. Soc. Hort. Sci.*, **34**:96 (1936)]

are able to survive extreme desiccation that would kill members of most species of plants. Also, during the long dry periods the stomata of desert plants are closed much or even all of the time, and it should be remembered that stomatal closure is the most important of all factors in reducing the diffusive capacity of a plant. Because most xerophytes, in contrast with most mesophytes, have little or no cuticular transpiration, transpiration essentially ceases when their stomata are closed.

ATMOSPHERIC HUMIDITY. We now turn our attention to the environmental factors that influence the rate of transpiration through their effect on the vapor pressure gradient between the intercellular spaces and the atmosphere; we shall begin with atmospheric humidity. It should not be necessary to stress the fact that the humidity of the atmosphere varies greatly from time to time and place to place. Humidity is usually low in dry regions, such as deserts and grasslands, and high in regions where there is a large amount of water evaporating into the air from bodies of water, moist soil, or by transpiration. The humidity of the air in any particular location varies from season to season, day to day, and even in some cases from hour to hour. This fluctuation of the water vapor concentration in the atmosphere is in contrast with the essential uniformity in the concentration of other gases, such as nitrogen, oxygen, and carbon dioxide, in the free atmosphere.

Although atmospheric humidity is most commonly expressed in terms of **relative humidity** (RH) this is not a useful unit in dealing with transpiration unless the temperature is also known, thus making it possible to determine the vapor pressure of the air at a particular temperature and RH by consulting suitable tables. Table 9-3 gives the vapor pressure at a few selected temperatures

Table 9-3 Vapor pressure at selected temperatures and relative humidities

Temper-ature, °C	Vapor pressure (mm Hg) at indicated relative humidity									
	10%	20%	30%	40%	50%	60%	70%	80%	90%	100%
0	0.46	0.92	1.37	1.83	2.29	2.75	3.21	3.66	4.12	4.58
5	0.65	1.31	1.96	2.62	3.27	3.92	4.58	5.23	5.89	6.54
10	0.92	1.84	2.76	3.68	4.60	5.53	6.45	7.37	8.29	9.21
15	1.28	2.56	3.84	5.12	6.40	7.67	8.95	10.23	11.51	12.79
20	1.75	3.51	5.26	7.02	8.77	10.52	12.28	14.03	15.79	17.54
25	2.38	4.75	7.13	9.50	11.88	14.26	16.63	19.01	21.38	23.76
30	3.18	6.36	9.55	12.73	15.91	19.09	22.27	25.46	28.64	31.82
35	4.22	8.44	12.65	16.87	21.09	25.31	29.53	33.74	37.96	42.18
40	5.53	11.06	16.60	22.13	27.66	33.19	38.72	44.25	49.79	55.32
45	7.19	14.38	21.56	28.75	35.94	43.13	50.32	57.50	64.69	71.88
50	9.25	18.50	27.75	37.00	46.26	55.51	64.76	74.01	83.26	92.51

and RHs. The RH is simply the actual vapor pressure of the air expressed as the per cent of the saturation vapor pressure. Since the saturation vapor pressure increases markedly with an increase in temperature, an increase in temperature results in a decrease in RH even though the vapor pressure remains the same. For example, at 15°C the saturation vapor pressure is 12.79 mm Hg and at this vapor pressure the RH is 100%, but with a rise in temperature to 30°C the saturation vapor pressure increases so much that a vapor pressure of 12.79 represents a RH of only slightly more than 40%.

Because the vapor pressure of the air in the intercellular spaces is usually near saturation and thus generally higher than that in the atmosphere, transpiration usually occurs as long as the stomata are open. The RH of the air in the intercellular spaces is usually at or near 100%. If both a leaf and the surrounding atmosphere are at 20°C and the intercellular space has a RH of 100%, the internal vapor pressure will be 17.54 mm Hg. If the RH of the atmosphere is 70% the vapor pressure will be 12.28 and the vapor pressure gradient (VPG) will be $17.54 - 12.28 = 5.26$. If, however, the RH of the atmosphere is only 40% the vapor pressure will be 7.02 and the VPG $17.54 - 7.02 = 10.52$, resulting in a transpiration rate about twice as fast as in the first example.

If a leaf were at a higher temperature than the air because it was exposed to direct sunlight the VPGs given above would be even steeper because, at the higher temperature, increased evaporation from the cell walls would maintain the vapor pressure near the new higher saturation point. It is even possible for transpiration to occur into a saturated atmosphere if the leaf is at a higher temperature than the air. For example, if a leaf is at 15°C and the air at 10°C and both are at 100% RH, the VPG would be $12.79 - 9.21 = 3.58$. Of course, the additional water would immediately condense into water droplets.

Vapor pressure is determined by two factors: the concentration of water vapor and the influence of temperature on the kinetic energy of the water molecules. The concentration of water vapor in the air is usually referred to as **absolute humidity,** which is expressed in units such as grams per cubic meter (g/m^3). As long as a leaf and the atmosphere are both at the same temperature the vapor pressure differences between them result entirely from concentration differences.

TEMPERATURE. A rise in temperature results in a marked increase in the rate of transpiration (Figure 9-13). The influence of temperature on the kinetic

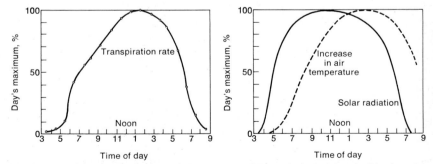

Figure 9-13. Rate of transpiration of alfalfa over an 18-hr period compared with changes in solar radiation and temperature. [Data of L. J. Briggs and H. L. Shantz, *J. Agr. Res.*, **5**:583 (1916)]

energy of water molecules makes little or no contribution to the increased VPG that brings about this increase in transpiration, but rather the temperature increase causes a marked increase in the concentration of water vapor in the intercellular spaces. The concentration of water vapor in the atmosphere increases little if at all. The marked increase in internal water vapor concentration occurs as a result of the large evaporating surface of the wet cell walls in comparison with the relatively small volume of air in the intercellular spaces. The temperature increase brings about an increase in the saturation vapor pressure of the intercellular spaces and also in the rate of evaporation, and enough water evaporates to saturate, or almost saturate, the intercellular spaces. Of course, the saturation vapor pressure of the outside air also increases, but the concentration of water vapor in the atmosphere increases little if at all in most cases because the external evaporating surfaces are small in comparison with the volume of the atmosphere and the atmosphere is expansive. Ordinarily the external atmosphere remains far from saturated.

Table 9-4 Examples illustrating the influence of temperature on the VPG between leaves and the air

		15°C	*25°C*	*Air 25°C; Leaf 30°C*
Leaf	RH	100%	100%	100%
	VP	12.79 mm Hg	23.76 mm Hg	31.82 mm Hg
Air	RH	70%	40%	40%
	VP	8.95 mm Hg	9.50 mm Hg	9.50 mm Hg
VPG		3.84 mm Hg	14.26 mm Hg	22.32 mm Hg

Some quantitative examples (Table 9-4) may serve to clarify the influence of temperature on the VPG and so on the rate of transpiration. If a leaf and the atmosphere are both at 15°C and we assume the RH to be 100% in the leaf and 70% in the air, the VPG will be 3.84 mm Hg. If the temperature of both then increases to 25°C later in the day, a not unlikely situation, we can assume that the internal RH will remain about the same, with a marked increase in vapor pressure, and that the external vapor pressure will increase only slightly while the RH decreases markedly. The VPG is now 14.26 mm Hg, and the rate of transpiration will be almost 4 times as rapid as it was at 15°C. If a leaf were 5°C warmer than the air because of exposure to direct sunlight, the VPG would be 22.32 mm Hg and its rate of transpiration would be about 1.6 times that of a shaded leaf of the plant.

AIR MOVEMENT. A third environmental factor that influences the rate of transpiration through its effect on the VPG is the rate of air movement over the surface of the leaves. A wind, or even a breeze, will blow away some of the moist air adjacent to the leaf surface and thus increase the steepness of the VPG. Also, if there is a difference between leaf and air temperature, convection currents are created even in an atmosphere with no other air flow, and these will also increase the VPG. Most of the increase in the rate of

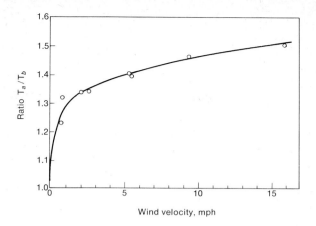

Figure 9-14. Influence of wind velocity on the rate of transpiration. [Data of E. V. Martin and F. E. Clements, *Plant Physiol.* **10**:613–636, 1935]

transpiration occurs at the lower wind velocities (Figure 9-14), probably because the steeper VPG at high velocities make the diffusive capacity of the plant the limiting factor. Also, the more rapid rates of transpiration may bring about a decrease in the internal vapor pressure and even stomatal closure because of their dehydrating influence and may reduce the VPG by cooling the leaves.

Soil factors. Any factor of the soil environment that brings about a reduction in the rate of water absorption by the roots may also result in a reduced rate of transpiration because of the reduced water potential in the plant. The reduced hydration of the plant may result in wilting and closing of the stomata. One soil factor that can bring about a reduction in the rate of transpiration is a decrease in the available soil water (Figure 9-15). Another is low soil temperature, which reduces the rate of water absorption and consequently the rate of transpiration (Figure 9-16). There are quite marked species differences as regards the influence of temperature on the rate of water absorption (Figure 9-17).

Wilting

A water deficit in a plant results in wilting (Figure 9-18) of the plant because of the reduced turgor pressure of its cells. Initially the loss of turgor may not be great enough to result in evident drooping of the leaves and stems, and

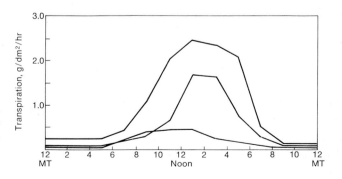

Figure 9-15. Daily periodicity of transpiration of *Phaseolus vulgaris* (bean) for three successive days during which the soil was becoming drier. [Data of Chung, Ph.D. Diss., Ohio State Univ., 1935]

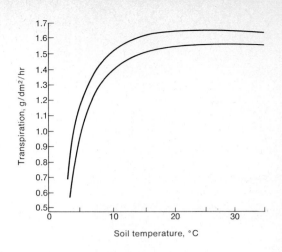

Figure 9-16. Influence of soil temperature of the rate of transpiration of sunflower plants (two different experiments). The decrease in transpiration at low temperatures resulted from low rates of water absorption. [Data of F. E. Clements and E. V. Martin, *Plant Physiol,* **9:**619 (1934)]

this condition is referred to as **incipient wilting.** However, in species such as magnolia with extensive mechanical supporting tissue in their leaves and with woody stems, marked loss of water and turgor may occur without wilting.

Two distinct types of wilting occur. **Temporary wilting** occurs when there is adequate soil moisture, but when the rate of transpiration exceeds the rate of water absorption (Figure 9-19). When the rate of transpiration is high the rate of absorption simply cannot keep up with it, and a water deficit develops. Such temporary wilting occurs commonly on the afternoons of hot summer days. Addition of water to the soil will not bring about recovery from temporary wilting because there is already adequate soil moisture, but anything that reduces the rate of transpiration sufficiently will be effective. The stomatal closure accompanying wilting may bring about recovery from temporary wilting, but after a short period of time the plant is likely to wilt again and may not recover until night.

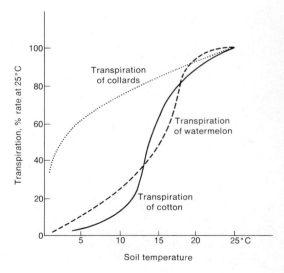

Figure 9-17. Species differences in rates of absorption and transpiration in relation to soil temperature. [Data of P. J. Kramer, *Amer. J. Bot.,* **28:**446 (1942)]

A   B

Figure 9-18. **(A)** Severely wilted leaves of squash plants in the late afternoon of a hot day. **(B)** The same plants the following morning. The wilting was temporary because no water was added to the soil. [From C. L. Wilson, W. E. Loomis, and T. A. Steeves, *Botany,* 5th ed., with the permission of the publisher, Holt, Rinehart and Winston, Inc.]

Permanent wilting results from a lack of available soil water. It is really permanent only in the sense that there will be no recovery from wilting until water is added to the soil. Most species can survive permanent wilting for several days to several weeks, but eventually death from desiccation results.

That a wilted plant has a very low level of metabolic efficiency is worth reemphasizing. The rate of photosynthesis may be severely limited by either stomatal closure, and the consequent carbon dioxide deficit, or the reduced

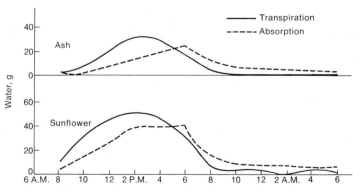

Figure 9-19. Rates of absorption of water and transpiration of two species of plants during a hot day. The excess of transpiration over absorption during the day results in temporary wilting. The plants recover from wilting when the rate of transpiration drops below that of absorption during the evening and night. [Data of P. J. Kramer, *Amer. J. Bot.,* **24:**10 (1937)]

Figure 9-20. Two cotton plants of the same age showing reduction of growth as a result of water stress and wilting. The plant at the right was kept well supplied with water whereas the soil in the pot at the left was allowed to dry progressively. [Courtesy of D. R. Ergle]

hydration of the protoplast (Chapter 3). The latter also reduces the rates of most other anabolic processes as well as membrane permeability. The growth of wilted plants is greatly reduced (Figure 9-20). The effects of growth inhibition by wilting may persist for some time (Figure 9-21).

Periodicity of Transpiration

The daily periodicity of transpiration is quite variable. It is influenced by species differences in resistance to water movement as well as by environmental factors. In general, the rate of transpiration is very low at night, rises rapidly during the morning to a peak around noon, and declines during the afternoon and evening (Figure 9-22). At night stomatal closure is the principal factor, whereas during the day both the degree of stomatal opening and changes in the vapor pressure gradient are involved. When the temperature is high there is likely to be a sharp drop in the transpiration rate as a result of desiccation and stomatal closure, often followed by a secondary increase and decrease resulting from the reopening and subsequent reclosing of the stomata.

In contrast with most plants succulents, such as Sedum, Bryophyllum, Agave, various cacti, and other xerophytes, have a very low rate of transpiration during the day and transpire mostly at night (Figure 9-22). This results from the fact that their stomata are open at night and closed during the day, the reverse of the usual situation. This, of course, essentially deprives photosynthesis of a carbon dioxide supply; however, survival is possible because succulents

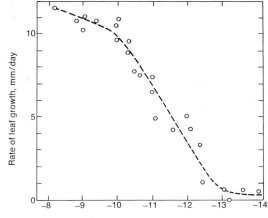

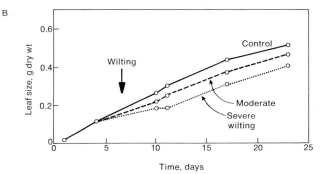

Figure 9-21. **(A)** Influence of the water potential of *Phaseolus vulgaris* (bean) leaves on their rate of growth. [Data of Brouwer, *Acta Bot. Neer.,* **12**:248 (1963)] **(B)** Influence of a single brief wilting period on the growth of young tomato leaves. Note the elapse of several days before the normal rate of growth resumed. [Data of D. Gates, *Aust. J. Biol. Sci.,* **8**:196 (1955)]

Figure 9-22. Daily periodicity of transpiration of sunflower and Opuntia (a cactus) plants, along with changes in rate of evaporation of water and the vapor pressure deficit, during parts of two hot summer days. Both plants were in soil at field capacity. The afternoon decrease in transpiration of sunflower probably resulted from temporary wilting and partial closure of the stomata. Note that the rate of Opuntia transpiration was low and the maximum was at night. See text for explanation. [Data of P. J. Kramer, *Amer. J. Bot.,* **24**:10 (1937)]

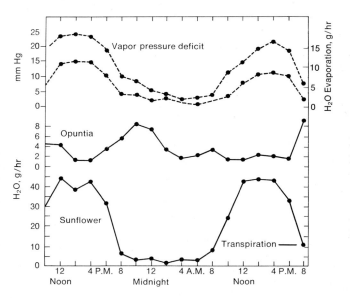

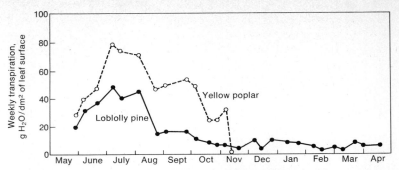

Figure 9-23. Seasonal periodicity of transpiration of potted seedlings of yellow poplar (deciduous) and loblolly pine (evergreen) at Durham, N. C. [After P. J. Kramer and T. T. Kozlowski, *Physiology of Trees,* McGraw-Hill Book Company, New York, 1960]

engage in extensive nonphotosynthetic carbon dioxide fixation during the night, primarily by carboxylation of phosphoenol pyruvic acid to oxalacetic acid and its conversion to malic and citric acids. The following day these acids can be photosynthetically reduced to sugars. This metabolic complex has obvious survival value in hot and dry environments.

There is also a marked seasonal periodicity of transpiration, as a result of either seasonal temperature changes or wet and dry seasons. In deciduous species there is also a marked decrease in diffusive capacity with leaf abscission, but even in evergreens the rate of transpiration is much lower in the winter than in the summer, spring, and fall (Figure 9-23). This results partly from the reduced steepness of the VPG as temperature decreases and partly because the low temperatures may induce stomatal closure. Also, low soil temperatures cause a considerable reduction in the rate of water absorption even when the soil is not frozen.

A period of warm weather during the winter increases the rate of transpiration of evergreens to a considerable degree. Since the soil warms up much less as well as more slowly than the air, the rate of water absorption remains low and the plants become desiccated. This may result in the death of part or all of the plant, a phenomenon known as **winter killing.** This is, of course, quite different from freezing injury.

Magnitude of Transpiration

Both the rate of transpiration per unit area of epidermis and the total amount of water vapor lost by a plant per unit of time vary so greatly from species to species, time to time, and place to place that it is difficult to give any meaningful quantitative data on the magnitude of transpiration in general. However, it may be noted that the rate of transpiration of broad-leaved temperate-zone plants during the day generally ranges between 0.5 and $3.5 \text{ g/(dm}^2)(\text{hr})$ and only rarely gets as high as 5 or $6 \text{ g/(dm}^2)(\text{hr})$. At night the rate is usually $0.1 \text{ g/(dm}^2)(\text{hr})$ or less.

A corn plant may lose over 2 liters of water per day. It has been estimated that the transpirational water loss by a field of corn during a growing season is enough to fill a pool the area of the field to a depth of 20 to 40 cm. Comparable figures for water loss by the plants of an eastern deciduous forest range between 43 and 56 cm per year. The quantity of water lost by transpiration is sometimes expressed as the transpiration ratio, which is the weight of water lost by transpiration divided by the weight of dry matter produced by photosynthesis during the growing season. This ratio varies greatly with the species and environmental conditions, ranging from 200 to 2000, although it is usually between 300 and 500.

Significance of Transpiration

Botanists have for many years been interested in the question as to whether the immense loss of water vapor by transpiration plays any useful roles in the life of plants, or whether it is simply an unavoidable and often harmful loss that cannot be escaped because there must be pathways such as stomata through which carbon dioxide and oxygen can diffuse. There is often a high correlation between the rates of photosynthesis and transpiration, but this is evidently simply due to the fact that both processes are favored by such factors as a high diffusive capacity (including open stomata) and an adequate supply of readily available water.

Three principal roles have been suggested for transpiration: (1) that it is necessary for the flow of water upward through the xylem to the shoot; (2) that it is necessary for delivery of mineral ions to the cells of the shoot at an adequate rate; and (3) that the cooling resulting from the evaporation of water during transpiration is essential to prevent the heat injury of leaves and other organs exposed to direct sunlight when the temperature is high.

FLOW OF WATER. Although the rate at which water flows through the xylem is determined by the rate of transpiration most of the time, transpiration is definitely not essential for an adequate rate of water flow. As we shall see in the following section, the rate of transpiration does not have any significant influence on water flow through the xylem resulting from root pressure. However, when water is being pulled up the xylem by the cohesion mechanism, its rate of flow is determined primarily by the rate of transpiration. Nonetheless transpiration is not essential for the operation of the cohesion mechanism, since the use of water in processes such as photosynthesis, hydrolysis, and growth is also effective in inducing its flow. Without transpiration the rate of water flow would be reduced by some 99%, but this slow flow would still provide water just as fast as it is being used.

DELIVERY OF MINERAL ELEMENTS. There have been numerous investigations designed to determine whether or not transpiration is essential for delivery of mineral elements to the shoot at adequate rates, but the problem is complex and the results have been conflicting. Transpiration can influence the rate of delivery of mineral ions only when it affects their rate of absorption from the soil. This does occur when water tensions in the xylem induce passive absorption of water, that is, the mass flow of water (and its solutes) from the soil, through the root tissues, and into the xylem. Also, Crafts and Hoagland have suggested that rapid flow of water through the xylem promotes active transport of ions into the xylem by keeping the salt content of the water in the xylem low.

However, the rates of diffusion of water and of the various ions into roots from the soil are independent of one another. Active absorption of ions is not influenced by the rate of water absorption. If water is flowing through the xylem slowly its ion concentration will be relatively high, whereas if the water is flowing rapidly the ion concentration will be lower. However, in both cases the rate of ion delivery to the shoot is likely to be about the same. This is analogous to the fact that the rate at which boxes are delivered at the end of a conveyor belt is determined by the rate at which boxes are placed on the other end of the belt, not by the rate at which the belt is moving (as long as it is moving at all). What *does* vary with the rate of movement of the belt is the concentration of the boxes on the belt, assuming that they are placed on the belt at the same rate in each case.

Although high rates of transpiration may increase mineral element absorption and translocation under certain conditions, most studies have failed to establish any consistent correlation; it is likely that any effect transpiration may have is not important or essential. There is no evidence at all of any plant developing mineral deficiency symptoms because of a low rate of transpiration.

REDUCTION OF TEMPERATURE.   There is no doubt that the evaporation of water during transpiration has a cooling effect on leaves or other organs (Figure 9-24), but the question is whether this cooling is ever essential in preventing heat injury to the tissues. Evaporation is only one of several means of heat loss from plants. Most of the heat loss is by reradiation, and this is particularly effective in flat, thin structures such as leaves. Another means of heat loss is by convection currents, which are generated whenever there is a temperature differential between a leaf and the air.

For many years plant physiologists thought that the cooling effect of transpiration did not play an important role in heat loss by plants, and that

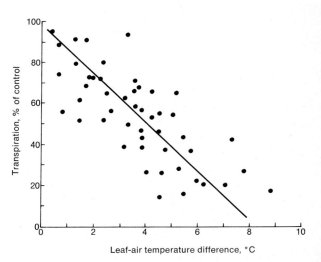

Figure 9-24. Effect of the rate of transpiration of cotton leaves exposed to the sun on the difference of temperature between the leaves and the air. The rates of transpiration of the experimental plants were reduced by application of transpiration-suppressing substances. [Data of R. O. Slayter and J. F. Bierhuizen, *Aust. J. Biol. Sci.,* **17**:131 (1964)]

reradiation and convection were capable of providing a rapid enough heat loss to prevent thermal injury or death even though there was no evaporation. Meyer and Anderson [11, p. 134] provided quantitative data in support of the idea that transpiration is not an essential factor in heat loss. However, in recent years a number of investigators, notably Gates [5, 6] and Leopold [2] and their coworkers, have reexamined this matter experimentally and have reached the conclusion that, under certain conditions, the heat loss by transpiration may indeed make an essential contribution toward prevention of thermal injury. Gates points out that a transpiration rate of 3 g/(dm²)(hr) will reduce leaf temperature by 15°C and that a rate about a tenth of this can dispose of about 29% of the absorbed heat, with 68% being lost by reradiation and only 3% by convection. At night there is essentially no transpiration and convection carries some heat from the air to the cooler leaves, so all heat loss is by reradiation. Leopold has found that leaves carrying on transpiration may be at least 5°C cooler than comparable leaves whose stomata have been closed experimentally.

On the other hand, other investigators point to the fact that when temperatures are the highest stomata are often closed and the transpiration rate is very low, but the plants are not often injured by heat. Of course, the drooping wilted leaves are generally not receiving as much solar radiation as turgid leaves that are more nearly at right angles to the radiation. Also, there has been little evidence of heat injury to plants that have had their stomata closed by chemicals or impermeable plastic films sprayed on the leaves. Nevertheless, it appears that transpiration does play a more important role in heat loss from plants than was once believed.

ADVERSE EFFECTS. Although there may be some doubt as to whether transpiration plays any essential roles in plants, and if so to what degree, there is no doubt that the water deficits developed in plants as a result of transpiration are definitely harmful. Kramer [in reference 17, p. 648] has stated that "It is probable that more plants are injured and killed by excessive transpiration than by any other cause." If there were no transpiration there would never be any winter killing of evergreens nor any temporary wilting of plants, and permanent wilting would be much less common because the soil water would not be depleted as rapidly. It is possible that land plants did not evolve in the best of all possible ways and that an unbroken cuticle permeable to carbon dioxide and oxygen but impermeable to water would have been better than the stomata, but it is obvious that the course actually taken by evolution has enabled plants to survive and flourish even in quite dry land habitats.

In recent years there has been considerable interest in the practical use of antitranspirants to reduce the rate of transpiration of plants being transplanted or of plants cultivated in dry or hot regions [22, 25]. These antitranspirants are of two principal types. One type brings about stomatal closure more or less permanently and includes a large variety of different chemicals. Even if such substances are not toxic to the plants they may be undesirable if the stomatal closure results in any significant decrease in the rate of photosynthesis. The other type consists of thin films of plastics or waxes which are sprayed on plants and then solidify into a sort of artificial cuticle. Such films should ideally be impermeable to water vapor and permeable to carbon dioxide and oxygen, but a number that have been used fall short in one respect or another. Both types can significantly reduce transpiration only if they can be applied to the greater part of the total leaf surface, and this is often difficult or impossible on a large field scale.

The Translocation of Water Through the Xylem

We now turn to a consideration of the forces that bring about the upward flow of water through the xylem from the roots to the cells of the stems, leaves, and reproductive organs of the shoot. This problem has interested botanists for many years and was investigated in considerable detail in the early 18th century by Stephen Hales [8]. To be generally applicable any satisfactory theory of water translocation must be able to account for the rise of water to the tops of the tallest trees, a height of some 120 m including the depth to which the roots penetrate the soil. A very large amount of energy is required to raise water to such heights, or even to the 30 m or so that is a more common height for trees.

Several mechanisms that might occur to a layman have never been considered seriously by botanists because of obvious theoretical inadequacies. The capillary rise of water through the lumens of tracheids or vessels could occur to a height of only a few centimeters in vessels and no more than a meter in the smaller tracheids. The capillary spaces within the cell walls could account for a higher rise, but it is known that most of the water flows through the lumens. Air pressure could account for a rise of no more than 10 m, and would also require evacuation of the xylem and exposure of the ends of the xylem in the roots to the air. At one time various vital theories were proposed, but these are now all considered discredited. Despite their differences, all these vital theories implied some participation by the living cells of the xylem, generally postulating some sort of pumping action. However, various experiments have made it clear that water can still flow through the xylem when all its cells have been killed in a section of a stem.

Only two of the various theories of water translocation through the xylem have proved to be both theoretically acceptable and supported by experimental evidence [13]. One is **root pressure,** and the other has been referred to variously as the **cohesion theory** (or **mechanism**), **transpiration pull,** the **transpiration stream,** the **transpiration-cohesion-tension mechanism,** and **shoot tension.**

Root Pressure

As the term implies, root pressure is the flow of water through the xylem as a result of pressure developed in the roots. The pressure is generated by the **active absorption** of water, which will be discussed later. Unlike the active absorption of mineral salts, the active absorption of water does not necessarily involve the expenditure of metabolic energy; the phrase implies only that the roots play an active role in absorption. Indeed, as we shall see later, active absorption appears to be largely, if not entirely, a matter of osmotic movement of water from the soil into and through the root tissues and then into the xylem. This is possible since the osmotic potential of water in the xylem of slowly transpiring plants is usually −2 bars or even somewhat less, whereas the osmotic potential of the soil water is generally on the order of −0.1 bar. In a general way, the tissues of the root can be considered to function as a multicellular differentially permeable membrane with a gradient of decreasing water potential (not osmotic potential) from the epidermis to the pericycle.

Root pressure can be demonstrated by decapitating a plant, attaching a piece of glass tubing, and observing the rise of xylem sap through the tubing. Attachment of a manometer permits the measurement of root pressure. The bleeding of xylem sap from cut or broken stems [21] and, in general, guttation result from root pressure, although somewhat dubious cases of active water secretion by hydathodes have been reported. Bleeding from the xylem may be quite profuse. Stephen Hales noted the rise of sap to a height of over 6.4 m in a tube attached to a cut vine stem over a period of several days. Crafts and Broyer [3] found that in 24 hr the exudate from squash plants ranged from 243 to over 550 ml, depending on the size of the root system.

Although root pressure undoubtedly can cause water to flow through the xylem, several facts seem to preclude it as being the primary and usual force causing the water flow. (1) The magnitude of root pressure is not great enough to force water to the tops of trees of any great height [9]. Root pressures of more than 2 bars have rarely been measured, and usually they are much lower. A root pressure of 2 bars can theoretically account for the rise of water to the height of about 20 m but, considering the energy required to overcome the friction of the flowing water against the walls of the vessels or tracheids, the actual rise would not be much more than half of this. (2) The rate of flow of water through the xylem resulting from root pressure is too slow, by a factor of 100 to 1000, to account for the usual rates of flow during the day. Rates as high as 4500 cm/hr have been measured and rates of around 1000 cm/hr are common in transpiring plants. Replacement of water by root pressure during periods of even moderate transpiration is not nearly rapid enough to make up the deficit. (3) Root pressure has never been found in many species of plants. (4) In species that do develop root pressure, it is generated only when the rate of transpiration is low as a result of such factors as stomatal closure or high soil moisture and high atmospheric humidity. (5) The water in the xylem is generally in a state of tension (negative pressure) rather than under positive pressure when transpiration is occurring.

Thus, although root pressure causes the flow of water through the xylem of some species part of the time, it apparently cannot be the sole, or even most general, cause of water flow through the xylem. Most students of water translocation have reached this conclusion, but a few such as White [23, 24] believe that root pressure plays a much more important role in the rise of water than is generally recognized. White has reported the development of root pressures of around 7 bars by isolated roots but, even if pressures of this magnitude existed in intact plants, they could raise water only to a height of about 35 m. Of course, most plants are no higher than this.

The Cohesion Mechanism

The literal pulling up of water columns in the xylem by tensions (negative pressures) generated in the shoot is made possible by the great cohesive strength of water when it is in airtight tubes of small diameter and contains a minimum of dissolved gases. This cohesiveness of water results from hydrogen bonding and perhaps other intermolecular attractive forces between water molecules. Estimates of the maximum tension to which a water column can be subjected without rupturing it have varied greatly, principally as a result of the different means of measurement. In 1948 Fisher estimated the theoretical maximum to be about -1275 bars, and earlier (1914, 1915) Dixon, Renner, and Ursprung all reported the measurement of maxima in the range of -200 to -350 bars in plant tissues, but in 1954 Greenidge [7] provided evidence that such values were too high and that the maximum tension that water columns in the xylem could stand without rupturing was only about -30 bars. However, even this would permit water columns to be pulled to the tops of the tallest trees without rupturing.

The fact that the water columns in the xylem are usually under tension rather than positive pressure has been demonstrated in various ways. If a carefully exposed xylem vessel viewed under a microscope is punctured the water column in it usually snaps apart, whereas the water would ooze or squirt out if it were under positive pressure. Micrometer measurements of vessel diameters have shown that they decrease with an increase in the rate of water flow, but the opposite would be true if the flow was caused by positive pressure. MacDougal was able to measure small daily variations in the diameter of tree trunks by use of dendrographs (instruments that magnified any changes in diameter by a system of levers and recorded them on a kymograph drum), and he found that the trunks were slightly smaller by day than by night (Figure 9-25). He ascribed this daily variation to the smaller aggregate diameter of the vessels when the water was flowing through them most rapidly and the tensions were greatest.

The operation of the cohesion mechanism of water rise has been demonstrated in a purely physical system (Figure 9-26) which consists of a long piece of glass tubing with a porous clay tube attached to its upper end, both being filled with water. The lower end of the tube is immersed in a container of

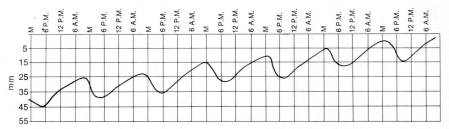

Figure 9-25. Dendrographic record of changes in diameter of a tree trunk for a week in early May when the midday rate of transpiration was high. Note the decrease in diameter during the day. The upward slope of the curve records the growth in diameter of the trunk during the week. The actual changes were 0.1 of those recorded by the kymograph. [Data of D. T. MacDougal, Carnegie Inst. Wash. Publ. 462, Washington, 1936]

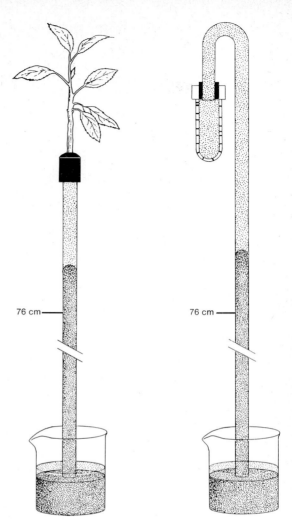

Figure 9-26. Demonstrations of the pulling up of water by the cohesion mechanism as a result of loss of water by transpiration from a branch of a plant and from a porous clay tube by evaporation. The water-filled glass tubes were immersed in mercury to avoid the use of tubes more than 10 m long, which would be needed to demonstrate the pull if only water had been used. The adhesive force between water and mercury is sufficient to pull the mercury up behind the water. Air pressure contributes toward the rise of the mercury to a height of 76 cm, but beyond that height only the cohesion mechanism is responsible for its rise. In demonstrations such as these the rise of mercury to heights of 100 cm or more has frequently been observed.

76 cm —

76 cm —

mercury. As water evaporates from the surface of the porous clay tube the water in the tube is pulled up, and there is sufficient adhesive force between the water and the mercury so that the mercury is also pulled up. Of course, to a height of about 760 mm atmospheric pressure is pushing the mercury up, but the rise of the mercury to a height of a meter or more is achieved when the apparatus is functioning properly. This experiment requires considerable care in setting up, including the use of water recently boiled to remove as much dissolved gas as possible, air-tight joints, and clean glassware. Use of a tree branch in place of the porous clay tube can also bring about the rise of the mercury to a height of a meter or more. This requires recutting the stem under water so that the air-filled vessels or tracheids at the end are removed and a continuous water column in the stem is provided. Recutting of stems under water also promotes the survival of cut flowers and results in much more rapid recovery from wilting by cut flowers or branches (Figure 9-27). Note that the

Figure 9-27. Two cuttings from Amaranthus (pigweed), both initially wilted severely, 30 min after being placed in water. The terminal 3 cm of the stem of the one on the left was cut off under water to remove the air-filled xylem. [Photograph by V. A. Greulach]

flow of water through such detached parts of plants cannot possibly be a result of root pressure.

The question remains as to whether great enough tensions can be developed to pull water to the tops of the tallest trees—about 120 m. A tension of about 12 bars is needed [120 m/(10 m)(bar) = 12 bars] but, at rapid rates of water flow, about an equal amount of energy is required to overcome friction, plus an additional 5 or 6 bar tension to pull the water through the tissues of the roots. Thus, a total tension of about −30 bars is needed [−12 + (−12) + (−6) = −30]. This means that if a leaf or stem cell is to obtain water from the xylem it must have a water potential of somewhat less than −30 bars, and indeed this appears to be the case. The Ψ_π of the leaves of trees, especially tall species, is commonly at least −30 bars and often considerably less, and Ψ_c will be even lower than Ψ_π in cases where there is a negative turgor pressure. Thus, the cohesion mechanism (in contrast with root pressure) is theoretically capable of raising water to the tops of the tallest trees.

It is interesting to note that a tension of 30 bars is essentially the maximum that water columns in the xylem can endure without rupturing, according to Greenidge, and it is possible that the maximum height of trees is limited by the height to which they can raise water. Of course, considerably smaller tensions are required to raise water to the tops of lower trees. A tension of 12 bars is theoretically adequate for a 30-m tree, although greater tensions than this are likely to exist at times. The Ψ_π of herbaceous plant leaf cells is generally in the range of −8 to −15 bars, in contrast with a usual range of −20 to −35 for tree leaf cells, but in each case tensions adequate for pulling water to the tops of the plants are possible.

As has been noted, the rate of transpiration is the principal factor determining the rate at which water is pulled up by the cohesion mechanism (Figure 9-28). The evaporation of water from the wet cell walls into the intercellular spaces reduces the water potential in the walls and causes water to diffuse into them from the protoplast or vacuole of the cell as well as from the walls of adjacent cells that may not be in direct contact with an intercellular space. Because of the great imbibitional forces in cell walls, almost any deficiency in their hydration will result in the movement of water into the walls from the protoplasts and vacuoles. The loss of water by evaporation and the resulting movement of water from cell to cell bring about the removal of water from the xylem of the veins. Because of the cohesiveness of the water in the xylem, the water is pulled up, and the tension is transmitted through the length of the xylem.

Although the loss of water by transpiration is primarily responsible for the rate at which water flows through the xylem, it should be noted again that any use of water in processes such as photosynthesis or hydrolysis or in the hydration of the protoplasts and walls of new cells will reduce the water potential and bring about removal of water from the xylem. Without transpiration water would flow through the xylem at about 1% of its usual rate but, since it would be delivered just as fast as it was being used, the slow rate of flow would be quite adequate.

Despite the evidence in support of shoot tension and its general acceptance by plant physiologists as the principal means of water rise, a few investigators continue to raise objections to it and cite evidence that they feel invalidates the mechanism or at least requires revision of the concept of how it operates. It has been pointed out that water under tension in a glass tube may rupture

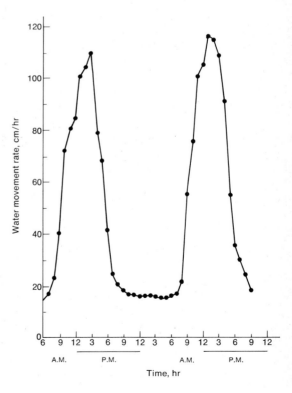

Figure 9-28. Daily periodicity of the rate of water flow through the stem of a cotton plant. The marked increases in the rate of flow during the daylight periods result primarily from the increases in the rate of transpiration. [Data of M. E. Bloodworth *et al.*, *Agron. J.*, **48**:222 (1956)]

with only slight jarring, whereas plant stems may be violently bent by winds without breaking all the water columns in the xylem. Actually, the difference in the diameter of the tubes in the xylem and the glass tubing and the fact that the walls of vessels and tracheids are impregnated with water could account for this difference.

It is also pointed out that extensive pruning of a tree will admit air into many vessels or tracheids but does not interfere with the flow of water to the remaining branches. Furthermore, when a tree trunk is cut horizontally more than halfway through from opposite sides at two different levels, water continues to flow through the trunk although all the vertical water columns have presumably been broken. In answer to such objections it may be noted that the conductive capacity of the xylem is generally much greater than actually required so that rupture of a substantial portion of the water columns is not critical. Also, lateral flow of water through the xylem can explain its movement around the horizontal cuts into the trunk.

Another difficulty is that during periods of rapid transpiration a considerable number of the vessels may contain gas that interrupts the water columns, this being particularly true in large vessels. The excess conductive capacity of the xylem may be great enough so that a considerable number of the water columns may be ruptured without producing a water deficit in the shoot. The breaks in the water columns may become filled with water by means that are not well understood, although this may not occur until the rate of transpiration has been greatly reduced, perhaps even by leaf abscission. Capillary action or root pressure might be effective in reestablishing water columns or, as Crafts has suggested, the gas in the breaks may be largely water vapor which may later condense. Also, it should be noted that the walls of the vessels or tracheids, unlike glass tubing, are permeable to gases and the gas may diffuse out through them.

Such questions about the operation of the cohesion mechanism constitute valid criticisms, but none of the critics has proposed a more acceptable theory as to the means of water translocation. It seems likely that further investigations will provide more adequate explanations than those given above as to how water can continue to be pulled up despite difficulties such as those that have been pointed out. At any rate, the evidence for shoot tension as the principal means of water rise in plants is so great that it will receive general acceptance until there is much more evidence against it and evidence for an acceptable alternative mechanism [14].

The Absorption of Water

Most water absorption by vascular plants is from the soil through roots, although epiphytes absorb water through aerial roots, leaves, or other organs and some submerged aquatics can absorb water through almost any surface. The present discussion considers only the absorption of water from the soil by roots.

Water absorption occurs most readily in the root hair zone, located a short distance back from the root tip. In this region the xylem is well differentiated and permeability of the root surface has not been reduced by suberization. Despite the fact that they greatly increase root surface, the importance of the root hairs in water absorption has probably been overestimated. In moist soils root hairs are probably not really essential and in dry soils they usually collapse. Although suberized root surfaces offer considerable resistance to water absorption, Kramer [10] found that much more water absorption occurs through them than had been believed. This is particularly true in large trees, where the suberized surfaces constitute a large portion of the total root surface, and in the winter, when the unsuberized root surface is at a minimum.

There are two principal means of water absorption by roots: **active absorption,** which results in root pressure, and **passive absorption,** which results from shoot tension. We shall consider both mechanisms, but they will be more intelligible if we first consider the ways that soils hold water.

Soil Water

Most soils are composed primarily of fine rock particles derived from the weathering of rocks mixed with variable portions of humus, which consists of the partially decayed remains of organisms. Soils also contain numerous living organisms of a wide variety, including bacteria, fungi, algae, protozoa, nematodes, and larger animals, but these are peripheral to our present discussion. We are, however, interested in the soil atmosphere, which fills the spaces between soil particles not occupied by water.

The primary classification of soil types is based on the size range of the rock particles. If these are predominantly less than 0.002 mm in diameter the soil is classified as clay, if 0.002 to 0.02 mm as silt, if 0.02 to 0.2 mm as fine sand, and if 0.2 to 2.0 mm as coarse sand. The particle size of a soil is of major importance in determination of its structure, texture, water holding capacity, and aeration.

Water is held in the soil by several different forces. A small amount of water is chemically bound within the soil particles as a component of soil minerals. A film of water (**hygroscopic** water) only a few molecules thick is held on the surface of each soil particle by strong adsorptive forces that reduce its potential to around -1000 bars, so this hygroscopic water is completely unavailable to plants. Since the small particles in a clay soil have a much greater total particle surface than the larger particles in an equal volume of sandy soil, the clay soil can hold much more hygroscopic water. In the spaces between soil particles that are small enough to hold water by capillary action, there will be **capillary** water as long as the soil is moist. Although the capillary forces can hold the water against the force of gravity, they are too weak to reduce water potential significantly. The principal factor reducing the potential of capillary water is the presence of molecules and ions of mineral salts, and the water potential is usually greater than -1 bar so that the capillary water is readily available to plants. Because its smaller particles are more numerous and fit together more closely, a clay soil has many more spaces of capillary size than a sandy soil and can hold more capillary water.

The larger soil spaces that cannot hold water by capillary forces are drained of water by gravity a short time after a rain or irrigation, unless drainage is blocked and there is a high water table. This temporarily held water is referred to as **gravitational** water. Like capillary water, its potential is reduced only by solutes and it is readily available to plants. However, retention of gravitational water in poorly drained soils creates an undesirable situation for most species of plants because of the displacement of the soil atmosphere. The poor aeration reduces the rate of root respiration and may even limit respiration largely to the anaerobic type. This may result in reduced accumulation of mineral salts, inadequate energy for other metabolic processes, reduced growth, and in some cases even death of the plants. Thus, for most species it would be desirable if a soil held all the capillary water it could but no gravitational water. Of course, some species such as rice, various aquatics, and plants that grow at stream or lake margins thrive when their roots are submerged. Since sandy soils have more spaces too large to hold capillary water than an equal volume of clay soil, gravitational water constitutes a higher per cent of the water supplied to a sandy soil and it is better aerated when well drained.

Two soil moisture values are of particular importance in relation to water absorption by plants. One is **field capacity** (FC), the percentage water content of a soil after drainage of gravitational water has practically ceased and the soil is holding all the capillary water it can. The other is the **permanent wilting percentage** (PWP), the percentage of water in a soil when plants growing in it undergo permanent wilting. This occurs approximately when all the capillary water has been exhausted. Whereas field capacity is a purely physical value, the PWP is basically a physiological value and can be expected to vary with the species of plant used in making the determination. Thus, a species with

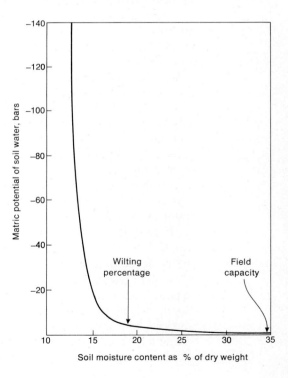

Figure 9-29. Relationship between the water content of a silty clay loam soil and the matric potential of the soil water. [Data of Magistad and Breazeale, *Univ. Ariz. Tech. Bull.* No. 25, 1929]

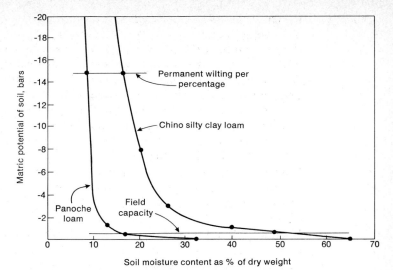

Figure 9-30. Relationship between the water content of two different soils and the matric potential of the soil water. Note that field capacity and PWP for the two soils are at quite different percentages of water content but are at the same matric potential in both cases. [After P. J. Kramer, *Plant and Soil Water Relations,* 2nd ed., McGraw-Hill Book Company, 1969]

a cell water potential of − 100 bars would remove more water from a soil by the time of permanent wilting than a species with only − 10 bars, and a species with a limited root system would leave more water in the soil at permanent wilting than would one with roots thoroughly permeating the soil sample being measured. However, because of the sudden marked decrease in soil water potential when the capillary water has been exhausted (Figure 9-29) the species used is of less importance than might be expected. The range of PWP obtained by use of different species is generally only a few per cent, but there are marked differences in the PWP from one type of soil to another. The difference between the FC and PWP is sometimes referred to as the **water supplying capacity** (WSC) of a soil, and this is approximately a measure of the capillary water held by the soil. (The percentage water content of a soil is based on the dry weight of the soil, whereas the percentage water content of a plant is based on its fresh weight.)

Because of their differing capacities for holding capillary and hygroscopic water, the FC and PWP of various kinds of soil differ (Figure 9-30). Clay soils have higher FC, PWP, and WSC than sandy soils, with loam soils intermediate (Table 9-5). It should be stressed that the figures in Table 9-5 are merely typical

Table 9-5 Typical soil water contents (%) of different types of soils

	Clay	Silt	Sand
Field capacity	40	20	10
Permanent wilting	20	10	5
Water-supplying capacity	20	10	5

values, and that each specific type of clay, silt, or sandy soil has its own characteristic values.

Although the FC and PWP of different soils may vary greatly, the potential of the soil water is essentially the same in all soils at FC and also at the PWP. Except in situations such as salt marshes or alkaline soils where there is an unusually high concentration of solutes, the soil water potential at FC is about -0.1 bars and at the PWP about -15 bars, although this varies with the species of plant. The water potential of air dry soil is about -1000 bars.

It is interesting to note that a soil cannot be wet to less than its FC and also that a soil will remain above FC only for a short time if there is dry soil below the moist layer so that a capillary equilibrium can be established. The amount of water supplied to a soil determines the depth of the moist layer of soil, not its percentage water content. A light rain wets only a thin layer of soil to FC, whereas a heavier rain wets a thicker layer of soil to FC. If a layer of soil is at FC, additional water will increase the percentage water content above FC only until the additional water has been drained through by gravity and has brought an additional layer of soil to FC. The line of demarcation between the moist and dry soil is sharp (Figure 9-31), and there is essentially no capillary movement of water into the dry soil.

Of course, as plants absorb water from the soil or as the soil water evaporates

Figure 9-31. A box with a glass wall was partially filled with dry soil and then some of the same soil at FC was added and corn was planted. Note that the roots of the seedlings did not grow into the dry soil any distance and also that there was no capillary movement of water into the dry soil. [Courtesy of Richard Böhning]

the water content is reduced below FC and may eventually reach the PWP. As roots absorb water from the soil, a layer of the soil near the root is depleted of capillary water and, since there is negligible capillary movement of water from moist to dry soil, the soil adjacent to the roots may be at the PWP while other portions of the soil are still at FC. One important result of continuing root growth is that it taps new regions of soil not yet depleted of water (or of mineral salts). Also, a highly branched root system that permeates the soil is advantageous from the standpoint of water availability and absorption.

Water may rise up through the soil by capillary action from an underground water table as well as moving down from the surface, and in either case the soil is brought to FC. Since the capillary spaces in clay soils are mostly smaller than those in sandy soils, water will rise higher in a clay soil than in a sandy soil as long as the underground water table remains. After periods of heavy rains the upper and lower layers of soil at FC may merge, and during such a period the roots of species with long roots may grow down even as far as the water table. Subsequent drying of the soil may result in a dry layer of soil between the moist layers, and such species may continue to thrive during a drought when all the upper part of the soil has been reduced to the PWP. Of course, the roots did not grow through the dry soil in a search for water. Actually, roots cannot grow through dry soil (Figure 9-31), and frequent light watering or rains may result in shallow root systems because only a thin layer of soil is wet to FC.

Active Absorption

Active absorption of water refers to absorption in which the roots themselves play an essential role. As we have seen, active absorption generates root pressure, so clarification of the forces involved in active absorption is essential for an understanding of root pressure as well as of the active absorption itself.

Numerous investigations over the years have resulted in the proposal of three principal sets of theories to explain the active absorption–root pressure complex: secretion theories, electroosmotic theories, and osmotic theories [9]. The secretion theories involve the secretion of water into the xylem at the expense of respiratory energy. The electroosmotic theories involve the influence of electrical potential gradients on the movement of water across membranes. Experimental evidence has been presented for and against all the theories, and there has been considerable controversy among proponents of the different theories. None of the theories has been completely discredited, but most of the students of the subject have now concluded that neither the secretory nor electroosmotic forces are adequate for explaining more than a minor movement of water and that simple osmotic movement of water can provide an adequate explanation.

Since the osmotic potential of the xylem sap is generally -2 bars or less whereas that of the soil solution is on the order of -0.1 bar as long as capillary water is available, and since the living cells of the root provide differentially permeable membranes across which a water potential gradient can be established from the epidermis to the pericycle, the osmotic movement of water from the soil solution into the xylem appears to be quite feasible.

The principal problem is to explain how a sufficient concentration of solutes is maintained in xylem sap to provide the necessary water potential gradient. This problem is complex and each of the various theories that has been proposed involves contradictions and obscurities. The extensive literature on the subject has been reviewed and evaluated by Kramer [9]. However, whatever the mechanism involved, the solute concentration of the xylem sap evidently is adequate.

A number of the theories, including the rather widely held one of Crafts and Broyer [3], involve the endodermis in an essential role in combination with the accumulation of ions by the cells of the epidermis and cortex by active transport. Crafts and Broyer proposed that the ions are carried through the endodermis and into the stele by diffusion through the protoplasts and cytoplasmic streaming and that the poorly aerated cells of the pericycle could provide less respiratory energy for accumulating and retaining ions which would then leak into the xylem. Back diffusion through the walls of the endodermal cells would be prevented by the suberized Casparian strips. Since water also cannot move through the walls of the endodermal cells but must pass through their protoplasts, the endodermis has been visualized as a particularly effective differentially permeable membrane in the osmotic system. However, the effectiveness of the endodermis as a barrier to the movement of water and ions through walls has been questioned on a number of bases, including the fact that water can easily flow through walls at the numerous places where branch roots puncture the endodermis.

Passive Absorption

The passive absorption of water does not require participation by the cells of the roots, which merely provide an absorbing surface. The water is absorbed through the roots rather than by them. Indeed, passive absorption can occur through a dead root system or when no roots are present, as when a shoot or a portion of a shoot is cut off and placed in a container of water. If a root system is killed (as by immersing the roots of a potted plant in boiling water) the rate of passive absorption usually increases, since the dead root cells offer less resistance to the passage of water through them than do the living cells with their differentially permeable membranes. Of course, a plant with a dead root system will not survive long because the roots will soon begin to decay and disintegrate.

As we have noted, passive absorption results from the development of tensions in the water columns of the xylem, and these tensions are transmitted across the root tissues to the root surfaces. This causes a mass flow of water through the roots and into the xylem. In living roots this mass flow of water occurs primarily through the free space of the root cells. The tensions may become quite great and, thus, cause a substantial reduction in water potential that results in more complete absorption of soil water than is possible by active absorption. Active absorption usually ceases when the soil water potential has been reduced to -2 bars (or even before this point), whereas passive absorption can occur even when the soil water potential is on the order of -15 to -20 bars.

Because the magnitude of the tensions developed in the water columns is determined primarily by the rate of transpiration, it would be anticipated that the rate of passive absorption would more or less parallel the rate of transpiration; and there is abundant experimental evidence that this is indeed the case. However, during periods of rapid transpiration there is commonly a lag of the absorption rate behind the rate of transpiration (Figure 9-18), primarily because of the resistance to water flow through the root tissues. This commonly results in temporary wilting, from which the plants recover later in the day when the transpiration rate drops and is exceeded by the rate of absorption. Of course, water absorption ceases when the permanent wilting percentage is reached and, under these conditions, any further transpiration results in the development of higher and higher tensions, possible rupture of the water columns, and perhaps eventual death of the plant by desiccation.

Since the flow of water through the xylem is usually brought about by the cohesion mechanism rather than root pressure, it seems evident that passive absorption is a much more common means of water absorption than is active absorption.

Factors Influencing the Rate of Water Absorption

As has already been noted, the rate of transpiration is generally the most important factor in determining the rate of passive absorption of water. Active absorption of water ordinarily occurs only when the rate of transpiration is low and, even though the rate of water absorption is also low under these conditions, it may exceed the rate of transpiration, resulting in guttation or bleeding.

A number of soil factors may limit the rate of water absorption under certain conditions. The solute concentration of the soil solution is in most cases low enough so that it is not a limiting factor in water absorption, but in a few situations such as salt marshes or the alkaline soils of dry regions solute concentration may be limiting. Soils that have been irrigated for years may also have high solute concentrations that limit water absorption because there is not enough rain to leach out the salts continuously brought in with the irrigation water. The burning of plants by excessive application of fertilizers is another example of inhibition of water absorption by a high solute concentration in the soil solution.

Of course, soil water content becomes limiting as the PWP is approached. At the other extreme, water absorption by many species may be limited by flooded soils because of inadequate aeration. Inadequate oxygen for respiration may reduce ion accumulation and, consequently, active absorption by reducing the steepness of the water potential gradient between the soil solution and the xylem sap. The limitation of water absorption by poor aeration has been cited as evidence for metabolic water absorption, but as we have noted there is really no evidence for nonosmotic active water absorption. Actually, the adverse effects of poor aeration on water absorption are still not well understood.

Another factor that may limit the rate of water absorption is low temperature (Figures 9-16 and 9-17). The decrease in the rate of water absorption with decreasing temperature is apparently brought about as a complex of factors.

The viscosity of water increases with a decrease in temperature, and this would be particularly important in reducing the rate of passive absorption. Also, the permeability of root cells decreases with temperature, thus increasing the resistance to water movement through the root tissues. A decrease in temperature may also bring about a decrease in the rate of root growth and the continuing growth of roots into soil not yet depleted of its available water. Of course, if the temperature becomes low enough to freeze the soil water, absorption essentially ceases but even temperatures well above freezing may greatly restrict water absorption.

Species native to warm climates are generally less capable of water absorption at low temperatures than those native to cool climates. Also, after a period of cold weather the rate of absorption may increase even though there is no change in temperature, possibly because of increased permeability of the roots. This is apparently one of the changes that occurs during the hardening of greenhouse plants prior to transplanting them outside. It has been found that active absorption of water generally rises to a maximum and then declines with a further increase in temperature. For both tomato and sunflower plants the optimal temperature for active absorption is 25°C. However, passive absorption continues to rise with temperature because of the marked increase in the rate of transpiration with temperature. Of course, there is a marked drop in both transpiration and absorption as the stomata close.

Mycorrhizae and Absorption

Most species of trees and shrubs and some herbs have mutualistic fungi of various species growing in or on their roots, and the association is referred to as a mycorrhiza. Endotrophic mycorrhizae occur within the cortical cells of the roots, whereas ectotrophic mycorrhizae grow over the root surface and often cover large parts of it (Figure 9-32). The mycorrhizal fungi are present in forest soil and become associated with seedlings that germinate there.

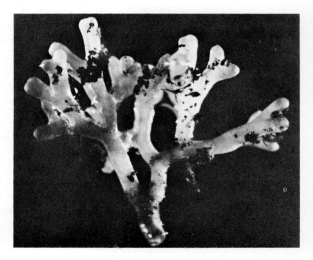

Figure 9-32. Photograph of mycorrhizae on *Pinus virginiana* roots. [Courtesy of Edward Hackskaylo, U. S. D. A.]

Seedlings growing elsewhere frequently do not thrive because of the lack of mycorrhizae.

The fungus secures food from the roots of its host while it contributes substantially to the absorption of water and mineral ions by the host plant [9], particularly in species with limited numbers of root hairs. It has also been suggested that the fungi may produce antibiotics that protect young roots from parasites. The growth of mycorrhizae is promoted by any factor that results in high carbohydrate levels in the roots, for example high rates of photosynthesis and moderate supplies of mineral elements, especially nitrogen and sulfur. High levels of these result in use of much of the carbohydrate in the synthesis of proteins and other substances.

References

[1] Branton, Daniel. "Membrane structure," *Ann. Rev. Plant Physiol.,* **20:**209–233 (1969).
[2] Cook, G. D., J. R. Dixon, and A. C. Leopold. "Transpiration: its effects on plant leaf temperature," *Science,* **144:**546–547 (1964).
[3] Crafts, A. S., and T. C. Broyer. "Migration of salts and water into xylem of roots of higher plants," *Am. J. Bot.,* **25:**529–535 (1938).
[4] Fischer, R. A. "Stomatal opening: role of potassium uptake by guard cells," *Science,* **160:**784–785 (1968).
[5] Gates, D. M. "Heat transfer in plants," *Sci. Amer.,* **213**(6):76–84 (Dec. 1965).
[6] Gates, D. M., R. Alderfer, and E. Taylor. "Leaf temperature of desert plants," *Science,* **159:**994–995 (1968).
[7] Greenidge, K. N. "Ascent of sap," *Ann. Rev. Plant Physiol.,* **8:**237–256 (1957).
[8] Hales, S. *Vegetable Staticks,* Innys & Woodward, London, 1727; Scientific Book Guild, London, 1961.
[9] Kramer, P. J. *Plant and Soil Water Relations,* 2nd ed., McGraw-Hill, New York, 1969.
[10] Kramer, P. J. "Absorption of water through suberized roots of trees," *Plant Physiol.,* **21:**37–41 (1946).
[11] Meyer, B. S., and D. B. Anderson. *Plant Physiology,* 2nd ed., D. van Nostrand, New York, 1952.
[12] O'Leary, J. W., and P. J. Kramer. "Root pressure in conifers," *Science,* **145:**284–285 (1964).
[13] Richardson, M. *Translocation in Plants,* St. Martin's Press, New York, 1968.
[14] Scholander, P. F., *et al.* "Sap pressure in vascular plants," *Science,* **148:**339–346 (1965).
[15] Slayter, R. O. *Plant Soil-Water Relationships,* Academic Press, New York, 1967.
[16] Stadelmann, E. J. "Permeability of the plant cell," *Ann. Rev. Plant Physiol.,* **20:**585–602 (1969).
[17] Steward, F. C. (ed.). *Plant Physiology: A Treatise,* Vol. II, Academic Press, New York, 1959.

[18] Sutcliff, J. *Plants and Water,* St. Martin's Press, New York, 1968.

[19] Ting, I. P., and W. E. Loomis. "Diffusion through stomates," *Am. J. Bot.,* **50:**866–872 (1963).

[20] Turrell, F. M. "Correlation between internal surface and transpiration rate in mesomorphic and xeromorphic leaves grown under artificial light," *Bot. Gaz.,* **105:**413–425 (1944).

[21] van Raalte, M. H., and R. J. Helder. "The bleeding sap of plants," *Endeavour,* **28:**35–39 (1969).

[22] Waggoner, P. E., and I. Zelitch. "Transpiration and the stomata of leaves," *Science,* **150:**1413–1420 (1965).

[23] White, P. R. "Vegetable Staticks," *Sigma Xi Quart.* (now *Am. Sci.*), **30:**119–136 (1942).

[24] White, P. R. *et al.* "Root pressure in gymnosperms," *Science,* **128:**308–309 (1958).

[25] Zelitch, I. "Stomatal control," *Ann. Rev. Plant Physiol.,* **20:**329–350 (1969).

Translocation and Mobilization of Solutes

10

Solutes, as well as water, are translocated over long distances through the vascular system of plants, but the translocation of solutes of various kinds is much more complicated and much less thoroughly understood than the translocation of water. It has been known for many years that solutes are translocated through the vascular tissues and that the velocity of solute translocation is far greater than can be accounted for by simple diffusion, but despite years of intensive research the mechanism of solute transport through the phloem has not been entirely clarified. Since the development of the nuclear reactor and cyclotron, the ready availability of radioisotopes for use as tracers has aided in the elucidation of a good many points that could not be settled by older research techniques, such as ringing the bark of woody plants. Nevertheless, there are still many aspects of solute translocation that are unexplored.

Associated with the translocation of solutes [6, 17] is the mobilization of solutes or the differential distribution of solutes among the various tissues and organs of a plant. In general, solutes move from regions where they are absorbed or synthesized to regions where they are utilized and also from older, senescent tissues such as older leaves to metabolically active tissues in young leaves, meristems, or developing flowers and fruits. Mobilization is often, but by no means always, from a region of higher to lower concentration of the particular solute. Solute mobilization will be considered in the latter part of this chapter.

Principal Substances Translocated

Although a considerable variety of inorganic and organic compounds may be translocated through the vascular tissues, most of the solutes translocated are of only a few different classes. One important group consists of the ions of mineral elements absorbed by the roots and translocated to the stems, leaves, and reproductive organs of the shoots. Many mineral ions are readily retranslocated from one part of a plant to another, particularly from older leaves to young leaves or other metabolically active tissues, and may be carried either up or down the vascular tissues. When minerals are applied as foliar sprays, the principal direction of translocation is likely to be toward the nearest meristem. Most mineral elements are translocated as ions rather than as constituents of organic compounds, but in many species of plants nitrate reduction occurs predominantly in the roots and there is very little translocation of nitrates or other inorganic nitrogen compounds unless excessive amounts are absorbed. For example, it was found that in apple trees 90% of the nitrogen compounds translocated from the roots to the shoots consisted of asparagine, glutamine, aspartic acid, and glutamic acid, and most of the remainder consisted of other amino acids rather than inorganic compounds.

Sugars constitute the great bulk of the organic compounds that are translocated and, as might be expected, the principal translocation of sugars is from the leaves or other photosynthetic organs to nonphotosynthetic tissues or metabolically active tissues that are using sugars faster than they can synthesize them. Such sugar translocation is predominantly toward the nearest meristem. There may also be substantial translocation of sugars from storage organs where starch or other foods are being hydrolyzed. At one time it was assumed that glucose and fructose were the principal sugars translocated, but there is now abundant evidence that monosaccharides are not translocated in quantity through the phloem. In most species sucrose is the principal sugar translocated. However, in some species, particularly certain trees such as ash, elm, and linden [12], there may also be substantial quantities of the galactose oligosaccharides translocated. These sugars are raffinose, stachyose, and verbascose, which consist respectively of one, two, and three galactose residues attached to the glucose residue of sucrose. It is possible that these tri- to pentasaccharides are important translocation sugars in more species than currently recognized. The phloem sap usually contains 10 to 25% solutes and the greater part consists of sugars.

Evidence as to the extent to which amino acids are translocated out of leaves is inconclusive, but the phloem sap rarely contains more than 1% of amino acids or other organic nitrogen compounds. The greatest movement of amino acids out of leaves probably occurs from senescent leaves prior to their abscission. Little is known about the translocation of fats, fatty acids, or glycerol, but it is likely that these are rarely if ever translocated. Sugar alcohols, particularly mannitol and sorbitol, are translocated in some species. However, by far the most important translocation forms of foods are sucrose and the galactose oligosaccharides.

Among the other substances translocated through the vascular tissues of plants, the auxins, vitamins, and other plant growth substances are probably of greatest importance. Although they constitute only a minute portion of the solutes translocated, they are effective in very low concentrations and the small quantities translocated are extremely important in the growth and development of the plant. The translocation of hormones will be considered in a later chapter, but it should be mentioned here that translocation of auxins through the phloem is polar and occurs only in a morphologically downward direction.

Pathways of Solute Translocation

In contrast with the fact that the xylem is clearly the pathway of water flow through the vascular system, determination of the pathway of solute translocation through the vascular system has been difficult. Although solutes are translocated through both the xylem and the phloem, there has been considerable uncertainty about the roles played by each one, particularly as regards upward and downward translocation. Over the years various investigators have reached conflicting conclusions from experimental data and have engaged in considerable controversy with one another.

One thing that contributes toward the problem is that there can be a substantial lateral interchange of solutes between the xylem and phloem throughout their length. This led to some uncertainty about the results of experiments involving surgical techniques, the principal procedure available to the earlier investigators. The most common technique was removal of a ring of bark from the stem of a woody plant (girdling), thus interrupting the phloem but leaving the xylem intact (Figure 10-1). Partial girdling was employed in some experiments as was removal of all or part of a short cylinder of xylem through a slit in the bark. The girdling of herbaceous stems is impractical, but phloem translocation can be interrupted by killing the cells in a short piece of the stem with a jet of steam or a ring of hot wax. This does not interfere with the flow of the xylem solution. Although such experiments provided a substantial amount of information, satisfactory clarification of the pathways of solute translocation was not achieved until radioisotopes became available for use as tracers. The tracers have been used in conjunction with surgical techniques and also without them.

In summarizing the present state of knowledge regarding the pathways of solute translocation we shall consider upward and downward translocation separately. Upward is used in a morphological sense, from the stem base toward the stem apex. Upward translocation is usually but not necessarily up in relation to the earth, one exception being a drooping branch of a weeping willow tree. In general, translocation into a leaf from the vascular cylinder of a stem can be considered as upward and translocation from a leaf to the vascular cylinder as downward.

Figure 10-1. A trunk of a young tree two years after it was ringed (girdled). The girdle prevented translocation of sugars and other substances from the leaves through the phloem, thus inhibiting growth below the girdle. Since the girdle did not prevent the upward flow of water through the xylem, the shoot above the girdle continued to thrive.

Upward Translocation of Solutes

The principal substances translocated upward through the roots and stems are the ions of mineral salts absorbed by the roots and the amino acids synthesized in the roots. It seems logical to assume that these would be carried upward in solution in the water that flows up the xylem, and this was generally considered to be the case until 1920. Then Curtis and his coworkers began publishing a series of papers providing experimental data they regarded as strong evidence (for example, Table 10-1) that the upward translocation of

Table 10-1 Comparative effects of cutting the xylem versus phloem on gain in dry matter and transport of minerals into defoliated shoot tips of sumach (*Rhus typhina*) during 7 days, average of 8 sets, weight in milligrams

	Controls, initial	Controls, 7 days	Xylem cut, 7 days	Phloem cut, 7 days
Dry weight/shoot	53.20	607.40	199.00	17.80
Ash/shoot	4.95	44.66	18.03	3.17
CaO/shoot	0.12	0.14	0.04	0.03
K/shoot	1.19	12.81	5.06	0.67

SOURCE: O. F. Curtis [7].

mineral ions occurred primarily, if not entirely, through the phloem [7, Chap. 15]. Other investigators soon published experimental data they believed to give evidence for upward translocation through the xylem. For example, when woody stems were girdled near the base, there was little or no reduction in the upward translocation of mineral ions. They pointed out that Curtis did his ringing experiments near the stem tip and that the mineral ions that passed through the phloem may have come primarily from leaves rather than roots. The controversy and uncertainty continued until 1939, when Stout and Hoagland [20] provided convincing evidence for the upward translocation of mineral ions through the xylem.

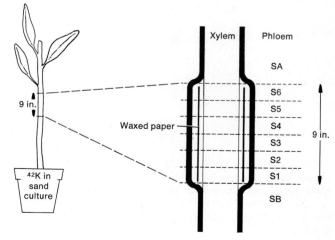

Figure 10-2. Experimental technique used by P. R. Stout and D. R. Hoagland [20] to determine the pathway of upward translocation of ^{42}K through the stem of a young willow tree. See Table 10-2 for results of the experiment.

Using young willow trees, Stout and Hoagland carefully separated the bark from the wood of a 9-in. section of stem and inserted waxed paper between them to prevent lateral movement of solutes between the xylem and the phloem (Figure 10-2). They then supplied the roots with a solution containing a ^{42}K salt and determined the radioactivity of various regions of the stem after 5 hr (Table 10-2). The data show that the translocation of ^{42}K was through the xylem

Table 10-2 Gain of ^{42}K from nutrient solution in sections (see Figure 10-2) of willow (*Salix lasiandra*) stem after 5-hr absorption period, ppm calculated from radioactivity

Section	Stripped plant		Unstripped plant	
	Bark	*Wood*	*Bark*	*Wood*
A	53	47	64	56
S6	12	119		
S5	0.9	122		
S4	0.7	112	87	69
S3	0.3	98		
S2	0.3	108		
S1	20	113		
B	84	58	74	67

SOURCE: P. R. Stout and D. R. Hoagland [20].

and also that, in intact stems, there is extensive lateral movement from the xylem to the phloem.

However, there is some upward translocation of mineral ions through the phloem, as shown by the experiments of Curtis and his coworkers and also subsequent investigators. There is a substantial translocation of mineral ions out of older leaves through the phloem, and when they reach the vascular cylinder of the stem they can move upward as well as downward through its phloem (Figure 10-3, Table 10-3). There is also a possibility that, when the

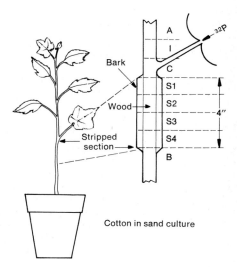

Figure 10-3. Experiment designed to determine the movement of ^{32}PO$_4$ applied to a leaf of a cotton plant through the stem. See Table 10-3 for results. [After O. Biddulph and J. Markle, *Amer. J. Bot.*, **31**:65 (1944)]

Table 10-3 Distribution of $^{32}PO_4$ in sections of a cottom stem after 1 hr of translocation from a leaf to which it was applied. Section I was stripped at the end of the hour, the other sections 16 hr before use. See Figure 10-3.

| | $^{32}PO_4$ in each section, µg calculated from radioactivity | | | |
| | Plant stripped | | Plant unstripped | |
Section	Bark	Wood	Bark	Wood
A		1.11		
I	0.458	0.100	0.444	
C		0.610		
S1	0.544	0.064	0.160	0.055
S2	0.332	0.004	0.103	0.063
S3	0.592	0.000	0.055	0.018
S4	0.228	0.004	0.026	0.007
B		0.653	0.152	

SOURCE: Biddulph, O., and J. Markle, *Amer. J. Bot.,* **31:**65 (1944).

rate of water flow through the xylem is slow, as in decidious trees during the winter, translocation of minerals upward through the phloem may play an important role.

The great bulk of sugar translocation is downward from the leaves where it is produced by photosynthesis. However, at times there is a substantial upward translocation of sugars, for example, from the endosperm or cotyledons in seedlings or from the storage roots of biennials when they are bolting and blooming. It would be logical to assume that such upward translocation of sugars would occur in the xylem, the sugars being dissolved in the upward-flowing water. That sugars reach the xylem in quantity is indicated by the characteristically high starch content of xylem parenchyma cells. However, the concentration of sugars in the water flowing through the xylem is generally low, and ringing experiments (for example, Table 10-4) indicate that the upward translocation of sugars is predominantly through the phloem.

Table 10-4 Effect of ringing at different distances from the terminal buds of *Crataegus sp.* upon subsequent growth*

Treatment	Average shoot elongation, mm
Control, not ringed	26.8
Ringed, second internode from tip	6.1
Partially ringed, second internode from tip	26.2
Ringed, fourth internode from tip	8.1
Partially ringed, fourth internode from tip	28.0
Ringed, base of 1-year wood	17.0
Ringed, base of 4-year wood	22.0

*Ringed April 8, measurements taken May 8. In the partially ringed branches one-half of the xylem and three-quarters of the phloem were removed. From 4 to 17 branches per treatment were used.

SOURCE: O. F. Curtis [6].

Downward Translocation of Solutes

The principal solutes translocated downward are sugars, but there is also downward translocation of other solutes, primarily from leaves, including mineral ions (Figure 10-3). There is little question about the fact that such downward translocation of solutes is through the phloem. The xylem obviously provides an improbable pathway for the downward translocation of solutes because of the upward mass flow of water through it, often at high rates (frequently at around 100 cm/hr and sometimes at 1000 cm/hr or more). As noted previously, the sugar concentration of the xylem sap is very low, usually almost 0 in the summer and from 0.02 to 0.05% in the winter and early spring. [The sugar maple is a special case. See the papers by Marvin, for example, *Plant Physiol.*, **31**:57–61 (1956).]

More positive evidence has come from numerous ringing experiments, which have conclusively demonstrated that the downward translocation of sugars is through the phloem (Figure 10-4). There have been several practical applications of stem girdling. For example, the trunks of trees that were to be cut down have sometimes been girdled to prevent the growth of water sprouts from the stumps. The girdle prevents the translocation of sugars to the roots, and in a year or two the roots exhaust the food accumulated in them and die of starvation.

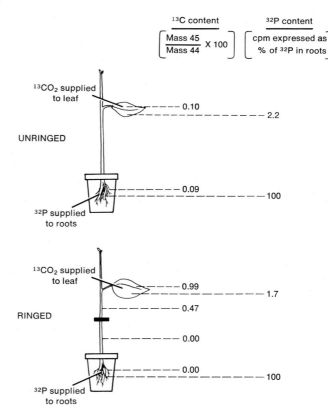

Figure 10-4. Experiment showing that ringing of the phloem prevents the downward translocation of ^{13}C-labeled sugars from the leaves but not the upward translocation of ^{32}P from the roots. [After G. S. Rabideau and G. O. Burr, *Amer. J. Bot.*, **32**:349 (1945)]

Xylem Translocation of Solutes

The mechanism of solute translocation upward through the xylem does not require extensive consideration here, since it is highly probable that the solutes are simply carried along with the flowing water. Thus, as we have seen in Chapter 9, shoot tension would provide the motive force most of the time and root pressure part of the time. We commonly speak of the translocation of water through the xylem, but what is actually flowing is a solution of various salts and also at times amino acids and sugars.

It should not be assumed that the solution originating in a root will be delivered intact and unchanged to a leaf, as would be the case if the solution were pumped through a plastic tube. As we have seen, the vessels or tracheids of the xylem provide a leaky system of tubes, and throughout their length various solutes diffuse out and into the phloem, the vascular rays, or other tissues. Thus, the solution reaching the stem tips and the leaves may be less concentrated than it was in the roots, and there may also be a change in the proportions of the various solutes resulting from losses and additions along the way. Furthermore, there is a possibility (although not a certainty) that some of the solutes may be adsorbed by the walls of the tracheids or vessels and that they may move upward through them rather than with the mainstream of water through the lumens. If such solute adsorption occurs the solutes would move more slowly than the water and, besides, some solutes might move more rapidly than others, as in a chromatogram.

Phloem Translocation of Solutes

The mechanism of solute translocation through the phloem will require much more extensive consideration than the mechanism of solute translocation through the xylem. Several different hypothetical mechanisms of phloem translocation have been proposed. Each one is supported by certain experimental evidence, but there is also experimental evidence against each one. No completely satisfactory hypothesis has yet been proposed, despite extensive research over many years, and students of phloem translocation have been unable to agree on which hypothesis is most acceptable. Indeed, the mechanism of phloem translocation has been one of the most refractory problems in the whole field of plant physiology. The problem will be solved eventually, possibly by some combination and modification of existing hypotheses or possibly by an entirely new hypothesis, but perhaps not for many more years.

We shall first outline the principal established facts about phloem translocation and then consider the pros and cons of each of the principal hypotheses.

Some Facts about Phloem Translocation

Certain facts about the translocation of solutes through the phloem have been established beyond any reasonable doubt and must be taken into consideration when any hypothesis of the mechanism of phloem translocation is proposed. We shall first present these facts, without any attempt to review the mass of experimental data on which they are based. Later on in this section we shall present still other information on phloem translocation that seems to be highly probable but for which the experimental evidence is less convincing.

1. Both the rate and velocity of phloem translocation are high. The rate of translocation is the total weight of solutes translocated per unit time and the velocity is a measure of the distance per unit time. On the basis of dry weight increases of pumpkin fruits, Colwell [3] calculated that translocation to the fruits occurred at rates of up to 1.7 g/hr and at velocities of from 20 to 155 cm/hr (Figure 10-5). These figures were not corrected for respiration or photosynthesis in the fruit. Crafts and Laurenz [5] have calculated that translocation to growing pumpkin fruits would have to occur at a velocity of 110 cm/hr if a 10% sugar solution was being translocated or 55 cm/hr if the solution was 20%. Actually, the sieve tubes often contain a 20% sugar solution. The calculation for squash was 134 cm/hr for a 10% solution. Calculated velocities of phloem translocation in various species of plants range from 10 to 500 cm/hr, with velocities of 50 to 150 cm/hr being most common. Since the velocity of diffusion of sugar from cell to cell is less than 2 cm/hr (generally much less), it is obvious that phloem translocation cannot be explained on the basis of simple diffusion. Phloem translocation is up to 40,000 times as fast as diffusion. Swanson [21] points out that Zucca melon fruits gain about 5 lb dry weight in a month. Assuming the

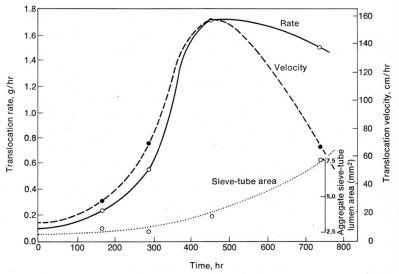

Figure 10-5. Rate and velocity of translocation of sugars and other solutes into pumpkin fruit as a function of time from anthesis (blooming). The velocity calculations were based on an assumed 20% solution in the phloem. The sieve-tube area is the aggregate lumen area of the sieve-tube elements of the fruit stalk. [Data of R. N. Colwell, from A. S. Crafts and O. A. Lorenz, *Plant Physiol.*, **19**:131 (1944)]

substance translocated is sucrose, the number of molecules would be 4.5×10^{24}. If this number of sucrose molecules could be placed end to end they would extend through a distance of 2.5 trillion miles.

There is evidence that not all substances are translocated at the same velocity. Different amino acids may be translocated simultaneously at different velocities. When ^{14}C-sucrose, ^{32}P, and tritiated water ($^{3}H_2O$) were supplied simultaneously, the water and phosphate were both translocated at 87 cm/hr but the sucrose moved at 107 cm/hr. It is possible, however, that the sucrose was simply introduced into the sieve tubes more rapidly. Also, there is usually a diurnal fluctuation in the velocity of translocation; the rate is generally higher during the day than at night, although the reverse has also been observed.

2. Exudation from the phloem usually occurs when it is punctured or cut. This exudation of sap may be both rapid and extensive, although it frequently ceases after a short time if the phloem is cut, possibly as a result of the plugging of the sieve plates by callose. In Italy a manna is produced commercially from the sieve tube exudate of two species of ash (*Fraxinus ornus* and *F. excelsior*). The trees are tapped as are rubber trees, and the exudate flows throughout the summer and fall, in contrast with the rather brief flow from many trees.

Perhaps the most satisfactory means of securing phloem exudate for experimental purposes is by utilization of aphids [23, 24]. These insects always sink their stylets into a sieve tube and, if the aphid is cut from its stylet, the latter serves as an effective drain. The stylet is so small that it apparently does not damage the sieve-tube element or cause callose formation, and exudation through it may continue for long periods of time. Observed rates of exudation through aphid stylets range from 0.05 to 3 mm^3/hr. The latter rate corresponds with the volume of 100,000 sieve-tube elements or a velocity of 500 cm/hr. Hill and Weatherly found that in willow a section of stem 16 cm long was necessary for maximum exudation and that this contained between 800 and 1000 sieve-tube elements in each tube, with 1600 to 2000 sieve plates to be traversed. Zimmerman calculated that to account for the observed rates of exudation through stylets the tapped sieve-tube element would have to be refilled from 3 to 10 times per second.

Phloem exudation implies that there is a substantial positive pressure in the sieve tubes, and this has been shown to be the case; turgor pressures can be up to 30 bars or even more. There is evidence that the turgor pressures in sieve tubes are generally higher than in other cells of the plant and that sieve-tube elements may still be turgid even when the leaves are wilted.

3. There is generally, if not always, a positive gradient of solute concentration in the sieve tubes from the supplying organs (usually leaves) to the receptor organs (roots, fruits, and so on). Huber found the gradient to be 0.012 M/m in red oak, and Zimmermann found it to be 0.01 M/m in white ash (Figure 10-6).

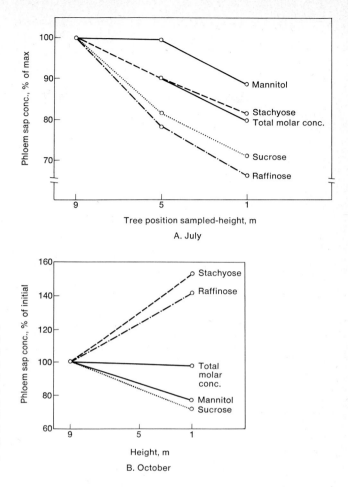

Figure 10-6. Gradients in sugar content down the stem of a white ash tree in July **(A)** and October **(B)**. Note the decreases in all sugars in the phloem sap in July when phloem translocation is active. In October (after leaf abscission) the total molar concentration gradient is slight, but note downward decrease in sucrose and mannitol and the increase in stachyose and raffinose. [Data of M. H. Zimmerman, *Plant Physiol.*, **32:**399 (1957)]

4. There is substantial evidence that there may be simultaneous bidirectional movement of solutes through the phloem (for example, Figure 10-7, Table 10-5) and even that such bidirectional movement may occur within a single

Table 10-5 Simultaneous bidirectional translocation of ³²P- and ¹⁴C-labeled solutes through the phloem of geranium stems during a 15-hr period. See Figure 10-7

Section of phloem analyzed	^{14}C (cpm/100 mg bark)	^{32}P (μg KH$_2$ ^{32}PO$_4$/100 mg bark)
	Radioactivity	
SA (above waxed paper)	44,800	186
S1	3,480	103
S2	3,030	116
SB (below waxed paper)	2,380	125

SOURCE: Chen, S. L., *Amer. J. Bot.*, **38:**203 (1951).

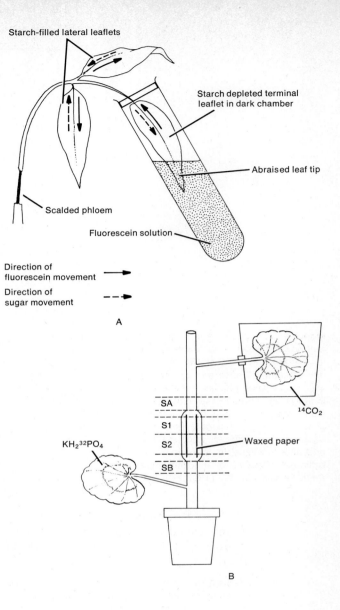

Starch-filled lateral leaflets

Starch depleted terminal leaflet in dark chamber

Abraised leaf tip

Scalded phloem

Fluorescein solution

Direction of fluorescein movement →

Direction of sugar movement --→

A

SA
S1
S2
SB

$^{14}CO_2$

Waxed paper

$KH_2^{32}PO_4$

B

Figure 10-7. Two experiments demonstrating the simultaneous bidirectional translocation of solutes through the phloem. **(A)** Translocation of sugars and fluorescein in opposite directions in a compound leaf of a bean plant. [After Palmquist, *Amer. J. Bot.*, **26**:665 (1939)] **(B)** Translocation of labeled sugars and KH_2PO_4 in opposite directions through the phloem of a geranium stem. See Table 10-5 for the results of the experiment. [After S. L. Chen, *Amer. J. Bot.*, **38**:203, (1951)]

vascular bundle. However, there is no clear evidence as to whether or not bidirectional movement of different solutes can occur within a single sieve tube. Information on this would be valuable in discriminating between the two principal hypotheses of phloem translocation.

5. Although most substances can be translocated in either direction through the phloem, a few can be translocated in only one direction. The outstanding example of such polar translocation is auxin, which can move through the phloem only in a morphologically downward direction. Calcium ions can not move downward through the phloem, but there is a possibility that they

may not be able to move in either direction because they precipitate out and that all the upward movement of calcium is through the xylem.

6. Translocation through the phloem can occur only as long as its cells are living. If the cells in a short section of stem are killed by a jet of steam or by other means, phloem translocation through this section ceases; this is in contrast with xylem translocation, which proceeds without interruption. Presumably the companion cells and perhaps also the phloem parenchyma as well as the sieve-tube elements must be living if translocation is to occur, although there is no definite information on this.

7. Associated with the preceding point is the fact that factors that reduce the rate of respiration in the phloem also reduce the rate of translocation. The factors include reduced oxygen and temperature and various metabolic inhibitors. However, some investigators have found that inhibitors are more effective when applied to the supplying or receiving organs and suggest that the inhibitory effect is more a matter of interference with the movement of substances into and out of the sieve tubes than with movement through the sieve tubes. The effect of temperature on translocation is rather complex, but in general there seems to be an optimum between 20 and 30°C (Figure 10-8).

8. Certain information about the nature of the sieve-tube elements and about the phloem in general must be considered in evaluating the various hypotheses regarding the mechanism of phloem translocation. At this point it would be helpful to review the descriptions of the sieve cells, sieve-tube

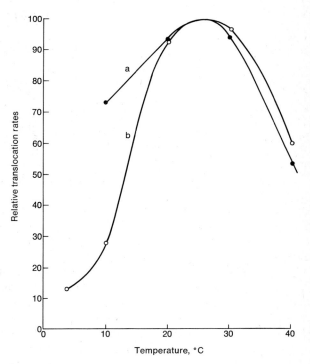

Figure 10-8. Rate of sugar translocation from bean leaves as a function of temperature. Curve a: Temperature of petiole varied. Curve b: Temperature of entire plant varied. [Data of S. P. Hewitt and O. F. Curtis, *Amer. J. Bot.,* **35:**746 (1948). Graph after C. A. Swanson and R. H. Böhning, *Plant Physiol.,* **26:**557 (1951)]

elements, and companion cells in Chapter 2, and the structure of the phloem as described in Chapter 8.

Very young sieve-tube elements are not greatly different from parenchyma cells, except for their elongated shape, but as they differentiate and mature marked changes occur. The golgi bodies, vacuolar membrane (tonoplast), and nucleus disintegrate and disappear, although nucleoli may persist for a while. The number of mitochondria is reduced and those remaining are smaller than usual with fewer cristae. The endoplasmic reticulum becomes appressed against the wall and appears vesicular or as stacks of cisternae, and no ribosomes are present. Endoplasmic reticulum has been reported as being present in the sieve pores. Proteinaceous **slime bodies** appear and may later coalesce into strands or form slime plugs in the sieve plates. The normally dense cytoplasm becomes restricted to a very thin layer against the walls, the greater part of the cell becoming filled with **mictoplasm,** or watery cytoplasm, resulting from the mixture of cytoplasm and cell sap. The plasma membrane apparently retains its differential permeability, although it has been suggested that the membranes at the ends of the cells are more permeable than those on the sides. The cell walls become thickened and hyaline, and the **sieve plates** with their **sieve pores** differentiate. The sieve pores resemble plasmodesmata but generally have a much greater diameter: 2 to 5 μ versus 0.02 to 0.2 μ for plasmodesmata, although sieve pores down to 0.2 μ have been observed on rare occasions.

Prepared sections of sieve elements almost always show the sieve plates plugged with **callose,** a polysaccharide made from glucose. However, it is now believed that these callose plugs are not present in functional sieve tubes and that they form rapidly when the sieve cells are injured, as during killing the tissue in its preparation for slides or when a sieve element is ruptured. Ruptures of sieve-tube elements as well as those of sieve plates are apparently plugged with callose, but in each case slime plugs are thought to precede the callose plugs. There is evidence that the sieve plates are normally plugged with callose each fall in woody plants, probably making them incapable of carrying on translocation. For this reason Zimmermann and other investigators believe that only the current year of phloem is functional, contrary to the general idea that several years of phloem may be functional in perennials.

Sieve elements do not contain chloroplasts, but there are a few small, poorly developed leucoplasts that synthesize a low molecular weight starch.

The morphology of mature sieve-tube elements suggests that, although the cells are living, they probably have very limited metabolic capabilities. The few, poorly developed mitochondria probably have a low rate of respiration, and there is probably little or no protein synthesis. However, the cells have some acid phosphatase activity, are capable of some starch synthesis, and can synthesize and hydrolyze callose. The watery mictoplasm that occupies most of the cell volume may well facilitate translocation through the cell, as do the sieve pores through the sieve plates. The fact that sieve-tube elements are more alkaline than most cells, perhaps because of the unusually high concentration of potassium ions in them, may be of significance in relation to translocation.

It has generally been assumed that the companion cells play some role, perhaps in the maintenance of life of the associated sieve-tube elements if not in translocation itself, but essentially nothing is known as to what this role might be. Less attention has been paid to the phloem parenchyma cells, but they may also play some role in translocation.

The Cytoplasmic Streaming Hypothesis

The first hypothesis regarding the mechanism of phloem translocation to be proposed was the cytoplasmic streaming (cyclosis) hypothesis of DeVries in 1885, and a number of other botanists soon became proponents of the hypothesis. During the thirties Curtis [6] and his students became vigorous supporters, but the majority of students of the subject regarded the hypothesis as unacceptable.

The proposed mechanism is quite simple: As the cytoplasm of the sieve cells or sieve-tube elements is streaming, it carries the solutes the length of the cell. These solutes would then diffuse through the sieve plates into the next cell in the series (or perhaps be moved across by an active transport mechanism) and would again be carried by the streaming cytoplasm.

One thing offered in support of this hypothesis is that it could account for bidirectional translocation even within a single sieve tube. Also, it is pointed out that this mechanism requires living, metabolically active cells, that treatments that retard metabolism and streaming also retard translocation, and that gradients of turgor pressure or solute concentration are not necessary. The continuity of cytoplasm through the sieve pores is also offered in support, and it has been suggested that streaming may be continuous from cell to cell.

One kind of evidence offered against the hypothesis is that the observed rates of cytoplasmic streaming are much too slow to account for the known velocities of translocation. Usually cytoplasmic streaming occurs at velocities of 0.02 to 0.04 cm/min, although in the large cells of Chara and Nitella velocities of 0.5 to 0.8 cm/min have been found. Since translocation commonly occurs at velocities of 2 to 8 cm/min, even the highest observed rates of streaming are far too slow. Translocation is 10 to 400 times more rapid than observed streaming velocities. An even more serious objection has been that most investigators have been unable to observe any cytoplasmic streaming at all in mature sieve cells or elements, although there is streaming in immature ones.

In response to these objections the proponents of the hypothesis have pointed out that: (1) measurements of the velocity of streaming have been made by utilizing chloroplasts and other relatively large microscopic bodies as tracers, and it is possible that solute particles may be carried much more rapidly by streaming; (2) immature sieve tubes may actually be the site of translocation rather than mature ones; and (3) streaming in mature sieve elements may be very sensitive to injury, even to adjacent companion cells so that the inability of most investigators to observe streaming has resulted from injury in the course of preparation of the tissue for microscopic examination.

The Pressure Flow Hypothesis

The pressure flow hypothesis is also called the **Münch hypothesis,** for the German botanist who proposed it in 1930 [16], and the **mass flow hypothesis,**

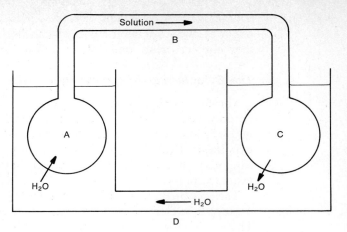

Figure 10-9. Diagram of a sectional view of a physical system that illustrates the operation of the pressure-flow theory of phloem translocation as proposed by Münch. See the text for discussion of this model.

although the latter name is not so suitable since cytoplasmic streaming is also mass flow. The pressure flow hypothesis has had a number of staunch supporters, notably Crafts [4] and, at least until recently, has been the most favored hypothesis despite some valid objections to it.

The pressure flow mechanism is more complicated and difficult to explain than the cytoplasmic streaming hypothesis and can best be approached by description of a mechanical model (Figure 10-9), which actually operates as described when set up in a laboratory. A and C in the diagram are osmometers permeable to water but not to the solutes used and correspond respectively to cells of leaves and roots. B is a glass tube corresponding to the sieve tubes of the phloem. D corresponds to the xylem, and the two vessels to the water supply to the cells.

Suppose that a solution with a water potential of −25 bars is placed in A, one of −20 bars in C, water in the connected vessels surrounding the osmometers, and solution (from A and C) in tube B. If A and C were independent rather than being connected by B and both were sealed, water would, of course, diffuse into A until a turgor pressure of 25 bars developed and into C until 20 bars was attained. However, when A and C are connected by B the turgor pressure is transmitted throughout the system and, because of this, the water potential in C will become 5 bars (25 + −20). Thus, water will diffuse out of C into the vessel and the solution will flow from A to C through tube B, carrying solutes from A to C. An equilibrium will be attained when the solute concentration, the turgor pressure, and so the water potential become equal throughout the system (A, B, C).

However, if additional solutes are continually added to A (as would occur when sugar is produced in a leaf cell by photosynthesis) and solutes were removed from C (as would occur when a root cell uses sugar in respiration or starch synthesis), the flow of solution through tube B would continue. It should be noted that while a solution is flowing through B there must be a gradient of solute concentration between A and C, but this is true only of the *total* solutes. For any individual substance in a mixture of solutes there could be a gradient, no gradient, or even a negative gradient (Figure 10-6).

It should not be too difficult now to visualize how the pressure flow mechanism would work in a plant. The mesophyll cells are supplying sugar by photosynthesis (and perhaps by starch hydrolysis), and this sugar along with other solutes would move into the phloem of the leaf veins by diffusion or active transport, thus maintaining a high solute concentration in the sieve tubes. The cells of the root (or other receptor such as the stem or fruits) remove sugar from solution by respiration, starch synthesis, or other processes, thus maintaining a solute concentration gradient and a turgor pressure gradient from the top to the bottom of the phloem and, consequently, a mass flow of solution through the sieve tubes. In his original hypothesis Münch postulated pressure flow as including the cells of the donor and receptor tissues, but proponents of the hypothesis now generally consider that the mechanism is limited to the sieve tubes themselves.

Several points have been presented in support of the pressure flow hypothesis. It is based on sound physical principles, as indicated by the successful functioning of the analogous mechanical model. Phloem exudation can be explained better on the basis of pressure flow than any other mechanism. Substances present in low concentration in leaves, such as 2,4-D and viruses, have been found to be translocated out only at the times that sugars are being translocated, and this fits in with the idea that a gradient of total solute concentration is essential.

The points presented in opposition to the hypothesis have been more numerous. It is difficult to conceive of bidirectional flow in a vascular bundle, although flow in opposite directions in different bundles is somewhat more plausible. Exudation from the phloem is not necessarily evidence for pressure flow translocation. It may be an artifact that operates only when sieve tubes are ruptured. In support of this view it is pointed out that exudation also occurs when laticifers are ruptured and that no one has ever suggested that these are translocation channels. The structural features of sieve tubes, particularly the end walls and the cytoplasmic contents, do not seem to provide a suitable pathway for pressure flow and offer a high resistance to flow. The pressure gradient required to account for observed rates of translocation is on the order of 20 bars/m, whereas the actual gradients appear to be less than 1 bar/m.

The requirement for living, metabolically active cells in the phloem is more difficult to explain than on the basis of the cytoplasmic streaming hypothesis. It has also been suggested that the phloem is incapable of transporting the quantity of water required as a solvent in the pressure flow hypothesis. Curtis and other investigators have pointed out that the supplying cells do not always have a higher solute content and turgor pressure than the receiving cells. Schumacher found that reduced leaf turgidity had no effect on the translocation of fluorescein or nitrogen compounds from the leaves, and others have pointed out that wilting may actually initiate removal of carbohydrates, nitrates, and minerals from leaves. Sprouting potato tubers have a lower turgor pressure than the sprouts to which sugars are being translocated, and the same is true of cotyledons of germinating seeds and seedlings as compared with the tissues to which foods are being translocated.

Some of these objections have greater validity than others, but the rather numerous proponents of the pressure flow hypothesis feel they have answers to most of them. We cannot pursue the controversy further, except to note that most of the time the necessary solute concentration and turgor pressure gradients between the supplying and receiving tissues are present. The Ψ_π in the shoots of herbaceous plants is usually on the order of -10 to -15 bars, in woody plants -20 to -30 or more, and in roots it is around -5 bars. Also, it has been pointed out that the turgor pressure gradient within the sieve tubes themselves is the important requirement and that the sieve-tube elements are often highly turgid even when leaves are wilted.

The Activated Diffusion Hypothesis

In the decade beginning in 1928 Mason, Maskell, and their coworkers published a considerable number of papers [14] on their studies of translocation in cotton and other plants and proposed their activated diffusion hypothesis of phloem translocation. They found that sugar moves in a positive concentration gradient, that the rate of translocation is proportional to the steepness of the gradient, and that, although partial ringing reduces the amount of sugar translocated, the amount per unit cross-sectional area of phloem increases. All these would be expected if sugar translocation were a matter of diffusion, although they recognized that the rate of sucrose diffusion is far too slow to account for observed rates of translocation. They proposed that respiratory energy was in some way utilized in increasing the rate of diffusion to the necessary level, either by increasing the rate of movement of the molecules or by reducing the resistance to diffusion. In regard to the latter possibility, they pointed out that, if sucrose could be vaporized so it could diffuse through air, its diffusion constant would be great enough to account for the observed rates of sugar translocation.

The activated diffusion hypothesis has received essentially no support from other investigators because it does not seem to have a sound theoretical basis. Increase in the kinetic energy of the diffusing molecules, even if it could be obtained, would involve an increase in temperature far out of line with any existing in plants, and there seems to be no conceivable way in which the resistance within the sieve tubes could be reduced to a level comparable with that of air.

The Interfacial Flow Hypothesis

In 1932 van den Honert [22] proposed an interfacial flow hypothesis of phloem translocation, pointing out that the reduction in surface tension at interfaces between two substances permits the rapid movement of solutes along the interfaces. He set up a physical analogy in the laboratory, placing water and then ether in a long trough. He then introduced some potassium oleate into the water-ether interface at one end and found that it moved to the other end at a rate 68,000 times as fast as it would have diffused. He suggested that sugars and other solutes might move along the interfaces between the cytoplasm and walls of the sieve-tube elements or between the cytoplasm and vacuoles. Unlike the pressure flow hypothesis this mechanism would not require any appreciable flow of water through the sieve tubes.

Despite the fact that the velocity and rate of flow would be adequate to account for those known to exist, this hypothesis has received little attention or support although it has not been invalidated. The principal objections have been that sugars, amino acids, and most other substances translocated are so soluble in water that they would not be present in the interfaces in adequate quantities and that the interfacial areas within sieve tubes are not great enough.

Bioelectrical Potential Hypotheses

In 1957 Fensom [9] proposed that phloem translocation involved electro-osmosis, at least across the sieve plates, and claimed that the necessary bioelectrical potential differences were present. He suggested that the sieve plates were charged by the production of carbonic acid by respiration and that the H^+ diffused through the barrier faster than the HCO_3^-, thus creating a potential difference.

In 1958 Spanner [19] proposed a similar theory except that he ascribed the potential differences to the rapid circulation of K^+ through the sieve plate in the direction of transport, out of the sieve-tube element (possibly into a companion cell), and then back to the original side of the sieve plate.

Neither hypothesis has received any particular attention or support and they will not be discussed further.

Summary

It should be evident by now that there is no completely acceptable hypothesis of phloem translocation, and even the major theories have experimental evidence against them as well as supporting them. There is no doubt about the fact that rapid translocation of solutes can and does occur through the phloem, but a satisfactory explanation of the mechanism by which it occurs has so far eluded plant physiologists despite extensive and intensive research over many years. The problem is an extremely difficult one, but we can be confident that it will eventually be solved. Perhaps future investigations will remove all objections to one hypothesis or another. Perhaps one of the minor hypotheses that have so far received little consideration or some entirely new hypothesis will prove to be the valid one. Perhaps the actual mechanism will involve features of two or more hypotheses, since they are really not mutually exclusive. At any rate, the pressure flow and cytoplasmic streaming hypotheses currently appear to be most strongly supported, and each one has staunch proponents and opponents. In general, the evidence for pressure flow appears to be stronger.

Mobilization of Solutes

The translocation of solutes in plants is from supplying organs to receptor organs, or sinks as they are sometimes called. There is often, but by no means always, a positive concentration gradient for any particular solute from the supplying organ to the receptor organ. Such a concentration gradient may be maintained over an indefinite time period by continued production of the solute in the supplying organ and its continued use in the receptor organ. For example, the sugars produced by photosynthesis in leaves are translocated to roots or fruits where they are used in respiration, assimilation, or accumulation after conversion to substances such as starch; this sequence maintains a sugar concentration gradient. When translocation is not along a positive translocation gradient from the supplying to the receptor organs, continued movement into the receptor no doubt involves the expenditure of metabolic energy in active accumulation or perhaps some other way. The term **mobilization** is frequently applied to the translocation of solutes to receptor organs, particularly when the translocation is against a concentration gradient for any particular solute or when a solute is retranslocated from one organ to another. The term may also be applied when there is differential translocation, such as movement of higher concentrations of a mineral element into younger leaves than into older ones. Mobilization may involve what Leopold [12] refers to as "yielding forces" of the supplier as well as "pulling forces" of the receptor.

Mobilization is rather complex and still not thoroughly understood, but it involves a variety of factors such as age of an organ, level of metabolic activity of an organ, phytohormones, adequacy of mineral nutrition, water availability, light, and temperature. Common mobilization pathways are from older to younger leaves (Figure 10-10); from leaves to apical meristems, flowers, fruits, and roots; from woody stems to nearby buds during their early growth in the spring; from petals to developing ovularies; and in germinating seeds from the endosperm or fleshy cotyledons to the rapidly growing tissues of the embryo or seedling.

As one example of the influence of light on mobilization, Mellor found that movement of carbohydrates from tobacco leaves was predominantly to the roots at night but mostly to the shoot tips when the light intensity was over 750 ft-c.

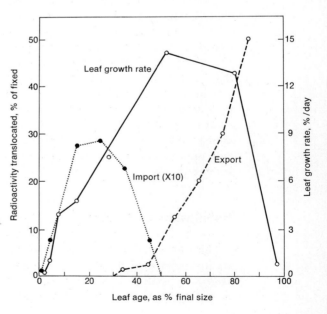

Figure 10-10. Translocation of ^{14}C-labeled sugars into or out of soybean leaves as a function of their ages. The plants were provided with $^{14}CO_2$ for 2 hr of photosynthesis, and the final determinations of radioactivity were made 6 weeks later. [Data of S. L. Thrower, *Aust. J. Biol. Sci.,* **15**:629 (1962)]

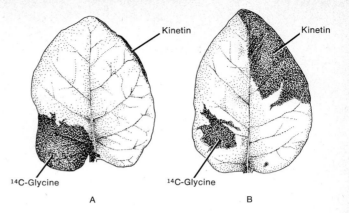

Figure 10-11. Drawings of radioauto-graphs of **(A)** a young leaf and **(B)** an old leaf of *Nicotinia rustica,* showing the mobilization of radioactive glycine by kinetin. Radioactive glycine was applied to lower left part of each leaf and kinetin to the upper right of each. Note the strong mobilization in the old leaf but not in the young one. [After radioautographs in K. Mothes, *Naturwiss.,* **47:**337 (1960)]

Kinetin

Kinetin

^{14}C-Glycine

^{14}C-Glycine

A

B

Vickery and others found that tobacco leaves lost their nitrogenous contents rapidly when kept in continuous darkness but not when exposed to normal diurnal light. Engelbrecht and Mothes found that high temperatures resulted in loss of amino acids from tobacco leaves.

One of the best examples of the mobilizing effects of phytohormones was provided by Mothes [15] who found that, when older tobacco leaves were excised, radioactive amino acids supplied to the lower left-hand side were strongly mobilized to the upper right-hand quadrant of the blade when that quadrant was treated with kinetin (Figure 10-11). Such mobilization was not evident in young leaves. When leaves of tobacco and most other species are excised and placed in water, protein hydrolysis becomes much more rapid than protein synthesis and the amino acids move to the midrib. Mothes found that, if kinetin was supplied to half of such a leaf, a high protein level was maintained and that the usual yellowing and scenescence did not occur. It is interesting that the leaves of species such as African violet and begonia, which can form adventitious buds and roots and can be used in propagation, do not lose protein and become chlorotic after abscission. It is possible that auxins and other phytohormones may also have a mobilizing influence, as in developing fruits or seeds, but there is no particular evidence for this.

There is extensive mobilization of solutes from older leaves to younger ones, including carbohydrates, amino acids and other nitrogenous compounds, and a variety of mineral elements, notably potassium, phosphorus, and sulfur. This may involve a reduced capability of the older leaves for holding solutes and for active accumulation. Old leaves lose up to 60% more solutes by leaching than do young ones. Biddulph has found that prior to abscission scenescent leaves lose a substantial portion of their solutes to the stems or younger leaves and that for some substances up to 90% of the total is lost. The most highly mobilized elements are nitrogen, phosphorus, potassium, sulfur, and chlorine, and sometimes magnesium and iron, whereas calcium, boron, manganese and silicon are translocated out little if at all.

Biddulph [2] has also conducted extensive experiments on the distribution and redistribution of various radioactive ions after their absorption by the roots of young bean plants. Although the distribution patterns are somewhat different

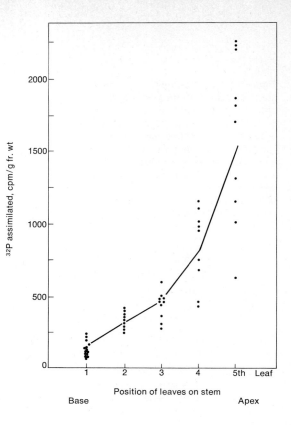

Figure 10-12. Accumulation of ^{32}P in the leaves of bean seedlings as a function of the age of the leaves. The radioactivity was determined 4 days after a ^{32}P-labeled salt was added to the nutrient solution. [Data of O. Biddulph, pp. 261–278 in E. Truog (ed.), *Mineral Nutrition of Plants*, University of Wisconsin Press, Madison, 1951]

for different elements, there is in general a greater initial movement to the younger leaves and apices than to the older leaves followed by extensive retranslocation from the older to younger leaves (Figure 10-12). Biddulph suggests that this may result from a more rapid transpiration rate of the younger leaves as well as from their greater metabolic activity and ion accumulation. Water deficiency, as well as age, may bring about translocation of solutes out of leaves.

The high mobilizing capacities of developing fruits and seeds have been demonstrated by numerous experiments (Figure 10-13). McCollum and Skok [13] found that mobilization of ^{14}C-labeled compounds from tomato leaves to fruits up to 20 days old was extremely great but then dropped to a very low level in older fruits, except for a slight rise at the mature green stage (Figure 10-14). There was essentially no translocation into red ripe fruits. Marré has reported an interesting example of mobilization from one fruit of Calonyction to another. The receptacle and peduncle of the flowers of this plant become greatly enlarged, fleshy, and filled with starch as the ovulary develops into a fruit. If the ovulary of one of the several flowers is removed or damaged the starch is hydrolyzed, the sugar is translocated to other flowers, and the peduncle and receptacle shrivel up. However, if all but one flower is removed from a cutting this mobilization does not occur (Figure 10-15). The developing ovulary evidently has a mobilizing effect on the carbohydrate in the peduncle and receptacle.

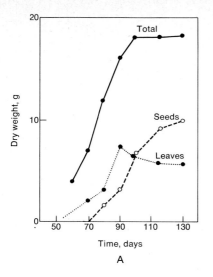

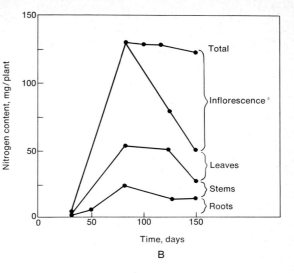

A

B

Figure 10-13. Data from three different experiments showing mobilization from leaves into developing reproductive organs. **(A)** Mobilization of foods from leaves to the developing grains of barley. [Data of H. K. Archbold and B. N. Mukerjee, *Ann. Bot., N.S.6:*1 (1942)] **(B)** Mobilization of nitrogen compounds from the leaves and stems of oat plants into the inflorescences. [Data of R. T. Williams, *Aust. J. Exptl. Biol. Med. Sci.,* **16:**65 (1938)] **(C)** Mobilization of nitrogen compounds from leaves to flowers and bolls (fruits) of cotton plants. The experiment was conducted in South Africa, where November is midsummer. [Data of F. Crowther, *Ann. Bot.,* **48:**877 (1934)]

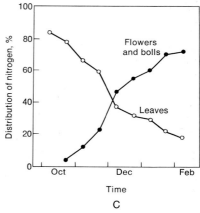

C

Figure 10-14. Mobilization of ^{14}C-labeled compounds from the leaves to the fruits of tomato plants during the development of the fruits. [Data of J. P. McCollum and J. Skok, *Proc. Amer. Soc. Hort. Sci.,* **75:**611 (1960)]

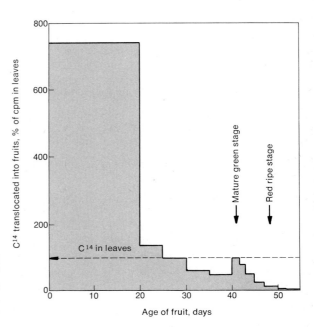

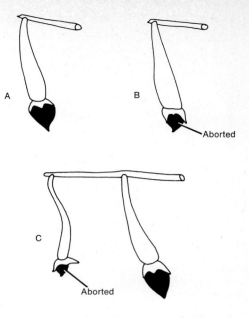

Figure 10-15. Mobilization by developing fruits of Calonyction. **(A)** intact fruit; **(B)** fruit with an aborted embryo on a cutting by itself; **(C)** fruit with an aborted embryo on a cutting with an intact fruit. The withered fleshy peduncle shows mobilization from it to the intact fruit. The peduncles normally have a high starch content. [After E. Marré, *Boll. Soc. Ital. Biol. Sper.*, **24:**602 (1948)]

Hay and others [10] found that 60% of the total nitrogen used by developing corn grains was mobilized from other tissues, and only 40% was absorbed from the soil and translocated directly. Of the mobilized nitrogen 60% came from the leaves, 26% from the stems, and 14% from the husks and shanks. Gregory found that in oats 93% of the phosphorus used in the developing inflorescence was mobilized from other parts of the plant if the external supply was abundant, but only 30% was mobilized when phosphorus was deficient. He also found that in cereal grasses in general about 90% of the total nitrogen and phosphorus was absorbed by the time the plants had attained only 25% of their final dry weight, and much of this was mobilized, first from the older to younger leaves and then to the inflorescences.

It been known for some time from the work of Curtis and others that mineral ions mobilized into buds of woody plants during their rapid development in early spring come principally from only a few centimeters of stem just below the bud, and essentially the same thing is true of sugars. The idea of extensive rise of sap from the roots at this time is largely a misconception. It has been noted that there is generally a mobilization of solutes from older leaves and stems to the apices, but if the apices are excised there may be an extensive change in the pattern of mobilization. Skok found that removal of sunflower apices reduced the translocation of ^{14}C-labeled compounds up the stem by about 50%. When tobacco plants are topped the nitrogenous compounds ordinarily mobilized by the apices are largely retained within the leaves.

Although the term mobilization is generally limited to solutes, it might also be applied to water. For example, there may be internal redistribution of water as a water deficit is created when the rate of transpiration exceeds the rate of absorption. The usual pathway of water distribution is from leaves to meristematic regions such as stem and root tips, the cambium, and developing fruits. Water also appears to move into the phloem from adjacent tissues. However, Bartholomew found that water moved out of lemon fruits during daily temporary wilting into other organs, and back in again at night, and this may be generally true of fleshy fruits that are mature or nearly so. Internal water redistribution is apparently in the direction of the water potential gradient and does not involve any expenditure of metabolic energy.

Circulation of Solutes in Plants

It should be evident that vascular plants do not have a circulatory system at all comparable to that of vertebrate animals, despite the misguided efforts of some authors of biology textbooks to draw unjustifiably close parallels between plants and animals. The vascular system of plants, in contrast with that of animals, consists essentially of parallel one-way pathways: water, mineral elements, and some organic solutes flow upward in the xylem and various solutes predominantly sugars downward through the phloem. All but a small fraction of the water carried up is lost by transpiration. However, despite these essentially one-way pathways various botanists have pointed out that solutes may undergo a type of circulation within plants (Figure 10-16).

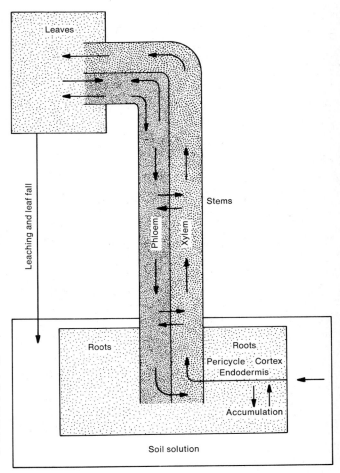

Figure 10-16. Diagram outlining the circulation of the ions of salts in plants. Downward translocation is through the phloem, but phloem translocation may also be upward and both into and out of leaves. Note also the lateral interchange between the xylem and phloem. Ions may be returned to the soil by leaching and leaf fall.

As early as 1850 Hartig suggested that sugars may circulate in plants. They would go from the leaves down into stems and roots where they would accumulate during the winter and would then be translocated up through the xylem in the spring. Later Mason and Maskell suggested that this could also occur without an intervening storage period. Currently, the most active proponent of the circulation concept is Biddulph, whose writings [1, 2] should be consulted for further details.

Biddulph has conducted extensive experiments on the translocation and mobilization of ^{32}P, ^{35}S, and other radioisotopes, primarily in young bean plants. He reports an especially rapid and prolonged circulation of phosphorus, up through the xylem from the roots, into leaves, and out of them through the phloem in a predominantly downward direction, followed by a series of similar circulations. The circulation apparently involves mobilization of the phosphorus from the older leaves to the younger leaves and meristems, although there may be a circulation of phosphorus into and out of these. Biddulph found that some ^{32}P circulated for as long as 96 hr, the longest experimental period, and estimated that any single phosphate ion might make at least three or four circulations in a day. He ascribes the mobility of the phosphorus to its rapid metabolic turnover. He found that there was considerable initial circulation of ^{35}S, including mobilization from the older to younger leaves and to meristems, but that after a short time most of the sulfur was tied up by incorporation into proteins in the younger leaves and meristems.

Although the concept of circulation in plants has been related primarily to solutes, it may be noted that the pressure flow hypothesis implies the circulation of some water down through the phloem and up again through the xylem. However, it should be stressed again that such circulation as exists in plants is not really comparable with circulation in animals.

References

[1] Biddulph, O. "Translocation of inorganic solutes," in F. C. Steward (ed.), *Plant Physiology: A Treatise,* Vol. II, Academic Press, New York, 1959, pp. 553–603.
[2] Biddulph, Susann, and O. Biddulph. "The circulatory system of plants," *Sci. Amer.,* **200**(2):44–49 (Feb. 1959).
[3] Colwell, R. N. "The use of radioactive phosphorus in translocation studies," *Amer. J. Bot.,* **29**:798–807 (1942).
[4] Crafts, A. S. *Translocation in Plants,* Holt, Rinehart, & Winston, New York, 1961.
[5] Crafts, A. S., and O. A. Lorenz. "Fruit growth and food transport in cucurbits," *Plant Physiol.,* **19**:131–138 (1944).
[6] Curtis, O. F. *The Translocation of Solutes in Plants,* McGraw-Hill, New York, 1935.
[7] Curtis, O. F., and D. G. Clark. *An Introduction to Plant Physiology,* McGraw-Hill, New York, 1950.
[8] Esau, K. "Explorations in the food conducting system of plants," *Amer. Sci.,* **54**:141–157 (1966).

[9] Fensom, D. S. "The bio-electrical potentials of plants and their functional significance. I. An electrokinetic theory of transport," *Can. J. Bot.,* **35:** 573–582 (1957).

[10] Hay, R. E., E. B. Earley, and E. E. DeTurk. "Concentration and translocation of nitrogen compounds in the corn plant (*Zea mays*) during grain development," *Plant Physiol.,* **28:**606–621 (1953).

[11] Kursanov, A. L. "Recent advances in plant physiology in the U.S.S.R.," *Ann. Rev. Plant Physiol.,* **7:**401–436 (1956).

[12] Leopold, A. C. *Plant Growth and Development,* McGraw-Hill, New York, 1964, Chaps. 3–5.

[13] McCollum, J. P., and J. Skok. "Radiocarbon studies on the translocation of organic constituents into ripening tomato fruits," *Proc. Amer. Soc. Hort. Sci.,* **75:**611–616 (1960).

[14] Mason, T. G., E. J. Maskell, and E. Phillips. "Further studies on transport in the cotton plant. III. Concerning the independence of solute movement in the phloem," *Ann. Bot.,* **50:**23–58 (1936).

[15] Mothes, K. "Über das Altern der Blätter und die Moglichkeit ihrer Wiederverjungung," *Naturwiss.,* **47:**337–350 (1960).

[16] Münch, E. *Die Stoffbewegungen in der Pflanze,* Fisher, Jena, 1930.

[17] Richardson, M. *Translocation in Plants,* St. Martin's Press, New York, 1968.

[18] Roach, W. A. "Plant injection as a physiological method," *Ann. Bot.,* **3:**155–226 (1939).

[19] Spanner, D. C. "The translocation of sugar in sieve tubes," *J. Exp. Bot.,* **9:**332–342 (1958).

[20] Stout, P. R., and D. R. Hoagland. "Upward and lateral movement of salt in certain plants as indicated by radioactive isotopes of potassium, sodium, and phosphorus absorbed by roots," *Amer. J. Bot.,* **26:**320–324 (1939).

[21] Swanson, C. A. "Translocation of organic solutes," in F. C. Steward (ed.), *Plant Physiology: A Treatise,* Vol. II, Academic Press, New York, 1959, pp. 481–551.

[22] Van den Honert, T. H. "On the mechanism of transport of organic materials in plants," *Koninkl. Akad. Wet. Amsterdam Proc.,* **35:**1104–1112 (1932).

[23] Zimmermann, M. H. "Transport in the phloem," *Ann. Rev. Plant Physiol.,* **11:**167–190 (1960).

[24] Zimmermann, M. H. "Movement of organic substances in trees," *Science,* **133:**73–79 (1961).

Introduction to Plant Growth and Development

11

The remainder of this volume will be devoted to a discussion of plant growth and development, including the reproductive processes that give rise to new individuals that then proceed to grow and develop in a pattern characteristic of their species. Despite the fact that growth and development are generally accepted without question as natural phenomena by laymen, biologists find growth and development to be extremely complex and puzzling. The development of an undifferentiated unicellular zygote into a mature individual is truly remarkable and still only poorly understood by biologists, despite an immense amount of research that has provided a continually increasing body of information.

The sequence of events in the development of various plants and animals has been described in considerable detail and the influences of various factors on growth and development have been determined, but any complete explanation of the behavior of cells of a developing organism remains elusive. What determines how many times a cell in an apical meristem divides before it begins to elongate, or how many cell divisions there will be in each of several planes, thus determining the size and shape of an organ? Why do the cells cut off to the inside of the vascular cambium develop into vessel elements and other cells of the xylem, whereas those cut off to the outside develop into sieve-tube elements or other cells of the phloem? Why does the zygote of a fern develop into a sporophyte plant, whereas the spores develop into strikingly different gametophyte plants? At present we have only a few scattered clues that would aid in answering such questions.

Although many unsolved problems remain in many areas of biology, few of these areas have as many problems or as complex ones as does development [5]. As might be expected, development is currently one of the most active areas of biological investigation. The problems are being approached from a variety of directions—biochemical, physiological, cytological, and genetic as well as the older morphological approach. James Bonner [2] believes that development is now the "new biology," rather than molecular biology, and that the solution of problems of development is now "doable." The spectacular advances in molecular biology and biochemical genetics during the past several decades are contributing greatly to this doability along with a better understanding of metabolic processes and the fine structure of cells. In addition new instrumentation and new or improved techniques, such as histochemical and cytochemical procedures, cell and tissue culture methods, and chromatography, have become available.

Growth Versus Development

The title of this chapter implies that growth and development are different things and, as a matter of fact, perhaps the majority of biologists consider this to be the case. The term **growth** is often used to refer only to an increase in the size or weight of an organism. Growth is thus subject to quantitative measurement and, from a series of measurements over a period of time, growth curves can be constructed. Growth is commonly considered to be an irreversible increase, which would rule out such things as the changes in size and weight of a leaf as the turgidity of its cells changes. Some biologists also prefer to restrict growth to an increase in organic matter, but this can lead to difficulties. For example, a seedling growing in the dark obviously increases greatly in both size and fresh weight, but its dry weight (predominantly organic matter) is less than that of the seed from which it grew because some of the accumulated food of the seed was used in respiration. A more suitable criterion of growth is an increase in the quantity of cytoplasm in the organism. This occurs even in the seedling just mentioned but, since measurement of an increase in cytoplasm is difficult, it is seldom used in measuring growth. Increases in size, fresh weight, and dry weight are all commonly used indicators for measuring growth. One may be selected as most suitable for a particular experiment, or in some cases two or even all three indicators may be used. Dry weight determinations are used least frequently, since they involve sacrificing the organism or tissues used.

In contrast to growth, **development** is more a qualitative than a quantitative change. It cannot be measured in any simple numerical way and is generally either described or illustrated by means of photographs or drawings. Development involves the differentiation of cells, tissues, and organs, as well as other accompanying changes that will be discussed in subsequent chapters, but it does not necessarily involve an increase either in size or weight. For example, the gastrula stage of a frog embryo has no greater size or fresh weight than the zygote from which it developed whereas its dry weight is less because of the food used in respiration, but it is made up of many cells and the obviously differentiated germ layers.

Although growth and development often can be differentiated quite clearly, some biologists consider development one aspect of growth rather than a separate phenomenon. Botanists in particular tend to use the terms in this sense, probably because in plants development from the meristems is a continuing process intimately associated with growth. In animals development is restricted largely to the embryonic and juvenile stages, and thus can be distinguished more easily from subsequent growth to the final adult size. It is, therefore, necessary to know whether the term growth is being used in the broader or more restricted sense.

Another term used in connection with development, and sometimes also growth, is **morphogenesis** (the origin of form). It is used with various shades of meaning, perhaps most commonly as an essential synonym for development. It is used more by those who approach the study of development from a morphological standpoint, rather than a physiological or genetic standpoint, and even by a few as a synonym for descriptive developmental morphology. Among zoologists, there is an increasing tendency to use the term as a substitute for **embryogeny.** In this textbook we will use morphogenesis and development essentially as synonyms.

Growth and Development at the Cellular Level

At the cellular level three rather distinct, though integrated, aspects of growth and development can be identified: cell division, cell enlargement, and cell differentiation. These will be considered briefly here, with further consideration at appropriate places in the following discussions.

Cell Division

Without cell division it would be impossible for a zygote to grow into a multicellular individual; however, in such cases as the early development of frog embryos and some stages in the development of plant embryos, cell division does not result in growth but rather in the production of more and progressively smaller cells. In the apical meristems of plants the meristematic cells generally enlarge between divisions but only to the approximate size of the parent cell, and this results in only a minor amount of growth in size.

The plane of the cell divisions plays an important morphogenetic role, since it is one factor determining the shape of an organ. If the cell divisions are most numerous in one plane, an elongated organ such as a stem or root results; if in two planes, a flat organ such as a leaf results; and if in three or more planes, a bulky and more or less isodiametrical organ such as a fruit results. The other factor determining organ shape is differential cell enlargement. The size of an organ is determined by the number of cells in it (that is, the number of cell divisions that occurred during its development) and by the size of the cells (Figures 11-1 and 11-2). Irregularly shaped organs like lobed leaves result when cell division ceases earlier in some regions than others. All these aspects of cell division are important morphogenetically and are obviously under genetic control, since they result in structures of a size and shape characteristic of their species.

In most organisms cell division (Figure 2-4) involves two more or less distinct stages: the division of the nucleus (mitosis or karyokinesis) and the subsequent division of the cytoplasm (cytokinesis). Since most if not all readers of this

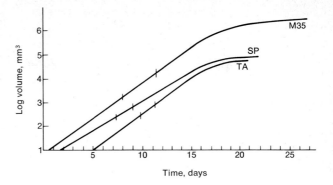

Figure 11-1. Contribution of cell division and cell enlargement to the growth of fruits of three races of Cucurbita differing in fruit size. In the period between the vertical bars cell division was ceasing and cell elongation was beginning. Before this period, growth was by cell division; after it, growth was by cell enlargement. Note that despite the shift the growth rate remained constant for about 15 days. [Data of Sinnott, from E. W. Sinnott, *Plant Morphogenesis,* McGraw-Hill Book Company, New York, 1960]

text will have studied the details of cell division previously, no attempt will be made to describe them here. However, it may be well to point out some differences of cell division in plants and animals. Although most of the mitotic events are strikingly similar in both groups of organisms, mitotic cells in the great majority of plant species do not have the centrioles and asters that are so characteristic of mitosis in animals. There are even greater differences in cytokinesis. In animal cells cytoplasmic separation is accomplished by the invagination of the plasma membrane from the sides toward the center, but in most higher plant cells cytokinesis begins with the formation of a pectic cell plate that extends toward the sides and later beocomes the middle lamella. Plasma membranes then form on each side of the cell plate and primary cell walls are laid down between the membranes and the middle lamella. These differences are related principally to the presence of cell walls in plants and their absence in animals.

There are probably also physiological differences between animal and plant cell division, or at least differences in the mechanisms that trigger cell division.

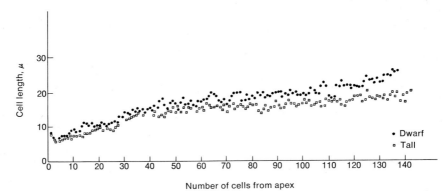

Figure 11-2. Lengths of successive cells along the apical meristem of a dwarf (upper) and tall (lower) variety of tomato. Note that the greater height of the tall variety resulted from the production of many more cells by division, not from greater cell length. The cells of the dwarf variety are somewhat longer in the older part of the meristem because they stop dividing and begin elongating sooner than those of the tall variety. [Data of E. Bindloss, *Amer. J. Bot.,* **29:**179 (1942)]

As will be seen in Chapters 12 and 13, several plant hormones including the auxins, cytokinins, and gibberellins play roles in cell division, but as far as is known these have little influence on cell division in animals, although cytokinins may affect the division of animal cells in culture. Several synthetic growth inhibitors such as maleic hydrazide inhibit cell division in plants but not in animals.

Cell Enlargement

As has been noted, cell division by itself does not result in growth, which requires also subsequent cell enlargement. The development of a parenchyma cell from a meristematic cell involves quite a spectacular enlargement. It is not unusual for cells such as cortical parenchyma or palisade cells in leaves to be around five times as wide and 20 times as long as the approximately cubical meristematic cells from which they were derived, which represents an increase in volume on the order of 500 times. Since the increase in length is often much greater than the increase in width, cell enlargement is often referred to as cell elongation. However, some cells such as pith parenchyma may be more or less isodiametric, and their enlargement may involve an even more marked increase in volume.

A prerequisite for cell enlargement in plants is plasticization of the cell wall so that it becomes stretchable. This requires the presence of a suitable concentration of auxin, a plant growth hormone. The turgor pressure of the cell then pushes the walls out, causing enlargement. As a result of the increased volume the turgor pressure decreases, bringing about a decrease in water potential and the diffusion of more water into the cell. As this process continues water enters the vacuoles, which become larger and merge so that finally one large vacuole comes to occupy most of the cell volume. The protoplast then constitutes a rather thin layer against the walls (sometimes with strands through the vacuole).

Because the walls of parenchyma cells are generally thicker than those of meristematic cells, it is obvious that cell enlargement also involves the incorporation into the walls of substantial quantities of cellulose, pectic compounds, and other cell wall materials as well as additions to the middle lamella. Despite the fact that the cytoplasm occupies only a thin peripheral layer, its volume is considerably greater than in the meristematic cell so that cell enlargement involves a considerable amount of assimilation. Other events include the development of chloroplasts or other plastids from proplastids and the differentiation of other cell organelles. Cell enlargement is, thus, a rather complex process.

Cell Differentiation

Although cell division plus cell enlargement result in growth and may represent the total of growth and development in various simpler plants, such as some of the filamentous algae, the complete development of vascular plants and indeed most other plants also involves differentiation into all of the various kinds of cells that make up the organism. There is, of course, a considerable amount of differentiation during the enlargement of meristematic cells, but cell differentiation is a term usually applied to the development of more specialized cells such as fibers, tracheids, vessel elements, sieve-tube elements, cork cells, collenchyma cells, and guard cells.

Among the visible events during cell differentiation (Chapter 16) are the deposition of secondary cell walls, sometimes in a differential pattern as in vessel elements and collenchyma cells; changes in the shapes of cells; synthesis and deposition of substances such as cutin and suberin in epidermal and cork cells; loss of the nucleus and vacuolar membranes in sieve-tube elements; complete loss of the protoplast, as in vessel elements and fibers; dissolution of end walls, as in vessel elements (Figure 11-3); and changes in the number, size, and kind of various cell organelles. Before such changes occur the differentiating cells generally pass through a parenchymalike stage as a result of the enlargement of the original meristematic cells. It is, of course, much easier to describe what happens during cell differentiation than to determine the biochemical changes that cause certain kinds of cells to differentiate at certain times and in the appropriate places in a plant. One thing that apparently is involved is the differential production of mRNA and enzymes, which will be discussed later.

Interaction of Factors in Growth and Development

The pattern of growth and development of any organism is determined by two primary sets of factors: the hereditary potentialities of the organism and the environment in which it is growing. The interaction of these factors can be illustrated by use of a simple diagram.

Hereditary
potentialities
$\searrow$
$\rightarrow$ Internal biochemical and   Observable
biophysical conditions $\longrightarrow$ growth and
and processes   development
$\nearrow$
Environmental
factors

The diagram is somewhat oversimplified because heredity and environment may influence one another. Environmental factors such as ionizing radiation may promote mutations, thus changing hereditary potentialities; and from an evolutionary standpoint environment is involved in natural selection, thus influencing the hereditary potentialities that are transmitted to offspring. However, the diagram does indicate that both heredity and environment operate by determining the nature of the biochemical and biophysical processes and conditions within the organism, and that these in turn determine the pattern of growth, development, and behavior of the organism.

It should be stressed that *everything* an organism is or does requires both a hereditary potentiality and an environment suitable for the expression of the potentiality. The question is not whether any particular characteristic is hereditary or environmental but rather whether a variation is a result of differences in heredity or environment (Figure 11-4). Since geneticists are

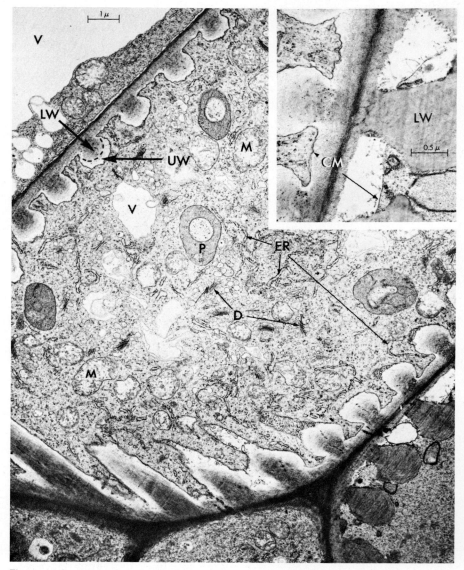

Figure 11-3. Two stages in the differentiation of vessel elements in the xylem of a bean leaf. In the earlier stage the cell is depositing secondary wall thickenings that are not yet completely lignified. Note the abundance of cell organelles at this stage. D, dictyosomes; ER, rough endoplasmic reticulum; M, mitochondria; LW, lignified wall; P, plastid; UW, unlignified wall; V, vacuole. In a later stage (lower right and, upper right, more highly magnified insert) the cytoplasmic structures are collapsing and disintegrating and the cell membrane (CM) is withdrawing from the wall. [Reprinted with the permission of The Macmillan Company from *Plant Structure and Development* by T. P. O'Brien and Margaret E. McCully. Copyright © 1969 by The Macmillan Company.]

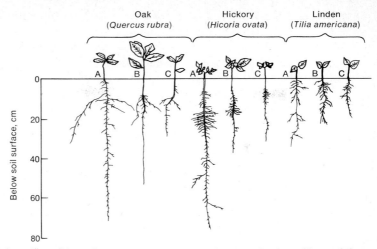

Figure 11-4. Interaction of heredity and environment on the growth of seedlings of three species of trees in three different environments: **(A)** Open prairie; **(B)** oak forest; **(C)** deep shade of a moist linden forest. (The linden seedlings in the prairie were watered to prevent death from desiccation.) [Modified from A. E. Holch, *Ecology*, **12**:259 (1931)]

interested primarily in hereditary variation, we sometimes lose sight of the fact that the basic and generally nonvariable capabilities of plants, such as those for developing xylem, phloem, roots, stems, leaves, and flowers of a particular type, are also dependent on hereditary potentialities. Heredity determines what an organism *can* do; environment determines which of these things the organism actually does and to what degree. All the hereditary potentialities of an embryo plant in a seed are worthless and can never be expressed unless the environment is suitable for the germination of the seed and the subsequent growth of the plant.

Some hereditary potentialities can be expressed in almost any environment that permits survival of the organism, whereas others require specific environmental conditions for expression. For example, a short-day plant may have all the hereditary potentialities for producing flowers of a certain characteristic shape and color, but these potentialities may never be expressed if the plant is kept under long days continuously. Most plants with the potentiality for synthesizing chlorophyll cannot do so if kept in the dark. However, a plant that lacks this hereditary potentiality, such as albino corn, Indian pipe, or any fungus, is incapable of synthesizing chlorophyll under any environmental conditions.

Hereditary Potentialities

As a result of major advances in molecular biology and biochemical genetics it now appears evident that the hereditary potentialities (genes) consist of coded information provided by specific sequences of nucleotides in DNA molecules, that the DNA codes are transcribed into the complementary codes of messenger

RNA (mRNA), and that the mRNA code, with the cooperation of transfer RNA (tRNA) and ribosomes, is translated into the specific type and sequence of amino acids in protein molecules (Chapter 5). Thus, the hereditary potentialities operate by determining the specific kinds of enzymes (and other proteins) an organism can synthesize and, thus, the specific biochemical reactions that can occur in the organism [35].

Of particular importance as regards differentiation and development is the fact that any particular cell of an organism at a specific time is not synthesizing all of the kinds of mRNA (and therefore not all the kinds of enzymes) for which it has DNA codes [15, 30]. Every vegetative cell of a plant normally contains the same set of DNA codes (a complete set of hereditary potentialities), but the specific kinds of mRNA it is producing varies with its stage of development and with its location in the plant. Thus different cells of the plant may have different enzymes and different biochemical capabilities. For example, a mesophyll cell of a leaf can form chloroplasts and synthesize chlorophyll but may not synthesize anthocyanin. A cell of a petal of the same plant may synthesize anthocyanin but may not develop chloroplasts or synthesize chlorophyll, even though both cells contain all the DNA codes for anthocyanin and chlorophyll synthesis and chloroplast development.

Differential enzyme synthesis may also be controlled at the level of protein synthesis [4], even though the necessary mRNA is present. Also, metabolic feedback mechanisms, such as accumulation of the end product of a series of biochemical reactions, may inactivate enzymes that *are* present in a cell.

The question now arises as to what determines whether a specific kind of mRNA is being synthesized by a cell at a specific time. One theory is that in eucaryotic organisms a gene is repressed if it is bound to histones, a type of protein found in chromosomes, and that chromosomal RNA complexed with histones provides the coding needed to attach the histone to specific genes [4]. In any cell at a particular time it seems likely that many, if not most, of the genes are repressed. Other mechanisms of gene repression and derepression (activation) have been proposed, particularly for microorganisms [35].

A still more fundamental question is what brings about the removal or addition of the histone-RNA complex, assuming that this is the mechanism involved in gene derepression and repression. There is increasing evidence that hormones are one of the controlling factors in both plants and animals and that gene repression and derepression is one of the ways in which hormones exert their influences [31]. Gibberellins produced by the embryos of germinating seeds bring about the synthesis of α-amylase in the endosperm, evidently by inducing the production of the mRNA coding for this enzyme. This and other examples will be considered in the next two chapters. Environmental factors may play a role in gene activation or repression by influencing hormone synthesis. For example, James Bonner and his coworkers [3] found increased RNA production in the buds of photoperiodically induced plants prior to their differentiation into flower primordia, and this was presumably brought about by the arrival of a flowering hormone from the leaves. Application of inhibitors of RNA synthesis to the buds prevented the formation of flower primordia.

Hormones are not the only substances that control gene activation. More localized regulating substances are probably involved when a certain type of differentiation is restricted to one or a few cells. The production of adaptive enzymes by microorganisms [35] is induced by the presence of the substrate. For example, lactase may be produced only when lactose is present in the medium. It has been reported that, if iron is added to a culture medium previously deficient in iron, Chlorella begins producing new kinds of mRNA that presumably code for some of the enzymes essential for chlorophyll synthesis.

Environmental Factors

Even if a cell contains all the enzymes essential for a particular metabolic process, the process can occur only if the environment is suitable. For example, photosynthesis can proceed only when essential environmental factors such as light and carbon dioxide are available. Furthermore, the rates of the various metabolic processes are generally determined by limiting environmental factors, as we have seen in previous chapters, and the rates of various processes (for example, photosynthesis and respiration) may have important influences on growth if not development. Some environmental factors, particularly light and temperature, also have marked morphogenetic influences. Finally, the environment determines what species can grow and thrive in any particular habitat or locality. Both the physical and biological environment have important influences on plant growth and development. The factors of the physical environment may be classified in various ways, but here we shall separate them into chemical factors, those that involve matter, and physical factors in a stricter sense, those that involve energy.

Biological Factors

An extensive consideration of the biological environment is beyond the scope of this book, but it should be noted that any ecological interaction between individuals of two species influences the life processes of both and in many cases alters the normal pattern of growth and development. Parasites, if they do not kill their host, frequently inhibit growth or, as in the bakanae disease of rice caused by the fungus *Gibberella fujikuroi,* they may cause an abnormal increase in growth. Some parasites, notably insects and fungi that induce plant galls [20], bring about spectacular morphogenetic changes in the host (Chapter 17).

The symbiotic (mutualistic) relationships between the algae and fungi of lichens [23, 25], mycorrhizal fungi and roots [16] (Chapter 9), and nitrogen-fixing bacteria and legumes (Chapter 5) all influence both organisms. Plants that are dependent on bees or other animals for pollination are in essence dependent on them for the continued growth and development of members of the species.

In some cases one organism influences another by altering its physical environment, as in the influence of forest trees on the smaller plants growing under them or the alteration of the environment of roots by earthworms as they burrow through the soil.

Chemical Factors

The growth and development of autotrophic plants require only a limited number of simple inorganic substances from the environment—several atmospheric gases, a dozen or so mineral elements, and water—but these are absolutely essential. The gases are carbon dioxide, used in photosynthesis, oxygen, used in aerobic respiration (Figure 11-5), and nitrogen, used in nitrogen fixation. The essential mineral elements have been considered in Chapter 6, as have been the various roles of water in Chapters 7 and 9. Fungi and other heterotrophic plants, like animals, must also obtain foods and in some cases vitamins from their environment.

Other substances present in the environment may also influence plant growth and development, generally in an adverse manner. Among these are atmospheric pollutants [36], such as sulfur dioxide, carbon monoxide, and various hydrocarbons, as well as toxic elements such as lead or arsenic or even essential elements such as copper and boron if too concentrated. Agricultural chemicals such as herbicides, insecticides, and fungicides added to the environment by man can influence plant growth and development, particularly if used improperly [13].

Figure 11-5. Influence of oxygen concentration on the growth of tomato roots in solution culture. The plants grew in separate solution cultures, each with a different oxygen concentration in the air above the solution, and then were assembled for this photograph. From left to right, the concentration of oxygen in the air was 1, 3, 5, 10, and 20%. [Courtesy of L. C. Erickson, University of California, Riverside]

Figure 11-6. Inhibition of growth of other species by volatile toxins (antibiotics) emitted by *Salvia leucophylla* (left), a shrub that has invaded a California grassland. A 2-m zone (A–B) is essentially bare of plants and some growth inhibition extends as far as C. [Courtesy of C. H. Muller]

Antibiotics are produced by some vascular plants, as well as by microorganisms, and may inhibit or reduce the growth of other vascular plants (Figure 11-6) as well as certain parasites [1]. Antibiotics provide another example of alteration of the physical environment by organisms.

The droplets of salt water swept off the oceans by the wind are carried to land plants and may be absorbed by them. The salts may contribute toward the mineral nutrition of the plants, but some elements such as chlorine or high total salt concentration can cause impaired or distorted growth [10]. Most of the droplets are not carried far inland, but the smaller ones (some of colloidal dimensions) may be carried considerable distances and may enter the stomata.

Physical Factors: General

The physical factors of the environment include the various manifestations of energy, including heat as it influences temperature, light and other electromagnetic radiation, sound, electricity, magnetism, gravity, and kinetic energy. Of these, temperature and electromagnetic radiation are the only ones that will be considered in any detail here. Their influences on plant growth and development are more extensive, more varied, and generally better understood than those of most of the other physical factors.

There have been several studies of the effects of audible sound waves, including various types of music, on plant growth, some of them conducted by reputable botanists. Although there have been some reports of measurable effects on plant growth, most botanists remain skeptical. However, there is no question about the fact that ultrasonic vibrations can be lethal to microorganisms and that, at high intensities, they can affect plant growth adversely and cause severe injury.

Gravity has marked influences on plant growth and development. The geotropic responses of plants to gravity, that is, the downward growth of roots and the upward growth of stems, is an important factor in plant growth and survival. Since geotropism involves the redistribution of auxin, a plant hormone, it will be considered in Chapter 12. Gravity also has other effects on plants, including the size and shape of some leaves (Figure 11-7) and the amount of secondary growth on the upper and lower sides of tree branches [29].

In nature, electricity influences plants primarily as lightning, which contributes toward nitrogen fixation but may also damage or kill trees. Electrostatic fields influence the potential differences within cells and organs but apparently do not constitute a differential environmental factor of any importance. The influence of high intensity magnetic fields on organisms has been rather extensively investigated and various effects have been reported, but the earth's magnetic field is probably not a differential factor in plant growth.

The principal types of kinetic energy, aside from the kinetic energy of molecules, that influence plants are the flow of air and water. Air flow influences plant growth through its effects on the rates of photosynthesis and transpiration. Air flow adjacent to leaves and stems reduces the concentration of water vapor and increases the concentration of carbon dioxide and, at high velocity, may induce stomatal closure. A more spectacular effect of air flow on plant growth is the slow and distorted growth of trees subjected to strong prevailing winds from one direction (Figure 11-8). Water flow influences aquatic organisms, primarily by altering the concentration of oxygen, carbon dioxide, and mineral salts adjacent to their surfaces.

Although energy factors such as these have minor influences on plant growth and development as compared with light and temperature, they cannot be ignored and, in some cases, may eventually prove to be more important than now realized.

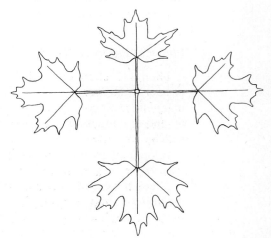

Figure 11-7. Effect of gravity (and probably light to a lesser degree) on the size and shape of sugar maple leaves on horizontal branches. The difference between the leaves of the vertical pair is evident, but not so evident is the fact that the lower halves of the horizontal leaves are slightly larger than the upper halves. In maple there are two leaves at a node and the leaves at one node are at a 90° angle to the pair at the next node. The drawing was based on the averages of 250 measurements of each of four types of leaves. [After E. W. Sinnott and K. S. Wilson, *Botany: Principles and Problems*, 6th ed., McGraw-Hill Book Co., New York, 1963]

A

B

Figure 11-8. Influence of strong prevailing winds on tree growth. **(A)** A pine tree (*Pinus flexilis*) on a mountain in Pike National Forest, Colorado. **(B)** Cypress trees (*Cuperessus macrocarpa*) near Monterey, California. Here salt spray carried by the wind is also a factor. [Both photographs courtesy of U. S. Forest Service]

Physical Factors: Temperature

INFLUENCES ON RATE OF GROWTH. Temperature has a pervasive influence on the rate of plant growth because it affects the rates of all the biophysical and biochemical reactions involved in metabolism, except the purely photochemical ones. In general, the rates of processes increase with temperature; the Q_{10} for physical processes such as diffusion are between 1.2 and 1.3 and for enzymatic reactions between 1.4 and 2.0. However, many processes such as photosynthesis and respiration involve a temperature-time factor, as we have noted earlier. The initial high rates at higher temperatures decline with time, sometimes to the extent that the process finally may occur more slowly than it would at a lower temperature (Figure 4-10). Furthermore, when temperatures rise much above about 40°C enzyme inactivation begins, and by 60°C most enzymes are denatured to such an extent that serious injury or death result. However, dry spores and seeds are often able to survive for some time at even higher temperatures, and thermophilic bacteria and blue-green algae that grow in hot springs at around 70°C obviously have enzymes that are not damaged at this temperature [7, 22]. Temperature may also influence the direction of reversible reactions such as the starch $\rightleftharpoons$ glucose-1-phosphate interconversion (Chapter 5).

The effects of temperature on the various metabolic processes of plants in turn influence the rate of growth by determining the quantity and kinds of foods that can be used in assimilation, the rates of translocation, the amount of ATP available from respiration, the rates of hormone synthesis, the degree of hydration of the cells, and a considerable complex of other internal conditions and processes. However, the rate of growth does not necessarily increase proportionally with the rates of the metabolic processes, nor does it necessarily have the same range of optimal temperatures as even the more important processes as photosynthesis. At the higher natural temperatures growth is generally slow because of such factors as desiccation, which is brought about by high rates of transpiration, and the fact that respiration increases more rapidly than photosynthesis with the rise in temperature.

The optimal temperature for plant growth and also the range of temperatures permitting growth vary considerably with the species, the stage of development, and the time of day. As might be expected, these temperatures are generally higher for plants native to temperate zones than for arctic plants and highest of all for tropical species. Arctic plants may have an optimum as low as 10°C and may grow appreciably even at a few degrees below freezing. Temperate zone plants usually do not grow below 5°C, have optima on the order of 25 to 35°C, and can grow up to about 35 or 40°C. Tropical plants generally have a minimum of around 10°C, optima of around 30 to 35°C, and a maximum of about 45°C.

THERMOPERIODICITY. Frits Went and other investigators have made it clear that plants have different optimal temperatures for growth and development during the night than they have during the day, a phenomenon he has called **diurnal thermoperiodicity** [37, 39]. For example, tomato plants were found to grow and develop fruits best when maintained at 26.5°C during the day and

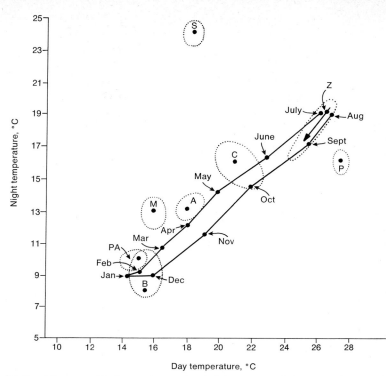

Figure 11-9. Optimal thermoperiodic requirements of eight genera of plants: A, Ageratum; B, English daisy; C, China aster; M, stock; P, petunia; PA, Iceland poppy; S, African violet; Z, zinnia. Note the markedly different optimal temperatures for African violet (Saintpaulia). The solid line shows the average monthly day and night temperatures in Pasadena, California, for a year. [After F. W. Went, *Amer. Sci.* **44**:378–398, 1956]

at 17 to 20°C at night, the favorable effect of the lower night temperature being cancelled by lighting at night. Most species that have been studied have similar thermoperiodic requirements, although the best day and night temperatures vary with the species, but a few species such as African violet thrive best when the night temperatures are somewhat higher than the day temperatures (Figure 11-9).

The change in optimal temperature with the stage of development of plants may be referred to as annual or seasonal thermoperiodicity (Chapter 15), and plants with such changing temperature requirements can thrive only in regions where the seasonal temperature changes approximate the changing optimal temperatures [38]. This phenomenon has been studied most extensively in bulbous plants such as tulips [39].

LOW-TEMPERATURE PRECONDITIONING. The most striking morphogenetic effects of temperature involve low-temperature preconditioning, where the low winter temperatures influence the growth and development of the plant during the following spring and summer. The phenomena influenced include the breaking of bud and seed dormancy and the bolting (internode elongation) and blooming of biennials (Figure 15-12). These phenomena will be considered in later chapters. As might be expected, low-temperature preconditioning is not required by species native to the tropics.

INFLUENCE ON PLANT DISTRIBUTION.   Temperature is one of the most important factors determining the regions in which any particular kind of plant can grow and thrive. This involves meeting not only the thermoperiodic and low-temperature preconditioning requirements of the species but also the temperature extremes that permit survival of the plants. As is well known, some species of plants can withstand the lowest temperatures on earth whereas others are killed by even a light freeze. The nature of freezing injury and of species differences in frost resistance are complex and still not thoroughly understood [26]. Species also differ in their capacity for surviving through periods of extremely high temperature.

Physical Factors: Electromagnetic Radiation

The spectrum of electromagnetic radiation is broad, ranging from cosmic rays with wavelengths as short as 10^{-14} cm to radio waves with wavelengths as long as 10^6 cm (Figure 11-10). Light, the portion of the electromagnetic spectrum visible to man, occupies a very narrow band of the spectrum but has more diverse and more important influences on the growth and development of plants in nature than any other part of the spectrum (Figure 11-11). For this reason most of this section will be devoted to light. Electromagnetic radiation influences an organism only when it is absorbed by one or more of the substances of which the organism is composed; transmitted or reflected radiation is ineffective.

The absorption of infrared (heat) radiation by plants influences them by increasing their temperature. The still longer radio waves are mostly not absorbed by plants (probably fortunately), but high intensity microwaves have adverse effects on plants [11] and animals [21]. However, these are not a component of the solar radiation received by the earth.

Radiation of shorter wavelengths than light (ultraviolet, x-, γ-, and cosmic rays) as well as the particles emitted by radioactive substances affect organisms primarily by causing ionization of various important substances. Ionizing

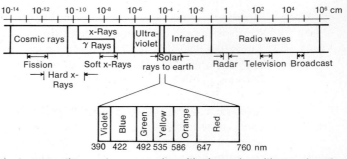

Figure 11-10.   The electromagnetic spectrum on a logarithmic scale, with wavelengths given in centimeters. The narrow band between ultraviolet and infrared is light (radiation visible to man), and is shown below enlarged with wavelengths in nanometers (nm).

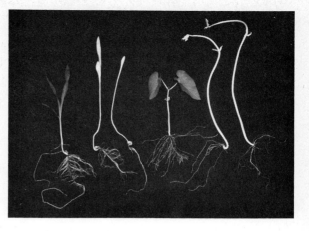

Figure 11-11. Seedlings of corn and bean that germinated and grew in light (left) and dark (right), showing various influences of light on plant growth and development. In the plants that had grown in complete darkness note the lack of chlorophyll synthesis, the tall and spindly stems, poor root growth, failure of the bean leaves to unfold and enlarge, and the greatly elongated bean hypocotyls. [Courtesy of Frank Salisbury]

radiation [28] has two principal types of effects: the induction of mutations and more direct morphogenetic effects (Figure 11-12). The latter will be considered in Chapter 17. Ultraviolet rays, particularly the shorter ones (that are fortunately largely absorbed by the earth's atmosphere) may be lethal to microorganisms, but they have little effect on larger plants and only on the superficial tissues. That ultraviolet radiation plays no essential roles in plants is indicated by the normal growth and development of plants under the glass of greenhouses, which absorbs essentially all of the ultraviolet.

PHOTORECEPTOR PIGMENTS. The substances in plants that absorb light of various wavelengths are referred to as pigments because the absorption of light

Figure 11-12. Influence γ-radiation from ^{60}Co on the growth of tobacco hybrids (*Nicotiana glauca* × *N. langsdorfii*). The figures in front of the pots give the daily dosage in roentgens (r). Note that 15 r/day induced blooming. [Courtesy of A. H. Sparrow, Brookhaven National Laboratory]

imparts a color to the substance. A substance that absorbs light throughout the visible spectrum looks black. There are some black plant pigments, but they are not known to play an important photochemical role. Most plant pigments absorb some wavelengths of light more completely than others, and thus are colored. For example, chlorophyll has its absorption maxima in the red and blue regions of the spectrum and absorbs only a small fraction of the light in the green region. Because most of the green light is reflected or transmitted, chlorophyll appears green to us. The colors of the various plant pigments range from one end of the visible spectrum to the other: violet, blue, green, yellow, orange, and red. The substances that absorb electromagnetic radiation with wavelengths either longer or shorter than those of light are colorless because they are beyond the range of man's vision.

As has been noted in Chapter 3, photosynthesis involves a variety of pigments other than chlorophyll, including carotenes, xanthophylls, phycocyanin and phycoerythrin. Photosynthesis, of course, plays an important role in the growth of green plants by providing the food that is essential for respiration and assimilation. Low rates of photosynthesis are frequently the limiting factor for plant growth. In addition to the light requirement for photosynthesis itself, light is essential for the conversion of protochlorophyll into chlorophyll in higher plants of most species.

The photoreceptor pigment in phototropism, that is, the bending growth of stems, coleoptiles, and leaves (Figure 11-13) toward light of higher intensity

Figure 11-13. Phototropic response of the leaves of a geranium plant. **Left:** Photograph taken from the direction of the incident light. **Right:** Photograph of the same plant taken at 90° from the direction of the incident light. [Courtesy of A. M. Winchester]

(Chapter 12), is a yellow substance. It has been difficult to determine whether it is riboflavin or a carotene because both substances have absorption spectra approximating, but not coinciding in either case, with the action spectrum of phototropism.

Several common plant pigments are not known to play any important photochemical role. Among these are the anthocyanins, anthoxanthins, and betacyanins, all of which are water soluble and occur in the cell sap of vacuoles. The color of anthocyanins ranges from violet through blue to pink or red, depending on the particular kind of anthocyanin and the pH of the solution, whereas anthoxanthins are generally a clear yellow. These substances are responsible for the colors of most flowers and may play a role in attracting pollinating insects. However, anthocyanins are present in many stems or leaves and can hardly play a similar role there. Certainly the betacyanin of redbeet roots does not play a photochemical role, nor does the abundant carotene in carrot roots. The carotenoids also occur in many above-ground organs where they can hardly be playing a role in either photosynthesis or phototropism.

A plant pigment that does have an important function in plant growth and development is **phytochrome,** a pale blue substance that is of widespread occurrence in plant cells but is present in too low concentrations to impart any visible color to them. Because of the numerous aspects of plant growth and morphogenesis in which phytochrome is the photoreceptor pigment, it will receive special consideration here and in subsequent chapters.

PHYTOCHROME IN PLANT GROWTH AND DEVELOPMENT. Phytochrome [18, 19] was discovered and characterized during the 1950s by Sterling Hendricks and his coworkers. It has proved to be a conjugated protein, the chromaphore being a straight-chain pyrrole compound chemically related to the phycocyanins and phycoerythrins of algae and the bile pigments of animals. Light absorption by phytochrome evidently influences a process or processes of basic metabolic importance since it is involved in a wide range of photomorphogenetic phenomena including photoperiodism (Chapter 15), seed germination, spore germination, leaf blade unfolding and expansion, hypocotyl and plumule hook opening (Figure 11-14), internode elongation, and the formation of some kinds of anthocyanin. A number of these will be discussed at appropriate points in subsequent chapters.

An interesting and important characteristic of phytochrome is that it exists in two interconvertible forms: one (P_R) has its maximum absorption peak in the red region of the spectrum at 660 nm and the other (P_{FR}) at 730 nm in the far-red region at the very edge of visibility for humans. When P_R is exposed to light of 660 nm it is converted to P_{FR} and, in turn, this is converted back to P_R when it is subsequently exposed to light of 730 nm (Figure 11-15). Of course, in nature plants are not exposed to such narrow wavelengths of light, but the broad spectrum light from the sun or electric lights converts P_R to P_{FR} even though it includes some 730 nm radiation, and P_{FR} slowly reverts to P_R that lasts until it again changes to P_R during the night. P_{FR} is apparently the enzymatically active form of the pigment, and the significance of this will be considered later on.

Figure 11-14. Influence of red and far-red light on the development of bean seedlings. **From left to right:** Continuous darkness; 2 min red; 2 min red + 5 min far-red; 5 min far-red (otherwise complete darkness until photographed). Note the effect of the red light on leaf expansion, straightening of the plumule hook, and hypocotyl and stem growth. [Courtesy of Crops Research Division, U. S. D. A., Beltsville, Md.]

Internal Conditions and Processes

A discussion of the internal conditions and processes of plants that are involved in growth and development could easily fill several large volumes, but here we shall make only a few comments about their general roles. All of the processes discussed in the first half of this book (except perhaps for the synthesis of such substances as alkaloids and latex that are of dubious metabolic significance in plants) plus many others are involved in plant growth and development in one way or another. Some additional processes will be considered in coming chapters. The plant growth substances are of particular importance internally and will be discussed in Chapters 12 and 13.

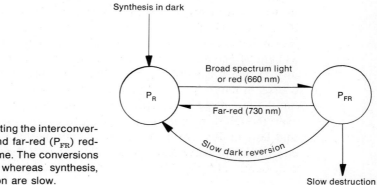

Figure 11-15. Diagram illustrating the interconversions between the red (P_R) and far-red (P_{FR}) red-absorbing forms of phytochrome. The conversions induced by light are rapid, whereas synthesis, dark reversion, and destruction are slow.

Elucidation of the internal conditions and processes involved in any biological phenomenon is a complex and difficult problem, and thus is almost always the last thing to be achieved. For example, the general nature and significance of photosynthesis have been known for over a century and much information on the influence of environmental factors on photosynthesis has been available for years, but the beginning of a detailed and reasonably satisfactory understanding of the photosynthetic processes dates back only to about 1940.

Photoperiodism was discovered in 1918 and within a few years a large body of information had been obtained on the hereditary potentialities of many plants in relation to the photoperiod, on the influence of various environmental factors, and on the observable growth and development of plants as influenced by the photoperiod. Furthermore, within a few years practical application of the information about photoperiodism was being made in the culture of horticultural plants. Yet even at the present time, and despite extensive research, we still have only a very sketchy understanding of the internal conditions and processes involved in photoperiodism.

Clarification of these internal factors will probably not give us any greater control over plant growth and development, but plant physiologists and biochemists and other botanists will not be satisfied until they have a reasonably clear picture of the nature of the internal factors involved in all aspects of growth and development.

Observable Growth and Development

In contrast to our incomplete understanding of the internal factors of growth and development, a vast amount of information has been accumulated over the years regarding the observable aspects of plant growth and development. The rates and periodicity of growth of plants of various species (that is, with varying hereditary potentialities) under different environmental conditions have been determined. The pattern of plant development has been described in great detail at the cellular and tissue levels as well as from the standpoint of external morphology; that is, the sequence of morphological changes during development as well as the structure of the cells, tissues, and organs that eventually result has been explicated. However, investigation of this aspect of plant growth and development is by no means exhausted. Plant development will be considered in subsequent chapters.

Growth Rates and Periodicity

Kinetics of Growth

If the growth of a plant is plotted on a cartesian (linear-linear) graph against time, an S-shaped or sigmoid growth curve usually results (Figure 11-16). This is generally the case regardless of whether the measurements are of increase in length, volume, fresh weight, or dry weight, or whether the growth of the plant as a whole or of individual organs such as stems, roots, leaves, or fruits

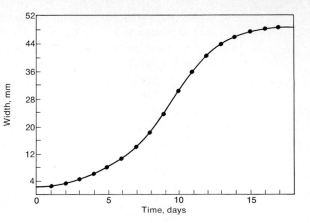

Figure 11-16. Growth in diameter of fruits, illustrating a typical sigmoid growth curve. [After E. W. Sinnot, *Plant Morphogenesis,* McGraw-Hill Book Co., New York, 1960]

is being measured. However, the growth rates of specific organs may differ from one another and from the growth rate of the plant as a whole, and the period of time covered by the sigmoid curve may be quite different. The growth of a plant as a whole is, of course, an integration of the growth of all its various organs.

The rate and duration of growth are controlled by both genetic and environmental factors. One does not need to be a biologist to be aware of the fact that the rate of growth is a species, or even a varietal, characteristic that varies greatly from one kind of plant to another or that the period of growth ranges from a few weeks or months for annuals to hundreds of years for trees. It is also common knowledge that growth rates are greatly influenced by such environmental factors as light, water availability, and the concentration of the essential mineral elements. The shape of a growth curve may be markedly altered by a severe environmental deficiency, such as lack of one or more essential elements, and the sigmoid curve may be almost obliterated when plants are treated with growth inhibitors such as maleic hydrazide (Figure 11-17).

There has been extensive mathematical analysis of growth curves [29] and a number of differential equations that characterize the sigmoid growth curve have been devised. Although these are generally oversimplified in view of the complex of factors influencing growth, equations such as the following one provide a reasonable approximation to actual sigmoid growth curves.

Growth per unit time = Constant × Present size ×
(Final size − Present size)

$$\frac{dG}{dt} = K \times G \times (G_f - G)$$

The equation states in essence that the growth increment per unit time is proportional to the size already attained (that is, that as organisms get larger they grow faster) and also to the difference between present and final size (that is, that as organisms approach their final size there is a progressive decrease in the rate of growth). When growth is plotted as the daily increment of growth (Figure 11-18) rather than as total cumulative growth these changes in growth

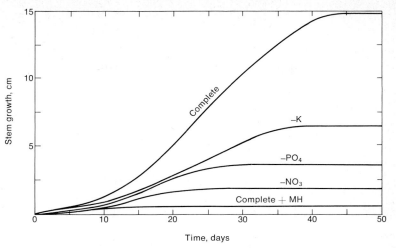

Figure 11-17. Growth of stems of plants provided with a complete mineral nutrient solution in comparison with plants provided with solutions lacking nitrogen, phosphorus, or potassium, and plants whose growth was inhibited by maleic hydrazide (MH). Note the marked deviation from the usual sigmoid growth curve when growth was inhibited by MH or a mineral deficiency. [Data of V. A. Greulach]

rate with time become quite evident, and such a curve is usually somewhat bell-shaped.

When growth is plotted on semilog paper with time on the linear axis, a different shape of curve results (Figure 11-19) that provides more insight regarding growth rates. (Figures 11-16, 11-18, and 11-19 were all plotted from the same data.) Note that, when expressed logarithmically, the initial rapid growth rate is constant so that its plot is a straight line with the slope of the curve indicating the growth rate. During this period growth essentially follows the compound interest law if the interest is compounded continuously; the growth per unit time is a constant percentage of that attained at the beginning of the time period, even though the actual increments per unit time (here a day) keep increasing.

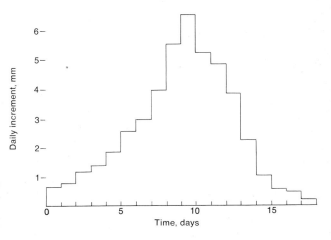

Figure 11-18. Daily increment of growth in diameter of fruits, plotted from the same data as Figure 11-16. [Data of E. W. Sinnott, *Plant Morphogenesis,* McGraw-Hill Book Co., New York, 1960]

Introduction to Plant Growth and Development 317

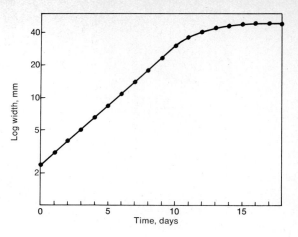

Figure 11-19. Growth in diameter of fruits, plotted as the logarithm of the fruit width from the same data as Figures 11-16 and 11-18. [After E. W. Sinnott, *Plant Morphogenesis,* McGraw-Hill Book Co., New York, 1960]

Although mathematical treatments of growth may be interesting and illuminating, the rather marked differences in the daily increments of growth demand metabolic explanations and these are still rather obscure. The rapid initial growth rate may be primarily a result of the progressive increase in the abundance of metabolic substances such as foods, enzymes, and hormones, but the subsequent slowing of growth is more obscure. Such things as exhaustion of metabolic substrates, increasing difficulty of adequate translocation of water and solutes with the increase in size, and an increase in catabolic processes in proportion to anabolic ones may be involved. Genetic factors are obviously important because the hereditary potentialities of any species set a rather definite limit on the size it can attain, and this may involve a progressive change in the kinds of enzymes being produced.

Range of Growth Rates

Although environmental factors have considerable influence on the rate of growth of any particular plant, the members of some species grow much more rapidly than those of other species. This means, of course, that the hereditary potentialities of a plant play a major role in determination of its rate of growth.

Among the most rapid rates of plant growth that have been measured are 60 cm a day for bamboo shoots, 30 cm a day for asparagus stems, and 15 cm a day for the peduncles of the century plant. Corn plants may grow as much as 13 cm a day. Other examples of rapid growth include the elongation of the internodes of biennials during the bolting period, the elongation of dandelion scapes, and the growth of large fruits such as pumpkins [17]. At the other extreme are such species as English boxwood, which grows only a few centimeters a year. Most species of plants, however, have intermediate rates of stem growth on the order of 1 to 4 cm a day.

Determinate and Indeterminate Growth

The growth of animals and of plant organs such as leaves, fruits, and seeds is **determinate,** that is, it proceeds for a certain period of time which is to a large degree under genetic control and then ceases. In contrast, growth from the apical meristems of stems and roots and from the cambium is **indeterminate,** that is, growth continues indefinitely, with the result that the size of the plant continues to increase with age. The effects of indeterminate growth are most strikingly evident in woody perennials, which may continue to increase in size for hundreds or even thousands of years, but even in these growth eventually slows down and then ceases when the genetically determined size is attained. One of the changes involved when a bud primordium is converted from the vegetative to reproductive condition is that growth becomes determinate. Thus, the stem elongation of normally unbranched plants with terminal inflorescences such as the sunflower ceases once the flower primordia have been formed. The same thing occurs in branched plants if all the buds become flower buds.

Differential Growth

The growth of vascular plants, unlike that of some other organisms, is limited to certain regions. Growth in length of stems and roots is from the apical meristems, and once cell enlargement is complete growth in diameter of stems and roots is from the cambium. Root elongation occurs largely in a zone a few millimeters long (the region of cell elongation) located just back of the apical meristem (Figure 11-20). The nodes of stems generally do not elongate,

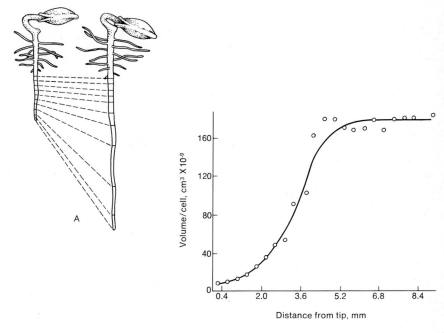

Figure 11-20. (A) Amount of growth in various regions of the primary root of a seedling. Ink marks were made at equal intervals on the root of a young seedling, and their spacing after further growth shows the amount of growth in each segment. **(B)** Average volume of pea root cells at increasing distances from the root tip. [B after R. Brown, W. S. Reith, and E. Robinson, *Symp. Soc. Expt. Biol.* **6:**329–347, 1952]

Figure 11-21. The growth of a leaf blade is essentially uniform throughout its area, as shown by inking a grid on a young leaf and noting the spacing of the lines when the leaf reaches its final size.

and the pattern of internode elongation varies with the species. Elongation may be uniform throughout the internode, may be more extensive in the central portion of the internodes, as in Polygonum, or may be restricted to the base of the internodes, as in Tradescantia in which the internodes continue to elongate for a considerable period of time. In most trees and many other plants elongation of a particular internode ceases after only a few days or weeks of growth, and all subsequent stem elongation is from the younger internodes. A widespread misconception among laymen is that the trunks of trees continue to elongate.

The growth of leaf blades is generally uniform throughout the blade (Figure 11-21), as well as determinate. However, the blades of grass leaves have a persistent basal meristem that results in continued growth in length of the blades. This makes grass suitable for lawns, but necessitates the periodic task of mowing.

Daily Periodicity of Growth

During the 24 hr of a day plants exhibit a quite marked periodicity in rate of growth. This periodicity results from diurnal changes in environmental factors, primarily light, temperature, and water, although it may also involve circadian rhythms [8]. The environmental factors influence the rate of growth through their effects on a wide variety of internal processes and conditions, notably photosynthesis, respiration, the hydration and turgidity of the cells, and somewhat more directly on cell division and cell elongation. The effects are complex and often difficult to analyze, and the factor that is limiting growth may change rather quickly.

The result may be rather marked and irregular fluctuations in the rate of growth, as in the case of the corn plants in Figure 11-22. The rapid increase in the growth rate in the morning was probably brought about both by an increase in temperature and the provision of increased substrates for growth by photosynthesis. The marked early afternoon decline in growth rate undoubtedly resulted from internal water deficits, and it should be noted that the decline was much less in the shaded plant than in the one exposed to full sunlight. The subsequent increase probably resulted largely from increased water content. The low rate during the night was probably primarily a result of substrate exhaustion, although decreased temperature may also have been a factor. Corn, in general, grows more during the day than the night, except when the internal water deficit is severe. Although light has an inhibiting influence on growth of stems, as shown by the rapid growth of stems in the dark (Figure 11-11), this is not evident in the growth of corn under field conditions. The effects of light on corn growth appear to be more a matter of its influence on the rate of photosynthesis and on increasing the temperature of the plant.

It should not be assumed that the growth periodicity of corn just described is typical for all plants at all times. The daily periodicity of growth varies with the particular set of environmental factors and the fluctuations of each one on any particular day. Note in Figure 11-22 that the growth periodicity of the second day was unlike that on the first.

Although environmental factors are primarily responsible for the daily periodicity of growth, genetic factors also play a role and two species under the same conditions may have different growth periodicities. For example, in tomatoes stem elongation occurs only at night. As has been noted, species differ as regards optimal temperatures and thermoperiodic requirements.

Seasonal Periodicity of Growth

There is also a marked seasonal periodicity in the growth of plants, at least in biennials and perennials, and in particular in trees and shrubs. The principal environmental factors influencing seasonal periodicity are temperature [24],

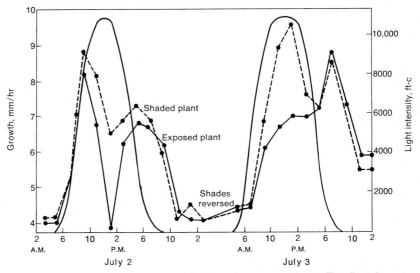

Figure 11-22. Daily periodicity of growth in height of two corn plants. The first day one was shaded and the other was exposed to full sunlight, whereas the second day each plant had light exposure opposite to that of the first day. The solid, bell-shaped curves show the intensity of full sunlight each day. See text for discussion. [Data of H. F. Thut and W. E. Loomis, *Plant Physiol.,* **19:**117 (1944)]

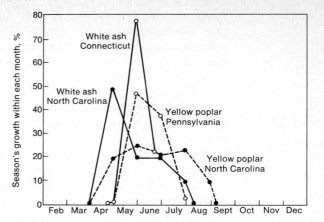

Figure 11-23. Seasonal periodicity of growth of white ash and yellow poplar trees in different latitudes. [After P. J. Kramer, *Plant Physiol.,* **18**:239 (1943)]

water availability, and the length of day [12]. Typically trees cease growth in the autumn or even during the summer (Figure 8-8), lose their leaves (if a deciduous species), and become dormant [27, 34]. During the winter growth is at a standstill, even in evergreen species, and the trees remain dormant until they have been exposed for a long enough period to low temperatures. In the spring, with the advent of higher temperatures, longer days, and/or greater water availability, growth is quite suddenly resumed and the growth rate soon attains its maximum for the year (Figure 11-23). By late spring or early summer terminal growth of many species ceases or slows down greatly, even though the environmental factors still seem to be favorable. Lateral diameter growth from the cambium generally continues through the summer, but at a much slower rate than in the spring, as evidenced by the spring and summer wood of an annual ring. Of course, there may be marked growth of flowers and fruits in some species during the summer.

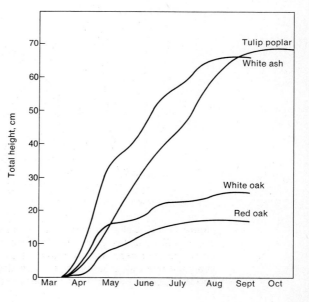

Figure 11-24. Growth in height of seedlings of four species as related to season. Note the species differences in the duration of the period of active growth and total amount of growth. [Data of P. J. Kramer, *Plant Physiol.,* **18**:239 (1943)]

There are genetic differences between species as regards both the maximum growth rates and the length of time growth continues before it ceases (Figure 11-24). In some species such as beech and the hickories stem elongation and leaf formation cease after only two or three weeks, in others such as dogwood and sumac stem elongation and leaf formation continue through the summer, and many other species are intermediate between these extremes. While growth is in progress it is by no means occurring at a constant rate. Aside from the daily growth periodicity, growth may occur in spurts covering a variable number of days, presumably because of periodic changes in temperature and other environmental factors.

References

[1] Bonner, J. "Chemical warfare among the plants," *Sci. Amer.,* **180**(3):48–51 (1949); "Chemistry in plant societies," *Nat. Hist.,* **68**:508–513 (1959).

[2] Bonner, J. "The next new biology," *Plant Sci. Bull.,* **11**(3):1–7 (Dec. 1965).

[3] Bonner, J., and J. A. D. Zeevaart. "Ribonucleic acid synthesis in the bud: an essential component of floral initiation in Xanthium," *Plant Physiol.,* **57**:43–49 (1961).

[4] Bonner, J. *The Molecular Biology of Development,* Clarendon Press, Oxford, 1965.

[5] Bonner, J. T. "The unsolved problem of development: an appraisal of where we stand," *Amer. Sci.,* **48**:514–527 (1960).

[6] Beerman, W., and U. Clever. "Chromosome puffs," *Sci. Amer.,* **210**(4):50–58 (April 1964).

[7] Brock, T. D. "Life at high temperatures," *Science,* **158**:1012–1019, (1967).

[8] Brown, F. A., Jr. "The rhythmic nature of plants and animals," *Amer. Sci.,* **47**:148–168 (1959).

[9] Brown, S. W. "Heterochromatin," *Science,* **151**:417–425 (1966).

[10] Cassidy, N. G. "Cyclic salt and plant health," *Endeavour,* **30**:82–86 (1971).

[11] Davis, F. S., J. R. Wayland, and M. G. Merkle. "Ultra-high frequency electromagnetic fields for weed control: phytotoxicity and selectivity," *Science* **173**:535–537 (1971).

[12] Downs, R. J. "Photoperiodic control of growth and dormancy in woody plants," in K. V. Thimann (ed.), *The Physiology of Forest Trees,* Ronald Press, New York, 1958, pp. 529–537.

[13] Egler, F. E. "Pesticides in our ecosystem," *Amer. Sci.,* **52**:110–136 (1954).

[14] Gibor, A. "Acetabularia: a useful giant cell," *Sci. Amer.,* **215**(5):118–124 (Nov. 1966).

[15] Gurdon, J. B. "The cytoplasmic control of gene activity," *Endeavour,* **25**:95–99 (1966).

[16] Harley, J. L. "The mycorhiza of forest trees," *Endeavour,* **15**:43–48 (1956).

[17] Haseman, L. "Some punkins," *Science,* **115**:526 (1952).

[18] Hendricks, S. B. "How light interacts with living matter," *Sci. Amer.,* **219**(3):174–186 (Sept. 1968).

[19] Hendricks, S. B., H. A. Borthwick, and R. J. Downs. "Pigment conversion

in the formative responses of plants to radiation," *Proc. Nat. Acad. Sci.,* **42**:19–26 (1956).

[20] Hovanitz, W. "Insects and plant galls," *Sci. Amer.,* **201**(5):151–162 (Nov. 1959).

[21] Jaski, T., and C. Susskind. "Electromagnetic radiation as a tool in the life sciences," *Science,* **133**:443–447 (1961).

[22] Kempner, E. S. "Upper temperature limit of life," *Science,* **142**:1318–1319 (1963).

[23] Kershaw, K. A. "Lichens," *Endeavour,* **22**:65–69 (1963).

[24] Kramer, P. J. "Thermoperiodism in trees," in K. V. Thimann (ed.), *The Physiology of Forest Trees,* Ronald Press, New York, 1958, pp. 573–580. See also Kramer, P. J., and T. T. Kozlowski. *The Physiology of Trees,* McGraw-Hill, New York, 1960.

[25] Lamb, I. M. "Lichens," *Sci. Amer.,* **201**(4):144–156 (Oct. 1959).

[26] Levitt, J. *Frost Killing and Hardiness of Plants,* Burgess Publishing Co., Minneapolis, 1941.

[27] Perry, T. O. "Dormancy of trees in winter," *Science,* **171**:29–36 (1971).

[28] Platzman, R. L. "What is ionizing radiation?" *Sci. Amer.,* **201**(3):74–84 (Sept. 1954).

[29] Sinnott, E. W. *Plant Morphogenesis,* McGraw-Hill, New York, 1960.

[30] Stebbins, G. L. "From gene to character in higher plants," *Amer. Sci.,* **53**:104–126 (1965).

[31] Thomas, D., and J. T. Mullins. "Role of enzymatic wall softening in plant morphogenesis: hormonal induction in Achyla," *Science,* **156**:84 (1967).

[32] van Overbeek, J. "The control of plant growth," *Sci. Amer.,* **219**(1):75–81 (July 1968).

[33] Veen, R., and G. Meijer. *Light and Plant Growth,* Macmillan, New York, 1959.

[34] Vegis, A. "Dormancy in higher plants," *Ann. Rev. Plant Physiol.,* **15**:185–224 (1964).

[35] Watson, J. D. *Molecular Biology of the Gene,* 2nd ed., W. A. Benjamin, New York, 1970.

[36] Went, F. W. "Air pollution," *Sci. Amer.,* **192**(5):62–73 (May 1955).

[37] Went, F. W. "The role of environment in plant growth," *Amer. Sci.,* **44**:378–398 (1956).

[38] Went, F. W. "Climate and agriculture," *Sci. Amer.,* **196**(6):82–94 (June 1957).

[39] Went, F. W. *Environmental Control of Plant Growth,* Ronald Press, New York, 1957.

Auxins and Ethylene

12

Among the important internal factors that play roles in plant growth and development are the plant hormones **(phytohormones).** These are sometimes referred to as plant growth substances (or regulators), a somewhat broader term that includes synthetic substances which have hormone activity but are not produced by plants, natural substances effective in low concentrations but not translocated, and the phytohormones. Hormones are organic substances that are biologically effective at very low concentrations (usually less than 1 ppm). They are usually synthesized in one part of an organism and are transported to other parts where they exert their influences. This does not necessarily exclude effects of a hormone on a cell that synthesizes it. Hormones provide an important means of chemical coordination of multicellular organisms. Unlike most animal hormones, plant hormones are not produced in well-defined endocrine glands, and they differ from animal hormones (largely if not entirely) as regards both chemical nature and their metabolic roles. The known plant hormones are all primarily involved in growth and development, whereas some animal hormones are not.

The discovery and study of plant hormones occurred at about the same time as that of animal hormones, or even earlier if the preliminary research that led to the discovery of auxin is included (Figure 12-1). Although Charles Darwin provided the first clues regarding the existence of plant hormones, active investigation of them did not begin until after 1926, when Frits Went provided proof that diffusible substance obtained from oat seedlings promoted their growth. In 1928 he devised a quantitative biological assay for this growth substance, which was named auxin.

Until about 1950 it was generally believed by western-world botanists that auxin was the only phytohormone of any consequence, but then in rapid succession the importance of others such as gibberellins, cytokinins, abscisic acid, and ethylene was discovered. These, as well as auxin, have been investi-

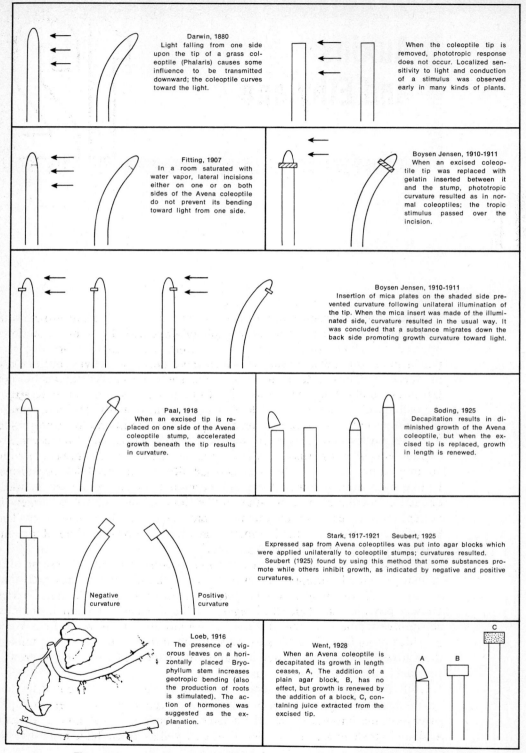

Figure 12-1. Outline of the early discoveries concerning auxin. [After P. Boysen-Jensen, *Growth Hormones in Plants,* 1936, with the permission of the publisher, McGraw-Hill Book Co., Inc. New York]

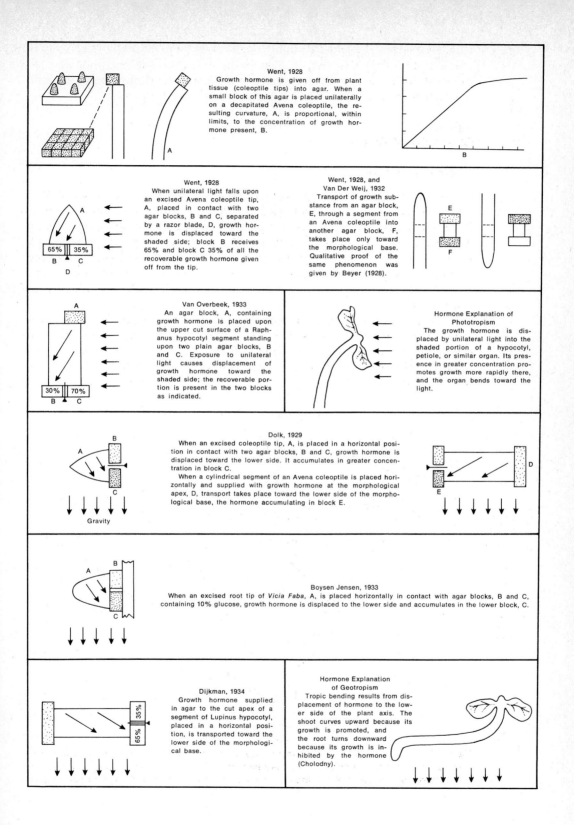

Went, 1928
Growth hormone is given off from plant tissue (coleoptile tips) into agar. When a small block of this agar is placed unilaterally on a decapitated Avena coleoptile, the resulting curvature, A, is proportional, within limits, to the concentration of growth hormone present, B.

Went, 1928
When unilateral light falls upon an excised Avena coleoptile tip, A, placed in contact with two agar blocks, B and C, separated by a razor blade, D, growth hormone is displaced toward the shaded side; block B receives 65% and block C 35% of all the recoverable growth hormone given off from the tip.

Went, 1928, and Van Der Weij, 1932
Transport of growth substance from an agar block, E, through a segment from an Avena coleoptile into another agar block, F, takes place only toward the morphological base. Qualitative proof of the same phenomenon was given by Beyer (1928).

Van Overbeek, 1933
An agar block, A, containing growth hormone is placed upon the upper cut surface of a Raphanus hypocotyl segment standing upon two plain agar blocks, B and C. Exposure to unilateral light causes displacement of growth hormone toward the shaded side; the recoverable portion is present in the two blocks as indicated.

Hormone Explanation of Phototropism
The growth hormone is displaced by unilateral light into the shaded portion of a hypocotyl, petiole, or similar organ. Its presence in greater concentration promotes growth more rapidly there, and the organ bends toward the light.

Dolk, 1929
When an excised coleoptile tip, A, is placed in a horizontal position in contact with two agar blocks, B and C, growth hormone is displaced toward the lower side. It accumulates in greater concentration in block C.
When a cylindrical segment of an Avena coleoptile is placed horizontally and supplied with growth hormone at the morphological apex, D, transport takes place toward the lower side of the morphological base, the hormone accumulating in block E.

Boysen Jensen, 1933
When an excised root tip of Vicia Faba, A, is placed horizontally in contact with agar blocks, B and C, containing 10% glucose, growth hormone is displaced to the lower side and accumulates in the lower block, C.

Dijkman, 1934
Growth hormone supplied in agar to the cut apex of a segment of Lupinus hypocotyl, placed in a horizontal position, is transported toward the lower side of the morphological base.

Hormone Explanation of Geotropism
Tropic bending results from displacement of hormone to the lower side of the plant axis. The shoot curves upward because its growth is promoted, and the root turns downward because its growth is inhibited by the hormone (Cholodny).

gated extensively. Indeed, for almost a half century a major part of the research in plant physiology has been on phytohormones. This chapter will be devoted to auxin, which is still the most extensively investigated phytohormone, and to ethylene, and Chapter 13 will discuss other phytohormones.

Chemistry of Auxin

Chemical Nature

The predominant, if not the only, natural auxin is indole-3-acetic acid (IAA). Other related indole compounds such as indole-3-pyruvic acid, indoleacetaldehyde, and indoleacetonitrile occur in plants. These generally have auxin activity when supplied to auxin-deficient tissues, but they may all be converted to IAA before they are effective. Various synthetic growth substances not produced by plants, such as α-naphthalene acetic acid and 2,4-dichlorophenoxy acetic acid (2,4-D), have auxin activity when supplied to plants. These will be considered later.

From time to time there are reports of naturally occurring growth substances other than indole compounds that have auxin activity, and it is possible that eventually a variety of auxins will be found. After all, only a few of the hundreds of thousands of species of plants have been checked for their phytohormone content.

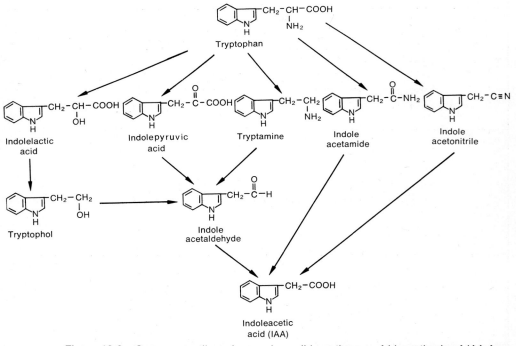

Figure 12-2. Summary outline of several possible pathways of biosynthesis of IAA from tryptophan. [After R. Phelps and L. Sequeira, p. 198 in F. Wightman and G. Setterfield (eds.), *Biochemistry and Physiology of Plant Growth Substances,* Runge Press, Ottawa, 1968]

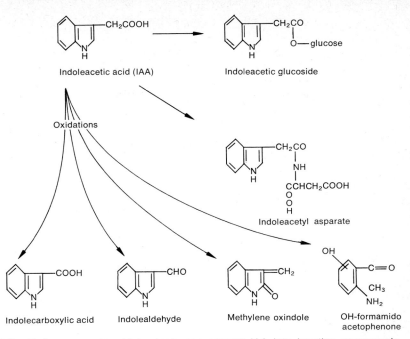

Figure 12-3. Various ways in which plants can convert IAA into inactive compounds. [After A. C. Leopold, *Plant Growth and Development,* McGraw-Hill Book Co., Inc., New York, 1964]

Synthesis and Degradation of IAA

It is generally considered that IAA is synthesized from tryptophan, with indolepyruvic acid and indoleacetaldehyde as intermediates, but other synthetic pathways may occur in at least some species (Figure 12-2).

Plants contain a variety of enzymes that catalyze the oxidation of IAA. These are referred to collectively as **IAA oxidase.** The oxidations are stepwise and the final products lack auxin activity. Polyphenols inhibit IAA oxidase activity, whereas monophenols increase it. Since the concentrations of these phenolic compounds change during plant development and are also influenced by environmental factors such as the photoperiod, they may well play important roles in the control of auxin levels [15]. There are also other means of IAA inactivation (Figure 12-3) that probably contribute toward the control of auxin concentration.

Quantitative Aspects of Auxin

Effective Auxin Concentrations

The concentrations of auxin effective in promoting growth are very low, as is characteristic of hormones. The optimal auxin concentration for growth is quite different for various organs and tissues and for various auxin effects. For stem growth it is about $10^{-5}\ M$ whereas for roots it is about $10^{-9}\ M$, or possibly somewhat higher. Concentrations of $10^{-3}\,M$ or more inhibit growth in all organs.

The reasons for the differential sensitivity of organs to auxin are not known. Various theories have been proposed in an effort to account for auxin growth inhibition at concentrations above the optimum, but in general good experimental validation is lacking. Descriptions of these theories can be found elsewhere [1, 18]. It is known that relatively high concentrations of auxin induce ethylene formation in plants, and Burg [4] and subsequently others have proposed that ethylene is actually the growth inhibitor, but this remains to be proved definitely.

Quantitative Determination of Auxin

Because of the extremely low concentrations of auxin in plants, very sensitive means of quantitative determination are essential. Since chemical methods have lacked the necessary quantitative sensitivity, at least until recently, bioassays have generally been employed.

THE AVENA CURVATURE TEST. The original and still the most precise and sensitive bioassay for auxin is the Avena (oat) coleoptile test devised by Went in 1928 and subsequently refined (Figure 12-4). The degree of curvature of coleoptiles following unilateral application of agar blocks containing auxin is proportional to the auxin concentration. Auxin from the plant tissue being assayed is allowed to diffuse into the agar. The coleoptile tips must be decapitated since they produce auxin, and the tests must be conducted in the absence of phototropically effective light. Quantitative calibration is provided by use of a series of agar blocks containing IAA solutions of different concentrations over the physiological range. To be reliable, the assay requires careful manipulations under controlled environmental conditions. To be classed as an auxin a substance must be effective in the Avena test. However, the assay is not effective for bound auxin.

OTHER BIOASSAYS. Of the other bioassays that have been devised, the most satisfactory and widely used are the slit-pea-stem test (Figure 12-5) and the Avena-coleoptile and pea-stem straight-growth tests. In all three the test tissues are placed in petri dishes containing the solutions to be tested for auxin activity. Although these assays are less precise and sensitive than the Avena curvature

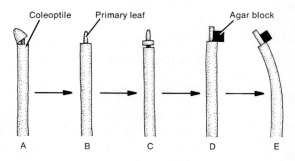

Figure 12-4. Steps in the Avena curvature test. **(A)** Auxin-producing coleoptile tip removed; **(B)** 3-hr interval; **(C)** second decapitation; **(D)** application of agar block containing auxin; **(E)** curvature after 90 min.

Figure 12-5. The split-pea bioassay for auxin. **Left:** sections of pea seedling stems are slit longitudinally about three-quarters of their length. **Center:** when placed in water the stem halves curve outward and downward. **Right:** when placed solutions containing auxin the stem halves curve upward and inward. The degree of curvature is proportional to the auxin concentration.

test, they are much easier to conduct satisfactorily. Since several milliliters of solution are required, these assays are more suitable for synthetic growth substances than for plant extracts. Details regarding the various bioassays are available elsewhere [16].

CHEMICAL DETERMINATIONS.   Colorimetric analyses and other methods for chemical determination of auxin have long been available, but they lack the sensitivity needed for quantitative determination of low concentrations of auxin and have been more useful for qualitative analysis. Newer techniques such as thin-layer chromatography and gas chromatography have now provided more sensitive means of chemical analysis.

Extensive use has been made of an assay combining biological and chemical methods. Two comparable paper chromatographs of an extract or solution are made. One is used for determining the location of the various compounds and for identifying them chemically. The other is cut into strips and bioassayed by placing coleoptile or stem sections over the moistened paper and later measuring their straight growth in various regions corresponding with the location of the chromatographically determined substances. The method is reasonably quantitative and also locates substances other than IAA that promote or inhibit growth.

Occurrence of Auxin in Plants

Species Distribution of Auxin

Auxin has been found in all species of vascular plants that have been analyzed for it, and it is presumably universal in this group of plants. It has also been found in various bryophytes, algae, and fungi. However, auxin may be of limited distribution among the algae and fungi, and in general applications of auxin to algae and fungi either have no effect or inhibit growth. The matter needs further study, since the endogenous auxin levels may be optimal already or the membranes of these plants may not be permeable to auxin. In any event, auxin can hardly play a hormonal coordinating role in unicellular or colonial plants and is probably ineffective in fungi with chitin walls. Some effects of fungus parasites on their host plants have been ascribed to auxin produced by the fungi, and auxin promotion of growth of yeasts and a few other fungi has been reported.

Libbert and his associates at the University of Rostock have made the interesting discovery that bacteria of at least four genera and 58 different strains that are epiphytic on higher plants synthesize IAA, which diffuses into the plants and influences their growth [29]. The plants may be dependent upon the bacteria for most of their IAA, since sterile plant sections or homogenates of plants can synthesize much less IAA than those including the bacteria. The bacteria obtain foods and perhaps tryptophan from the plants, thus resulting in a mutualistic relation. Libbert points out that his findings require reconsideration of previous experimental data on IAA synthesis by plants, and suggests that the possibility of production of other plant growth substances by epiphytic bacteria should be investigated.

Regions of Auxin Synthesis

In general, the regions of most active auxin synthesis in vascular plants are active meristems, in particular the apical meristems of buds, coleoptile tips, and the meristematic tissues of developing embryos. Considerable auxin is also synthesized in young tissues with elongating cells, including young leaves that are expanding. Leaves continue to produce auxin at a reduced rate after they have attained their final size, and the rate continues to decrease as the leaf ages. Pollen also produces considerable auxin, particularly during pollen tube growth.

Translocation of Auxin

The translocation of auxin is predominantly through the phloem, and is polar in a morphologically downward (basipetal) direction (Figure 12-6). The velocity of translocation is between 0.5 and 1.5 cm/hr. Auxin can also move from cell

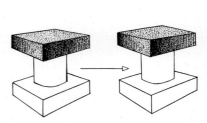

Figure 12-6. Demonstration of polarity of auxin transport. **Above:** A section of coleoptile or stem is placed on an auxin-free agar block in its normal position and an agar block containing auxin is placed on top of the section. After a period of time the auxin has been transported from the upper to lower block, as can be shown by the Avena curvature test. **Below:** A similar section is inverted and placed between two agar blocks. No auxin is transported through the inverted section.

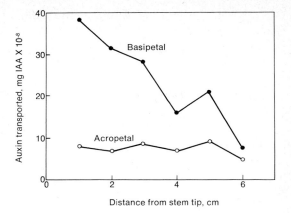

Figure 12-7. The basipetal (downward) transport of auxin by bean stems decreases with increased distance from the stem apex. The small amount of acropetal (upward) auxin transport is not influenced by distance from the tip. [Data of W. P. Jacobs, *Amer. J. Bot.,* **37:**248 (1950)]

to cell outside the phloem, and this probably involves active transport since the rate is too rapid for simple diffusion and since its movement is inhibited by low oxygen and metabolic inhibitors. IAA translocation is specifically inhibited by 2,3,5-triiodobenzoic acid, as is its movement from cell to cell.

The polar translocation of auxin is most pronounced near the stem tips, and farther down there may be slight upward (acropetal) auxin transport. In roots translocation is predominantly toward the tip (acropetal) and so continuous with the downward flow of auxin in the stem. When considered in detail auxin translocation involves a considerable number of complications [17, 18].

Distribution of Auxin in Plants

In general, the auxin concentration in stems decreases with increasing distance from the apex (Figure 12-7). Its concentration in roots is usually lower than in stems and may be higher just back of the tip than either at the tip or farther up the root. In oat seedlings auxin is most concentrated in the coleoptile tip, declines toward the coleoptile base, and again increases toward the root tip (Figure 12-8).

The auxin level also varies with time and age. Dormant plants have low auxin levels, and in general (as in leaves) auxin concentration decreases in aging organs or tissues. The differential distribution of auxin as regards both time and space within a plant plays an important role in growth and development.

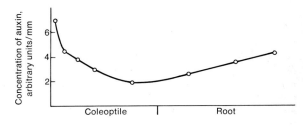

Figure 12-8. Distribution of auxin in the Avena (oat) seedling. [After data of K. V. Thimann, *J. Gen. Physiol.,* **18:**32 (1934)]

Effects of Auxin at the Cellular Level

Cell Elongation

Auxin is essential for cell elongation and cell enlargement in general. The auxin is evidently involved in the incorporation of new wall materials as well as in plasticizing the wall by loosening the crossbonding between cellulose microfibrils and other cell wall components. Despite extensive investigations, it is still not certain which bonds are broken or how this occurs. There is considerable evidence for the theory that auxin acts by promoting the methylation of pectic acid to pectin, including the fact that pectin methylesterase activity is increased by auxin. However, some investigators have failed to find a correlation between methylation and cell elongation. More recently evidence has been presented that auxin may operate through some sort of effect on the hydroxyproline-rich cell wall proteins, but the way in which auxin brings about plasticization of the walls remains dubious.

Good evidence that auxin is involved in the incorporation of wall materials of enlarging cells has been provided by the autoradiographic studies of Ray [22] and others. However, when elongation occurs by tip growth only, as in root hairs and pollen tubes, the role of auxin (if any) in either plasticizing the wall or incorporating new wall materials is not clear.

Cell Division

Auxin is often considered to be involved primarily in cell elongation, but there is a variety of evidence that it also influences cell division and, indeed, it may be found that auxin is generally essential for cell division in higher plants. In the spring, division of the cambial cells of woody plants does not start until the buds have broken dormancy and resumed auxin production. The division of cambial cells ceases if the terminal bud is removed but resumes if auxin is applied to the cut. Auxin presumably influences cell division as well

Figure 12-9. Differentiation of vascular tissue in a piece of lilac callus after grafting a lilac bud to the callus. Note the orientation of the scattered vascular strands and the one that is continuous with the bud. [After R. H. Wetmore and S. Sorokin, *J. Arnold Arboretum,* **36:**305 (1955)]

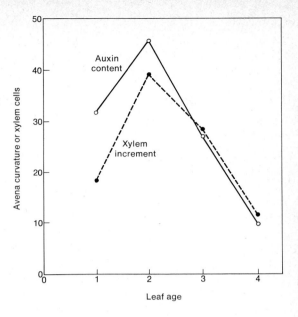

Figure 12-10. Correlation of xylem differentiation in growing petioles of Coleus leaves with the amount of auxin diffusing into agar blocks from the petiole bases. [After W. P. Jacobs and I. B. Morrow, *Amer. J. Bot.*, **44**:823 (1957)]

as cell enlargement when it promotes the development of adventitious roots. It is not clear how auxin influences cell division or how its roles differ from those of other phytohormones that promote cell division. However, it is possible that auxin is essential for the cell enlargement that occurs between cell divisions and perhaps for the laying down of the new walls between daughter cells.

Cell Differentiation

Auxin also plays a role in cell differentiation, at least as regards the vessel elements and fibers of the xylem. Wetmore and Sorokin [28] found that differentiation of xylem occurred in tissue cultures of lilac pith when a lilac bud was grafted onto the tissue (Figure 12-9) or when a localized application of auxin and sucrose was made, but not otherwise. Other investigators have also provided evidence for the role of auxin in xylem differentiation (for example Figure 12-10).

Thus, auxin can play a role in all three phases of cellular growth and development. Although auxin is probably involved in wall formation in all three cases, it may also function in other ways at the cellular level.

Effects on Observable Growth and Development

The effects of auxin on observable growth and development are numerous and diverse, and it may seem rather surprising that a phytohormone can have so many different effects. Some of the auxin effects involve interactions with other phytohormones, as will be seen in the next chapter. The observable effects of auxin are, of course, related in one way or another to its effects at the cellular level.

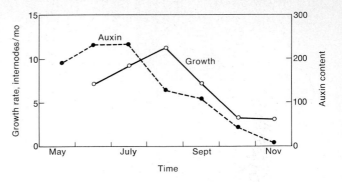

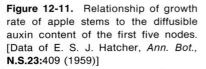

Figure 12-11. Relationship of growth rate of apple stems to the diffusible auxin content of the first five nodes. [Data of E. S. J. Hatcher, *Ann. Bot., N.S.***23**:409 (1959)]

Organ Elongation

The elongation of stems and coleoptiles is dependent on the presence of suitable concentrations of auxin because it is primarily a result of auxin-induced cell elongation (Figure 12-11). However, application of auxins to intact plants generally either has no effect on stem or coleoptile elongation or inhibits it. This is probably due to the fact that the endogenous auxin level is already optimal or, in the case of IAA, to its inactivation by IAA oxidase or other means. An interesting example of the influence of auxin on stem elongation is provided by the work of Gunckel and Thiman [8] on the Ginkgo tree. The Ginkgo has two kinds of branches: dwarf shoots with very short internodes and long shoots. These investigators found that the terminal buds of long shoots produce much more auxin over a longer period of time than those of the dwarf shoots (Figure 12-12).

The optimal auxin level for root elongation is lower than that for stem elongation, and some investigators have even concluded on the basis of evidence they obtained that the only effect of auxin on root elongation was inhibitory or that it had no effect. However, Scott [27], in a review of the literature on auxin and roots, has made it clear that auxin is essential for root elongation. He and other investigators have abundant evidence that auxin

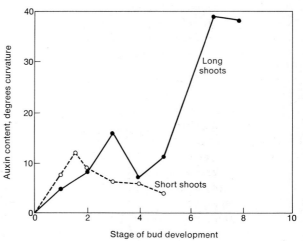

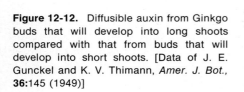

Figure 12-12. Diffusible auxin from Ginkgo buds that will develop into long shoots compared with that from buds that will develop into short shoots. [Data of J. E. Gunckel and K. V. Thimann, *Amer. J. Bot.,* **36**:145 (1949)]

promotes root growth, even at levels that were once considered to be inhibitory. Root elongation is promoted by auxin concentrations as low as 10^{-12} M to as high as 10^{-7} M. The latter is within the range that promotes stem elongation, even though the optimal auxin level for stem elongation is higher. As in stem elongation, auxin levels higher than the promotional ones inhibit growth. Burstrom [see 27] has pointed out that auxin enhances the *rate* of cell elongation in both roots and stems, but that in roots it exerts a secondary inhibitory effect by reducing the *time* during which elongation continues. He suggests that the inhibitory effect might occur at the point where the auxin moving upward from the root apex meets the auxin moving down from the shoot. This could explain the fact that the zone of cell elongation is shorter in roots than in stems.

Tropisms

Tropisms are growth curvatures of plant structures such as stems, roots, coleoptiles, petioles, and tendrils in response to environmental stimuli of unequal intensity on two sides of the structure. The curvatures result from the differential elongation of the cells on opposite sides of the organ. Tropisms do not include plant movements resulting from differential cell turgidity, such as the "sleep" movements of leaves, the closing of sensitive plant leaves, and the closing of Venus's fly-trap leaves, nor the locomotion of an organism toward a stimulus such as light (phototaxis).

Auxin is definitely involved in phototropism and geotropism and may also be involved in other tropisms. For example, it had been thought that auxin was not involved in **thigmotropism,** the coiling of a stem or tendril around a support as a result of a contact stimulus, but Reinhold [24] and several other investigators have evidence that auxin may play a role.

PHOTOTROPISM. When a terminal bud or coleoptile is exposed to light that is more intense on one side than on the other, the auxin concentration becomes higher on the less brightly lighted side. The increased auxin moving down to the cells on this side of the organ causes them to elongate more than the cells on the more brightly lighted side, so the organ bends toward the light (Figure 12-13).

When phototropism is considered in greater detail it becomes evident that it is a complex phenomenon that is incompletely understood [1, 7, 18]. It is not known whether the photoreceptor pigment is carotene or riboflavin, since their absorption spectra are so nearly the same (Figure 12-14). The action spectrum of phototropism fits the absorption spectrum of riboflavin much better at the 370 nm peak, which is absent in the naturally occurring *trans*-carotenes. However, *cis*-β-carotene, which is not known to occur in plants, does have a peak in the ultraviolet [7], and it is possible that such a carotene is present in plants but is isomerized during extractions. Using inhibitors of carotenoid biosynthesis, the carotenoids of plants can be reduced to less than 20% of the usual level without reducing phototropic sensitivity. Thus, the present evidence is slightly in favor of riboflavin as the photoreceptor.

The metabolic pathways leading from light absorption to the differential auxin redistribution are not known, but there is an initial induction period

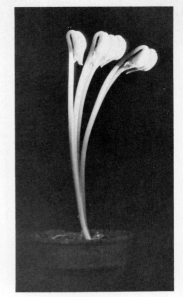

Figure 12-13. Multiple exposure photograph of an etiolated bean seedling showing phototropic bending after it had been removed from the dark and exposed to unilateral light. [Courtesy of John G. Haesloop]

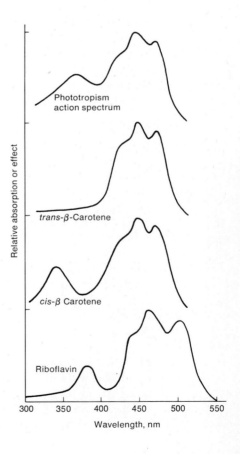

Figure 12-14. Action spectrum of phototropism and the absorption spectra of two carotenes and riboflavin. *cis-β*-Carotene is not known to occur in nature. [After A. W. Galston, *Amer. Sci.,* **55:**144 (1967)]

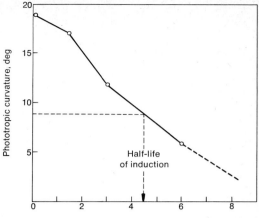

Figure 12-15. Decline in the induced geotropic state of sunflower seedlings. Two days after the seedlings were decapitated (terminal bud removed) they were exposed to unilateral light for 16 hr. At various times after completion of the light exposure groups of seedlings were supplied 10 mg/liter IAA each. The curvatures were measured 3 hr after auxin application in each case. [Data of R. Diemer, *Planta*, **57**:111 (1961)]

Phototropic curvature, deg

Half-life
of induction

Time from light stimulus to auxin application, hr

that requires oxygen and during which metabolic gradients are established even though auxin is lacking (as in decapitated coleoptiles). Subsequent terminal auxin application then results in bending, even though there is no longer unilateral illumination. There is a steady loss of the induced state with time (Figure 12-15).

Theoretically, light could bring about differential auxin distribution in at least three ways. It could inhibit auxin synthesis, promote auxin destruction or inactivation, or cause auxin to move laterally away from the brighter light. Each theory has had its proponents, but a variety of evidence [1, 18] now supports the lateral movement mechanism. For example, Briggs and his co-workers [3] found that if a corn coleoptile tip illuminated unilaterally had been completely partitioned vertically by a piece of microscope cover slip there was no significant difference between the auxin levels on the two sides, but if the partition stopped 0.5 mm below the tip auxin increased on the darker side and decreased on the lighter side (Figure 12-16). Also, there was no significant difference between the auxin levels of upartitioned tips in the dark and tips that were lighted unilaterally. Also, de la Fuente and Leopold [6] found that ^{14}C-IAA applied to corn coleoptiles moved laterally away from light.

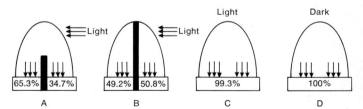

Light — Light — Light — Light — Dark

65.3% | 34.7%   49.2% | 50.8%   99.3%   100%

A   B   C   D

Figure 12-16. Experiments showing lateral movement of auxin in excised corn coleoptile tips induced by lateral light. **(A)** Tip partially separated by a piece of microscope cover-slip glass, which also separated the agar block on which the tip was placed. **(B)** Tip and agar completely separated by glass. **(C)** Intact tip in light. **(D)** Intact tip in dark. Note that B, C, and D show that light neither inhibited auxin synthesis nor promoted auxin destruction. [Data of W. R. Briggs, *Plant Physiol.*, **38**:237 (1963)]

There is considerable evidence that auxin transport on the less brightly lighted side of a stem or coleoptile is enhanced. This could, of course, be associated with any of proposed methods of lateral auxin redistribution in the tip.

GEOTROPISM. Geotropism is the orientation of plant parts in response to gravity. In general, stems and coleoptiles are negatively geotropic, that is, they grow vertically upward, although branches commonly grow at an angle to the earth and rhizomes grow horizontally. Roots are positively geotropic, that is, they grow downward in response to gravity. If a plant is placed in a horizontal

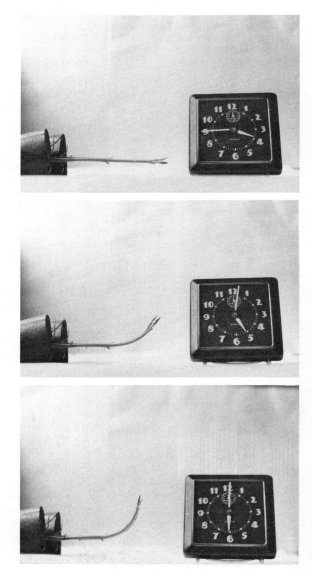

Figure 12-17. Geotropic response of young tomato plants. The leaves were removed just prior to the first photograph to make the stem curvature more evident. [Courtesy of John G. Haesloop]

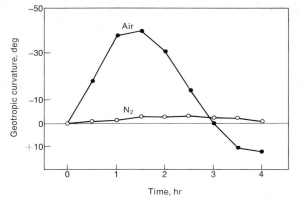

Figure 12-18. For 30 min sunflower seedlings were placed in a horizontal position, some in air and some in a nitrogen atmosphere. At time 0 both lots were returned to the vertical position and placed in air. Note the lack of geotropic curvature of the plants induced in a nitrogen atmosphere. [Data of L. Brauner and A. Hager, *Naturwiss.*, **15**:429 (1957)]

position the stems bend and grow upward, whereas the roots bend and grow downward (Figure 12-17). In both auxin becomes more concentrated on the lower side, the higher concentration promoting the growth of the lower side of the stem but inhibiting growth of the lower side of the roots.

Geotropism is an even more complex and poorly understood phenomenon than phototropism [7, 30], and the numerous investigations of it have often provided conflicting results. It has been suggested that auxin may not be the only growth substance displaced by gravity and also that some unknown growth inhibitor rather than auxin may be involved in root geotropism. The way in which gravity brings about auxin displacement is not understood. There is an induction period that does not require auxin but does require oxygen, indicating that gravity does not affect auxin redistribution directly (Figure 12-18).

Two possible components of the geotropic system were first identified years ago and have been intensively investigated up to the present time, but whether or how they participate in the geotropic response is still not clear. In 1858 Naegeli first found that starch grains settled to the lower side of a cell under the influence of gravity, and subsequently many investigators have attempted to implicate these as **statoliths** that provide the primary perception of gravity in geotropism. Cell organelles such as mitochondria and Golgi bodies have also been proposed as possible statoliths, but apparently these do not move down fast enough to account for the rather rapid time (1 to 5 min) for initiation of the geotropic response. It has also been suggested that perception of gravity might be accomplished by the rise of oil globules or vacuoles toward the upper side of the cell, thus resulting in a thicker layer of cytoplasm and presumably more auxin on the lower side. However, most experimental evidence supports the role of starch grains as statoliths. For example, Hertel and others [10] found that in a corn mutant with much smaller than usual starch grains the gravitational displacement was 30 to 40% less than that of the usual larger grains, the lateral displacement of auxin was 40 to 80% less, and the geotropic curvature was also significantly less. The attempted explanations of the connection between statolith movement and the differential distribution of auxin in response to gravity have been less than satisfactory.

In 1907 Bose first reported the development of an electrical potential difference between the upper and lower sides of plants in a horizontal position, and this has been investigated extensively since then. The lower side of a stem or root becomes positive to the upper side. The potential difference may be as great as 60 millivolts (mv) but is usually 10 mv or less. It has been proposed that this potential difference, which is probably generated by the downward movement of cations, could be directly responsible for attracting auxin cations to the lower side. However, it now seems clear that the potential gradient is a result, rather than the cause, of auxin displacement [2, 7]. The potential difference does not occur without auxin, and auxin displacement evidently occurs before the potential difference develops.

The geotropic responses of plants are of great importance in plant adaptation and survival. Without them the roots of seedlings would grow down and the stems up only if a seed happened to be oriented so that the embryo was vertical with the plumule at the top. The trunks of forest trees and the plants in a field of corn would be at every conceivable angle. Without geotropism agriculture, and so presumably civilization, would have been virtually impossible.

Apical Dominance

Long before auxin was discovered it was known that the terminal bud of a stem inhibited the growth of the lateral buds (apical dominance) and that removal of the terminal bud resulted in the development of one or more of the lateral buds into branches (Figure 12-19). Apical dominance generally becomes weaker with increasing distance from the terminal bud, so some of the lower lateral buds may develop into branches even if the terminal bud is present. However, some lateral buds never grow into branches, particularly in trees. There are marked species differences in apical dominance, for example,

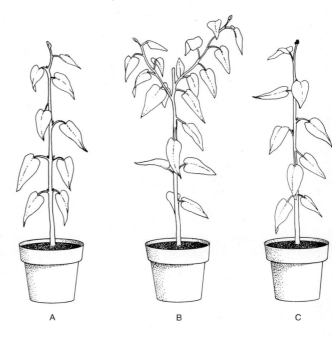

Figure 12-19. Role of auxin in apical dominance. **(A)** Intact plant with no branches. **(B)** Removal of terminal bud results in the loss of apical dominance and development of some lateral buds into branches. **(C)** Auxin in lanolin applied to tip of stem from which terminal bud was removed replaces the lost apical dominance and prevents branch development.

A   B   C

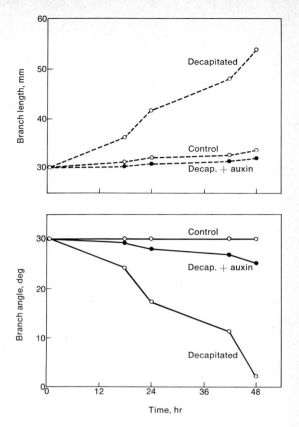

Figure 12-20. Removal of the terminal bud of *Mirabilis jalapa* resulted in both branch growth promotion and a decrease in the angle of the branch from the vertical. Application of auxin to the cut stem tip prevented both. [Data of Y. Vardar, *Rev. Fac. Sci. Forest. Univ. Istanbul.*, **20**:245–256, (1955)]

in some plants such as sunflowers the dominance is usually so strong that there are no branches, whereas in others like Coleus and many shrubs apical dominance is weak and branching is abundant.

In 1934 Skoog and Thimann found that a block of agar containing IAA could substitute for a decapitated terminal bud of broad bean in the inhibition of lateral bud growth, and numerous subsequent studies have supported the conclusion that auxin is responsible for apical dominance. Although the auxin concept of apical dominance has been challenged from time to time, it is still generally accepted. However, the situation is probably more complex than it was thought to be, and various interacting factors appear to be involved. For example, cytokinins can counteract the inhibition of bud growth by auxin, and it has been reported that auxin exerts apical dominance in flax plants only when the available nitrogen is rather low.

Auxin from a terminal bud is apparently responsible also for the growth of branches at an angle to the vertical (plagiotropism). If a terminal bud is removed, at least the uppermost lateral bud not only begins developing into a branch but also often becomes orthogeotropic and grows essentially vertically (Figure 12-20). Auxin applied to the main stem tip after the lateral branch has started growing generally prevents this shift in geotropic response. The reorientation of the uppermost branch to the vertical is particularly striking in pines and other gymnosperms with an excurrent (monopodial) branching pattern. The terminal bud of the vertical branch then exerts apical dominance.

Inhibition of Abscission

Numerous investigations have demonstrated that auxin plays a role in preventing the abscission [11, 25] of leaves, flowers, and fruits. That auxin from a leaf blade inhibits abscission is demonstrated by the fact that petioles generally abscise soon after the blade is cut off, but this does not happen if auxin is supplied to the cut end of the petiole. The abscission of the leaves of deciduous trees and shrubs in the autumn results from the decline in auxin level in the leaves as they age and become senescent. The premature abscission of leaves that have been damaged extensively by insects, subjected to water deficits, or have mineral deficiencies probably also involves reduced auxin levels. Flowers often abscise if they have not been pollinated, and fruits commonly abscise when mature and also sometimes prematurely.

The role of auxin in preventing abscission is by no means simple and involves various complications and interactions. Addicott and Lynch, Jacobs, and others have presented evidence that an auxin gradient is involved (Figure 12-21). If the auxin reaching the abscission zone from the blade is more concentrated than the auxin reaching the inner side of the zone (from the terminal bud and younger leaves), abscission is prevented; but, when the auxin concentration is the same on both sides or higher on the inner side, abscission occurs (Figure 12-22). The gradient theory has been questioned by some investigators who have data that does not support it. Auxin is not the only phytohormone involved in abscission. Abscisic acid promotes abscission as does ethylene, whereas gibberellins have been reported to both promote and inhibit abscission. Their interactions with auxin are not well understood.

Abscission zones (Figure 12-23) are formed in the petiole bases of most

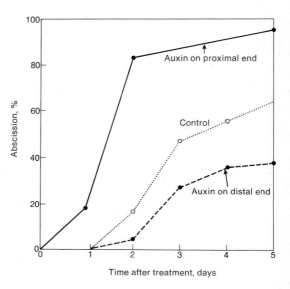

Figure 12-21. IAA solution (105 mg/liter) applied to petioles excised from bean plants (along with a section of stem) either inhibits or promotes petiole abscission, depending on where it is applied. [Data of F. T. Addicott and R. S. Lynch, *Science,* **114:**688–689 (28 Dec. 1951)]

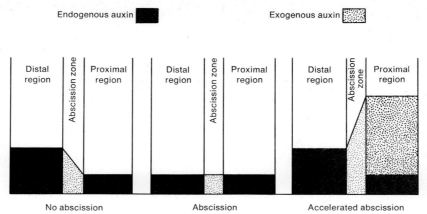

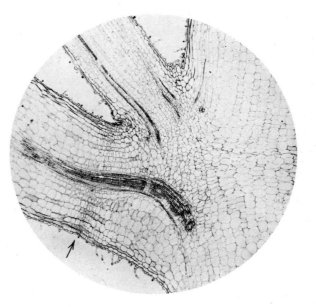

Endogenous auxin ■   Exogenous auxin ▨

Distal region	Abscission zone	Proximal region
Distal region	Abscission zone	Proximal region
Distal region	Abscission zone	Proximal region

No abscission   Abscission   Accelerated abscission

Figure 12-22. Diagram illustrating the influence of the auxin gradient across the abscission zone on abscission. [After F. T. Addicott and R. S. Lynch, *Ann. Rev. Plant Physiol.* **6:**211 (1955)]

species of dicotyledons, but they are lacking in a number including tobacco and some oaks. Monocotyledons generally lack abscission zones. The leaves of evergreens do not abscise until they are several years old. The abscission zone has less mechanical tissue than adjacent regions. Several new layers of cells are in some cases laid down across it prior to abscission. The abscission may result from dissolution of the middle lamella alone, the middle lamella and cell walls, or even the lysis of whole cells. These disruptions involve hydrolyses and perhaps other enzymatic processes and require metabolic energy, as indicated by the necessity of oxygen and readily available carbohydrate. Auxin apparently blocks these disruptive processes in some way that is still not clear.

Figure 12-23. Abscission layer developing at the base of the petiole of a Coleus leaf. Note the branch in the axil of the leaf. [Copyright, General Biological Supply House, Inc., Chicago]

Adventitious Root Development

If stems cut from certain species of plants are placed in water or moist sand, adventitious roots develop at the basal end of the cuttings. Horticulturists have long used the rooting of stem cuttings as a means of plant propagation. Some plants such as Coleus and willows root readily, whereas others form few or no adventitious roots. Removal of the leaves prevents or greatly reduces rooting, evidently because of lack of nutrients supplied by the leaves (Figure 12-24).

In the early 1930s IAA was found to function in adventitious root development. The degree of rooting was found to be proportional to the auxin content of the cutting and also adventitious roots could be induced by application of IAA, even in many species that never form adventitious roots naturally (Figure 12-25). IAA, and to a greater extent synthetic growth substances with auxin activity, are now widely used for the propagation of many horticultural plants by cuttings. However, cuttings from some species do not form adventitious roots even when treated with auxins. This fact provides one of several types of evidence that growth substances other than auxin are also involved in adventitious root formation.

The role of auxin in adventitious root formation is obviously different from its role in root growth, since the auxin concentrations effective in root formation inhibit root growth. In some cases the adventitious root primordia are not formed until after auxin application, whereas in other cases auxin simply induces the growth of previously differentiated primordia.

Flower Initiation

The possibility that auxin might be involved in flower formation has been investigated extensively, and the results show rather consistently that auxin levels that promote stem elongation inhibit flower initiation. Only in pineapple have auxin applications been found to promote flowering, and this results from auxin-induced ethylene synthesis. In many species there is a decrease in auxin

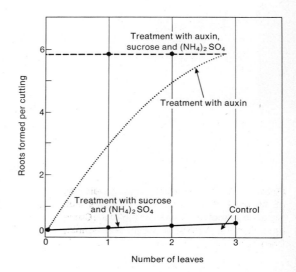

Figure 12-24. The number of roots formed by auxin-treated cuttings of hibiscus increases with the number of leaves left on the cutting, but sucrose applied to cuttings with no leaves replaces the promoting influence of even three leaves. [Data of J. van Overbeek, S. A. Gordon, and L. E. Gregory, *Amer. J. Bot.,* **33**:100–107, (1946)]

Figure 12-25. Formation of adventitious roots by holly cuttings treated with auxin (above) compared with lack of roots on the untreated controls. [Courtesy of K. V. Thimann]

prior to flower initiation, and this may reduce the auxin below inhibitory levels. However, auxin is not known to play any general decisive role in flower initiation, although it may possibly be involved in the growth of flower parts once they have been differentiated. Other phytohormones do play important roles in flower initiation, which will be discussed in greater detail later.

Fruit Development

The development of ovularies (and in some species accessory structures) into fruits usually does not occur without pollination and in some cases also fertilization and embryo development. In 1936 Gustafson found that, in many species of angiosperms, the development of fruits from unpollinated flowers could be induced by auxin applications, although the ovules did not develop into seeds. Many subsequent experiments have supported the concept that auxin is essential for fruit development and that it is, in most cases, supplied by pollen and developing embryos. Naturally seedless fruits, such as some grapes, citrus fruits, and bananas, apparently secure adequate auxin from other sources. The development of fruits without fertilization of the egg is known as **parthenocarpy** and as artificial parthenocarpy when it results from applications of growth substances.

Among the plants in which auxin induces artificial parthenocarpy are grapes, figs, okra, strawberries, and many members of the potato and cucumber families. However, applied auxin is ineffective in some other plants, including members of the rose family such as cherries, peaches, apricots, apples, and pears.

The Biochemistry of Auxin Action

The way in which auxin acts at the biochemical level is still not clear despite extensive efforts to elucidate it. Various hypotheses have been proposed, including a direct effect of auxin on cell wall components, effects on the permeability of the plasma membrane, the functioning of auxin as a coenzyme or coenzyme component, stimulation of respiration by auxin, or the derepression of genes by auxin and so the production of specific kinds of mRNA and enzymes.

Auxin does bring about an increase in the rate of respiration (Figure 12-26), but this appears to be a result of auxin action rather than one of the sequential events leading to auxin action.

There is increasing evidence that auxin does induce mRNA production and thus, presumably, specific enzymes that catalyze auxin-induced reactions, although unlike some other phytohormones auxin has never been linked to the induction of any specific enzyme. Some investigators think that hormones in general may act primarily by promoting the synthesis of specific kinds of mRNA and enzymes. As a sample of the evidence linking auxin to RNA production, Key and Shannon [14] found that IAA concentrations promoting cell elongation enhanced ^{14}C-nucleotide incorporation into RNA, whereas inhibitory IAA concentrations decreased it. Auxin induced a 25 to 30% increase in RNA, largely ribosomal RNA, and brought about a net transfer of labeled RNA from the nucleus to ribosomes. Key [13] reported that inhibitors of RNA and protein synthesis also inhibited cell elongation. Noodén and Thimann [21] secured similar results to varying degrees, particularly with inhibitors of protein synthesis, and found that all substances that inhibited auxin-induced growth also inhibited the incorporation of ^{14}C-leucine into proteins. DL-*p*-Fluoro-

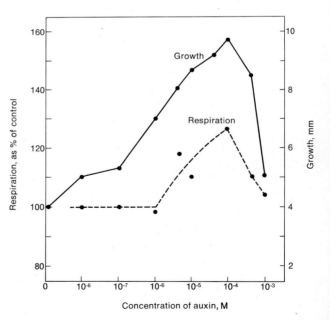

Figure 12-26. Influence of auxin on the growth and respiration of corn coleoptile sections. [Data of R. C. French and H. Beevers, *Amer. J. Bot.*, **40**:660 (1953)]

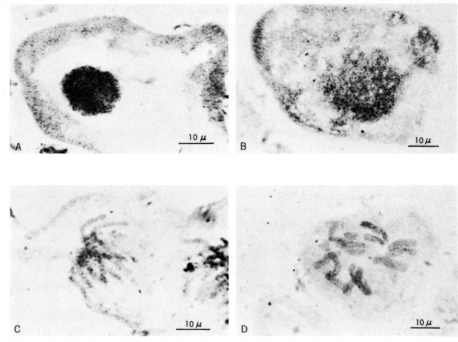

Figure 12-27. Autoradiographs showing localization of radioactive auxin in nuclei and chromosomes, as well as in the cytoplasm. **(A)** Onion root tip cell exposed to ^{14}C-IAA for 6 hr. **(B)** *Vicia faba* root tip cell exposed to ^{14}C-2,4-D for 1 hr. **(C)** *Vicia fiba* root tip cell exposed to ^{14}C-2,4-D for 1 hr and then grown for 72 hr. **(D)** Same as **C,** but grown for 120 hr before fixing. Note the retention of the 2,4-D by the chromosomes. [Courtesy of R. H. Hamilton and S-H Liao, *Science,* **151:**822 (18 Feb. 1966). Copyright 1966 by the American Association for the Advancement of Science]

phenylalanine inhibited the elongation of oat coleoptiles, whereas phenylalanine reversed this inhibition. They suggested that auxin might act by inducing the production of mRNA that codes for unstable enzymes involved in auxin-induced growth. Masuda and Kamusaka [20] reported that auxin can stimulate the biosynthesis of RNA within 10 to 15 min. Autoradiographs of cells supplied with ^{14}C-labeled auxin show radioactivity in the nuclei and chromosomes (Figure 12-27).

However, most investigators of the ways in which auxin promotes cell elongation believe that the auxin-induced production of mRNA and enzymes is not involved because cell elongation begins within just a few minutes after auxin is supplied. The time required for the production of mRNA and enzymes and transport of the enzymes to the wall is much too long, even if Masuda and Kamusaka's unusually brief period of time is accepted. Thus, it appears that auxin must promote cell elongation in a more direct way, either by interacting with cell wall constituents themselves or by influencing the permeability of the plasma membrane [23]. This does not exclude the possibility that some of the other effects of auxin at the cellular level involve auxin-induced RNA production.

Synthetic Growth Regulators with Auxin Activity

Commercial synthesis of substances with auxin activity has made available substantial supplies, not only for experimental purposes but also for horticultural use on a large-scale practical basis. In addition to IAA itself and other indole compounds such as indolebutyric acid, the principal synthetic growth substances with auxin activity fall into three chemical classes: derivatives of phenoxyacetic acid, naphthaleneacetic acid, or benzoic acid, or in some cases the acids themselves (Figure 12-28). It may seem strange that substances so different chemically from IAA have auxin activity, but in general it can be predicted that molecules with an aromatic ring having at least one free ortho position, a carboxyl group on a side chain of suitable length, and a certain spatial configuration between the ring and the side chain will have auxin activity.

The auxin activity of these synthetic growth regulators varies considerably. Some are less effective than IAA whereas others are substantially more effective, at least partly because they are not inactivated by IAA oxidase. Furthermore, their effectiveness as compared with one another and IAA varies considerably from one type of auxin effect to another (Table 12-1). Most of the

Table 12-1 Activity of some synthetic growth substances as percent of activity of IAA

Compound	Avena curvature test	Avena section test	Pea curvature test	Rooting of cuttings	Bud inhibition	Avena root inhibition
Idolebutyric acid	8	9	100	150	100+	100
Indene-3 acetic acid	1	7	20	100	14	—
α-Naphthalene acetic acid	2+	15	370	150	100+	4
β-Naphthoxyacetic acid	0	—	—	25	100	—
2,4-Dichlorophenoxyacetic acid	0	300	200–1200	1500	—	30

SOURCE: Bonner, James. *Plant Biochemistry,* Academic Press, New York, 1950, with permission of the publisher.

synthetic growth regulators cannot duplicate the growth correlations of IAA, at least partly because their translocation is generally not polar.

Because they are cheaper than IAA and effective at lower concentrations, several of the synthetic growth regulators including α-naphthaleneacetic acid (NAA), 2,4-dichlorophenoxyacetic acid (2,4-D), and 2,4,5-trichlorophenoxyacetic acid (2,4,5-T) are extensively used in horticultural applications such as the rooting of cuttings and artificial parthenocarpy [26]. However, the most extensive use of 2,4-D and 2,4,5-T is as selective herbicides, the latter generally being most effective in killing poison ivy and woody plants. Neither of these substances damages grasses, at least to any degree, making them particularly useful for the control of broad-leaved weeds in lawns and fields of cereal grains and other grasses. The reasons for the general immunity of grasses and some other monocotyledons to these herbicides are still not understood well. The concentrations lethal to dicotyledons are in the range of 500 to 1000 ppm, much higher than the very low concentrations that produce typical auxin effects. Sublethal concentrations on the order of 25 ppm may inhibit growth and induce extensive abnormal development, particularly of leaves (Figure 12-29).

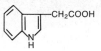

Indoleacetic acid

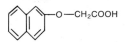

Indolebutyric acid

Naphthaleneacetic acid

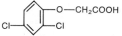

β-Naphthoxyacetic acid

2, 4, 5-Trichlorophenoxyacetic acid

2, 4-Dichlorophenoxyacetic acid

2, 4, 6-Trichlorobenzoic acid

2, 3, 6-Trichlorobenzoic acid

Figure 12-28. Structural formulas of IAA and several synthetic auxins.

Figure 12-29. Influence of sublethal concentrations of 2,4-D on the development of cotton leaves. A normal leaf is at the right. [Courtesy of Wayne McIlrath and D. R. Ergle]

Both 2,4-D and 2,4-5-T were once thought to be harmless to man and other animals at herbicidal concentrations, but now they have been reported to cause birth defects in experimental animals. These compounds are among the herbicides and defoliants used on a massive scale in the Viet Nam war. Many scientists and others have strongly criticized their use because of their disruptive effects on the ecosystem and the possibility of direct adverse effects on man. The low concentrations that have typical auxin effects are probably safe from both standpoints.

Ethylene

It is appropriate to consider ethylene at this point because of its numerous and intimate interrelations with auxin. Ethylene (an unsaturated hydrocarbon gas, $H_2C{=}CH_2$) has long been known to cause epinasty (downward bending) of leaf petioles, notably in tomato plants, which are sensitive to as little as 0.1 ppm. Since ethylene is a common constituent of commercial gas (particularly gas manufactured from coal), tomato plants have been used as sensitive indicators of gas leaks. As early as 1924 Denny found that ethylene promoted the ripening of lemon fruits, and by 1933 it was established that it is produced in increased quantities by fruits just prior to ripening. There is now abundant evidence that ethylene is involved in the ripening of fleshy fruits [19], with a few exceptions such as peaches and apricots, and ethylene is widely used commercially to promote fruit ripening. Also, ventilation of stored fruits to remove ethylene they produce is used to delay ripening.

More recently it has become evident that ethylene is synthesized in most plant organs and has varied effects on plant growth and development, and thus it is considered by many to be one of the phytohormones. It may seem strange that a gas can function as a hormone, but it meets the requirements for classification as such. In any event, it is definitely a plant growth regulator.

As early as 1935 Zimmermann and Wilcox found that IAA stimulated ethylene synthesis in plants, and this has now been well documented, particularly as regards inhibitory concentrations of auxin. It was noted earlier that Burg [4] and others have proposed that growth inhibition by auxin results from the auxin-induced ethylene. The role of ethylene in the auxin-induced inhibition of root growth is particularly well established. It has been found that many auxin effects can be duplicated by exposing plants to ethylene, and presumably in these cases auxin acts by inducing ethylene synthesis. Among these effects are the initiation of adventitious roots, the inhibition of flowering in most plants, the promotion of pineapple flowering [5], the inhibition of light-induced hypocotyl hook opening [12], and the increase in the proportion of pistillate flowers of cucumber plants. The leaf senescence and abscission induced by synthetic auxins such as 2,4-D can also be induced by ethylene [9], and ethylene may also stimulate abscission of petals. At 1 ppm ethylene prevents the opening of carnation flower buds.

Ethylene counteracts the normal polar translocation of auxin, and in pea stems at least it induces the immediate cessation of lateral auxin transport. This may explain why ethylene eliminates the normal geotropic responses of seedlings. Burg and Burg [4] have reported that ethylene, interacting with auxin, promotes transverse cell enlargement and inhibits the usual longitudinal elongation.

As the above examples indicate, auxin and ethylene evidently interact as a closely integrated pair of phytohormones, and it is likely that further research will reveal even more interrelations. Perhaps ethylene should be considered as an intermediate in certain reaction sequences induced by auxin rather than as a discrete plant growth substance.

References

[1] Audus, L. J. *Plant Growth Substances,* Interscience, New York, 1959.

[2] Brauner, L. "The effect of gravity on the development of electrical potentials in plant tissues," *Endeavour,* **28:**17–21 (1969).

[3] Briggs, W. R., R. D. Tocher, and J. F. Wilson. "Phototropic auxin redistribution in corn coleoptiles," *Science,* **126:**210–212 (1957).

[4] Burg, S. P., and E. A. Burg. "The interaction between auxin and ethylene and its role in plant growth," *Proc. Nat. Acad. Sci.,* **55:**262–269 (1966).

[5] Burg, S. P., and E. A. Burg. "Auxin-induced ethylene formation: its relation to flowering in the pineapple," *Science,* **152:**1269 (1966).

[6] de la Fuente, R. K., and A. C. Leopold. "Lateral movement of auxin in phototropism," *Plant Physiol.,* **43:**1031–1036 (1968).

[7] Galston, A. W., and P. J. Davies. *Control Mechanisms in Plant Development,* Prentice-Hall, Englewood Cliffs, N. J., 1970.

[8] Gunckel, J. E., and K. V. Thimann. "Studies of development in long shoots and short shoots of *Ginkgo biloba* L.," *Amer. J. Bot.,* **36:**145–151 (1949).

[9] Hallaway, Mary, and Daphne J. Osborne. "Ethylene: a factor in defoliation induced by auxins," *Science,* **163:**1067–1068 (1969).

[10] Hertel, R., R. K. de la Fuente, and A. C. Leopold. "Geotropism and the lateral transport of auxin in the corn mutant amylomaize," *Planta,* **88:**204–214 (1969).

[11] Jacobs, W. P. "Studies on abscission: the physiological basis of the abscission-speeding effect of intact leaves," *Amer. J. Bot.,* **42:**594–604 (1955); "What makes leaves fall?" *Sci. Amer.,* **193**(5):82–89 (1955).

[12] Kang, B. G. *et al.* "Ethylene and carbon dioxide: mediation of hypocotyl hook opening response," *Science,* **156:**958–959 (1967).

[13] Key, J. L. "Ribonucleic acid and protein synthesis as essential processes for cell elongation," *Plant Physiol.,* **39:**365–370 (1964).

[14] Key, J. L., and J. C. Shannon. "Enhancement by auxin of ribonucleic acid synthesis in excised soybean hypocotyl tissue," *Plant Physiol.,* **39:**360–364 (1964).

[15] Lang, A. "Plant growth regulation," *Science,* **157:**589–592 (1967).

[16] Leopold, A. C. *Auxins and Plant Growth,* University of California Press, Berkeley, 1955.

[17] Leopold, A. C. "The polarity of auxin transport," *Brookhaven Symp. Biol.,* **16:**218–234 (1963).

[18] Leopold, A. C. *Plant Growth and Development,* McGraw-Hill, New York, 1964.

[19] Mapson, L. W. "Biosynthesis of ethylene and the ripening of fruit," *Endeavour,* **29:**29–33 (1970).

[20] Masuda, Y., and S. Kamusaka. "Rapid stimulation of RNA biosynthesis by auxin," *Plant Cell Physiol.,* **10:**79–86 (1969).

[21] Noodén, L. D., and K. V. Thimann. "Evidence for a requirement of protein synthesis for auxin-induced cell elongation," *Proc. Nat. Acad. Sci.,* **50:**194–200 (1963); "Action of inhibitors of RNA and protein synthesis on cell enlargement," *Plant Physiol.,* **41:**157–164 (1966).

[22] Ray, P. M. "Radioautographic study of cell wall deposition in growing plant cells," *J. Cell Biol.,* **35:**659–674 (1967).

[23] Ray, P. M. "The action of auxin on cell enlargement in plants," in A. Lang (ed.), *Communication in Development,* Academic Press, New York, 1969.

[24] Reinhold, L. "Induction of coiling in tendrils by auxin and carbon dioxide," *Science,* **158:**791–793 (1967).

[25] Rubenstein, B., and A. C. Leopold. "The nature of leaf abscission," *Quart. Rev. Biol.,* **39:**356–372 (1964).

[26] Tukey, H. B. (ed.). *Plant Regulators in Agriculture,* John Wiley & Sons, New York, 1954.

[27] Scott, T. K. "Auxins and roots," *Ann. Rev. Plant Physiol.,* **23:**235–258 (1972).

[28] Wetmore, R. H., and S. Sorokin. "On the differentiation of xylem," *J. Arnold Arboretum,* **36:**305–317 (1955).

[29] Wightman, F., and G. Setterfield (eds.). *Biochemistry and Physiology of Plant Growth Substances,* Runge Press, Ottawa, 1968. See E. Libbert *et al.* pp. 213–230.

[30] Wilkins, M. B. "Geotropism," *Ann. Rev. Plant Physiol.,* **17:**379–408 (1966).

[31] Wilkins, M. B. (ed.). *The Physiology of Plant Growth and Development,* McGraw-Hill, New York, 1969.

Other Phytohormones

13

Since the discovery of auxin a considerable number of other phytohormones have been identified or postulated. In addition to auxin and ethylene, the most intensively studied ones are the gibberellins, the cytokinins, and absisic acid. This chapter will be devoted principally to these three.

Among the other phytohormones are the traumatins, or wound hormones. In 1939 English, Bonner, and Haagen-Smit isolated a substance from bean fruits that induced renewed meristematic activity in cells of the fruits (and also in potato tuber discs) and thus the formation of periderm wound tissue. The substance was found to be a dicarboxylic fatty acid ($HOOC \cdot CH{=}CH \cdot (CH_2)_8COOH$) and was named traumatic acid. It is apparently released from injured cells and moves to adjacent intact cells where it exerts its effects. Since it is ineffective in many species of plants, traumatic acid is evidently not a universal wound hormone. There has been little sustained interest in traumatins, although the healing of wounds occurs in most plants. It is not known to what extent traumatins and other phytohormones are involved.

Several of the B vitamins [41] are essential for root growth and, because they are generally synthesized only in the shoots, translocated to the roots, and are effective at low concentrations, they meet the requirements for hormones. They are probably also essential for shoot growth, but this is more difficult to demonstrate. Excised roots will not grow unless the essential vitamins are added to the culture medium. Vitamin B_1 (thiamin) is generally essential, and other B vitamins such as B_6 (pyridoxine) and nicotinic acid are also essential for some species. Various B vitamins, including pantothenic acid, thiamin, nicotinic acid, and biotin, are also essential for the growth of excised immature embryos in culture media and presumably also during their normal development.

In general, the vitamins or their derivatives function as electron-transporting coenzymes or cofactors in oxidation-reduction reactions, a role not ascribed to most phytohormones so that there is some question as to whether the vitamins should be considered as true phytohormones. This is particularly true of such vitamins as riboflavin (vitamin B_2), vitamin K, and vitamin C (ascorbic acid) which probably function primarily in the cells where they are synthesized.

The liquid endosperm of coconut seeds (coconut water) and the immature endosperms of other species such as corn and horsechestnut in the milk stage will frequently promote the growth of excised embryos and plant tissues in culture when these do not grow in the usual basic culture media that contain foods, essential minerals, vitamins, and several phytohormones. In addition to such substances, the liquid endosperms contain other substances including leucoanthocyanins, inositols, and sorbitol. These substances appear to be essential for the development of young embryos and perhaps the growth of plant tissues in general, but older embryos and most plant tissues presumably can synthesize them. They are no doubt growth substances, and whether or not they should be considered as phytohormones depends on the translocation factor.

Several phytohormones, including florigen, vernalin, and antheridogen, are involved specifically in reproductive development and will be discussed later on in the chapter on reproductive physiology. Although florigen was proposed in 1936 by the Russian botanist Cajlachjan as a phytohormone essential for flower initiation, it remains hypothetical because it has never been isolated definitively nor identified chemically. Vernalin, a hormone postulated to be involved in vernalization, also remains hypothetical, but there is considerable evidence for the existence of both.

From time to time there are reports on previously unknown plant growth substances. For example, Mitchell and his coworkers [31] isolated from pollen several chemically related lipids that promote internode elongation of intact plants when applied at the rate of 10 μg per plant. They were designated as brassins because they were first isolated from the pollen of rape, a member of the genus Brassica, although they have since been isolated from pollen of alders and other plants. It seems likely that other phytohormones and plant growth substances that are not phytohormones will continue to be discovered, and some of these may prove to be as important as the gibberellins, cytokinins, and abscisic acid that we will now consider.

Gibberellins (GA)

The discovery of the gibberellins [7, 25, 48, 54] began in 1926 when a Japanese plant pathologist secured an extract from *Gibberella fujikuroi*, an ascomycetous fungus parasitic on rice, that greatly stimulated the growth of rice and corn seedlings. The disease was called the bakanae (foolish seedling) disease because the infected rice plants grew at least 50% taller than uninfected ones. In 1935 Yabuta isolated a substance with marked growth promoting effects from extracts and named it gibberellin. Active investigation of gibberellin continued in Japan, but the botanists of the western world were completely unaware of it until 1950 because of poor scientific communication during World

War II. Since then there has been extensive study of the gibberellins throughout the world, particularly in England and the United States.

Species Distribution of Gibberellins

Despite the effects of gibberellin on plant growth, it was not considered a phytohormone until evidence demonstrated that it was synthesized by higher plants as well as by a fungus. It is now clear that the gibberellins are widely distributed through the plant kingdom. Gibberellins have been found in all species of seed plants that have been checked for them and also in many lower vascular plants, bryophytes, and algae. Gibberellins are apparently synthesized by only a limited number of species of bacteria and fungi, but several species of fungi in addition to *Gibberella fujikuroi* produce them in abundance.

Chemical Nature of the Gibberellins

In contrast with the predominance of IAA as the principal (if not the only) natural auxin, numerous closely related compounds with gibberellin activity have been isolated from fungi and higher plants. Over 30 different gibberellins have been identified, and more are likely to be found. In 1961 Cross and his coworkers determined the molecular structure of various gibberellins and found that all of them had the same complex skeletal construction, called a gibbane structure from the name of the fungus. The gibberellins differ from one another in the number and location of the —OH groups and in the location or absence of a double bond in the A ring (Figure 13-1). Since the gibberellins are acids, they are often referred to as gibberellic acid (GA). The neutral gibberellins

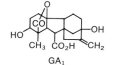
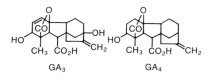
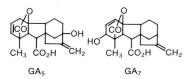

Figure 13-1. Structural formulas of five of the gibberellins.

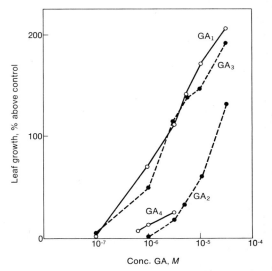

Figure 13-2. Influence of four gibberellins on the leaf growth of rice seedlings over a period of 6 days. [Data of T. Hashimoto and T. Yamaki, *Bot. Mag.* (*Tokyo*), **73**:64 (1960)

are evidently esters of GA, and can be converted readily to the acids. In this discussion GA and gibberellins will be used in interchangeably. The different kinds of GA are referred to as GA_1, GA_2, GA_3, and so on. All of the gibberellins have the same general phytohormone activity, but their effectiveness in inducing various specific growth responses varies considerably and some are ineffective in inducing certain responses (Figure 13-2).

Most plants contain several different gibberellins, but no plant has been found to contain all of them. So far GA_2 has been isolated only from fungi, but all the others, through GA_9 at least, have been found in higher plants. If the specific kind of GA is not indicated it is usually assumed to be GA_3.

Biosynthesis of Gibberellins

In 1958 Birch and his coworkers [6] and subsequently others worked out at least one pathway of gibberellin synthesis. The initial steps follow the normal path of terpene biosynthesis, beginning with the formation of mevalonic acid from acetate and then the conversion of mevalonic acid to isopentyl pyrophosphate. The remaining steps are specific for the gibberellins (Figure 13-3). Several synthetic growth inhibitors, including CCC, phosphon-D, Amo-1618, and B995, exert their effects by reducing the GA level in plants, evidently by inhibiting GA biosynthesis.

Enzymes that oxidize or otherwise inactivate GA have not been found, but several lines of evidence, including the decrease in GA in seeds as they mature, suggest the plants can inactivate gibberellins. They are sometimes bound to proteins and other substances, and such bound gibberellin may be inactive.

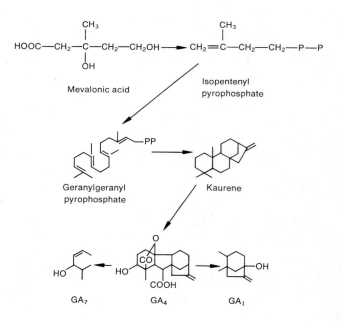

Figure 13-3. Pathway of gibberellin biosynthesis as deduced from experiments with *Fusarium moniliforme*. [After M. B. Wilkins (ed.), *The Physiology of Plant Growth and Development*, McGraw-Hill Book Co., New York, 1969]

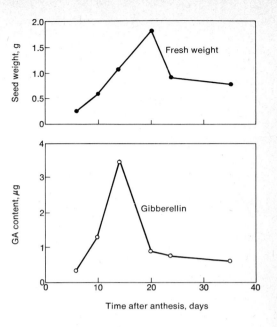

Figure 13-4. Weight of developing seeds of Pharbitis in relation to the gibberellin content of the seeds. [Data of Y. Murakami, *Bot. Mag.* (*Tokyo*), **74**:241 (1961)]

Distribution of Gibberellins in Vascular Plants

Gibberellins are found in all the organs of higher plants, but are most abundant in rapidly growing and developing tissues such as apical meristems, enlarging leaves, the endosperms and embryos of developing seeds (Figure 13-4), and growing fruits. The highest GA concentrations are in developing seeds. The young endosperm of Echinocystus has been found to have as much as 470 μg GA_3 equivalents per gram fresh weight compared with the usual 1 to 10 μg/g in vegetative tissues.

It has sometimes been assumed that the regions of high GA concentration are also the regions of most active GA synthesis, but GA can be translocated through the vascular tissues. Unlike the translocation of IAA, translocation of GA is not always polar. It can be carried up through the xylem and down (or perhaps both ways) through the phloem like other solutes. In general, gibberellins can diffuse from cell to cell, but GA_5 and perhaps some others cannot.

There is much greater variation in the gibberellin content from species to species and from variety to variety within a species than there is in the auxin content. In general, the highest GA concentrations are found in rapidly growing tall species. In species with tall and dwarf varieties the tall ones generally contain higher levels of GA (Figure 13-5), although dwarfness may result from other factors than low GA in some varieties.

Effective Concentrations of Gibberellins

Like other hormones, the gibberellins are effective at very low concentrations. A concentration of 2×10^{-11} M is effective in promoting seed germination.

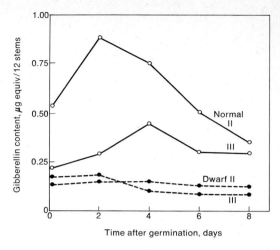

Figure 13-5. Gibberellin content of normal (tall) and dwarf varieties of *Pharbitis nil* (Japanese morning glory). [Data of Y. Ogawa, *Bot. Mag. (Tokyo)*, **75**:449 (1962)]

As little as 0.1 μg per plant doubles the height of dwarf pea plants in 3 weeks. Only 0.01 μg per plant has been found to bring about increased growth of dwarf corn plants. Unlike auxin, the gibberellins do not inhibit shoot growth at concentrations above the promotive ones. Superoptimal concentrations as high as 100 ppm or even more do not increase growth further, but neither do they inhibit shoot growth nor have toxic effects.

Bioassays for Gibberellins

The first bioassay for gibberellins was devised by Phinney [38] in 1957 and involves the use of dwarf corn plants. In this assay the length of the first leaf sheath is measured (Figure 13-6). Other assays involve the effects of GA on the length of dwarf pea epicotyls, the length of lettuce hypocotyls, the length of oat leaf sections, and the induction of α-amylase production by barley endosperm (Figure 13-7). The last assay, devised by Jones and Varner [22], is the most sensitive of these and requires only 1 day, in comparison with 3

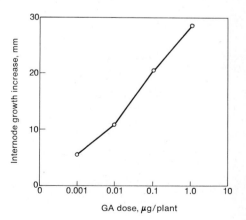

Figure 13-6. Influence of various concentrations of gibberellin on the growth of dwarf corn. [Data of O. E. Smith and L. Rappaport, *Adv. Chem. Ser.*, **28**:42 (1961)]

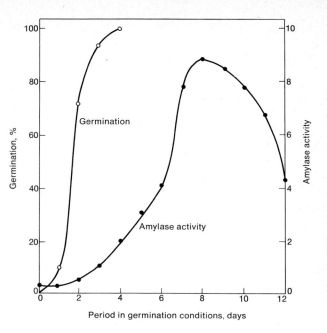

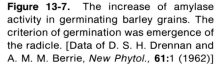

Figure 13-7. The increase of amylase activity in germinating barley grains. The criterion of germination was emergence of the radicle. [Data of D. S. H. Drennan and A. M. M. Berrie, *New Phytol.*, **61**:1 (1962)]

to 10 days for the other assays (Figure 13-8). A difficulty in the bioassay of gibberellins results from the varying activity of the different gibberellins. For example, some that are effective in flower initiation are inactive in the growth bioassays.

Because of the small quantities of gibberellins present in plants quantitative chemical analyses are not practical, but the different kinds of gibberellins present in a plant can be isolated and identified by chromatography.

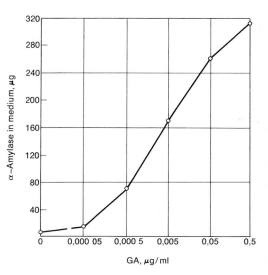

Figure 13-8. Relationship between GA_3 concentration and the release of α-amylase from barley half-seeds. [Data of R. L. Jones and J. E. Varner, *Planta*, **72**:155 (1967)]

Effects of Gibberellins on Observable Growth and Development

Like auxin, the gibberellins have a variety of effects on plant growth and development [7, 55]. Applied gibberellins can bring about extensive stem elongation, the elongation of grass leaves, the development of parthenocarpic fruits in some species, the initiation of flowers in some long-day or cold-requiring species, the breaking of dormancy, and the germination of light-requiring seeds. Presumably the naturally occurring gibberellins are also involved in these and other gibberellin effects. Early in the study of gibberellins there was a controversy as to whether or not they were just another kind of auxin, but it is now clear that they constitute a distinct group of phytohormones since most of their effects are different from those of auxin (Table 13-1).

Table 13-1 Comparison of some effects of auxin and gibberellins when applied to plants or plant parts*

	Auxins	Gibberellins
Cell elongation	+	+
Cell division	+	+
Stem elongation		
Intact plants, esp. dwarfs	0, −	+
Excised young stem sections	+	0
Bolting of rosette plants	0	+
Root elongation	−	0
Initiation of adventitious roots	+	0
Elongation of grass leaves	0	+
Cucumber hypocotyl growth	+	+
Epinasty	+	0
Induction of callus	+	0
Breaking of dormancy	0	+
Seed germination	0	+
Flower initiation		
Biennials	0	+
Some long-day plants	0, −	+
Short-day plants	0, −	0
Sex reversal		
Pistillate to staminate flowers	0	+
Staminate to pistillate flowers	+	0
Archegonia to antheridia	0	+
Artificial parthenocarpy		
Tomato	+	+
Apple	0	+

*+ = promotion, − = inhibition, 0 = no effect. In cases where the experimental results have been variable the usual or predominant effect is given.

STEM ELONGATION. Unlike auxin, the gibberellins are effective in bringing about stem elongation when supplied to intact plants. They are particularly effective in causing many dwarf varieties to grow as tall as the tall varieties of the same species (Figure 13-9) and in bringing about the internode elongation (bolting) of rosette plants in the absence of the long days or the low-

Figure 13-9. Influence of gibberellin on the growth of a dwarf (bush) variety of bean. Untreated control at left. [Courtesy of S. H. Wittwer]

temperature preconditioning naturally required for bolting (Figure 13-10). Biennials can thus be caused to bolt and bloom the first year by gibberellin treatments.

Gibberellins generally have little or no effect on stem elongation of tall species or varieties, probably because the natural gibberellin level is optimal. Also gibberellins do not promote the growth of all dwarf varieties, including several varieties of dwarf corn. Some other factor is evidently limiting the growth of such dwarfs. Gibberellin can prevent the light-induced inhibition of stem growth (Figure 13-11). Lockhart found that, in the dark, dwarf pea plants grew as high as the tall varieties and concluded that light reduced the GA level in the dwarfs. Later Gortner proposed that light causes the dwarfs to become less responsive to GA_5.

Gibberellins promote the growth of hypocotyls as well as of stems, but have little or no effect on root growth. The roots of dwarf varieties are usually as large as those of tall varieties of the same species.

LEAF GROWTH. GA generally promotes the elongation of grass leaves, probably through an effect on their basal intercalary meristems. It has little or no effect on the enlargement of the broad leaves of dicotyledons, and the leaves

Figure 13-10. Promotion of bolting and blooming of radish, a long-day plant, by gibberellin. Untreated control at right. Both plants were kept under short days. [Courtesy of S. H. Wittwer]

of plants treated with gibberellins generally are no larger than those of the untreated controls. However, GA does promote the growth of leaf discs in the dark, and Wheeler [53] has reported that when bean seedlings are moved from the dark to the light their gibberellin level rises sharply. Thus, gibberellins may be involved in the light-induced expansion of dicotyledon leaves even though it does not influence their final size.

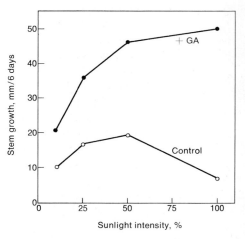

Figure 13-11. The inhibition of stem growth of pinto bean plants by sunlight is prevented by the application of gibberellin (4 μg per plant). [Data of J. A. Lockhart, *Amer. J. Bot.,* **48:**387 (1961)]

FRUIT DEVELOPMENT.   The induction of artificial parthenocarpy is one of the effects that auxin and gibberellins have in common. However, the gibberellins are ineffective in some species where auxin is effective and vice versa. For example auxin is ineffective in several members of the rose family, such as peaches, apples, and pears, whereas GA is quite effective. Both auxin and gibberellin can bring about artificial parthenocarpy in some plants including tomatoes. The generally high GA level in developing fruits suggests that it plays a role in natural fruit development. It is possible that both auxin and gibberellins are essential for fruit development in general, but that in specific cases the natural level of one or another is adequate. Jackson and Coombe [21] have found a strong correlation between the endogenous gibberellin level of fruits and their rate of growth.

FLOWER INITIATION.   Unlike auxin, the gibberellins are effective in bringing about flower initiation in many plants, in particular rosette plants that require either long days or vernalization for flower initiation as well as for bolting. Gibberellins are also effective in flower initiation in some, but by no means all, nonrosette long-day species, and there have been scattered reports that they promote earlier blooming in some day-neutral plants; however, they are ineffective in causing short-day plants to bloom under long days [24]. The various gibberellins differ greatly in their effect on flower initiation. GA_7 is apparently the most generally effective one and GA_3 is ineffective in many species. It is possible that in species where GA_3 applications promote flowering the plants can convert it into GA_7 or another gibberellin that promotes flower initiation.

   The gibberellins have still another effect on flowers. In plants such as hops and cucumbers [15, 37] that have stamens and pistils in separate flowers, the gibberellins promote the development of staminate flowers where pistillate flowers would otherwise have developed. Also, in fern gametophytes GA promotes the development of antheridia in place of archegonia.

BREAKING OF DORMANCY.   The buds of most trees, shrubs, and other perennials become dormant in the autumn and can not resume growth until dormancy has been broken by a sufficient period of cold. The seeds of some species have dormancy that is broken by low temperature, and other species have seeds that remain dormant until they have a brief exposure to light after they have imbibed water. Gibberellins can substitute for the natural environmental factors in breaking these types of dormancy, and it appears likely that the environmental factors may promote gibberellin synthesis or activation as well as possibly inactivating growth inhibitors.

Effects of Gibberellins at the Cellular Level

   Some investigators have presented evidence that gibberellins promote only cell elongation (Figure 13-12), others that they promote only cell division, and still others that they promote both. It now seems evident that growth promotion by gibberellins results primarily from increased cell division and that its effect on cell elongation is secondary and probably minor (Figure 13-13).

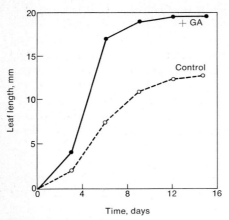

Figure 13-12. Wheat seedlings were given 612 kr of x-ray, which completely stops cell division, but GA still promoted leaf growth. [Data of A. H. Haber and H. J. Luippold, *Amer. J. Bot.,* **47:**140 (1960)]

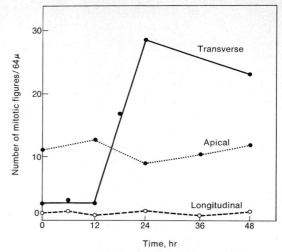

Figure 13-13. Effect of gibberellin (GA₃) on cell division in the apical meristem and in transverse and longitudinal planes in the subapical meristem. In all three cases the number of meristematic figures in the untreated controls remained at the 0 hr level throughout the 48-hr period. [Data of R. M. Sachs, C. F. Bretz, and A. Lang, *Amer., J. Bot.,* **46:**376 (1959)]

    The cells in gibberellin-treated plants are generally no longer than those in untreated controls. The immense increase in internode length of rosettes brought about by gibberellins could not possibly be accounted for by cell elongation alone. Auxin is effective in parthenocarpy where fruit growth is principally a matter of cell enlargement, whereas gibberellin is effective when the growth involves substantial cell division. However, light inhibits stem growth by reducing both cell division and cell elongation, and GA counteracts both of these light inhibitions. GA does not promote cell elongation of cells in isolated young stem sections whereas auxin does, but such sections are deficient in auxin and GA is ineffective in cell elongation in the absence of auxin.

    Gibberellins have essentially no effect on cell division in the apical meristems, except for their role in the breaking of dormancy. Their effect on cell division is primarily in the subapical or intercalary meristems. However, the possibility remains that their basic effect here is on cell elongation and that the increased cell division is a result of the cells attaining a greater than critical length. The promotion of grass leaf elongation by GA is related to the presence of an intercalary meristem, whereas their lack of effect on root growth may be due to the absence of a subapical meristem in roots.

    The developmental effects of gibberellin, such as the initiation of flowers and the conversion of pistillate to staminate flowers, suggest that the gibberellins may also have an influence on cell differentiation. GA has been found to promote the enlargement and differentiation of young sieve cells in pines [12] while they are dormant, but the differentiation is not complete and may not involve a direct influence of GA.

Biochemistry of Gibberellin Action

There is a substantial amount of evidence from various investigations that gibberellins increase the auxin content of plants substantially, often within a few hours after they are applied. There is evidence that the increase in auxin results both from an increased rate of auxin synthesis and a decreased rate of auxin inactivation, possibly because GA in some way reduces the activity of IAA oxidase and IAA peroxidase. This effect on auxin level could explain such promotion of cell elongation as is induced by GA, and the relatively few cases where auxin and gibberellin have the same effect, for example, in artificial parthenocarpy of tomatoes. However, the influence of gibberellins on auxin level cannot explain most of the effects brought about by gibberellins, because they are largely different from auxin effects and because added auxin is incapable of duplicating such gibberellin effects.

A more promising possibility for explaining gibberellin effects in general is that they bring about an increase in certain enzymes by derepressing genes that code for their synthesis, or perhaps in some cases by an effect at the level of protein synthesis. In 1960 Yomo and Paleg independently discovered that GA induces large increases in hydrolytic enzymes in germinating seeds, and subsequently Varner and his coworkers [9] have extensively investigated the role of gibberellins in germinating barley seeds. When GA is absent the aleurone cells of barley endosperm contain mere traces of α-amylase, but within 7 or 8 hr after excised endosperms are treated with GA or an extract from the embryos there is abundant *de novo* synthesis of α-amylase. In intact germinating seeds the embryos evidently produce gibberellin that diffuses into the endosperm. A variety of evidence indicates that the α-amylase synthesis results from derepression of the gene that codes for it. For example, actinomycin-D, an inhibitor of RNA synthesis, blocks α-amylase formation during the 7 hr or so before it appears but not after that, all the necessary mRNA apparently having been formed.

Other enzymes that are induced or increased by GA include proteases, ribonuclease, phosphatase, and pectin methyl esterase. To what extent the various GA effects result from its influence on enzyme synthesis is not known, but this could provide a theoretically plausible explanation of gibberellin action. The differing effects and activities of the various gibberellins could well be accounted for by the specific enzymes each one is capable of inducing.

Cytokinins

The discovery in 1954 of a new class of plant growth substances, now known as the cytokinins, resulted from tissue culture studies of tobacco pith that had been in progress in the laboratories of Folke Skoog at the University of Wisconsin. IAA induced substantial cell enlargement in the pith cultures, but there was no cell division unless vascular tissue was placed in contact with the pith or the medium was supplemented by the addition of coconut milk, malt extract, yeast extract, or autoclaved DNA. In 1954 Carlos Miller, a student of Skoog, isolated a crystalline substance from autoclaved herring sperm DNA that was capable of inducing cell division in the tobacco pith cultures at the amazingly low concentration of 1 part per billion (ppb) [30]. The substance was named **kinetin** and was later found to be 6-furfuryl amino purine (6-furfuryl adenine) (Figure 13-14).

Subsequently other adenine derivatives were found to have effects similar to kinetin, and these were referred to collectively as kinins. However, it was found that the term had already been used for a group of polypeptides that influence the contraction of involuntary muscles, so several other names were coined including cytokinins, which is now generally used. At first some plant physiologists questioned whether cytokinins should be classed as phytohormones, because the early ones were either derived from nonplant sources

Kinetin

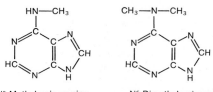

N⁶-Methylaminopurine   N⁶-Dimethylaminopurine

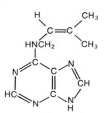

N⁶-(Δ²-Isopentenylamino) purine

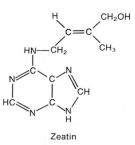

Zeatin

Figure 13-14. Structural formulas of kinetin and four naturally occurring cytokinins.

or synthesized, but a variety of natural cytokinins have now been isolated from plants and it is clear that they do constitute a distinct class of phytohormones [20].

Chemical Nature of the Cytokinins

The cytokinins differ from one another in the nature of the side chain attached at the N^6 position of adenine (Figure 13-14). A wide variety of side chains confer cytokinin activity, but the side chain must be nonpolar. However, even adenine itself has slight cytokinin activity. Some cytokinins are less effective than kinetin whereas others are more effective. Modification of the adenine nucleus, particularly at the number 1 position, generally results in loss or reduction of cytokinin activity. However, attachment of ribose or ribose phosphate at the number 9 position usually results in little change in activity. Also, 8-azakinetin has cytokinin activity, even though it is not a purine derivative.

Several other nonpurines, including *N,N'*-diphenyl urea, benzimidazole, 2-benzthiazolyloxyacetic acid, leucoanthocyanins, and inositol have some degree of cytokinin activity, and a few workers have suggested that they be included among the cytokinins. However, since auxins and gibberellins can also promote cell division it seems best to limit the cytokinins to purine derivatives instead of including all substances that promote cell division.

Although most known cytokinins are synthetic, several cytokinins have been isolated from plants and identified chemically since 1960. In addition, saps and extracts from many species of plants have been found to have cytokinin activity although specific cytokinins have not been isolated and identified chemically. Natural cytokinins are generally more active than kinetin and the synthetic cytokinins, notably zeatin which was isolated from corn grains and is at least ten times as active as any synthetic cytokinin.

Although the various cytokinins differ greatly in activity, they are all effective at very low concentrations. Applied concentrations on the order of 10^{-5} to 10^{-7} M or 10 to 0.01 mg/liter are quite effective, and the natural concentrations are also low. For example, the cytokinin content of the bleeding sap of grape vines has been reported to be between 0.05 and 0.1 ppm.

Occurrence and Distribution of Cytokinins in Plants

Although only a small percentage of the many species of plants have been checked for the presence of cytokinins, it now seems likely that cytokinins are widely if not universally synthesized by vascular plants and also by many nonvascular plants. They have been found in several bacteria and in yeasts and other fungi. In vascular plants they are most abundant in young and actively dividing tissues such as endosperms, embryos, growing fruits, seedlings, and apical meristems. These may also be the sites of most active cytokinin synthesis, although translocation from other regions of the plant is not excluded.

When synthetic cytokinins are applied to plants there is little or no movement from the region of application, except that they may be carried up the xylem when cut stems are placed in a cytokinin solution. However, there is now evidence that natural cytokinins are translocated upward in the xylem as solutes in the flowing water and that they may also be translocated in the phloem. Root tips have been found to be a rich source of cytokinins, and these are evidently translocated to the shoots through the xylem. There have been conflicting reports regarding the extent and polarity of cytokinin translocation through the phloem. Some of the discrepancies can perhaps be accounted for by the finding of Seth, Davies, and Wareing [43] that the translocation of kinetin through the phloem is greatly promoted if IAA is applied along with the kinetin, and similar results have been secured by others. Although phloem translocation of cytokinins is apparently predominantly basipetal (downward), there has been some evidence for translocation in both directions.

Effects of Cytokinins at the Cellular Level

The cytokinins were originally characterized as growth regulators that induce cell division in certain tissue cultures, and it now appears probable that they play a general role in cell division in intact plants. Although auxin and gibberellins also promote cell division in certain cases, they cannot substitute for a cytokinin requirement.

Although cytokinins are primarily cell division phytohormones (Figure 13-15), it has been found that they can also promote cell enlargement. As early as 1956 Miller [28] and also two other groups of investigators found that kinetin greatly increased cell enlargement in discs from etiolated leaves, an effect normally requiring light and not inducible by auxin. Kinetin has been reported to increase cell elongation in tobacco pith under certain conditions and, although it inhibits cell elongation in pea stem sections and sunflower hypocotyls, it promotes lateral enlargement of the cells. Arora and coworkers [3] found that kinetin causes the cortical cells of tobacco roots to enlarge up to four times their normal size.

The morphogenetic effects of cytokinins, including induction of bud initiation and the alteration of the sex expression of flowers, suggest that the cytokinins

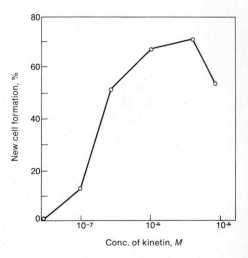

Figure 13-15. Influence of various concentrations of kinetin on induction of cell division in cultured tobacco pith during a 7-day period. Each culture medium contained also 2 mg/liter IAA. [Data of F. Skoog and C. O. Miller, *Symp. Soc. Exp. Biol.,* **11:**118 (1957)]

can also influence cell differentiation. Stetler and Laetsch [47] have reported that cytokinins play a role in the maturation of proplastids into plastids, and Bergmann [5] found that cytokinin promoted the differentiation of tracheids by activation of lignin synthesis.

Effects of Cytokinins on Growth and Development

Like auxin and gibberellins, the cytokinins have a variety of effects on observable growth and development. These include stimulation or inhibition of growth, bud initiation, breaking of dormancy and apical dominance, mobilization of metabolites, delay of senescence, and several rather limited effects on reproductive development.

EFFECTS ON GROWTH. Unlike the gibberellins, the cytokinins are ineffective in promoting the stem elongation of intact plants. However, they may cause the thickening of stems, hypocotyls, and roots, probably as a result of their promotion of lateral cell enlargement, or in some cases the stimulation of cambial cell division [54]. Cytokinin applications generally inhibit the elongation of a main root, but at the same time may stimulate the initiation of lateral roots. In addition to substituting for light in the expansion of the leaves of dicotyledons, they may also bring about increased enlargement of leaves in the light.

BUD AND ROOT INITIATION. In 1957 Skoog and Miller [44] reported an interesting influence of the auxin/kinetin ratio on the differentiation of buds and roots in tobacco pith cultures. At a ratio of about 100:1 only an undifferentiated callus developed. Decreasing the ratio, either by reducing the auxin or increasing the kinetin, resulted in the formation of buds that in some cases grew into complete tobacco plants. Increasing the ratio promoted the initiation of roots rather than buds. Detached African violet leaves and cuttings of begonia leaves, which normally form adventitious buds and roots, can be induced to form more buds by cytokinin applications. Schraudolf and Reinert [42] found that begonia leaf cuttings treated with kinetin developed buds all along the edge, whereas normally buds develop only at the basal end of the main veins. The leaves of some species, including certain kinds of begonias, that do not naturally form buds and roots on cuttings can be induced to do so by treatments with cytokinins. Torrey [49] found that root segments of bindweed formed adventitious buds more readily when treated with kinetin.

However, the capacity of cytokinins to promote adventitious bud initiation is not universal. Harris and Hart [19] found that cytokinins inhibited bud initiation in *Peperomia sandersii*, except in isolated cases where exceptionally high concentrations were used. Normally leaf cuttings of this species form buds only after roots are initiated, and presumably the cytokinin inhibited bud formation by preventing root initiation. Cytokinins have been completely ineffective in inducing bud initiation in tissue cultures and cuttings from some species, particularly those that never form adventitious buds naturally. Other factors than the cytokinins are evidently involved in adventitious bud initiation.

Although the buds that develop on moss protonemata are not homologous with the buds of vascular plants, several investigators have found that cyto-kinins stimulate their development and may also increase the number that develop into gametophores.

Auxin, rather than cytokinin, is usually effective in the initiation of adventi-tious roots, but Allsop and Szweykowski [1] induced abundant root initiation from young leaves of *Marsilea drummondii,* a water fern, with kinetin. Such adventitious root development in this species had never been observed, either naturally or after auxin treatments.

COUNTERACTION OF APICAL DOMINANCE. Cytokinins release lateral buds from apical dominance without decapitation of the terminal bud [11]. Sorokin and Thimann [45] have proposed that this results from cytokinin-induced differen-tiation of xylem joining the vascular tissue of the bud and stem, thus providing an increased supply of water and solutes that facilitate development of the bud into a branch. The numerous buds induced by cytokinins in tissue cultures and cuttings do not exert apical dominance on one another.

BREAKING OF DORMANCY. Cytokinins have been found to break the dormancy of buds and seeds of a number of different species [23]. Seeds of plants such as lettuce and tobacco that require light for germination will germinate in the absence of light if treated with cytokinins, which are also synergistic with light in further promoting germination. Cytokinins are also effective in breaking other types of physiological seed dormancy, including the upper dormant seed in the cocklebur fruit. The dormancy of duckweed plants is also broken by cytokinins.

Worsham and others have found that the seeds of witchweed (*Striga asiatica*), a root parasite of corn and some other grasses that has become a problem in North Carolina following its accidental introduction, can be induced to germinate by kinetin treatments. In nature the seeds germinate only when near the roots of the host plants. That this results from cytokinin diffusing from the roots is suggested by the fact that root extracts are effective in breaking the dormancy of the seeds.

EFFECTS ON REPRODUCTIVE DEVELOPMENT. Unlike the gibberellins, the cyto-kinins are not known to have any extensive or spectacular effects on repro-ductive development. There have been scattered reports of cytokinin-induced flower initiation in long-day plants or plants that require vernalization and of promotion of parthenocarpy in a few plants such as grapes and figs. Whether cytokinins have a more widespread influence on reproductive development or not remains to be seen. It may be that cytokinins are essential for various stages of reproductive growth and development but that the natural levels are gener-ally adequate. Young fruits are rich in cytokinins, and they may be essential for fruit growth, at least during periods of active cell division. Cytokinins also appear to be essential for embryo growth and development.

Cytokinins have been found to bring about the conversion of staminate flowers of grapes [34] and several other plants into pistillate flowers. Thus, their effect on sex expression is similar to that of auxin and opposite to that of gibberellins.

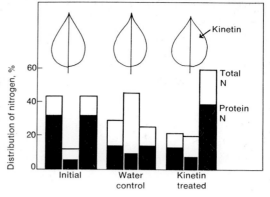

Figure 13-16. Changes in soluble and protein nitrogen in excised tobacco leaves. The three bars in each set represent the left half, midrib, and right half of the leaves. Nitrogen determinations of the water controls and the leaves with kinetin applied to the right halves were made 9 days after excision. [Data of K. Mothes, L. Engelbrecht, and O. Kulajewa, *Flora*, **147**:445 (1959)]

MOBILIZATION OF SOLUTES.   A unique and striking effect of cytokinins is their role in mobilization of solutes of a wide variety (Figure 13-16), including amino acids, auxin, and several mineral elements, notably phosphorus. Mothes and his coworkers [32] found that, when a radioactive amino acid was applied to one area of a leaf and a spot of kinetin was applied to another area, there was prompt and extensive movement of the amino acid to the kinetin-treated area (Figure 10-11). The kinetin remained localized in the treated area. The mobilizing influence of kinetin was pronounced in older leaves but not young ones. There is naturally extensive mobilization of solutes from older tissues to young and actively growing regions of the plant such as apical meristems, young leaves, and developing fruits and seeds, and it seems likely that cytokinins are involved in such mobilization.

DELAY OF SENESCENCE.   As leaves and other organs become senescent there is a marked decline in their content of chlorophyll, proteins, nucleic acids, and other important substances, probably as a result of both hydrolysis and a reduction in synthesis. This is accompanied by extensive translocation of amino acids, mineral elements, and other substances out of the leaves. Mere detachment of the leaves of many species of plants will bring about such senescence, even though they are adequately supplied with water (Figure 13-17). Richmond and Lang [39] found that the senescence of detached cocklebur leaves could be delayed for as long as 20 days if they were treated with cytokinins. The natural autumnal senescence of attached leaves can also be delayed by cytokinin applications.

The delay of senescence by cytokinins undoubtedly involves their effect on mobilization. For example, Leopold and Kawase [26] found that cytokinin treatment of some attached leaves of a plant induced earlier senescence of untreated leaves. However, other cytokinin influences than mobilization are also evidently involved. Isolated leaf discs have their senescence delayed by cytokinins, even when there is no region from which mobilization can occur, and Osborne [35] found that kinetin increased the ratio of RNA and protein to DNA, in contrast with the decline in the ratios during senescence. This suggests that cytokinins maintain high levels of biosynthesis through an effect on RNA synthesis and, consequently, enzyme synthesis.

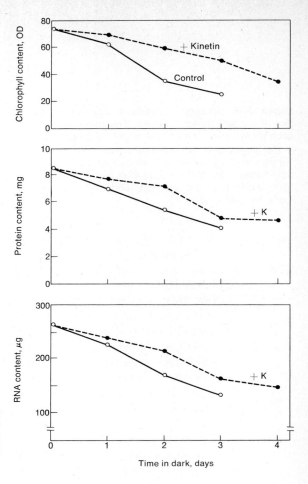

Figure 13-17. Delay of senescence of excised Xanthium leaves by kinetin, as indicated by the slower decrease of chlorophyll, protein, and RNA in the kinetin-treated leaves. [Data of D. J. Osborne, *Plant Physiol.*, **37:**595 (1962)]

Cytokinins reduce the symptoms of stress brought about by a wide variety of adverse environmental factors such as growth inhibition by drought, γ-rays, damage by cold, and injury by viruses, herbicides, and pesticides [24, 50]. Their role here may be similar to that in the delay of senescence.

Bioassays for Cytokinins

Cytokinin bioassays based on the effects on cell division, chlorophyll preservation, cell enlargement, germination, and differentiation have been devised [54]. Tests based on the first two are the most sensitive, specific, and widely used. Two cell division assays are used: the tobacco pith culture assay and the soybean callus assay devised by Miller [29] in 1963. Both are sensitive, reacting to as little as 2×10^{-8} kinetin, but the soybean callus assay is faster and more quantitative because it is proportional to the log of the cytokinin concentration up to 10^{-5} M or higher. However, for crude tissue extracts the tobacco tissue is better, since it is less sensitive to the toxic or inhibitory substances usually present.

In 1961 Osborne and McCalla devised an assay based on the influence of cytokinins on chlorophyll retention. Leaf discs from cocklebur are aged in dim light, treated with the test solution, and kept in the dark for 48 hr. The chlorophyll is then extracted and measured photometrically. Others have devised assays using detached leaves from wheat or oat seedlings. The chlorophyll assays are less sensitive than the cell division tests and also less specific because other substances such as auxin and sugars may also promote chlorophyll retention.

Biochemistry of Cytokinin Action

Extensive evidence has been accumulated indicating that the cytokinins exert their primary metabolic influences at the level of nucleic acid and protein synthesis, although their specific roles are still not well established. We can present only a few of the many types of evidence here, but more extensive discussions are available elsewhere.

In 1967 Guttmann [18] found that kinetin treatment of onion roots quickly brought about an increase in the RNA content of the nuclei, and subsequently others reported similar results. Gunning and Barkley [17] found that cytokinins greatly increase the incorporation of ^{32}P into the nucleic acids of detached oat leaves. In 1966 Fox [14] reported that cytokinins were incorporated by soybean tissue cultures almost exclusively into some kinds of tRNA, and in the same year Zachau and his coworkers found that a cytokinin was incorporated into yeast serine tRNA adjacent to the anticodon. There have been several other reports that cytokinins are incorporated in tRNAs of several species in this position. It has been suggested that the incorporated cytokinin may be essential for the binding of the tRNA to the mRNA on the ribosomes.

As noted previously, cytokinins increase both the RNA and protein content of senescing leaves, and several investigators have found that cytokinins markedly increase the synthesis of some enzymes but not others. For example, Anderson and Rowan [2] found that cytokinin increased a tRNA synthetase, whereas Srivastava and Ware [46] reported that cytokinin decreased ribonuclease activity. Such findings suggest that cytokinins may act as gene derepressors or repressors, or at least that they affect the synthesis of specific enzymes in some way.

Abscisic Acid (ABA)

Except for superoptimal concentrations of auxin, or perhaps more accurately the ethylene they induce, the phytohormones we have considered all promote growth and development in one way or another. However, as early as 1934 it was proposed that plants also produce growth inhibitors, and since then there have been numerous reports of a wide and bewildering variety of natural plant growth inhibitors. Many of these were never isolated or identified. Those that have been are mostly phenols, flavinols, or sesquiterpenes. The growth inhibitors have been investigated primarily from the standpoint of their induction of dormancy of buds and seeds (Figure 13-18). Germination inhibitors, which

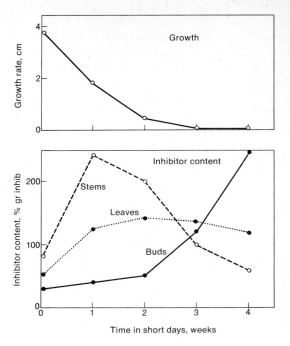

Figure 13-18. Relationship between the growth rates of *Betula peubescens* stems kept under short (10 hr) days and the content of growth inhibitors in the stems, leaves, and buds. [Data of M. Kawase, *Proc. Amer. Soc. Hort. Sci.*, **78**:532 (1961)]

are responsible for some types of seed dormancy, include a much larger and more diverse variety of substances than inhibitors of bud growth [13] and will be considered in a later chapter. Here we will consider only one of the growth inhibitors, but discussion of growth inhibitors in general can be found elsewhere [25]. It should be noted, however, that none of the synthetic growth inhibitors such as Amo-1618, phosphon-D, chlorocholine chloride (CCC), B995, and maleic hydrazide (MH), which have been used widely for practical as well as experimental purposes, is known to occur naturally in plants (Figure 13-19).

Since 1960 one of the multitude of reported plant growth inhibitors—abscisic acid (ABA)—has emerged as probably the most widespread, important, and predominant of the group. Several of the reported inhibitors, including abscisin II, dormin, and β-inhibitor have proved to be abscisic acid, and it is possible that some of the unidentified inhibitors that were studied were also abscisic acid. Whether or not other plant growth inhibitors comparable in importance with abscisic acid will emerge from the multitude in the future remains to be seen.

The discovery of abscisic acid may be considered to date from 1964 when three independent groups of investigators working with different species reported on the extraction of substances that promoted abscission or bud dormancy. Following through on a 1961 report by Liu and Carns [27], H. R. Carns, F. T. Addicott, and their coworkers at the University of California at Davis reported the isolation from young cotton bolls (fruits) of substances that promoted leaf abscission. These substances were named abscisin I and abscisin II. K. Rothwell and R. L. Wain of the University of London, following up earlier work by R. F. M. van Steveninck in New Zealand, isolated from lupins a substance that promoted flower abscission. At the University College of Wales

P. F. Wareing and his students isolated a substance from sycamore leaves that caused dormancy when applied to the buds of the tree and named it dormin. When abscisin II, the lupine factor, and dormin were identified chemically it became evident that they were all the same compound, which is now known as abscisic acid [52].

Chemical Nature of Abscisic Acid

Abscisic acid is an acidic sesquiterpene (Figure 13-19) and forms esters with sugars (mostly glucose) which are also effective growth inhibitors. ABA is biosynthesized from mevalonic acid, like other sesquiterpenes, but the later steps in the biosynthesis are still in doubt. ABA has also been synthesized by chemists, and this synthetic ABA provides the major source for experimental purposes since only small quantities can be extracted from plants. Natural ABA is strongly dextrorotary, but synthetic ABA is made up of equal proportions

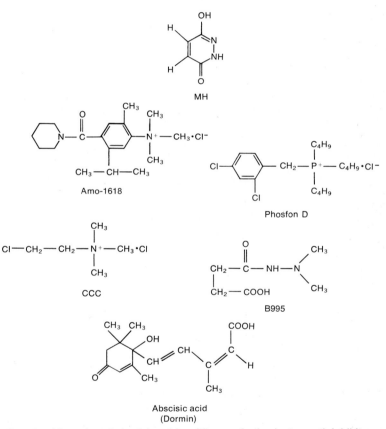

Figure 13-19. Structural formulas of abscisic acid and five synthetic plant growth inhibitors. Note the diversity of molecular structure. Amo-1618, Phosfon D, and CCC inhibit gibberellin biosynthesis. MH (maleic hydrazide) inhibits cell division, probably by interfering with nucleic acid synthesis. Abscisic acid is the only one of these substances known to occur naturally in plants.

of the dextrorotary and levorotary isomers. In contrast with the rather numerous gibberellins and cytokinins, only one abscisic acid has been identified so far.

Like hormones in general, abscisic acid is effective at very low concentrations, and as little as 0.25 ppm is active.

Occurrence and Distribution of Abscisic Acid

ABA has been isolated from numerous species of angiosperms and from several species of gymnosperms and pteridophytes including fir, yew, and bracken fern. It seems likely that it will be found to be of essentially universal distribution among vascular plants. So far little is known about its occurrence in nonvascular plants. ABA has been extracted from essentially all organs of vascular plants: roots, stems, leaves, fruits, and seeds, and has been found in the sap of both the xylem and the phloem. The natural concentrations are generally between 0.01 and 1 ppm of fresh tissue, although 4 ppm were found in rose hips (fruits) collected in December.

Effects at the Cellular Level

Like the growth-promoting phytohormones, ABA influences both cell division and cell enlargement, inhibiting both processes. As a growth inhibitor it cannot be expected to have any very marked effects on cell differentiation. However, bud dormancy induced by ABA in woody plants involves the formation of typical bud scales, and ABA presumably has at least an indirect effect on the cell differentiation involved in this.

Effects on Plant Growth and Development

The known effects of abscisic acid on plant growth and development are somewhat less extensive and varied than those of the other established phytohormones. They include the induction of dormancy, promotion of abscission and senescence, and inhibition of growth.

INDUCTION OF DORMANCY. Although ABA is by no means the only substance that induces dormancy of buds and seeds, it has been found effective in a considerable number of woody perennials and seeds and is probably one of the most widespread inducers of dormancy in nature. ABA has also been found effective in inducing bud dormancy in duckweeds and in some herbaceous annuals that normally do not have dormant buds.

ABA may also play a role in apical dominance. Dörffling [see 54, p. 180] found that, when apical buds of sycamore and pea were decapitated, the concentration of β-inhibitor (ABA) in the lateral buds declined and that application of ABA directly to the lateral buds would then inhibit their growth.

In general, application of ABA to seeds inhibits their germination (Figure 13-20). ABA has been found in the seeds of various plants including ash, avocado, pea, grape, yellow lupine, peach, and rose, and in at least some of these it is apparently responsible for seed dormancy.

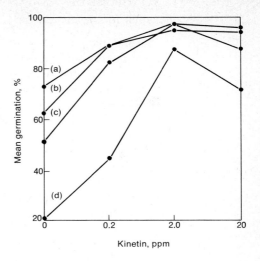

Figure 13-20. Effect of various concentrations of ABA and kinetin on the germination of lettuce seeds. ABA concentrations are (a) 0, (b) 0.2, (c) 2.0, and (d) 20 ppm. [Data of P. F. Wareing *et al.*, from P. F. Wareing and G. Ryback, *Endeavour*, **29**:84 (1970)]

INHIBITION OF GROWTH. Of course, the induction of dormancy in buds and seeds inhibits growth, but several other types of growth inhibition have been reported. ABA inhibits the growth of oat coleoptiles, even in the presence of adequate IAA. It inhibits flowering in some long-day plants under long days. Whether or not it is involved in preventing internode elongation of rosette plants has not been established, but it apparently does not inhibit growth of plants to the degree that various synthetic growth inhibitors do when applied to plants. Goodwin and Cansfield [16] found that ABA inhibited the growth of potato stems and that it was involved in the apical dominance of tuber buds, but it promoted the growth of the tubers. This is the only report of growth promotion by ABA, but it may be an indirect result of the inhibition of stem growth.

PROMOTION OF SENESCENCE AND ABSCISSION. Although the promotion of leaf and flower abscission were among the first established effects of ABA, its role in natural abscission has not been as well documented as its role in the induction of dormancy. ABA applications have been found to promote abscission in various species, but they are less effective on intact leaves than when applied to debladed petioles. Thus, the hope that ABA might be used on a practical basis to induce abscission of cotton leaves to facilitate mechanical harvesting of the bolls has not materialized and other defoliants are employed. The reduced effectiveness of ABA on intact leaves may be related to the abscission-inhibiting effects of IAA and cytokinins in the blades. It should be stressed that abscission is a complex phenomenon [40, 56] involving not only inhibition by IAA and cytokinins and promotion by abscisic acid and gibberellins [8] but also various nonhormonal factors.

ABA applications hasten the senescence of detached leaves and leaf discs but are less effective in promoting the senescence of attached leaves. This difference probably also results from the availability of other phytohormones in attached leaves and perhaps from other factors.

Assays for Abscisic Acid

Several bioassays have been devised for ABA, including the use of oat coleoptile sections and explants of stem sections with petiole bases from cotton seedlings. The first determines the effect of ABA on growth and the second on the promotion of abscission.

Chemical and physical assays for ABA are more sensitive and effective than those for most other phytohormones, and these have been used for quantitative as well as qualitative determinations. ABA can be measured by ultraviolet spectrophotometry, by gas-liquid chromatography, or by its strong dextrorotary effect. If ABA is treated with a formic-hydrochloric acid mixture and then treated with sodium hydroxide, a strong but transient violet color results. Because the test appears to be very specific for ABA and its esters, it can be used for qualitative determination of ABA in plant extracts.

Biochemistry of Abscisic Acid Action

Various effects of ABA are counteracted by different phytohormones. For example, IAA counteracts the effect of ABA on abscission, gibberellins its promotion of dormancy in buds of woody plants, cytokinins its promotion of duckweed bud dormancy and also its promotion of lettuce seed dormancy. These diverse interactions with other phytohormones suggest that the basic effect of ABA is at a more fundamental level, and indeed there is considerable evidence that it influences nucleic acid and protein synthesis or degradation.

Chrispeels and Varner [10] found that ABA counteracts the effect of GA in inducing α-amylase synthesis in barley seeds. van Overbeek and his co-workers found that ABA inhibits the synthesis of both DNA and RNA by duckweed [51], whereas Wareing and his group reported that ABA-induced leaf disc senescence is accompanied by a decrease in the synthesis of both RNA and protein within 4 hr but that no specific DNA fraction was selectively inhibited [52]. These and other results suggest a definite effect of ABA at the level of nucleic acid and protein synthesis.

Summary

In this and the preceding chapter an attempt has been made to stress the importance of phytohormones in plant growth and development and to present a few of the important facts and concepts about phytohormones that have emerged from the extensive and intensive research on them since the beginning of this century. It should be clear that, although much has been learned about phytohormones, much more remains to be learned and clarified by future investigations, particularly as regards their basic modes of action. It is likely that the list of well-established plant hormones will be increased in the future by addition of at least some of the currently hypothetical ones, some that so far have received little attention, and some that are now completely unknown.

One point that should be stressed is that there are extensive interactions among the various phytohormones, and their effects are not as independent of one another as may have been suggested by some of the preceding discussions. Another point that should perhaps be evident is that plant hormones have more numerous and diverse effects than animal hormones, many of which play quite specific roles. At the cellular level all the established phytohormones influence both cell division and cell enlargement, and most of them have certain influences on cell differentiation. For the most part, however, their interactions at the cellular level remain to be elucidated.

Although we have now completed our basic consideration of phytohormones, their roles in plant development will receive considerable attention in the subsequent chapters.

References

[1] Allsop, A., and A. Szweykowska. "Foliar abnormalities, including repeated branching and root formation, induced by kinetin in attached leaves of *Marsilea*," *Nature*, **186**:813–814 (1960).

[2] Anderson, J. W., and K. S. Rowan. "Activity of aminoacyl-transfer-ribonucleic acid synthetases in tobacco leaf tissue in relation to senescence and to the action of 6-furfurylaminopurine," *Biochem. J.*, **101**:15–18 (1966).

[3] Arora, N., F. Skoog, and O. N. Allen. "Kinetin-induced pseudonodules on tobacco roots," *Amer. J. Bot.*, **46**:610–613 (1959).

[4] Audus, L. J. *Plant Growth Substances*, Interscience, New York, 1959.

[5] Bergmann, L. "Der Einfluss von Kinetin auf die Ligninbildung und die Differenzierung von *N. tabacum*," *Planta*, **62**:221–254 (1964).

[6] Birch, A. J., R. W. Ricards, and H. Smith. "Biosynthesis of gibberellic acid," *Proc. Chem. Soc. (London)*, **1958**, pp. 192–193.

[7] Brian, P. W. "The gibberellins as hormones," *Int. Rev. Cytol.*, **19**:229–266 (1966).

[8] Chatterjee, S. K., and A. C. Leopold. "Kinetin and gibberellin actions on abscission processes," *Plant Physiol.*, **39**:334–337 (1964).

[9] Chrispeels, M. J., and J. E. Varner. "Gibberellic acid-enhanced synthesis and release of α-amylase and ribonuclease by isolated barley aleurone layers," *Plant Physiol.*, **42**:398–406 (1967).

[10] Chrispeels, M. J., and J. E. Varner. "Hormonal control of enzyme synthesis: on the mode of action of gibberellic acid and abscisin in aleurone layers of barley," *Plant Physiol.*, **42**:1008–1016 (1967).

[11] Davies, C. R., A. K. Seth, and P. F. Wareing. "Auxin and kinetin interaction in apical dominance," *Science*, **151**:468–469 (1966).

[12] DeMaggio, A. E. "Phloem differentiation: induced stimulation by gibberellic acid," *Science*, **152**:370–372 (1966).

[13] Evenari, M. "Germination inhibitors," *Bot. Rev.*, **15**:153–194 (1949); "The physiological action and biological importance of germination inhibitors," *Sym. Soc. Exp. Biol.*, **11**:21–43 (1957).

[14] Fox, J. E. "Incorporation of a kinin, N^6-benzyladenine, into soluble RNA," *Plant Physiol.*, **41**:75–82 (1966).

[15] Galun, E. "Effects of gibberellic acid and naphthaleneacetic acid on sex expression and some morphological characters in the cucumber plant," *Phyton*, **13**:1–8 (1959).

[16] Goodwin, P. B., and P. E. Cansfield. "The control of branch growth on potato tubers III. The basis of correlative inhibition," *J. Exp. Bot.,* **18:**297–307 (1967).

[17] Gunning, B. E. S., and W. K. Barkley. "Kinin-induced directed transport and senescence in detached oat leaves," *Nature,* **199:**262–265 (1963).

[18] Guttman, Ruth. "Alterations in nuclear ribonucleic acid metabolism induced by kinetin," *J. Biophys. Biochem. Cytol.,* **3:**129–131 (1967).

[19] Harris, G. P., and E. M. H. Hart. "Regeneration from leaf squares of *Peperomia sandersii* A. DC: a relationship between rooting and budding," *Ann. Bot.,* **N.S. 28:**509–526 (1964).

[20] Helgeson, J. P. "The cytokinins," *Science,* **161:**974–981 (1968).

[21] Jackson D. I., and B. G. Coombe. "Gibberellin-like substances in developing apricot fruit," *Science,* **154:**277–278 (1966).

[22] Jones, R. L., and J. E. Varner. "The bioassay of gibberellins," *Planta* **72:**155–161 (1967).

[23] Kahn, A. A. "Cytokinins: permissive role in seed germination," *Science,* **171:**853–859 (1971).

[24] Lang, A. "Effect of gibberellin on flower formation," *Proc. Nat. Acad. Sci.,* **43:**709–717 (1957); "Plant growth regulation," *Science,* **157:**589–592 (1967).

[25] Leopold, A. C. *Plant Growth and Development,* McGraw-Hill, New York, 1964.

[26] Leopold, A. C., and M. Kawase. "Benzyladenine effects on bean leaf growth and senescence," *Amer. J. Bot.,* **51:**294–298 (1964).

[27] Liu, W., and H. R. Carns. "Isolation of abscisin, an abscission accelerating substance," *Science,* **134:**384–385 (1961).

[28] Miller, C. O. "Similarity of some kinetin and red light effects," *Plant Physiol.,* **31:**318–319 (1956).

[29] Miller, C. O. "Kinetin and kinetin-like compounds," *Mod. Methods Plant Anal.,* **6:**194–202 (1963).

[30] Miller, C. O., F. Skoog, M. H. Von Saltza, and F. M. Strong. "Kinetin, a cell division factor from deoxyribonucleic acid," *J. Amer. Chem. Soc.,* **77:**1329 (1955).

[31] Mitchell, J. W., N. Mandava, J. R. Plimmer, and J. F. Worley. "Brassins—a new family of plant hormones from rape pollen," *Nature,* **225:**1065–1066 (1970).

[32] Mothes, K. "Uberdas Altern der Blatter und die Moglichkeit ihrer Wiederverjungung," *Naturwiss.,* **47:**337–350 (1960).

[33] Naylor, J., G. Sander, and F. Skoog. "Mitosis and cell enlargement without cell division in excised tobacco pith tissue," *Physiol. Plantarum,* **7:**25–29 (1954).

[34] Negi, S. S., and H. P. Olmo. "Sex conversion in a male *Vitis vinifera* L. by a kinin," *Science,* **152:**1624–1625 (1966).

[35] Osborne, D. J. "Effect of kinetin on protein and nucleic acid metabolism in *Xanthium* leaves during senescence," *Plant Physiol.,* **37:**595–602 (1962).

[36] Osborne, D. J. and D. R. McCalla. "Rapid bio-assay for kinetin and kinins using senescing leaf tissue," *Plant Physiol.,* **36:**219–221 (1961).

[37] Peterson, C. E., and L. D. Anhder. "Induction of staminate flowers on

gynoecious cucumbers with gibberellin A$_3$," *Science,* **131:**1673–1674 (1960).

[38] Phinney, B. O. "Growth response of single-gene dwarf mutants in maize to gibberellic acid," *Proc. Nat. Acad. Sci.,* **43:**398–404 (1957).

[39] Richmond, A. E., and A. Lang. "Effect of kinetin on protein content and survival of intact Xanthium leaves," *Science,* **125:**610–651 (1957).

[40] Rubenstein, B., and A. C. Leopold. "The nature of leaf abscission," *Quart. Rev. Biol.,* **39:**356–372 (1964).

[41] Schopfer, W. H. *Plants and Vitamins,* Chronica Botanica, Waltham, Mass., 1943.

[42] Schraudolf, H., and J. Reinert. "Interaction of plant growth regulators in regeneration processes," *Nature,* **184:**465–466 (1959).

[43] Seth, A. K., C. R. Davies, and P. F. Wareing. "Auxin effects on the mobility of kinetin in the plant," *Science,* **151:**587–588 (1966).

[44] Skoog, F., and C. O. Miller. "Chemical regulation of growth and organ formation in plant tissues cultured *in vivo*" *Symp. Soc. Exp. Biol.,* **11:**118–131 (1957).

[45] Sorokin, H., and K. V. Thimann. "The histological basis for inhibition of axillary buds in *Pisum sativum* and the effects of auxin and kinetin on xylem development," *Protoplasma,* **59:**326–350 (1964).

[46] Srivastava, B. I. S., and G. Ware. The effect of kinetin on nucleic acids and nucleases of excised barley leaves," *Plant Physiol.,* **40:**62–64 (1965).

[47] Stetler, D. A., and W. M. Laetsch. "Kinetin-induced chloroplast maturation in cultures of tobacco tissue," *Science,* **149:**1387–1388 (1965).

[48] Stowe, B. B., and T. Yamaki. "Gibberellins: stimulants of plant growth," *Science,* **129:**807–816 (1959).

[49] Torrey, J. G. "Endogenous bud and root formation by isolated roots of Convolvulus grown *in vitro,*" *Plant Physiol.,* **33:**258–263 (1958).

[50] van Overbeek, J. "Plant hormones and regulators," *Science,* **152:**721–731 (1966); "The control of plant growth," *Sci. Amer.,* **219**(1):75–81 (July 1968).

[51] van Overbeek, J., J. E. Loeffler, and M. I. R. Mason. "Dormin (abscisin II), inhibitor of plant DNA synthesis?" *Science,* **156:**1497–1499 (1967).

[52] Wareing, P. F., and G. Ryback. "Abscisic acid: a newly discovered growth-regulating substance in plants," *Endeavour,* **29:**84–88 (1970).

[53] Wheeler, A. W. "Changes in leaf growth substances in cotyledons and primary leaves during the growth of dwarf bean seedlings," *J. Exp. Bot.,* **11:**222–229 (1960).

[54] Wilkins, M. B. (ed.). *The Physiology of Plant Growth and Development,* McGraw-Hill, New York, 1969.

[55] Wittwer, S. H., and M. J. Bukovac. "The effects of gibberellin on economic crops," *Econ. Bot.,* **12:**213–255 (1958).

[56] *Plant Physiol.,* **43**(9B) (1968). This entire issue is devoted to papers on abscission.

Plant Reproduction

14

Since every plant is the product of some kind of reproduction, the reproductive processes are prerequisite to plant growth and development if not an inherent part of them. The capacity for reproduction is, of course, one of the prime and essential characteristics of organisms. Without it a species could exist for only one generation, and an increase in the population of the species would be impossible. The assortment and recombination of genes in the course of sexual reproduction is one of the major sources of the hereditary variations that are subject to natural selection. In plants reproduction also plays an important role that is not essential in mobile animals—dispersal of the species. Although water plants may float or be carried by currents from one place to another and some land plants invade adjacent areas by means of stolons, rhizomes, or horizontal roots, plant dispersal is accomplished primarily by means of spores or seeds that may be carried considerable distances by such agents as wind, water, or animals.

Reproduction is essentially a matter of the detachment of one or more cells from the parent or parents, and the detached portion then develops into a mature individual if conditions are suitable. Sexual reproduction involves the fusion of two cells, or in some cases only two nuclei, and is widespread in the plant as well as in the animal kingdom. Another characteristic of sexual reproduction is that meiosis occurs at some stage in the life cycle. Asexual reproduction involves neither fusion nor meiosis, so there is only one parent and, barring mutations, an organism reproduced asexually has hereditary potentialities identical with those of the parent. Asexual reproduction is much more widespread among plants than among animals. Most species of plants can reproduce sexually, but the majority of them also have some means of asexual reproduction.

Asexual Reproduction of Nonvascular Plants

Binary Fission

The principal means of reproduction (see Figure 14-1) of bacteria, blue-green algae, and some of the other unicellular algae and fungi is by binary fission. This is cell division with the separation of the daughter cells from one another. Here cell division results in reproduction and population growth rather than contributing to the growth of an individual as it commonly does in multicellular organisms. Under favorable conditions reproduction and population growth by binary fission can be very rapid. The cells may divide as frequently as every 20 min, so that theoretically a single cell could have 68 billion descendents in 12 hr. In actuallity, such a rapid rate of population growth could not continue for as long as 12 hr because of such factors as the depletion of food or water or the accumulation of toxic substances in most natural situations.

Budding is a modified kind of fission characteristic of yeasts (Figure 14-1). One of the cells resulting from division is considerably smaller than the other and may be regarded as the offspring cell, so in budding the parent cell does not lose its identity as it does in binary fission. The small cells later attain full size, but before they become detached from the parent cell they may in turn bud and results in a temporary filament composed of cells of decreasing size.

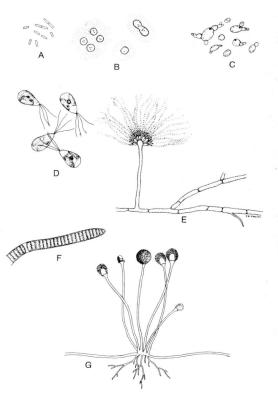

Figure 14-1. Examples of different types of asexual reproduction in lower plants. **(A)** Reproduction of bacteria by fission. **(B)** Reproduction of Gleocapsa, a unicellular blue-green alga, by fission. **(C)** Reproduction of yeast by budding. **(D)** Zoospores of Ulothrix, a green alga. **(E)** Reproduction of Aspergillus, a mold, by conidia. **(F)** Reproduction of Oscillatoria, a filamentous blue-green alga, by fragmentation at the point where a heterocyst (clear area) has formed. **(G)** Reproduction of bread mold (Rhizopus) by spores produced on aerial sporangia. [After V. A. Greulach and J. E. Adams, *Plants: An Introduction to Modern Botany,* John Wiley & Sons, Inc., New York, 1967]

Sporulation

Asexual reproduction by means of spores is quite common among the algae and fungi, but true asexual spores are not found in other plant groups. Spores are characteristically unicellular, but some species produce bicellular or multicellular spores. The spores may be produced within a **sporangium** as in the black bread mold (Rhizopus), within one or more cells of a filament or other thallus as in various species of green algae, or as chains of cells that separate from one another when mature as in the blue and green molds (Figure 14-1). The latter are called **conidiospores** or, more commonly, simply conidia. The spores of aquatic algae and fungi are in most species flagellated, and are designated as **zoospores.**

Unfortunately, the term **spore** has been used for a variety of cells other than the true asexual spores. All of the spores produced by bryophytes and vascular plants and many of those produced by algae and fungi are products of meiosis and are properly referred to as **meiospores.** Although meiospores have sometimes been referred to as a means of asexual reproduction, they are actually an essential component of the sexual life cycle of plants and will be discussed in connection with sexual reproduction. The **basidiospores** of mushrooms and bracket fungi and the **ascospores** of such fungi as the morels and Neurospora are meiospores, not true asexual spores. The **zygospores** of various fungi are really zygotes and not spores. The thick-walled resting spores produced by some bacteria, algae, and fungi are not reproductive cells at all but provide a means of survival during periods of unfavorable environment such as desiccation.

Vegetative Propagation

A few kinds of nonvascular plants produce specialized vegetative propagules. Among these are the **soredia** of lichens. A soredium is a small cluster of both fungal hyphae and algal cells that becomes detached and can develop into a new lichen. Soredia are an interesting means of reproducing these composite organisms but hardly surprising since they are probably the only kind of reproductive structure that could have evolved. Liverworts have evolved a unique type of reproductive propagules known as **gemmae** (Figure 14-2). These are born in small clusters in cups located on the upper surface of the gametophytes and are multicellular budlike structures. When mature the gemmae become detached and may grow into a new gametophyte plant.

However, the vegetative propagation of nonvascular plants is principally just a matter of fragmentation of a plant into two or more parts. Fragmentation is usually accidental, resulting from such things as animal activity, currents, waves, or the death of intermediate portions of a thallus or mycelium. However, some filamentous blue-green algae fragment at **heterocysts,** which are specialized thick-walled cells that die at maturity (Figure 14-1). In general, the filaments or other thalli of algae and the mycelia of fungi are quite unspecialized; thus, regeneration is usually not necessary for the formation of a complete plant from a fragment. Cultivated mushrooms are propagated vegetatively by man by fragmentation of their mycelia. The pieces of culture

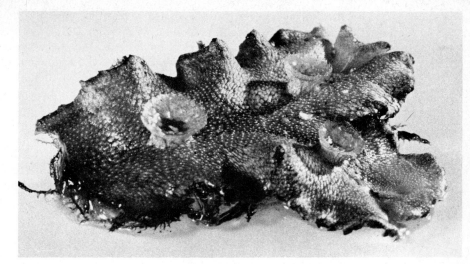

Figure 14-2. Marchantia, a liverwort, with three gemma cups. A number of gemmae, which are clusters of cells, develop in each gemma cup. The gemmae become detached and, after dispersal, each one may develop into a new plant. [Courtesy of Carolina Biological Supply Company]

medium containing mycelium are known as **spawn.** When liverworts or mosses become fragmented, some degree of regeneration may be required for the development of complete new plants.

Asexual Reproduction of Vascular Plants

The various means of asexual reproduction among vascular plants can all be considered as vegetative propagation, although specialized and normally produced propagules such as the seedlike axillary buds of some lilies and the plantlets borne on Kalanchoë leaves are natural asexual reproductive structures as distinctive as the asexual spores of algae and fungi. Vegetative propagation of vascular plants is rather limited in nature, but man has made extensive use of it in propagating cultivated plants [9]. A considerable number of economic species or varieties are sterile and can be propagated only vegetatively, as in the case of bananas, seedless grapes and citrus fruits, and hydrangeas. Other species such as Bermuda, St. Augustine, and Zoysia grasses are propagated vegetatively because of their limited seed production and the more rapid establishment of a lawn from pieces of turf.

Fruit trees are almost always propagated vegetatively even if they are not sterile because of their extreme heterozygosity and their long generation time. These factors make the breeding of essentially homozygous individuals with all the characteristics of the variety impractical if not impossible as well as unnecessary. Similarly, since plants propagated vegetatively are exact genetic copies of the parent, vegetative propagation is routinely used instead of propagation by seeds whenever it is economically feasible, even in annuals such as potatoes. Sexual reproduction is used only for the production of new varieties.

All the organs of vascular plants—stems, roots, leaves, and even flowers—may be involved in vegetative propagation, and the following discussion of types of vegetative propagation will be based on the organ involved.

Propagation by Stems and Buds

A number of different vegetative propagules consist of specialized stems with buds or simply buds alone. Among these are the **bulbs** of various members of the lily family, including onions, tulips, and lilies. Bulbs are usually underground, but onions may form aerial bulbs in place of the usual flowers of the inflorescence (Figure 14-3). Bulbs commonly develop branch bulbs that develop into new individuals after detachment. Some species of lilies have compact, spherical axillary buds that superficially resemble seeds. These abscise readily and develop into a new individual if they fall in a suitable habitat. Like bulbs, the **corms** of species such as gladiolus and crocus branch and serve as vegetative propagules. The numerous **tubers** produced by some plants such as Irish potato and Jerusalem artichoke provide a means of vegetative propagation because each tuber can give rise to offspring following the death of the parent plant.

Less highly specialized stems may also provide natural means of vegetative reproduction, even though they hardly qualify for designation as propagules. The brittle stems of various species such as willow trees, elodea and some other water plants, and cacti may break off from the parent plant (fragment) and develop adventitious roots, thus giving rise to new individuals. Stolons and rhizomes may only contribute toward the lateral growth and spread of a plant, but when they become detached from the plant by accident or the death of an older portion they provide a means of vegetative propagation. Dewberries, raspberries, currants, and other species with arching stems frequently develop upright shoots and adventitious roots from nodes near the stem tip that are

Figure 14-3. An onion inflorescence in which aerial bulbs developed in place of some of the flowers. [Courtesy of J. Arthur Herrick]

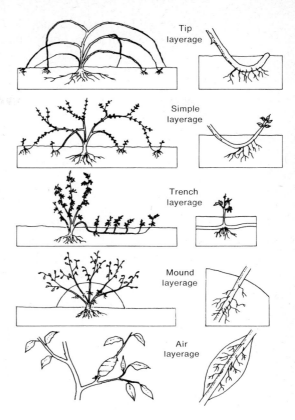

Figure 14-4. Several types of layering used in the propagation of plants. [After E. L. Denisen, *Principles of Horticulture,* The Macmillan Company, New York, 1958]

in contact with the soil. If these become detached from the parent plant vegetative propagation results.

Man utilizes all these natural means of vegetative reproduction for propagation of economic plants and, in addition, uses cuttings, layering, or grafting for the propagation of many species that rarely if ever reproduce vegetatively in nature. **Stem cuttings** provide the usual means of propagating many shrubs, vines, greenhouse plants, and a few trees. The use of cuttings has been greatly facilitated by treatment of the cut stem ends with auxin, since this promotes the development of the adventitious roots. Although some plants such as coleus and willows root readily without applied auxins, the treatment promotes rooting in most species, and many species will not root at all without applied auxin. Only a few species fail to root even with treatment.

Some species that cannot readily be propagated by cuttings are propagated by **layering,** which differs from cutting in that the development of adventitious roots is induced before instead of after the stem is severed (Figure 14-4). Mound layering is like natural tip layering and involves placing stem ends under small mounds of soil until they root. This procedure may be used for species that do not naturally propagate by tip layering as well as those that do. Air layering (Figure 14-4) involves partial cutting and slitting of a stem, treatment of the cut surfaces with auxin, and covering the cut portion with moist sphagnum moss and polyethylene.

Grafting, like layering, may be regarded as a modified type of cutting. In grafting the stem cutting of the plant being propagated (the **scion**) is attached (grafted) to a young plant that has had all its branches removed, leaving only the main stem and roots (the **stock**). Grafting is used for those species whose cuttings do not root readily, give rise to weak root systems, or to roots subject to diseases. Grafting is the standard means of propagating most cultivated fruit trees, roses, and some shrubs like lilac. Seeds of different varieties, or even different but related species, than the scions are usually planted to provide the stocks; these often are wild relatives. It is essential that the graft union should heal properly and that the rate of growth of the cambiums of the stock and scion be essentially the same. Some herbaceous plants, as well as woody ones, may be grafted successfully, but this is done for experimental purposes (for example reciprocal grafts of tobacco and tomato plants, Chapter 5) rather than for commercial propagation. **Budding** is a modified form of grafting in which only a single bud with adjacent bark is used as the scion, the bud being inserted under the bark of the stock through a cut. Budding is the standard means of propagating some trees such as pecan.

Propagation by Leaves

Although leaves are not involved in vegetative propagation as much as stems, a number of species have interesting means of leaf propagation. A number of species of Kalanchoë (Bryophyllum) produce well-formed plantlets in the notches of their leaf margins (Figure 14-5). These abscise and fall to the ground where they may take root and develop into mature plants. In some species plantlets develop only after the fleshy leaves have been detached from the plant, but in others they regularly form on attached leaves. Duckweeds, which are small, floating angiosperms, reproduce principally by the formation of bulblets on their leaves in the autumn. These abscise and sink to the bottom, rising to the surface the next spring and developing into plants. Duckweeds rarely bloom and produce seeds. The walking fern has long, pointed leaves that develop adventitious buds and roots near their tips when the leaves are in contact with the ground, and these give rise to new plants.

The leaves of a number of species develop adventitious buds and roots when detached from the plant. African violets are commonly propagated in this way. Some species of begonia are propagated by removing leaves, cutting them in pieces, and placing them on moist soil or sand in high humidity. Each piece

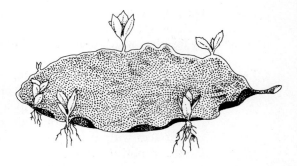

Figure 14-5. Plantlets that have developed from shoot and root primordia located in the notches of the leaf margin of a leaf of *Bryophillum calycinum* (*Kalanchoë pinnata*). [After J. P. Mahlstede and E. S. Haber, *Plant Propagation,* John Wiley & Sons, Inc., New York, 1957]

gives rise to adventitious buds and roots and thus to a new individual. Many species can develop roots but not buds on detached leaves, and even more species are incapable of developing either roots or buds so their leaves cannot be used for propagation. Leaves that can develop adventitious roots and buds, unlike those of most species, do not have a large decrease in proteins after detachment. Apparently they have a higher level of cytokinins and perhaps other phytohormones. Leaf cuttings and detached leaves rarely provide a means of propagation in nature, although they are widely used by man for plant propagation.

Propagation by Roots

Roots can provide a means of vegetative propagation if they have the capacity of developing adventitious buds. This capacity is much much more marked in some species than others, and bud formation is generally much more extensive in, or even restricted to, roots that have been detached from the shoot or from the rest of the root system. When placed in water the fleshy roots of sweet potatoes form numerous branch roots at the lower end and numerous shoots at the upper end. The latter may be separated, and each one will then develop into a plant. Although dahlia and peony roots are commonly used for propagation, they must have a piece of stem attached to them, since the roots themselves cannot develop adventitious buds.

Vegetative Propagation by Flowers

Despite the fact that flowers are the site of the sexual reproductive organs of angiosperms, it is interesting to note that they may also be involved in the vegetative propagation of plants. In a few cases, such as the aerial bulbs of onions already mentioned, flower primordia develop into vegetative propagules instead of flowers. More commonly, asexually produced embryos develop within the ovules and are present in seeds, a condition known as **apomixis.** In some cases one or more apomictic embryos occur in a seed along with the usual sexually produced embryo, whereas in other cases normal embryos never develop.

One rather common type of apomixis in plants is **parthenogenesis,** the development of an unfertilized egg into an embryo. The parthenogenetic embryo may be haploid but more commonly it is diploid or even polyploid because meiosis has failed to occur. Nitsch and Nitsch [15] have suggested, on the basis of the development of tobacco pollen grains into embryos in their tissue culture experiments (Chapter 17), that embryos may even develop naturally from pollen nuclei after their introduction into the ovule, thus possibly explaining cases where plants had only the genetic traits of the pollen parent. Both of these are actually cases of asexual reproduction, even though they involve alteration of the sexual reproductive processes.

Apomictic embryos may also develop from cells of the embryo sac other than the egg or from cells of the nucellus or even the inner integument. Among the plants in which apomixis is common are dandelions, blackberries, and hawthorns. An interesting example of apomixis was found in a lily where apomictic embryos were formed generation after generation (as indicated by lack of any genetic traits from a pollen parent) but where pollination and fertilization were essential for formation and development of the endosperm.

Sexual Reproduction of Plants

Although asexual reproduction, including the various means of vegetative propagation, is widespread throughout the plant kingdom, the vast majority of plants reproduce sexually, either as their only means of reproduction or in combination with one or more means of asexual reproduction. Of the major plant groups, only the blue-green algae and bacteria lack sexual reproduction, and even they may have rudimentary means of gene transfer that might be considered sexual in a way [19]. Although a considerable number of angiosperms are seedless or otherwise sterile and can reproduce only by some kind of vegetative propagation, they continue in most cases to develop flowers and often fruits. In some other species or varieties seed production is scanty and vegetative propagation provides the principal means of reproduction. However, these situations are more common among cultivated than wild plants, and in general the principal means of plant reproduction is sexual.

In plants, as in animals, the essential features of sexual reproduction are the formation of **gametes** that fuse, and thus form a zygote, and the occurrence of meiosis at some stage in the life cycle. The normally haploid (*n*) gametes give rise to a diploid (*2n*) **zygote.** Meiosis results in production of haploid from diploid cells, thus preventing a doubling of the number of chromosomes each generation. Although plants and animals both share the basic essentials of sexual reproduction, there are many marked differences as regards such things as structure and functioning of the reproductive organs, the means of bringing gametes together, and the time at which meiosis occurs in relation to the time of gamete fusion in the life cycle. Indeed, there are marked differences in these and other respects among plants of one major division and another, even though there are a number of similarities common to plants in general.

In the following pages the life cycles of a number of representative plants will be outlined in an effort to elucidate the general nature of sexual reproduction among plants. Since the scope of this book does not permit consideration of representatives of all the major groups of plants, the life cycles will be restricted to the green algae, the bryophytes, and the vascular plants. It should be made clear, however, that the other groups of algae and fungi also reproduce sexually, and in many cases their means of sexual reproduction are quite unique, interesting, and complex. Discussions of their life cycles may be found elsewhere [3]. The green algae were selected for inclusion because they are generally considered to be on the main phylogenetic line leading to the higher plants.

Selected Plant Life Cycles

In this section the sexual life cycles of eight different plants will be outlined. They include three genera of green algae (Ulothrix, Ulva, and Oedogonium), a moss, and four vascular plants (a fern, a clubmoss, a conifer, and an angiosperm). Each successive life cycle includes one or more features that may be regarded as advances from a primitive type of sexual reproduction (as exemplified by Ulothrix) to the most complex type found in angiosperm plants.

Ulothrix

The various species of Ulothrix are fresh-water algae commonly found in streams. They are filamentous, the unbranched filaments being composed of varying numbers of cells. The basal cell is modified as a holdfast that attaches the filament to a rock or other substrate. All the other cells contain a single large chloroplast shaped like a snap-on bracelet. However, any of these cells may undergo internal divisions that usually result in the formation of four or eight flagellated **zoospores.** A break in the wall of the original parent cell enables the zoospores to swim out of the filament, and each zoospore may develop into a new filament, thus providing a means of asexual reproduction. Each cell of the filament as well as each zoospore has the haploid number of chromosomes.

Under suitable conditions a Ulothrix cell may divide into what appears to be more (often 32 or 64) and smaller zoospores, but these are actually gametes (Figure 14-6). Gametes differ from zoospores in having only two flagella each instead of four and in being unable to develop into a new Ulothrix plant until after fusion with another gamete. All the gametes of Ulothrix are structurally the same and are referred to as **isogametes.** However, there are apparently some physiological differences among the isogametes because fusion commonly occurs between gametes from different plants. The zygote resulting from the fusion of the gametes is not flagellated. It develops a hard, thick wall that makes

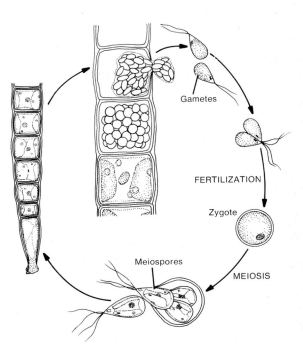

Figure 14-6. Life cycle of Ulothrix, a green alga. Only the zygote has the diploid (2n) chromosome number; all other stages are haploid (n).

it resistant to unfavorable environmental factors, such as desiccation, and may remain dormant for several months before it germinates.

When the zygote germinates it undergoes meiosis, resulting in the production of four tetraflagellate meiospores. Although the meiospores look like zoospores, they are an essential stage in the sexual life cycle rather than a means of asexual reproduction like zoospores. After a free-swimming period, each meiospore normally develops into a filamentous plant composed of haploid cells. Occasionally, however, a meiospore divides by mitosis into two daughter meiospores which then develop into filaments. The filamentous plants later reproduce sexually by producing gametes and asexually by producing zoospores.

Since in Ulothrix, and a considerable number of other algae and fungi, the zygote undergoes meiosis it is the only diploid stage in the life cycle. The meiospores, filaments, gametes, and zoospores are all haploid. This contrasts strongly with the usual situation in the animal kingdom (Figure 14-7) where the zygote develops into a diploid individual and meiosis does not occur until gametogenesis begins. Thus, all stages in the life cycle of an animal except for the gametes are diploid. The difference between animals and Ulothrix and many other algae is in the time of meiosis in relation to the time of fusion. In Ulothrix meiosis occurs just after fusion whereas in animals it occurs just before fusion.

Ulva

Ulva is a marine green alga that is found in intertidal zones. It has a puckered leaflike thallus that is two cells thick and may be up to 30 cm long and almost as wide. Its bright green color and leaflike thallus have given rise to its common name—sea lettuce.

Instead of undergoing reduction division the zygote of Ulva germinates, first into a filament (Figure 14-8) that then develops into the characteristic leaflike thallus. The cells of this thallus are diploid but, within some of the marginal cells, meiosis occurs, resulting in the formation of four meiospores per cell. The meiospores are discharged into the water where they may develop into new leafy thalli by the way of a filament stage. These plants look like the plant derived from the zygote, but they differ in that their cells are haploid and the marginal reproductive cells produce isogametes rather than meiospores. Although the gametes are all structurally the same, they have a physiological sex differentiation so that only + and − gametes can fuse. Furthermore, the sexual plants that produce the gametes as well as the meiospores from which the plants developed are differentiated into the + and − strains. Fusion of a + and a − gamete results in a diploid zygote, which can then develop into a diploid plant of the next generation.

The principal difference between the life cycles of Ulothrix and Ulva is that in Ulva both the zygotes and the meiospores develop into multicellular individuals. The result is a life cycle with alternating diploid and haploid generations, in contrast with the haploid individuals of Ulothrix and with the diploid individuals of animals. The diploid zygote of Ulva gives rise to a diploid plant (**sporophyte**) that produces meiospores by meiosis. These haploid meiospores

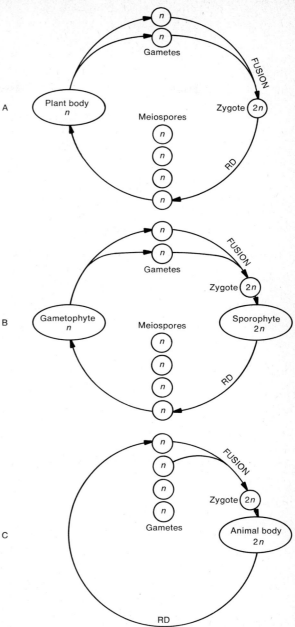

Figure 14-7. Generalized and simplified life cycles of **(A)** algae such as Ulothrix; **(B)** vascular plants, bryophytes, and many lower plants with alternation of gametophyte and sporophyte generations; and **(C)** most animals.

develop into haploid plants **(gametophytes)** that produce gametes, which may then fuse and form a diploid zygote, thus completing a life cycle. This **alternation of generations** (Figure 14-7) is characteristic of many algae and fungi other than Ulva and also of all higher plants, including both bryophytes and vascular plants.

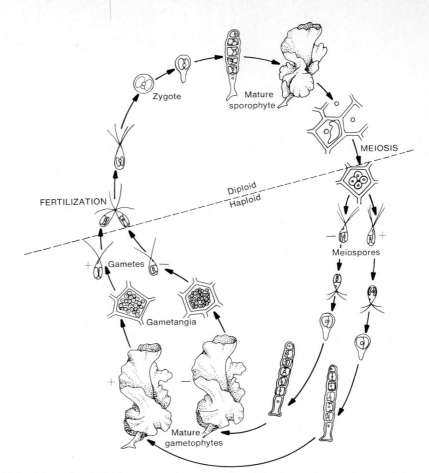

Figure 14-8. Life cycle of Ulva (sea lettuce), a green alga with gametophyte and sporophyte generations.

Oedogonium

Oedogonium is a filamentous, fresh-water green alga. Its sexual life cycle differs from that of Ulothrix and Ulva in that it produces **heterogametes,** that is, gametes clearly differentiated into **eggs** and **sperm** (Figure 14-9). The large, nonmotile egg is borne within an enlarged cell known as the **oogonium.** The small, motile sperm are borne in a series of short cells that are called **antheridia;** two sperm usually are produced in each antheridium. At maturity the sperm swim from the antheridia but the eggs remain in their oogonia and are fertilized there by the sperm, which swim through a pore in the oogonium wall. Once the zygote is formed it, like the zygotes of Ulothrix, undergoes reduction division when germinating, and the meiospores develop into haploid filaments. Thus, neither Oedogonium or Ulothrix has a real alternation of generations.

The advance in sexual reproduction exhibited by Oedogonium is, of course, heterogamy. The fact that the vast majority of both plants and animals are heterogamous rather than isogamous is probably not just an accident but is probably a matter of survival value. The large egg cells contain an abundance of food, but since they are nonmotile they do not waste it in locomotion. The small sperm provide the motility that brings the gametes together without the expenditure of the substantial energy needed to move a larger cell.

Some species of Oedogonium are **homothallic,** that is, one filament produces both eggs and sperm, whereas other species are **heterothallic** and have male plants that produce sperm and female plants that produce eggs. Thus, the three algae we have considered illustrate varying degrees of sexual differentiation: isogamy and heterogamy, homothallism and heterothallism, and haploid plants or alternation of diploid and haploid generations. In the bryophytes and tracheophytes heterogamy and alternation of generations are universal, but homothallism as well as heterothallism is found among the bryophytes and lower vascular plants.

A Moss: Polytrichum

The life cycles of mosses and liverworts involve a typical alternation of generations that can be represented by the same sort of outline as was used for Ulva. The diploid sporophyte plants produce haploid meiospores by reduction division, and the meiospores develop into gametophyte plants that produce gametes, which may then fuse and form diploid zygotes that develop into sporophyte plants (Figure 14-10). However, there are a number of important differences. In contrast with Ulva and a variety of other algae, the sporophyte and gametophyte plants are not identical or even similar in their external form.

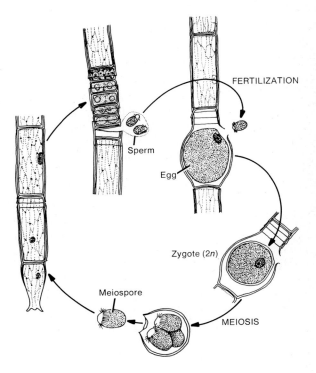

Figure 14-9. Life cycle of Oedogonium, a heterogamous green alga. As in Ulothrix, only the zygote is diploid.

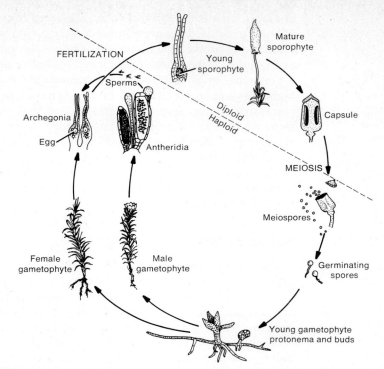

Figure 14-10. Life cycle of a moss. [After E. W. Sinnott and K. S. Wilson, *Botany: Principles and Problems,* 6th ed., McGraw-Hill Book Co., New York, 1963]

The gametophytes of mosses have upright, leafy stalks that are generally closely packed together and filamentous rhizoids that penetrate the substrate and perform the functions of roots. It should be noted, however, that these structures are not true roots, stems, or leaves and are not homologous with those structures of vascular plants. Most of the photosynthesis of mosses is carried on by the gametophytes. The eggs and sperm are produced within multicellular gametangia, the **archegonia** and **antheridia** respectively, rather than in single cells. Each antheridium produces numerous flagellated sperm, but each flask-shaped archegonium* produces only a single egg. Despite the fact that mosses and liverworts are land plants, their sperm (like those of algae) must swim to the eggs through water, so fertilization of the eggs is dependent on the presence of a surface film of water through which the sperm can swim. Some species of mosses develop both antheridia and archegonia at the tip of each stalk, but more commonly they are borne on separate stalks or even separate plants.

The moss zygotes develop into hairlike sporophyte plants while still inside the archegonia, and the base (foot) of the sporophyte becomes embedded in the tissue of the gametophyte. The sporophyte secures its water, mineral elements, and most of its food from the gametophyte. The upper end of the sporophyte stalk **(seta)** terminates in a multicellular **sporangium,** often called the **capsule.** The cells in the central region of the sporangium differentiate into

---

* Archegonia are multicellular, in contrast with the unicellular oogonia.

numerous separated cells called **spore mother cells.** Each one of these undergoes reductional division, forming four meiospores. Unlike the meiospores of algae, but like the meiospores of vascular plants, the moss meiospores are not flagellated. At maturity the sporangium opens by dehiscence of a lid **(operculum)** and the meiospores come out and are dispersed by air currents. When the meiospores germinate they first develop into a branched filament **(protonema)** that later forms budlike structures that develop into the upright leafy stalks. The protonemata, like the Ulva filaments that later develop into the broad mature thalluses, are commonly considered to be examples of ontogeny recapitulating phylogeny and to provide clues to the evolutionary ancestory of the species.

The gametophytes of liverworts, like those of mosses, are larger and more complex than the sporophytes and are the principal photosynthetic plants, but both the gametophytes and sporophytes of liverworts differ structurally from those of mosses to varying degrees [3].

A Fern: Polypodium

The life cycle of ferns [2, 5, 18] can be outlined by the same sort of general diagram used for mosses and even algae like Ulva, but ferns differ from bryophytes in a number of important details characteristic of vascular plants in general and considered to represent phylogenetic advances (Figure 14-11). The sporophytes, rather than the gametophytes, are the larger and more complex plants. The sporophytes, like those of vascular plants in general, have true roots, stems, and leaves, although the stems of most species of ferns (other than the tree ferns) are underground rhizomes. The sporangia are borne on leaves, appropriately designated as **sporophylls.** In Polypodium and many other fern genera the sporangia are borne on the undersides of ordinary photo-

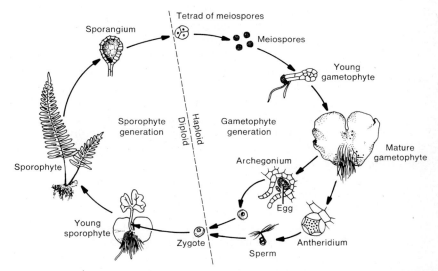

Figure 14-11. Life cycle of a fern. [Modified from E. W. Sinnott and K. S. Wilson, *Botany: Principles and Problems,* 6th ed., McGraw-Hill Book Co., New York, 1963]

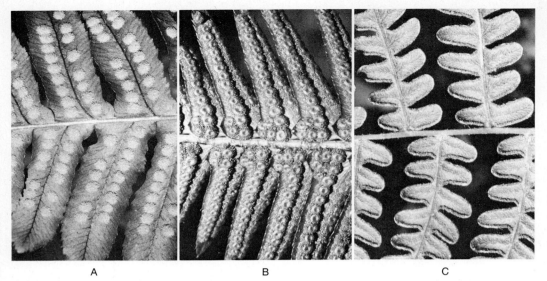

A   B   C

Figure 14-12. Three types of sori on the under side of foliage leaves of ferns. **(A)** Sori without indusia covering the sporangia (Polypodium). **(B)** Sori with umbrellalike indusia (*Polystichum munitum*). **(C)** Continuous marginal sori (*Pteridium aquilinum*). [Courtesy of the Carolina Biological Supply Company]

synthetic foliage leaves, usually in clusters called **sori** (Figure 14-12). However, in some species of ferns the sporophylls are distinct from the foliage leaves, are branches of the foliage leaves, or are just the terminal portions of foliage leaves [2, 3]. All but the most primitive vascular plants bear their sporangia on sporophylls, and the higher vascular plants all have sporophylls that are highly modified and quite unlike the foliage leaves.

As in mosses, spore mother cells are produced within the sporangia. Each spore mother cell undergoes meiosis and produces four meiospores, the sporangium dehisces at maturity, and the meiospores are dispersed by air currents. In Polypodium and related genera the sporangium suddenly snaps open, shooting the meiospores out. This results from the drying out of the cells of the **annulus,** which have unusually thick walls except on the outer surface (Figure 14-13). In suitable habitats the meiospores germinate and develop into independent gametophytes (often called **prothallia**).

In Polypodium and its relatives the gametophytes are small (less than 1 cm in diameter), heart-shaped, green plants that are perdominantly only one cell thick. Rhizoids are developed on the under surface as are the antheridia and archegonia, the latter being nearer the notch of the prothallus. The antheridia are globose whereas the archegonia (Figure 14-14) are flask-shaped like those of mosses, except that the neck of the archegonium is usually bent. As in mosses, a film of water is essential if the sperm are to swim to the eggs and fertilize them. The zygote begins developing in the archegonium and the young sporophyte is attached to the gametophyte (Figure 14-15).

In other orders of ferns the gametophytes may be larger (up to 5 cm long), fleshy, variously shaped (cylindrical, stellate, branched), and even subterranean, but in all ferns the gametophytes are small and inconspicuous compared with the large, highly developed sporophytes.

A few species of ferns have a feature found universally among the higher vascular plants but not in the bryophytes or green algae: **heterospory.** Heterosporous sporophytes produce two different kinds of meiospores, the **megaspores** and the **microspores,** produced in **megasporangia** and **microsporangia** respectively. As the names imply, megaspores are often, but not always, larger than the microspores. However, the important difference is that megaspores develop only into female gametophytes whereas microspores develop only into male gametophytes. Thus, the gametophytes of heterosporous species are always heterothallic (and, of course, heterogamous). In effect, heterospory extends sexual differentiation from the gametophytes into the sporophytes, even though the sporophyte generation is sometimes considered to reproduce itself asexually. It should be evident by now, however, that sporophytes are just as essential

Figure 14-13. Photomicrograph of a fern sporangium containing microspores. The sporangium has not opened, but the lip cells at which dehiscence occurs are visible at the right, somewhat below center of the sporangium. [Courtesy of the Carolina Biological Supply Company]

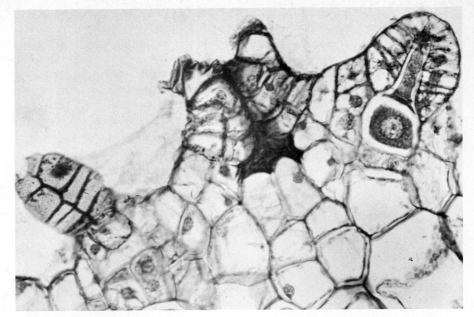

Figure 14-14. Photomicrograph of a section through a fern gametophyte showing (upper right) an immature unopened archegonium with the egg. The neck canal cells and ventral canal cell have not yet disintegrated. [Courtesy of the Carolina Biological Supply Company]

a part of the sexual life cycle of plants as are gametophytes because they are the site of meiosis.

Ferns and mosses have been favorite subjects for morphogenetic investigations for such reasons as the ease of culturing numerous gametophytes in a limited space, the ready accessibility of the reproductive organs (sporangia, antheridia, and archegonia) and of the two unicellular stages in the life cycle

Figure 14-15. A young fern sporophyte still attached to the parent gametophyte. Note the cotyledon, foot (region of attachment), and radicle of the sporophyte. [Courtesy of the Carolina Biological Supply Company]

(meiospores and zygotes), and the marked morphological differences between the gametophytes and sporophytes [4, 11]. Although the extensive research has provided considerable interesting and valuable morphogenetic information, such problems as why meiospores develop into gametophytes and zygotes into sporophytes are still not well resolved. However, it has been evident for some time now that the number of sets of chromosomes is not controlling as was once thought, since it has been found that zygotes develop into sporophytes whether diploid or tetraploid whereas diploid meiospores still develop into gametophytes.

Like plants in other major groups, ferns do not always follow the standard alternation of generations life cycle. Farrar [8] has found four species of ferns (in the vicinity of Highlands, N. C.) that rarely have sporophytes; the gametophytes reproduce themselves asexually by means of gemmae. Earlier Wagner and Sharp [17] had found a greatly reduced variety of the shoestring fern in the southern Appalachians that lacked the sporophyte generation and had gametophytes that reproduced asexually by gemmae. Evans [7] found a fern (*Polypodium dispersum*) that has both gametophyte and sporophyte generations but lacks both meiosis and fertilization; thus, it has an asexual life cycle. The gametophytes lack antheridia and archegonia, but produce outgrowths that develop into sporophytes. The sporophytes produce sporangia and spores, but the spore mother cells fail to undergo meiosis. Thus, both the sporophytes and gametophytes are diploid. In still other ferns there are also diploid sporophytes and gametophytes without sexual reproduction, but the spore mother cells undergo meiosis. This is possible because of the fact that, just prior to the formation of the spore mother cells, there is a doubling of the chromosomes **(syndiploidy).** This transient 4n chromosome number is then reduced back to the 2n number by meiosis.

A Clubmoss: Selaginella

Unlike the ferns, the clubmosses are not on the evolutionary line leading to the seed plants; however, the life cycle of Selaginella is included here because its reproductive organs have several features that are not present in the ferns but have evolved to an even more marked degree in the seed plants.

As in all vascular plants, the clubmoss sporophyte (Figure 14-16) is the conspicuous generation. The leaves of clubmosses are small and numerous and the internodes are very short, so the leaves essentially cover the stem and in some species overlap one another. The leaves have a single, unbranched midrib vein and, unlike the ferns and seed plants, there is no leaf gap above the point where it leaves the vascular tissue of the stem. In Selaginella (but not all clubmosses) the sporophylls are distinct from the foliage leaves and are borne in terminal cones **(strobili)** that are homologous with the cones of gymnosperms. However, unlike the cones of gymnosperms and the flowers of angiosperms, the apical meristem generally continues its growth, resulting in growth in length of the cone or renewed vegetative growth with foliage leaves. The sporophylls are similar in size and shape to the foliage leaves, but they may lose their chlorophyll while the foliage leaves are still green.

Selaginella is heterosporous and the megaspores and microspores are borne

Figure 14-16. Portions of the sporophytes of four species of Lycopodium, a genus of clubmosses. **Left to right:** *L. obscurum, L. lucidulum, L. alopecuroides,* and *L. flabelliforme.* All have their sporophylls in terminal strobili (cones) except *L. lucidulum,* in which some of the foliage leaves along the stem are also sporophylls. [Courtesy of the Carolina Biological Supply Company]

in separate sporophylls: **megasporophylls** and **microsporophylls** (Figure 14-17). In most species both kinds of sporophylls are borne in a single cone, the megasporophylls being below the microsporophylls. A single sporangium is formed near the base of each sporophyll or sometimes in the axil of the leaf.

Each microsporangium contains numerous microspore mother cells. Some of these may degenerate, but most of them undergo reduction division, giving rise to a tetrad of haploid microspores. The microspores develop into greatly reduced male gametophytes. This development generally begins before the microspores emerge from the microsporangium, but it may continue afterward. Emergence is through a slit in the apex of the microsporangium, and the dehiscence may be sudden and violent enough to eject the microspores (or male gametophytes) explosively. The male gametophyte consists of only a few cells, generally only nine (a single vegetative cell and an eight-celled antheridium), and remains within the wall of the microspore. It is, thus, a microscopic structure and also it lacks chlorophyll. Numerous flagellated sperm are produced in the antheridium. The male gametophyte could well be called a pollen grain since it is homologous with the pollen grains of seed plants, which are also male gametophytes of microscopic size. However, as we shall see later, the male gametophytes of seed plants are even more highly reduced.

The megasporangium contains a number of potential megaspore mother cells, but usually all but one of these degenerate. The surviving one undergoes reduction division, producing four megaspores, some of which may not survive. The megaspores are considerably larger than the microspores—on the order of 0.2 to 0.8 mm in diameter in various species. At maturity the megasporangium dehisces and the megaspores generally fall to the ground, but they may remain within the megasporangium until after they have developed into mature female gametophytes. A female gametophyte is composed of several hundred cells, but like the male gametophyte it develops within the spore wall and it is essentially the same size as the megaspore from which it developed. Each female gametophyte has several small but recognizable archegonia, each containing a single egg. The upper side of the megaspore wall eventually ruptures, exposing the archegonia. Some of the exposed vegetative cells may give rise to rhizoids, or they may develop chlorophyll and carry on photosynthesis. However, most of the food used by the female gametophyte is that which was in the megaspore.

If a male gametophyte falls on or near a female gametophyte, the sperms swim around and at least one usually enters each archegonium and reaches an egg. Although all the eggs are usually fertilized, the first one fertilized generally inhibits development of the other zygotes; usually only one sporophyte embryo develops. The clubmoss embryo is strikingly similar to a dicotyledon embryo, with two cotyledons that are folded over the apical meristem and a primary root. However, the clubmoss embryo also has a small foot, a structure present in ferns but not seed plants.

The female gametophytes with the enclosed embryo sporophytes is almost, but not quite, a seed. Gymnosperm seeds also contain embryos surrounded

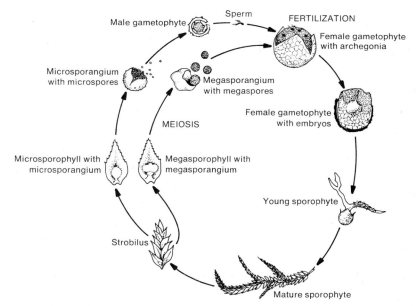

Figure 14-17. Life cycle of Selaginella, a clubmoss. [After E. W. Sinnott and K. S. Wilson, *Botany: Principles and Problems,* 6th ed., McGraw-Hill Book Company, New York, 1963]

by female gametophyte tissue but have one other important component—the seed coat that developed from the megasporangium. If the fertilized clubmoss gametophytes remained within the megasporangium and the megasporangium were capable of developing into seed coats, then clubmosses would also have seeds. However, clubmoss embryos lack the protection of a seed coat and must continue their development into a mature sporophyte without interruption if they are to survive.

A Conifer: Pine

The sporophytes of conifers are the conspicuous and widely known trees and shrubs that make up this group of plants. At least as far as size and age are concerned, the sporophyte reaches its maximum development in conifers such as redwood, sequoia, Douglas fir, and the bristlecone pine. In pines and other conifers the sporophylls are distinctly different in size and shape from the foliage leaves, lack chlorophyll at least when mature, and are borne in cones with determinate apical growth. The female cones (Figure 14-18) are made up of megasporophylls and the smaller and more ephemeral male cones (Figure 14-19) are composed of microsporophylls. Both kinds of cones generally develop in the spring.

DEVELOPMENT OF MEGASPORES AND FEMALE GAMETOPHYTES. Two megaspo-woody microsporophylls bears on its under side two microsporangia that occupy most of the surface of the microsporophyll (Figure 14-20). The numerous spore mother cells in each microsporangium undergo reduction division, each one thus producing four haploid microspores. The microspores begin developing into male gametophytes while still in the microsporangium (Figure 14-21). As in the clubmosses, the male gametophytes develop within the microspore wall. The two membraneous wings of a microspore remain after it has developed

Figure 14-18. Ovulate (female) cone of a pine at the time of pollination. After pollination such cones enlarge greatly and develop into the familiar pine cones with large, woody scales on which the seeds are borne. [Courtesy of the Carolina Biological Supply Company]

Figure 14-19. Staminate (male) cones of a pine near maturity. After the pollen is shed these cones abscise. [Courtesy of the Carolina Biological Supply Company]

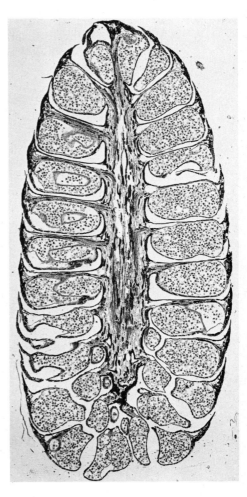

Figure 14-20. Photomicrograph of a longitudinal section through a staminate pine cone. Note the numerous pollen grains in the microsporangia, which are borne on the lower side of each microsporophyll. [Courtesy of the Carolina Biological Supply Company]

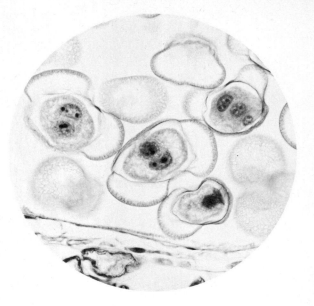

Figure 14-21. Photomicrograph of pine pollen grains in several stages of development. Note the winged extensions from the walls, which facilitate dispersal of the pollen by the wind. [Courtesy of the Carolina Biological Supply Company]

into a male gametophyte (pollen grain) and make more effective the wide dispersal of the pollen by wind. The first mitotic division gives rise to an antheridial cell and a prothallial cell, and the latter then divides again. These two prothallial cells then disintegrate, and the antheridial cell divides into a tube cell and a generative cell. The tube cell does not divide further, but the generative cell divides into a stalk cell and a body cell (generally after dispersal of the pollen), and the body cell then divides, forming two sperm. Thus, although there are five cell divisions during the development of the male gametophyte, it has only four cells at maturity or, perhaps more accurately, four nuclei since there are no cell walls between the nuclei. The sperm are not flagellated. It is evident that the male gametophytes of pines are even more reduced than those of clubmosses, and there is no trace of an antheridium.

DEVELOPMENT OF MEGASPORES AND FEMALE GAMETOPHYTES. Two megasporangia develop on the upper surface of each megasporophyll. Since the megasporangia of seed plants are also called ovules, the part of the megasporophyll bearing the megasporangia is often referred to as the **ovuliferous scale.** Each megasporangium produces only one large, central megaspore mother cell (Figure 14-22), which is surrounded by a tissue several cell layers thick (the **tapetum**), the one-cell thick megasporangium wall (the **nucellus**), and in turn by the integument that constitutes the outer tissue of the ovule. The integument is adjacent to the nucellus except at the end opposite the stalk where they are separated, forming a cavity known as the **pollen chamber.** A small pore in the integument (the **micropyle**) provides an outside opening through which pollen can enter. The megaspore mother cell undergoes reduction division, producing four haploid megaspores which are arranged in a linear (rather than spherical) tetrad. However, three of the four megaspores disintegrate, leaving only one to develop into a female gametophyte. It is interesting to note that this also occurs in angiosperms and in the meiosis of females of higher animals. (In animals the process results in the production of one egg and three nonfunctional polar bodies.) In pines the meiosis occurs about 1 month after pollination. Following pollination the ovuliferous scales thicken at their tips and become appressed against one another, closing the cone.

About 5 months after it was produced, the functional megaspore begins developing into the female gametophyte. For about 6 months this development consists of a series of slow mitotic divisions without any cell wall formation, resulting in a coenocytic female gametophyte with a thousand or so nuclei. There is a large central vacuole, the cytoplasm and nuclei being appressed against the megaspore wall. During this time there is a substantial increase in the size of the female gametophyte, the ovule (megasporangium), and indeed of the entire female cone. Accompanying the enlargement of the female gametophyte is a progressive disintegration of the cells of the tapetum and nucellus and eventually these tissues disappear entirely, leaving the female gametophyte surrounded only by the integument. About 13 months after pollination the central vacuole is invaded by the cytoplasm and the nuclei and cell walls are formed around each nucleus, resulting in a true multicellular female gametophyte. Two to four somewhat rudimentary archegonia differentiate at the micropylar end of the gametophyte, each containing a single egg, and the female gametophyte is now finally mature.

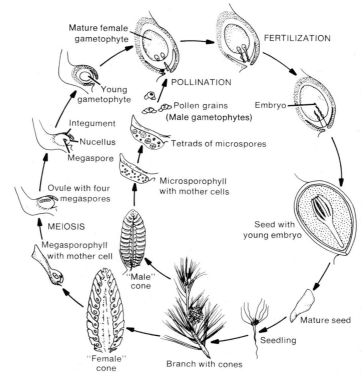

Figure 14-22. Life cycle of a pine. [After E. W. Sinnott and K. S. Wilson, *Botany: Principles and Problems,* 6th ed., McGraw-Hill Book Company, New York, 1963]

POLLINATION AND FERTILIZATION.   At the time of pollination (over 13 months earlier) the male gametophyte consisted of only a tube cell and a generative cell. At least some of the pollen grains (male gametophytes) that were carried to the female cones by wind happen to land near the micropyle. At that time the micropyle has a projecting droplet of mucilagenous fluid derived from some of the nucellar cells that disintegrated. Those pollen grains trapped in this fluid are actually drawn through the micropyle by the fluid as it becomes partially dehydrated. Once in the pollen chamber growth of the pollen tube begins promptly, and the tube slowly grows into the female gametophyte, the tube nucleus remaining close to the tip of the tube. However, the generative cell does not divide into the stalk and body cells until about a year after pollination. Shortly before the pollen tube completes penetration of the megasporangial tissue and reaches the female gametophyte, the body cell divides into the two sperm, which move toward the end of the tube. The tube wall ruptures and the sperm and other nuclei enter the archegonium, where one of the sperm fertilizes the egg and produces the diploid zygote that is the beginning of a new sporophyte generation.

Thus, because of the slow development and maturation of the gametophytes fertilization occurs over a year after pollination. This should make it clearly evident that pollination [14] and fertilization are two distinct phenomena, although they are unfortunately sometimes confused. Although the eggs in all the archegonia may be fertilized and the zygotes may begin developing into embryos, usually all but one abort.

DEVELOPMENT OF SEEDS.   As the zygote begins mitosis four nuclei are usually produced before any cell walls are formed. Then a sixteen-celled, four-tiered proembryo develops, the lower tier of four cells giving rise to one to four true embryos. The cells of the second tier elongate greatly, pushing the embryos into the gametophyte tissue, and then the other cells also elongate, pushing the embryos even deeper into the gametophyte. These elongated cells are called **suspensors** and eventually disintegrate. All four cells of the embryonic tier may begin to develop into embryos, although generally three of them abort. Thus, pine seeds generally contain only one embryo, although occasionally they are polyembryonic either because more than one zygote developed into an embryo or because more than one embryo from a single zygote survived.

The fully developed embryo consists of a hypocotyl, an epicotle with an apical meristem, and several cotyledons. It is completely surrounded by the haploid cells of the female gametophyte. These cells contain abundant accumulated food that can be used by the growing embryo when the seed germinates. The integument develops into a hard seed coat. Thus, the pine seed consists of tissues from three generations: the seed coat from the parent sporophyte, the female gametophyte, and the embryo sporophyte of the new generation. Maturation of the seed requires about 4 months after fertilization. When the seeds are mature the cone opens by reflexing of its scales, and the winged seeds are dispersed by autumn winds. The seeds generally germinate the following spring. Thus, it takes about 18 months from pollination to seed maturation and 2 full years to seed germination.

An Angiosperm

In describing the life cycle of an angiosperm (Figure 14-23) we shall depart from our previous practice of selecting a particular genus or species. Instead, we shall put together a generalized hypothetical angiosperm that illustrates the essential features of angiosperm life cycles as simply and typically as possible. There is such diversity in the structural details of angiosperm reproductive organs, including gametophytes, flowers, fruits, and seeds, that it is difficult to select any one species that illustrates all stages in the life cycle both simply and typically.

As in other vascular plants, the sporophyte generation is the large and conspicuous one and the gametophytes are even more reduced than in the gymnosperms. Angiosperm sporophytes have an amazing range and variety of sizes and structural patterns, from large trees and shrubs, grasses and herbs of numerous kinds, to small floating duckweeds. The sporophylls of angiosperms are all markedly different from the foliage leaves and are always borne

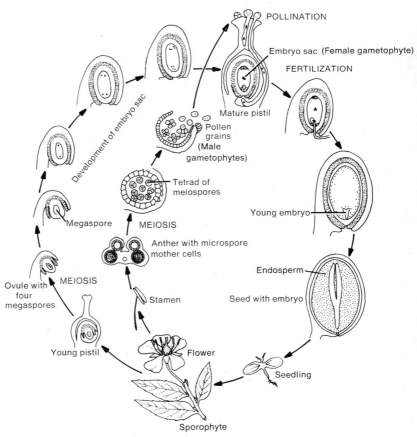

Figure 14-23. Life cycle of an angiosperm. [After E. W. Sinnott and K. S. Wilson, *Botany: Principles and Problems,* 6th ed., McGraw-Hill Book Company, New York, 1963]

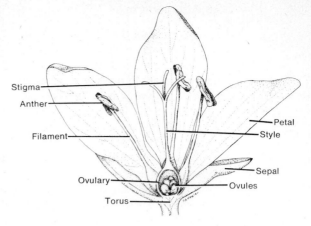

Stigma
Anther
Filament
Ovulary
Torus
Petal
Style
Sepal
Ovules

Figure 14-24. Generalized drawing of a complete flower showing the flower parts and their arrangement. [After V. A. Greulach and J. E. Adams, *Plants: An Introduction to Modern Botany,* John Wiley & Sons, Inc., New York, 1967]

in **flowers** (Figure 14-24), which may be defined as determinate sporophyll-bearing axes, although this definition does not exclude the cones of gymnosperms. The microsporophylls of the angiosperms are the **stamens,** and each **pistil** is composed of one or more megasporophylls (also called **carpels**).*

In addition, a **complete flower** also contains **sepals** and **petals,** although the flowers of many species lack one or another or both of these. Although they are not sporophylls, the sepals and petals are also modified leaves and differ from the foliage leaves in size, shape, and structure. Sepals are usually green and petals are usually white or colored, but in some species the sepals are colored whereas in others the petals are green.

DEVELOPMENT OF MICROSPORES AND MALE GAMETOPHYTES. Stamens, which are quite fragile, consist of a **filament** and an **anther** (Figure 14-25). They do not appear leaflike but, if one considers the filament to be a midrib and imagines the addition of a blade to either side of it, the stamen becomes easier to visualize as a modified leaf. Indeed, in some plants such as waterlilies there are intergradations between petals and stamens, and some of the intergrading forms have both blades and anthers.

The anthers typically contain four pollen chambers, which are actually microsporangia. Numerous microspore mother cells develop in each microsporangium (Figure 14-26), and these undergo reduction division forming spherical tetrads of microspores (Figure 14-27) that then separate. The development of a microspore into a male gametophyte is a simple matter, since the mature male gametophyte has only three nuclei (Figure 14-28). The first mitotic division of the microspore nucleus results in the production of a tube cell and a generative cell and occurs within the microsporangium (Figure 14-29). The only other mitosis is the division of the generative nucleus into two sperm. In some species this occurs within the microsporangium, in others it occurs after pollination and during the growth of the pollen tube.

* Meeuse [13] strongly criticizes and challanges the generally held belief that stamens and carpels are sporophylls.

Figure 14-25. Cherry flowers, one of which is sectioned to give a better view of the flower parts. Note that the bases of the sepals, petals, and stamens are fused, forming a cup-shaped structure that surrounds the lower part of the pistil. Such a flower is perigynous. [Courtesy of the Carolina Biological Supply Company]

As the microspore develops into a male gametophyte (pollen grain) there is further development of the wall, resulting in the formation of characteristic ridges, spines, or grooves in many species (Figure 2-36). This sculpturing of the wall [10], along with other characteristics such as size, makes possible the identification of pollen [6] under a microscope, often to the species. The outer sculptured wall of pollen is called the **exine,** and the thinner wall is called the **inine.** The exine has one to three germ pores through which the inine emerges when it forms a pollen tube. During maturation of the pollen the partition between the two microsporangia on each side of the anther generally breaks down, thus forming a single pollen sac on each side. At maturity each pollen sac dehisces (Figure 14-30) and the pollen is then dispersed by the wind or by insects [1] or other animals.

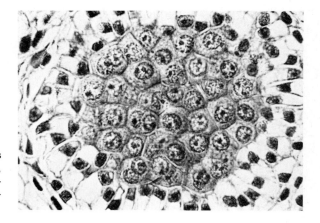

Figure 14-26. Photomicrograph of a cross section of a microsporangium in a lily anther, showing the microspore mother cells. [Courtesy of the Carolina Biological Supply Company]

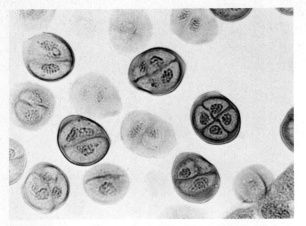

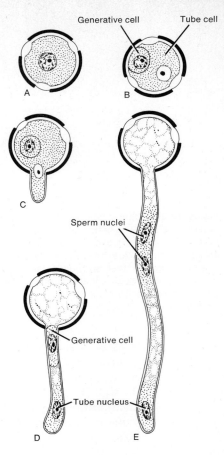

Figure 14-27. Photomicrograph of lily microspores that have developed from microspore mother cells. Note the two cells produced by the first meiotic division and the tetrads of microspores resulting from the second meiotic division. [Courtesy of Ripon Microslides]

Figure 14-28. Semidiagrammatic drawings showing the development of a male gametophyte (pollen) from a microspore **(A).** Growth of the pollen tube begins **(C)** after the pollen grain has been transported to the stigma of a pistil (pollination).

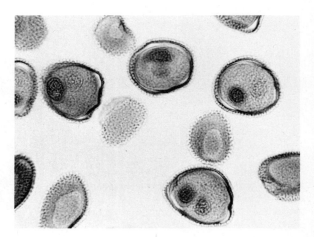

Figure 14-29. Lily pollen grains in the two-nucleate stage. [Courtesy of Ripon Microslides]

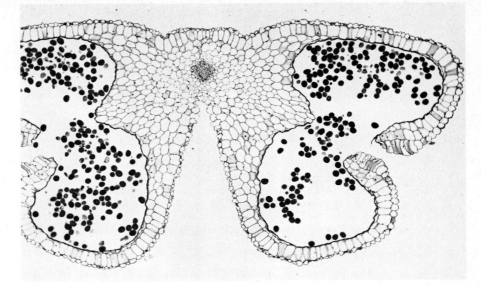

Figure 14-30. Photomicrograph of a cross section of a mature lily anther still containing some of the pollen grains. Note that the walls between the adjacent microsporangia have broken down, creating two pollen sacs, and that each pollen sac has dehisced, permitting the dispersal of the pollen. [Courtesy of the Carolina Biological Supply Company]

CARPELS AND PISTILS.   Meanwhile, within the pistil megasporangia (ovules) have developed, and these in turn have formed megaspore mother cells, megaspores, and then female gametophytes. The pistil, consisting of **stigma, style,** and **ovulary (ovary),** is a structure characteristic only of angiosperms and the basic difference between them and gymnosperms. It is difficult to conceive of the pistil as being made up of megasporophylls (carpels) since its structure is so different from that of the megasporophylls of lower vascular plants, which are still more or less leaflike.

A pistil composed of a single carpel, as in the pod of legumes, can be thought of as having been formed by the rolling of the megasporophyll into something of a tube, with the lateral margins growing together (Figure 14-31). Thus, the cavity of the ovulary within which the ovules develop would be formed, though the cavity does not extend into the style and stigma. If the ovuliferous scales of the pine had curved up in the course of evolution and the margins had grown together enclosing the ovules, a similar situation would have resulted— but this did not happen, so the pine is a gymnosperm. In species with pistils composed of several carpels one can visualize a ring of carpels growing together at their lateral margins, the space within forming the cavity or cavities of the ovulary. Some compound pistils formed from more than one carpel have the same number of ovulary cavities as carpels. Other species have only one cavity but the number of carpels involved in formation of the pistil can usually be determined from such things as the number of sutures on the outside of the ovulary, the number of placenta on which the ovules are borne, or the number of divisions of the stigma or style.

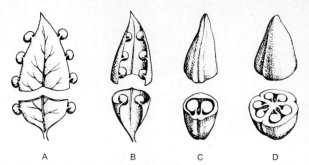

Figure 14-31. Diagrammatic drawings illustrating how pistils may have arisen from the **(A)** leafy megasporophyll (carpel) of a primitive ancestor of the angiosperms in the course of evolution. **(A–C)** A single megasporophyll with its megasporangia (ovules) could roll together and its margins could fuse, forming a simple (monocarpelate) pistil. **(D)** A compound pistil could be formed by the fusion of two or more carpels (three in this case). [After W. H. Muller, *Botany,* 2nd ed., The Macmillan Company, New York, 1969]

The presence of multiple stigmas, and less commonly stigmas and styles, in various species can be attributed to the failure of the carpels to fuse with one another the entire length of the pistil. The **placentas** are the places in the ovulary where ovules develop and they are generally found at the point where carpel margins have met. The number, location, shape, and size of the placentas varies greatly from species to species but is always consistent for any particular species. There is also great variation from species to species in the number of ovules per ovulary (Figure 14-32), for example, from one in peaches and plums, to several in peas and beans, on up to dozens, hundreds, or even thousands in some species.

Ovules begin development as a hemispherical protuberance of cells from the placenta. The base of this elongates somewhat forming a stalk **(funiculus)** while the upper portion enlarges into a spherical to ovoid structure—the nucellus or the megasporangium proper. Two circles of cells at its base then start growing over the nucellus and eventually cover it except at the terminal micropyle, thus forming the outer and inner integuments.

DEVELOPMENT OF MEGASPORES AND FEMALE GAMETOPHYTES. Within the megasporangium a single large megaspore mother cell differentiates, and it promptly undergoes reduction division, forming a linear tetrad of four haploid megaspores. As in pine, three of the megaspores disintegrate, leaving a single functional megaspore (Figure 14-33). This then develops into a female gameto-phyte, or **embryo sac** as it is often called (Figure 14-34). There are varying numbers of cells and varying arrangements of these in the embryo sacs of various species of angiosperms, but the number of cells is always small in contrast with the substantial number in the female gametophytes of gymno-sperms. In the type we are describing there are only eight nuclei present, and the fusion of two of these later reduces the number to seven.

The first mitotic division of the megaspore results in the formation of an embryo sac with two nuclei. Each of these gives rise to four nuclei by subse-quent mitoses. One nucleus from each group moves toward the center of the embryo sac, leaving three at each end. Next walls form around the six terminal

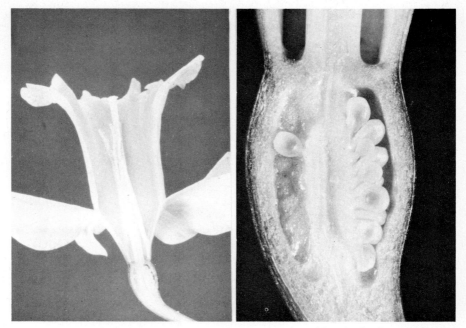

Figure 14-32. **Left:** Photograph of a jonquil flower sectioned longitudinally, showing the various flower parts. Note that the sepals, petals, and stamens appear to be attached to the top of the ovulary. Actually, their fused bases are also fused with the ovulary wall. Such flowers are epigynous. Fruits that develop from epigynous flowers have tissues derived from the fused bases of the other flower parts as well as from the ovulary wall. **Right:** Enlarged view of the ovulary, showing the ovules on a central placenta. [Courtesy of the Carolina Biological Supply Company]

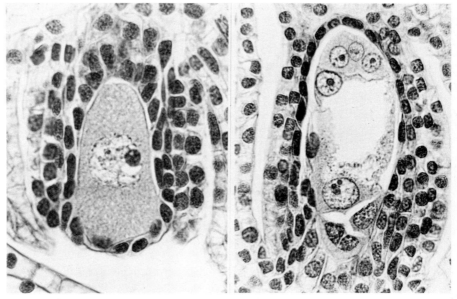

Figure 14-33. **Left:** Megaspore of lily. **Right:** Embryo sac (female gametophyte of lily). [Courtesy of Ripon Microslides]

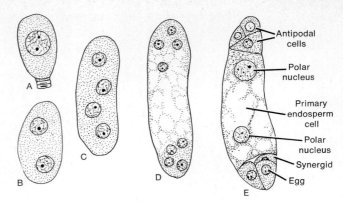

Figure 14-34. Semidiagrammatic drawings showing stages in the development of a female gametophyte (embryo sac) from a functional megaspore **(A).** Note also the three small nonfunctional megaspores of the tetrad resulting from meiosis.

nuclei, but not around the two central ones. At this stage the embryo sac may be considered as consisting of seven cells—the six terminal ones and a large central binucleate cell. These two central nuclei are called the polar nuclei. The middle cell at the micropylar end of the embryo sac is the **egg,** and the two cells flanking it are the **synergids.** The three cells at the opposite end are called the **antipodal cells.** The embryo sac (female gametophyte) is now mature and does not undergo any further development until after fertilization occurs.

POLLINATION AND FERTILIZATION.   After pollen grains have been transferred to the stigma of a pistil (pollination) they begin germinating by the formation of pollen tubes that grow through the tissues of the stigma, style and ovulary, into the cavity of the ovulary, and then into the ovules. The growth of pollen tubes is restricted to a short region near the tip. The direction of pollen tube growth is influenced by chemotropic substances in the various regions of the pistil, and these apparently vary from species to species [16]. There is evidence that growing pollen tubes secure nutrients and probably also growth substances from the tissues of the pistil. Pollen tubes growing through the pistil generally attain a greater length than when they are cultured in vitro. During pollen tube growth the generative cells divides into two sperm, if it has not done so previously. The number of pollen tubes growing through the pistil is generally considerably larger than the number of ovules, so there is little chance that one will fail to reach each ovule. Although pollen tubes usually enter ovules through the micropyle, they can also enter at other places.

Once the pollen tube has reached the embryo sac, the male and female gametophytes are in contact and the stage is set for fertilization. The end of the pollen tube ruptures, and the three cells of the pollen enter the embryo sac. The tube nucleus, which has apparently played an important role during pollen tube growth, disintegrates. One sperm fertilizes the egg, resulting in a diploid zygote that is the beginning of the new sporophyte generation. The other sperm and the two polar nuclei fuse, resulting in the production of the primary endosperm nucleus.

In most species each of the three nuclei is haploid and so the endosperm nucleus is triploid. However, in some species one or both of the polar nuclei may have a higher ploidy so the endosperm nucleus may be $4n$, $5n$, or even more. For example, in lilies one of the polar nuclei is triploid, resulting in a 5n endosperm nucleus and subsequently a 5n endosperm. The antipodals and synergids disintegrate, leaving only two nuclei in the embryo sac: the diploid zygote and the usually triploid endosperm nucleus. Although the antipodals and synergids have no future, at least the latter have played roles of some importance, including nutrition and possible chemotropic attraction of the pollen tube. The angiosperms are unique among all organisms in having double fertilization (Figure 14-35).

SEED DEVELOPMENT. Soon after fertilization the primary endosperm divides, and mitosis occurs repeatedly until numerous endosperm nuclei without walls have formed. These become distributed throughout the embryo sac (which is no longer a female gametophyte). While the endosperm nuclei are dividing the zygote begins division, first forming a filament of four to eight cells. Then the terminal cell of the filament divides several times, forming a spherical cluster of eight cells that will develop into the embryo proper (Figure 14-36). At the same time the terminal cell at the micropylar end of the filament enlarges greatly, thus pushing the embryo into a more central position in the developing endosperm. This enlarged cell and the remaining filament cells constitute the suspensor, which disintegrates as the embryo continues to develop. The suspensor has, however, played an important role in synthesizing proteins, RNA, and other substances and in nutrition of the embryo. In the meantime, walls

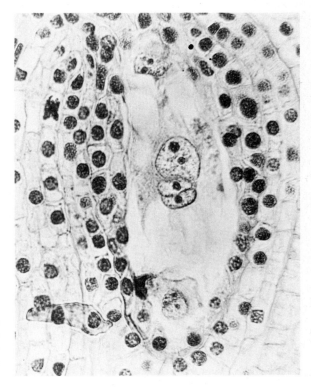

Figure 14-35. A lily embryo sac at the time of double fertilization. Note the fusion of the sperm and egg nuclei (below) and of a sperm and the fusion nuclei (center of embryo sac). [Courtesy of Ripon Microslides]

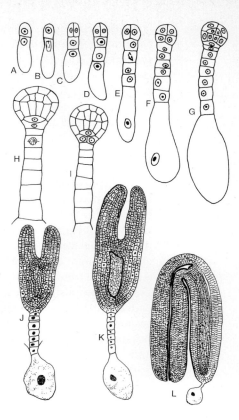

Figure 14-36. Development of the embryo of *Capsella bursa-pastoris* (shepherd's purse). (Stages A–I are drawn to a larger scale than stages K–L.) The embryo proper develops from the spherical cell mass at the upper end of the filament. The remainder of the filament, including the large terminal cell, constitutes the suspensor. [After Souèges and Schaffner, from E. W. Sinnott, *Plant Morphogenesis,* McGraw-Hill Book Company, New York, 1960]

have been forming around each endosperm nucleus, thus forming a multi-cellular endosperm that completely surrounds the embryo and fills all the space between the embryo and nucellus.

While the embryo and endosperm are developing there is a substantial increase in the size of the ovule, and the integuments differentiate into seed coats. The latter involves secondary wall thickening in most species, often including the deposition of substances such as cutin and lignin. The seed coats of most species are either leathery or hard and bony and often quite thick, but some seeds such as those of peanuts have thin and papery seed coats. Seed coats of many species are pigmented, and some such as milkweed and cotton develop filamentous hairs whereas others develop spines or hooked projections. Once differentiated, the cells of seed coats generally die.

Since the nucellus usually disintegrates by the time an ovule has developed into a seed, the seeds of most species consist only of the seed coats, the endosperm, and the embryo. However, in a few plants including beets the nucellus develops into a substantial tissue known as the **perisperm,** which plays a nutritive role comparable with that of the endosperm. In some plants including peas and beans the endosperm disintegrates and is used as food by the developing embryo before the seed is mature. Such seeds consist only of the seed coats and embryo. The cotyledon of such seeds are usually thick and fleshy and contain abundant food that is hydrolyzed and used by the embryo and seedling during germination.

The fully differentiated embryos of angiosperms consist of a primary root initial (the **radicle**), a stem like **hypocotyl** that extends from the radicle to the cotyledons, one or two **cotyledons** (in monocotyledons and dicotyledons respectively), and above the cotyledonary node the embryonic stem or **epicotyl** with its terminal bud (the **plumule**).

FRUIT DEVELOPMENT. While the ovules are developing into seeds, the ovularies are developing into **fruits.** The universal presence of fruits in angiosperms and their absence in gymnosperms is, of course, a result of the fact that the ovules of angiosperms are borne in the ovulary cavity created by the fusion of the carpel or carpels making up the pistil, whereas the ovules of gymnosperms are borne on the surface of the ovuliferous scale or megasporophyll. The term **gymnosperm** literally means "naked seed," that is, a seed not enclosed within a fruit. It should be stressed that the seeds of angiosperms are *always* found within fruits, at least until such a time as a ripe fruit may dehisce or disintegrate and so expose the seeds, despite some superficially apparent exceptions. For one thing, fruits are not restricted to the fleshy, succulent, and flavorful structures called fruits by laymen; they include also dry fruits such as pods, capsules, and nuts. For another thing, "seeds" such as those of the strawberry and sunflower that are apparently not enclosed are really fruits. Dissection of these fruits reveals a seed inside each fruit. The grains of corn and other grasses consist of fruits each enclosing a single seed, and in such cases the fruits and seed coats are tightly adherent. In some species other structures than the ovulary also contribute toward the formation of fruits—these are referred to as accessory fruits. Only the core of apples and pears is derived from the ovulary. The rest of the fruit develops from the fused bases of the sepals, petals, and stamens that have grown over the ovulary and covered it. The fleshy part of a strawberry is derived from the expanded stem tip (receptacle) on which the flower parts are borne, and thus is also an accessory fruit, although it is quite distinct from the true hard and dry seedlike fruits derived from the numerous ovularies of the flower.

In a later section of this chapter we shall consider the structure of fruits, seeds, and flowers in somewhat greater detail, with some consideration of their great diversity of structure in different families of angiosperms.

Comparison and Summary of Sexual Life Cycles

All species of vascular plants and bryophytes and many species of algae and fungi have a sexual life cycle involving alternation of gametophyte and sporophyte generations. In its simplest and most generalized form this alternation of generations can be outlined as follows: the normally haploid gametophytes produce gametes that fuse, thus forming a normally diploid zygote that is the beginning of the sporophyte generation; the zygote develops into a sporophyte plant that produces spore mother cells that then divide into four haploid meiospores by meiosis (reduction division), and the meiospores then develop into gametophyte plants. There are two unicellular stages in the life cycle: the zygote and the meiospore, not considering the gametes and spore mother cells that give rise to these. Algae such as Ulcthrix and Oedogonium might even be considered as having alternation of generations if it is assumed that the zygote (which immediately undergoes reduction division) is the total sporophyte generation.

Although the generalized life cycle is applicable to most plants, it should be evident that it does not present the differences among the representative life cycles that we have outlined. These differences are generally phylogenetic advances that appear as one goes from the more primitive to the more advanced plants, but it should be emphasized again that the representative species discussed above do not represent a single evolutionary line. The important phylogenetic advances, in the approximate order of their appearance are

1. From isogamy to heterogamy (eggs and sperm).

2. From homothallic (bisexual) to heterothallic (male and female) gametophytes.

3. Development of sporophylls in vascular plants.

4. Differentiation of sporophylls from foliage leaves.

5. From homospory to heterospory, including the appearance of megasporangia and microsporangia and also megasporophylls and microsporophylls.

6. Seed formation, resulting from the retention of the embryo sporophyte within the ovule (megasporangium) and the development of the integuments of the ovule into seed coats.

7. Progressive increase in the size and complexity of the sporophyte and progressive decline of the gametophytes, including eventual disappearance of antheridia and archegonia and all but a few vegetative cells.

8. From gymnosperm endosperms composed of female gametophyte tissue to angiosperm endosperms derived from a triple fusion nucleus.

9. Appearance of fruits derived from ovularies (and sometimes associated structures) that enclose the seeds, as a result of the formation of pistils from one or more megasporophylls (carpels).

It will be noted that these trends lead toward progressive sexual differentiation, first in the gametophytes and later in the sporophytes, as well as a progressive decline of the gametophytes.

It is interesting to note that it would not take a great deal more evolutionary simplification of the gametophytes of angiosperms (that is, functioning of the microspores as sperm and megaspores as eggs rather than their development into rudimentary gametophytes) to produce the animal type of life cycle in plants. Before leaving this summary, it should be made clear that there is by no means a perfect correlation between the evolutionary advancement of a species and the phylogenetic trends outlined above. For example, fern gametophytes are commonly homothallic whereas moss gametophytes may be heterothallic. The more advanced angiosperms are likely to be relatively small herbaceous plants, in contrast with the generally large and woody sporophytes of many more primitive species, as well as more likely to be monoecious (or have perfect flowers) than to be dioecious, although there is no consistent trend in either respect.

Angiosperm Reproductive Structures

In view of the great diversity of the reproductive structures of angiosperms it seems well at this point to consider briefly a few of the many different types of flowers, fruits, and seeds produced by different species of angiosperms. As has been noted earlier, there is also considerable variation in the female gametophytes (embryo sacs) as regards size, number of cells or nuclei and their arrangement, and details of endosperm formation. The species differences in male gametophytes (pollen) are less extensive and are largely a matter of size, shape, wall sculpturing, and the time at which the nuclear divisions occur. These gametophytic differences will not be discussed further here.

Flowers

One of the fundamental differences in flowers relates to whether they are **perfect** (stamens and pistils in the same flower) or **imperfect,** with separate staminate and pistillate flowers. If a species has staminate and pistillate flowers they may both be on the same plant as in corn **(monoecious)** or on separate staminate and pistillate plants (Figures 14-37 and 14-38) as in willows **(dioecious).** A few species, notably of the squash family, may have perfect and imperfect flowers on one plant, a condition referred to as **polygamy.**

Another basic difference of flowers of various species relates to the presence

Figure 14-37. Pistillate inflorescence (catkin) of a willow. Each individual flower consists of only a single pistil. [Courtesy of Louise Keppler]

Figure 14-38. Staminate inflorescences (catkins) of willow. [Courtesy of Louise Keppler]

or absence of other flower parts. A **complete** flower has all four sets of parts: sepals, petals, stamens, and pistils, and many angiosperm species have perfect flowers. Sometimes, as in tulips, this is not obvious at first glance since the petals and sepals are both the same size, shape, and color. The flowers of a particular species may lack one, two, or even three (in species with imperfect flowers) of the flower parts, but of course if all four were lacking there would be no flower. Petals are more likely to be lacking than sepals, and in some species without petals the sepals are colored like petals.

The absence of both petals and sepals is most likely to occur in wind-pollinated flowers (Figure 14-39), and even if one or another of these is present they are likely to be small, inconspicuous, and greenish. Such flowers are commonly not even recognized as flowers by laymen. The flowers of grasses are greatly reduced and generally lack both sepals and petals, consisting only of stamens and a pistil or, in monoecious species such as corn, of only stamens *or* a pistil. However, grass flowers are generally partially enclosed within small, dry, papery modified leaves known as **bracts,** and a spikelet of several flowers may be enclosed within still other bracts. Bracts are modified leaves associated with flowers, but are not considered as flower parts either in grasses or other species where they occur. Perhaps the ultimate in reduction of flowers is in the pistillate flowers of corn, each one consisting only of a single pistle; its ovulary is the immature grain of corn, its style the long silk with a stigma at its end.

There is also an immense range of variation from species to species in the size, shape, number, color, and arrangement of the various flower parts. We cannot consider many of these differences here, but it should be noted that they are of great importance as taxonomic criteria in the classification of angiosperms and also as key characteristics in the identification of plants by the use of manuals. Each order, family, genus, and species of angiosperms has certain flower characteristics in common, and obviously the number of these common characteristics increases as one goes to smaller and smaller taxa. Even the monocotyledons and dicotyledons have distinguishing flower characteristics. The monocotyledons have their flower parts in threes or sixes and the dicotyledons in fours or fives or some multiple of these in most cases.

Among the more obvious structural differences in flowers are the shapes of the **corollas** (the petals as a group). The corolla may be regular (radially symmetrical) as in buttercups and strawberries, or irregular (bilaterally symmetrical) as in peas and snapdragons. In either kind of flower the petals may be separate (buttercup, pea) or joined laterally into corolla tube either part or all of their length (morning glory, snapdragon). In some species other flower parts including sepals or stamens may be joined laterally. Compound pistils which are composed of several carpels are extreme examples of such lateral fusion of a set of flower parts, since it is really a carpel (rather than a pistil) that is comparable with the other flower parts. In some species one or more sets of flower parts may be fused with others, at least at their bases. Fusion of petals and stamens is perhaps most common.

In some species such as cherries the bases of the sepals, stamens, and petals are fused into a tube that surrounds the ovulary but is not attached to it. This

Figure 14-39. A single staminate flower of willow with its subtending bract. The flower consists of only two stamens. [Courtesy of Walter E. Rogers]

makes it appear that they arise higher on the receptacle than the pistil, which is not the case. Such flowers are called **perigynous** (Figure 14-25). In other plants such as apples and jonquils the tube is also fused with the ovulary of the pistil and results in what is called an inferior ovulary. Such flowers are **epigynous** (Figure 14-32). The usual, less modified type of flower without either such kind of fusion is called **hypogynous** (Figure 14-24). Epigynous flowers are particularly interesting from the developmental standpoint, since the fused bases of the sepals, petals, and stamens as well as the ovulary contribute to the formation of the fruit.

In many species the flowers are constructed in ways that reduce the probability of self-pollination or even prevent it [12]. Several species have **dimorphic** flowers, that is, the flowers on some individuals have long pistils and short stamens whereas those on other individuals have short pistils and long stamens. Thus, the part of an insect's body bearing pollen from one plant will be in the right position to contact the stigma of the opposite type of flower on another plant. In a number of other species pollen liberation from the anthers of a flower occurs before the stigma of the flower is receptive, thus favoring cross-pollination with a somewhat older flower. There are also varied types of self-sterility that are generally physiological rather than morphological, the

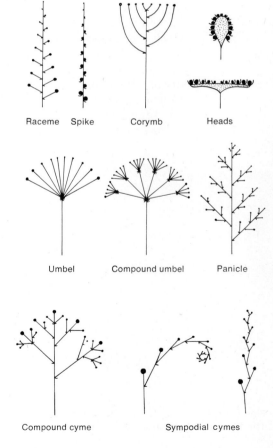

Figure 14-40. Diagrams representing the principal types of inflorescences. Of those shown, only the cymes (bottom) are determinate. The sizes of the black dots (representing individual flowers) indicate the sequence of flower formation (the largest are the first).

Figure 14-41. The inflorescences of larkspur (Delphinium) are racemes. [Courtesy of W. Atlee Burpee Seed Company]

pollen of an individual plant being ineffective in eventually bringing about fertilization of the eggs in its own pistils. This may result from inability of the pollen tubes to reach the ovules or from more complex factors. Of course, in dioecious species self-pollination is impossible.

Inflorescences

The flowers of some species of angiosperms are borne singly, but more commonly the flowers are borne in clusters called inflorescences. The inflorescences of various species are extremely diverse, but most of them belong to one of the types illustrated in Figure 14-40. (See also Figures 14-41 and 14-42.) Inflorescences can be classed into two main groups on the basis of the sequence of flower bud opening. In the **indeterminate** or **racemose** type the sequence is from the lower (or outer) flowers toward the upper (or inner) ones, whereas in the **determinate** or **cymose** type the sequence is from the upper (central) flower toward the lower (or outer) flowers. The modified stems on which the individual flowers of an inflorescence are borne are called **pedicels.** The main stems of inflorescences as well as the branches that do not terminate in a flower are called **peduncles.**

Figure 14-42. The inflorescences of Verbena are compound cymes. [Courtesy of W. Atlee Burpee Seed Company]

The various kinds of racemose inflorescences differ from one another primarily in the pattern and degree of branching of the peduncle and, in the cases of the spikes and heads, the essential absence of pedicels. The staminate inflorescences of a good many species, particularly trees such as oaks, hickories, and chestnuts, and both the staminate and pistillate inflorescences of trees such as willows and poplars are modified spikes called **aments** or **catkins.** These are characterized by having flowers without petals, bracts below each flower, and by being pendulous rather than erect.

Laymen frequently regard an inflorescence, particularly when it is rather compact, as a single flower rather than the cluster of flowers it really is. This is particularly true of members of the composite family, such as sunflowers,

Figure 14-43. The inflorescence (head) of a member of the composite family, with enlarged drawings of a ray flower and a disc flower. [After W. H. Muller, *Botany,* 2nd ed., The Macmillan Company, New York, 1969]

Figure 14-44. An inflorescence of *Cornus florida* (flowering dogwood). Only a few of the flower buds have opened. The four large white structures are bracts, not petals. Several other species of the genus either lack bracts or have only very small ones.

zinnias, marigolds, and chicory, that have heads in which all of the flowers or just the outer ones have strap-shaped corollas that, at first glance, appear to be the petals of a single flower (Figure 14-43). The leafy structures at the base of the head, which are often incorrectly assumed to be sepals, are actually bracts. Other inflorescences commonly mistaken for single flowers are the headlike cymes of flowering dogwoods and poinsettias. The white (or sometimes red or pink) "petals" of dogwoods (Figure 14-44) are bracts and not petals at all, and the same is true of the red (or white) "petals" of poinsettias.

The kinds of inflorescences most likely to be confused with one another are umbels, round-topped corymbs, and round-topped or spherical cymes. Queen Ann's lace and other members of the carrot family have umbels, usually compound ones, but the cymes of plants such as onions and hydrangeas strongly resemble umbels until they are examined carefully.

Fruits

As we have already pointed out, the seeds of angiosperms are always borne within fruits, which are enlarged and ripened ovularies plus, in some species, accessory structures that enlarge and ripen along with an ovulary and form a unit with it. The term fruit, like other terms such as work, force, and energy, has a technical scientific meaning as well as a popular meaning, and the two meanings overlap only partially. Although cherries, grapes, peaches, apples, and pineapples are fruits in either sense, many fruits in the botanical sense such as bean pods, nuts, grains of wheat or corn, maple keys, sunflower "seeds," poppy capsules, tomatoes, cucumbers, and pimentos are not considered to be fruits in the popular sense. Also, a few things like rhubarb (the large, fleshy petioles of the leaves) that are not fruits at all are so considered in the popular sense.

The enlarged and ripened ovulary wall of a fruit is called the **pericarp,** and it may be fleshy or dry, soft or hard. The pericarp may consist of as many as three distinct layers: the **exocarp** (skin), **mesocarp,** and **endocarp.** In peaches, plums, and cherries the mesocarp is fleshy but the endocarp is hard and stony and is commonly mistaken for the seed; in fact the seed can be seen only by breaking open the endocarp. A simple fruit is derived only from a single ovulary, and is the most common type (Table 14-1). Simple **accessory** fruits such as the pomes of apples, pears, and quince are derived partly (inside the core line) from a single ovulary and partly from the fused bases of the other flower parts.

Aggregate fruits are clusters of individual fruits that develop from the several pistils of a single flower, the most common examples being raspberries, blackberries, and strawberries. A raspberry or blackberry is actually a cluster of small drupes (Table 14-1), each derived from a different ovulary of the flower. In raspberries the ripe fruits easily detach from the hemispherical receptacle, but in blackberries they remain tightly attached and the receptacle is eaten along with the fleshy fruits. In strawberries the receptacle becomes greatly enlarged and fleshy whereas the ovularies become hard, dry pericarps that are commonly called seeds. The seeds are, of course, inside of these fruits. Thus, blackberries and strawberries (but not raspberries) are accessory as well as aggregate fruits.

Multiple fruits develop from the various flowers of an entire compact inflorescence rather than from a single flower, and the fleshy parts of the fruits are frequently accessory structures. The best known multiple fruits are mulberries, pineapple, and figs. Each flower of the pistillate inflorescence develops an **achene** (Table 14-1), while the **calyx** (sepals) of each flower develops into a fleshy accessory fruit that surrounds the achene. The core of a pineapple is the enlarged peduncle of the inflorescence, whereas the outer part is derived from the ovularies and fused bases of the numerous flowers of the inflorescence. The fleshy part of a fig is derived from the peduncle of the inflorescence which has invaginated and enlarged, making it appear that the flowers are borne inside although the cavity has a pore at the tip. The so-called seeds of figs are actually achenes, each one derived from the ovulary of a separate pistillate flower of the inflorescence. This multiple, accessory fruit of figs is called a **synconium.**

It will be noted from Table 14-1 that several of the types of fruits are limited, largely if not entirely, to a single family of plants. Also, it will be noted that many fruits are misnamed. For example, strawberries and raspberries are really not berries, whereas most true berries are usually not referred to as such.

Table 14-1 Outline of principal types of simple fruits

1. Fleshy fruits
   1.1 Berry—Pericarp all fleshy, except perhaps for a hard or leathery exocarp, and a number of seeds
      1.1.1 Berry proper—With a thin skin (e.g. grape, tomato, blueberries, currents, peppers)
      1.1.2 Pepo—With a hard rind (e.g. watermelon, squash, and other members of cucumber family)
      1.1.3 Hesperidium—With a leathery skin and radial partitions (e.g. lemons and other members of citrus family)
   1.2 Drupe—With a hard endocarp and usually a single seed (e.g. peach, plum, cherry, olive)
   1.3 Pome—Endocarp forming a core, rest of fruit accessory (e.g. apple, pear)
2. Dry fruits
   2.1 Dehiscent (Opening when mature thus releasing the seeds; with several to many seeds)
      2.1.1 Legume—One carpel, usually with two dehiscing sutures (e.g. pea, bean, and other members of legume family)
      2.1.2 Follicle—One carpel, usually with one dehiscing suture (e.g. larkspur, milkweed, magnolia)
      2.1.3 Capsule—Two or more carpels, dehiscing by several slits or pores or by separation of top (e.g. tulip, lily, cotton, azalea, portulaca, poppy)
      2.1.4 Silique—Two carpels, pericarp dehiscing in two halves, leaving membranous portion with seeds on it (e.g. honesty, shepherd's purse, and other members of the mustard family). Short siliques are sometimes called silicles.
   2.2 Indehiscent fruits (Usually with one seed, sometimes a few)
      2.2.1 Achene—One carpel, one loose seed (e.g. buckwheat, marigold, sunflower, and members of composite family)
      2.2.2 Samara—Essentially a winged achene, sometimes with two carpels (e.g. maple, ash, elm)
      2.2.3 Nut—Essentially a large achene with a hard, thick pericarp (e.g. chestnut, buckeye, acorn of oaks, hazelnut)
      2.2.4 Caryopsis or grain—One carpel, one seed, seed and pericarp united at all points (e.g. corn, wheat, rice, and other members of the grass family)
      2.2.5 Schizocarp—Two or more carpels, each usually with one seed, carpels separating at maturity but each still enclosing its seed (e.g. carrot, parsnip, and other members of carrot family)

Seeds

The seeds of the various species of angiosperms differ from one another markedly as regards size, shape, nature of the seed coats, and other characteristics. It is essentially impossible to classify them into a limited number of major structural types of any basic significance. In size seeds range from the immense seeds of coconuts (the coconuts themselves after removal of the fibrous pericarp) to the almost microscopic seeds of orchids, although the seeds of most species range between 1 and 10 mm in diameter. The embryos of different species vary as regards size, shape (whether straight or folded into a U-shape), and size and thickness of the cotyledons, but the most fundamental difference is in the presence of one cotyledon in monocotyledons and two in dicotyledons. The structural features of embryos are generally the same: cotyledon(s), epicotyl or plumule, hypocotyl, and radicle. Most embryos are fully developed at the time of seed dispersal, but in a few plants such as orchids the embryos are rudimentary at the time of dispersal and reach their full development only weeks or months later.

The endosperm develops in all species, but in some, such as shepherd's purse, pea, and bean, it disintegrates during seed development and is lacking in mature seeds. The embryos of such species generally have thick cotyledons that contain considerable accumulated food. Endosperms are generally solid tissues, although they may pass through a milky stage early in their development. However the inner portion of the endosperm of coconuts is a liquid at maturity—the coconut water or milk. Coconut water (milk) is a complex water solution of various nutrients and growth substances and also contains floating endosperm cells. The outer part of the endosperm makes up the meat of the coconut. In a few species, such as red beets, the nucellus of the ovule persists in the mature seed and forms a diploid tissue called the **perisperm** which, like the endosperm, is a site of food accumulation.

The seed coats are perhaps more variable from species to species than any other seed structure. In thickness and texture they range from the thin, papery seed coats of peanuts to the hard, thick seed coats of water lotus and coconut. The seed coats of a few species such as bittersweet are soft and fleshy **arils** and, like a variety of fleshy fruits, are eaten and dispersed by birds. The seed coats of some species are impermeable to either water or oxygen or both and thus provide seed coat dormancy, which will be discussed in the next chapter. Thick, hard seed coats may also confer seed coat dormancy by mechanical resistance to the imbibition of water and growth of the embryo. The seed coats of species with dehiscent fruits frequently have features such as tufts of hairs, wings, or hooked spines that favor seed dispersal by the wind or animals, whereas indehiscent fruits may have comparable modifications of the pericarp. In grasses the seed coats lose their identity by fusion with the pericarp, which assumes most of the protective functions of seed coats.

As examples for illustrating seed structure we are selecting three rather large seeds of common cultivated plants: bean, castorbean, and corn (maize). Castorbean seeds and corn seeds have endosperms at maturity, whereas bean seeds do not (Figure 14-45). Corn seeds are included as an example of the rather unusual seeds of grasses not only because their seed coats and pericarps are fused but also because their embryos have several unusual structures. One is the shield-shaped cotyledon, **(scutellum),** which does not emerge at germination but remains in contact with the endosperm and absorbs food from it. Another is the tubular leaf sheath or **coleoptile** that encloses the plumule in the seed and continues to do so through most of the germination period. The coleoptile is regarded as being a modified leaf. Finally, the radicle is also enclosed by a tubular **coleorhyza.**

Bean seeds nicely illustrate several common features of seeds. One is the readily visible persistent micropyle and another is the **hilum,** the scar at the point where the ovule was attached to the funiculus. The cotyledons of the castorbean, like those of most seeds with endosperms, are much more leaflike

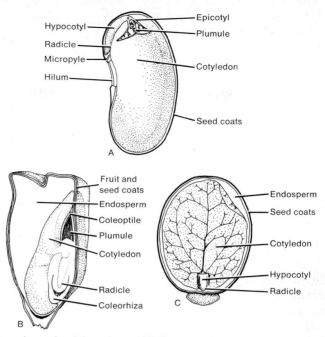

Figure 14-45. Sections through seeds of three types. **(A)** Bean seed, showing one of the two cotyledons. Endosperm is lacking in mature bean seeds. **(B)** A corn grain (caryopsis), characteristic of the grasses. The pericarp (matured ovulary wall) is fused with the seed coats, so a grain is both fruit and seed. The single cotyledon is also called the scutellum. **(C)** A castor bean seed, showing one of the two leafy cotyledons. Most of the accumulated food is in the endosperm. [After V. A. Greulach and J. E. Adams, *Plants: An Introduction to Modern Botany,* John Wiley & Sons, Inc., New York, 1967]

than those of corn and bean and make the fact that cotyledons are modified leaves more evident. During and after germination castorbean cotyledons enlarge considerably, develop chlorophyll, and function much like ordinary foliage leaves during the days or weeks before they abscise. The castorbean seed has a structure not found in most kinds of seeds: the rather irregular and somewhat spongy excrescence near the micropyle called the **caruncle.** The caruncle absorbs most of the water prior to and during germination.

References

[1] Becker, H. F. "Flowers, insects and evolution," *Natur. Hist.,* **72**(2):38–45 (Feb. 1965).
[2] Bierhorst, D. W. *Morphology of Vascular Plants,* Macmillan, New York, 1971.
[3] Bold, H. C. *Morphology of Plants,* 2nd. ed., Harper & Row, New York, 1967.
[4] DeMaggio, A. E. "Ferns as a model system for studying polyploidy and gene dosage effects," *BioScience,* **21**:313–316 (1971).

[5] Doyle, W. T. *The Biology of Higher Cryptogams,* Macmillan, New York, 1970.

[6] Echlin, P. "Pollen," *Sci. Amer.,* **218**(4):80–90 (April 1968).

[7] Evans, A. M. "Ameiotic alternation of generations: a new life cycle in the ferns," *Science,* **143**:261–263 (1964).

[8] Farrar, D. R. "Gametophytes of four tropical fern genera reproducing independently of their sporophytes in the southern Appalachians," *Science,* **155**:1266–1267 (1967).

[9] Hartmann, H. T., and D. E. Kester. *Plant Propagation: Principles and Practices,* Prentice-Hall, Englewood Cliffs, N. J., 1968.

[10] Heslop-Harrison, J. "Pollen wall development," *Science,* **161**:230–237 (1968).

[11] Klekowski, E. J., Jr. "Ferns and genetics," *BioScience,* **21**:317–322 (1971).

[12] Mather, K. "Mating discrimination in plants," *Endeavour,* **2**:17–21 (1943).

[13] Meeuse, A. D. J. *Fundamentals of Phytomorphology,* Ronald Press, New York, 1966.

[14] Meeuse, B. J. *Story of Pollination,* Ronald Press, New York, 1961.

[15] Nitsch, J. P., and C. Nitsch. "Haploid plants from pollen grains," *Science,* **163**:85–87 (1969).

[16] Rosen, W. G. "Cellular chemotropism and chemotaxis," *Quart. Rev. Biol.,* **37**:242–259 (1962); "Ultrastructure and physiology of pollen," *Ann. Rev. Plant Physiol.,* **19**:435–462 (1968).

[17] Wagner, W. H., Jr., and A. J. Sharp. "A remarkably reduced vascular plant in the United States," *Science,* **142**:1483–1484 (1963).

[18] Whittier, D. P. "The value of ferns in an understanding of alternation of generations," *BioScience,* **21**:225–227 (1971).

[19] Wollman, E. F., and F. Jacob. "Sexuality in bacteria," *Sci. Amer.,* **195**(1):109–118 (July 1956).

Reproductive Development of Angiosperms

15

Although there have been many interesting and important investigations of the reproductive development and physiology of nonvascular plants and the lower vascular plants, space limitations necessitate the restriction of this chapter to the reproductive development and physiology of the angiosperms, except for a number of incidental references to other plants at points where they are pertinent. Discussions relating to the lower vascular plants and nonvascular plants can be found elsewhere [5, 33, 39].

Initiation of Flowering

Botanists have long been intrigued by the fact that each species of flowering plant naturally initiates flowers and blooms at a characteristic and quite precise time of the year. They have attempted to find out why this is so, but significant progress in clarifying the matter has been made only since the early years of the present century [15].

We now know that there are at least three or four important morphological or environmental conditions that control or influence the initiation of flowers. One of these is **ripeness to flower,** that is, the necessity of a plant to attain a certain age or a certain stage of morphological development (such as the differentiation of a certain number of nodes) before flower initiation can occur. A second phenomenon is the influence of the relative lengths of the days and nights on flower initiation, a response referred to as **photoperiodism.** Biennials, winter annuals, and some perennials require the inductive influence of a period of low temperature that makes possible the initiation of flowers at a subsequent time, a process referred to as **vernalization.** Less important for flower initiation, but quite important for other aspects of plant growth, is the phenomenon called **thermoperiodism,** that is, the influence of alternation of different day and night temperatures. Some species are influenced by only one of these, some by two or more.

The initiation of flowering is by no means a simple matter. Although an immense amount of information has resulted from the extensive investigations of the phenomenon there are still extensive gaps in our knowledge, particularly as regards the internal processes and conditions that are involved.

Ripeness to Flower

In a rather vague way botanists had long been aware that plants generally attained a certain age, size, or stage of development before they bloomed and, before the discovery of photoperiodism and vernalization, it was assumed that this was the principal factor determining the time of flower initiation. However, one of the first experimental studies of flowering in relation to the stage of development was reported as late as 1918 by the German plant physiologist Georg Klebs [23], who coined the term Bluhreife (ripeness to flower).

The early investigators of vernalization found that both spring rye and vernalized winter rye did not develop inflorescences until at least seven leaves had appeared, and they stressed the importance of attainment of a certain stage of development in contrast with a certain size or age. Although four leaf primordia are already present in a rye embryo within an ungerminated seed, manipulations such as premature harvesting of the seeds or keeping the plants under continuous light would not reduce the prerequisite number of leaves below six. Similar ripeness to flower requirements appear to be common in most plants that must be vernalized for flower initiation.

The period required for ripeness to flower varies greatly from one species or variety to another and is entirely lacking in some plants. For example, the short-day plants *Pharbitis nil* (Japanese morning glory) and *Chenopodium rubrum* (pigweed) will bloom in the cotyledonary stage if kept under short days from the beginning. Of course, in nature they grow into large plants before blooming because they germinate and grow under long days for a considerable time. However, most photoperiodic species will not bloom, even if kept under suitable photoperiods from the beginning, until they have attained a larger size or greater age and have developed one or more true leaves. Such plants may be said to have a ripeness to flower requirement. It seems likely that day-neutral plants that do not have any vernalization or thermoperiodic requirement have initiation of flowering delayed only by a ripeness to flower requirement.

Perennial plants and in particular trees usually have a long juvenile stage (Chapter 18) during which no flowers are formed. For most trees this period is 5 years or more. Ripeness to flower is evidently not attained until the end of the long juvenile stage, although the term is often not applied to perennial plants. Attention is called again to the rather unusual situation in English ivy where the juvenile stage may last for years and the shrubby, flowering mature stage is rarely attained. Although most perennials bloom year after year once they have attained ripeness to flower, a few perennials like bamboo are **monocarpic,** as are annuals and biennials, that is, they bloom only once during their life and after the formation of seeds and fruits they become senescent and die. The various species of bamboo require from 5 to 50 years to attain ripeness to flower, and usually all individuals of a species in a large area will bloom and die the same year. That neither size, as such, or environmental factors initiate flowering is suggested by the facts that neither cutting or other destruction of the old shoots nor transplanting to another climate alter the time of blooming.

Despite the fact that the attainment of ripeness to flower varies so greatly from species to species and that it is still a rather vague and poorly understood concept, there seems to be little doubt that it is a real and important factor in the initiation of flowering in many species of plants. Attempts have been made to explain ripeness to flower in various ways including the requirement of production of a certain minimum quantity of food, attainment of a certain prerequisite hormonal balance, slowing down of vegetative growth (which often preceeds flowering), and in photoperiodic plants formation of enough leaves (or sensitive enough leaves) to permit adequate photoperiodic induction. There is some evidence that one or more such things may be involved in certain species, but there is not enough information to determine the degree to which they are generally involved or the complex of environmental and physiological conditions required for attainment of ripeness to flower in any particular species. A few specific examples that may provide some clues will be cited.

Robbins found that gibberellic acid would cause reversion of the mature, flowering form of English ivy to the juvenile, nonflowering form. Longman and Wareing found that they could cause birch trees to bloom when 1 year old (rather than when 5 years old as usual) by keeping the young trees under continuous long days. Birch is not photoperiodic as regards flowering, but the long days resulted in much larger trees and they concluded that a certain size rather than a certain age was prerequisite to flower initiation. Girdling of branches of trees sometimes induces flowering several years earlier than usual on the girdled branch, presumably by preventing the translocation of food out of the branch. However, despite these and other findings the physiology of attaining ripeness to flower still remains essentially unclarified.

Photoperiodism

The announcement in 1920 by W. W. Garner and H. A. Allard of the United States Department of Agriculture at Beltsville, Md., of their discovery of photoperiodism [17] provided the first really substantial explanation for the seasonal blooming of plants and also identified for the first time a previously unrecognized environmental factor of great biological importance—the length of the daily light and dark periods. However, despite the thorough and extensive experimental documentation of their discovery by Garner and Allard, some botanists were at first cautious about accepting such a revolutionary finding. The discovery of photoperiodism, like many other important basic scientific discoveries, was an accidental byproduct of other research. Also, as in many other cases of an important discovery, previous investigators had all the experimental evidence they needed to make the discovery but failed to understand the full significance of their data and to draw the critical conclusions.

A number of botanists working on factors influencing flower initiation, such as Klebs in Germany and Tournois and Bonnier in France, had investigated the influence of light on blooming but attempted to explain their results on the basis of total quantity of light energy rather than the mere length of the daily light and dark periods. Murneek and Whyte [29] reviewed the early work on photoperiodism, including further work by Garner and Allard and other investigators, which soon dispelled all doubts about the reality of photo-periodism and its biological importance.

Garner and Allard's discovery of photoperiodism was an incidental and essentially accidental byproduct of an entirely different investigation of Maryland Mammouth tobacco, a new variety that had been bred only a short time earlier. Unlike most tobacco varieties, it failed to bloom in the field at the usual time in the summer and continued vigorous vegetative growth into the early fall. When there was impending danger of freezing, the plants were moved into a greenhouse because of their prospective economic importance. As the autumn progressed the plants finally began blooming. Garner and Allard were also working with Biloxi soy beans; they found that plantings made at 2-week intervals from May through July all bloomed at essentially the same time in September, despite age differences ranging from 60 to 120 days. Garner and Allard now began systematic investigations in an effort to identify the environmental factor or factors involved in the timing of flower initiation in these plants, but none of the known environmental factors proved to be effective. As a last resort they then checked on what seemed to be a highly improbable idea—the length of the daily light period—and found that it was indeed the effective factor. Maryland Mommouth tobacco and Biloxi soy beans would bloom only when the length of the daily light period became short enough.

Instead of rushing into print with their discovery, Garner and Allard then checked on the influence of day length on the blooming of a considerable number of other species (Figure 15-1). They identified three groups of plants as regards their response to day length: **short-day plants** (SDP), **long-day plants** (LDP), and **day-neutral plants** (DNP). This first report on photoperiodism, as Garner and Allard named the response of plants to the relative lengths of day and night, was long and thoroughly documented and provided a sound basis for subsequent investigation of the phenomenon (Figure 15-2).

Although the investigation of photoperiodism has been related principally to its role in flower initiation, Garner and Allard (and later others) found that it is also involved in a number of aspects of vegetative development. For example, long days promote onion bulb formation, the growth of the fibrous roots of dahlia, the growth of strawberry runners, and vigorous vegetative growth in general. Long days also promote the development of plantlets on Bryophyllum and Kalanchoë leaves (whereas these are SDP as regards flower initiation). Short days promote leaf senescence and abscission and dormancy of trees and other perennials and the development of underground storage organs, such as the tubers of potatoes and Jerusalem artichokes and the roots of radishes, yams, and dahlias.

Not long after the discovery of photoperiodism zoologists found that it was also involved in a variety of seasonal phenomena in arthropods and vertebrates. For example, long days promote the pupation of midges, laying of diapause

Figure 15-1. The dark house used by Garner and Allard to provide short days for plants in their 1919–1920 experiments is in striking contrast to the elaborate controlled environment growth chambers and phytotrons currently used for research on photoperiodism. The trucks holding the plants were on tracks to facilitate moving the plants into the dark house at a certain time. [Courtesy of the U. S. D. A., Beltsville, Md.]

eggs by silkworms, production of parthenogenetic aphids, maturation of fish eggs, the growth and maturation of gonads in various lizards, birds, and mammals, and egg laying by young chickens and turkeys. Short days promote the initiation of diapause in mites and midges, the development of aphids from fertilized eggs, and the induction of oestrous in a number of mammals. Photoperiodism has also been implicated in the seasonal molts and changes in feather or hair color in birds and mammals, in hibernation, and in bird migration.

Day length provides the perfect seasonal timing device because, on any particular date in a certain location, it is always the same year after year and thus is precisely predictable, in contrast with seasonal variations in other environmental factors such as temperature, humidity, and light intensity. Although extremely dense clouds may reduce the photoperiodically effective day length somewhat, this is a small complication compared with the marked variability of other seasonal environmental factors.

Photoperiodism also plays a role in latitudinal plant distribution. From the equator toward either pole the days become progressively longer in the summer and shorter in winter. At and near the equator the day length is 12 hr throughout the year. At the poles there is continuous light in midsummer and continuous darkness in midwinter. Thus, plants that require more than 12 hr of light for flower initiation (most LDP) cannot propagate themselves sexually at or near the equator, whereas SDP cannot reproduce at far northern (or southern) latitudes because long days prevail throughout the growing season. It has been

Figure 15-2. Garner and Allard's photographs of some of their experimental plants. **(A)** Maryland Mammouth tobacco plants under long summer days. **(B)** Maryland Mammouth plants of same age kept under short days. **(C)** Biloxi soybean plants kept from germination under long days (left) and short days (right). **(D)** Spinach plants kept under long days (left) and short days (right). [From W. W. Garner and H. A. Allard, *J. Agr. Res.,* **18:**553–606 (1920). Courtesy of U. S. D. A., Beltsville, Md.]

found that some species of grasses and other plants with a wide latitudinal distribution have a number of varieties with different critical day length requirements, that is, the farther they are from the equator the longer their critical day length.

It should be noted that the photoperiodically effective day length cannot be determined just from the almanac times for sunrise and sunset, since a considerable part of the two daily twilight periods (roughly a half-hour each in temperate zones) provide light of sufficient intensity to be photoperiodically effective. However, the twilight period is very short at the equator and becomes longer toward the poles.

There have been a number of investigations of the genetics of the photoperiodic response, and in tobacco and most other species studied it appears to be controlled by a single pair of genes. However, in some species such as sorghum the interaction of a number of gene pairs is involved. Generally, in any one species the genetic variation involves either short-day and day-neutral varieties or long-day and day-neutral ones, but not long-day and short-day ones.

Finally, it is worth noting that extensive practical applications of photoperiodism have been made, beginning soon after photoperiodism was discovered. These applications relate largely to bringing commercial flowers into bloom out of season, using electric lights to provide long days in the fall and winter and dark shades or cabinets to provie short days in summer. For example, shortened days insure chrysanthemums for early football games and poinsettias before Christmas, whereas lengthened days provide Easter lilies before Easter no matter how early it may come. This provides an illustration of the fact that practical application of biological information does not involve the need for understanding the physiological processes and conditions involved for, as we shall see later, even today the physiology of photoperiodism remains largely obscure.

PHOTOPERIODIC CLASSES OF PLANTS.   In their 1920 paper Garner and Allard identified three basic groups of plants as regards photoperiodic flowering response: long-day plants (LDP) (Figure 15-3), short-day plants (SDP) (Figure 15-4), and day-neutral plants (DNP). LDP initiate flowers only when the day length exceeds a certain critical minimum, SDP only when the day length is less than a certain critical maximum, and DNP can initiate flowers under any natural day length. The critical day length varies considerably from species to species, and indeed days of intermediate length may induce flowering in both some LDP and some SDP (Table 15-1). Continued investigation of the flowering responses of many species of plants by Garner and Allard and many others soon made it evident that the situation was not as simple as it first appeared and that some species did not fit into any of the three basic classes. A few plants, such as some sugarcane varieties, proved to be **intermediate day plants** (IDP) that remained vegetative if the days were either too long or too short. The reverse of this has been found by the French botanists Methon and Stroun in *Media elegens,* which initiates flowers when the days are either shorter or longer than two critical times but not when the days are of intermediate length.

A few species are either **short-long day plants** (SLDP) or **long-short day plants** (LSDP), the first day length presumably being required for flower initiation and the second for continued flower development. The SLDP, such as some varieties of wheat and rye, are plants that bloom in the spring or early summer when the required sequence of day lengths occurs, whereas the LSDP, such as night-blooming jasmine (*Cestrum nocturnum*), bloom in late summer or fall. Some SDP and some LDP have a strictly qualitative photoperiodic response, remaining vegetative indefinitely when kept under noninductive photoperiods,

A

B

C

D

whereas others have a quantitative response, eventually initiating some flowers at day lengths beyond the critical. Some plants may be photoperiodic when young and day neutral when older, such as certain sunflower varieties that are first SDP and then DNP later. Also, temperature may modify the photoperiodic response of plants. For example a number of SDP, such as Maryland Mammouth tobacco, jimson weed, orange flare cosmos, and some strawberries, are day neutral at low temperatures on the order of 13°C. Temperature may also influence the critical day length for some plants or the number of photoperiodic cycles necessary for the induction of flowering.

Finally, it should be noted again that within many species there are varieties or strains with differing photoperiodic responses. Some may be either SDP or LDP whereas others are day neutral, some may be influenced by temperature more than others, or there may be a range of different critical day lengths within a species. It is rather interesting to note that many economic plants are day neutral although their wild or less intensively cultivated relatives may be either SDP or LDP. This may have resulted from unintentional selection of day-neutral strains because of their longer periods of flowering and fruiting.

This discussion of the photoperiodic responses of different species and varieties of plants may suggest that there is such a diversity that the situation is rather chaotic, but it should be noted that the majority of photoperiodic plants are more or less regulation SDP or LDP.

SHORT DAY | Ca. 60 days COCKLEBUR | LONG DAY

Figure 15-4. Photoperiodic response of Xanthium (cocklebur), a SDP. Note the fruits on the plant under short days. Xanthium has probably been used more extensively in investigations of photoperiodism than any other plant. [Courtesy of Frank B. Salisbury]

Table 15-1 Photoperiodic classification of selected plants, with critical day length in hours

Short-day plants	Critical length less than
Bryophyllum pinatum	12
Chrysanthemum, most species	15
Cosmos sulphureus, Klondyke cosmos	14
Euphorbia pulcherrima, Poinsettia	12
Fragaria chiloensis, strawberry (most)	10
Glycine max, Biloxi soybean	14
Nicotiana tabacum, Maryland Mammoth tobacco	14
Oryza sativa, winter rice	12
Viola papilionaceae, violet	11
Xanthium strumarium, cocklebur	15
Long-day plants	**Critical length more than**
Anethum graveolens, dill	11
Avena sativa, oats	9
Hibiscus syriacus, althea	12
Hordeum vulgare, winter barley	12
Hyoscyamus niger, henbane	10
Lolium italicum, Italian ryegrass	11
Rudbeckia bicolor, coneflower	10
Spinacia oleracea, spinach	13
Trifolium pratense, red clover	12
Triticum aestivum, winter wheat	12
Day-neutral plants	
Cucumis sativa, cucumber	
Fagopyrum tataricum, buckwheat	
Fragaria chiloensis, everbearing strawberry	
Gardenia jasminoides, Cape jasmine	
Ilex aquifolium, holly	
Impatiens balsamina, balsam	
Lycopersicon esculentum, tomato	
Phaseolus vulgaris, string bean	
Poa annua, annual bluegrass	
Zea mays, corn	

SOME BASIC FACTS ABOUT PHOTOPERIODISM. Although our knowledge of the internal processes and conditions involved in photoperiodism is very incomplete and unsatisfactory, a considerable amount of information about plant photoperiodism has resulted from years of research. In this section we shall summarize briefly some of the more important information and its implications. Later on we shall consider some theories regarding the nature of the photoperiodic processes.

1. A variety of evidence has shown that leaves are the photoreceptor organs in photoperiodism. In Xanthium (cocklebur) and a number of other plants providing even a single leaf with the required short day–long night photoinductive cycle (by placing it in a light-proof bag part of a day while the rest of the plant is in long days) will bring about flower initiation. Defoliated

plants do not initiate flowers even when kept under suitable photoperiods. The most sensitive leaves are generally the youngest fully expanded ones, and the very young leaves still in a bud are not effective. Since it is the leaves that are sensitive to light and the morphogenetic changes occur in buds (conversion of vegatative to flower buds) it appears evident that the transmission of a hormone or some other stimulus is involved.

2. Although exposure of a single leaf to suitable photoperiods can induce flower initiation, the effect is generally more marked if the other leaves are re-moved, and in some species removal of the other leaves is essential if flowering is to occur. Removal of the mature leaves of henbane (*Hyoscyamus niger*), a LDP, permits it to bloom in short days. Results such as these suggest that leaves in noninductive photoperiods may produce flower inhibitors, at least in some species.

3. When a cocklebur plant has been photoperiodically induced by short days and is then transferred to long days and approach-grafted to another cocklebur plant (by slicing a segment from the side of each stem and binding the stems) that has been under long days continuously, the latter also becomes induced and blooms under long days. Indeed, the graft union can then be severed and the receptor plant can then be grafted to still another plant kept under long days continuously; in this case the original receptor plant will induce the new receptor. This process has been repeated success-fully up to seven times [42]. Such results suggest not only that there is a flowering hormone that can be translocated through graft unions but also that induction results in a persistent change in the induced plants. It is unlikely that sufficient flowering hormone was produced in the original donor to induce the entire series of plants.

   Plants of different species but the same family can often be approach-grafted successfully, and an induced donor of one species can cause the blooming of a receptor of the other species that has been kept continuously under noninductive photoperiods. This occurs even when the donor is a LDP and the receptor a SDP, vice versa, or even when the donor is a DNP. For example, if a Maryland mammoth tobacco plant (SDP) and a henbane plant (LDP) are grafted together and are both kept continuously under short days, the tobacco plant becomes induced and in turn causes the henbane plant to become induced and bloom. Under continuous long days both plants also bloom if grafted together, but in this case the henbane is the donor and the tobacco is the receptor. Such grafting experiments indicate that the flowering hormone is the same in all species of plants, regardless of their response to day length. The differences are in the conditions under which hormone synthesis can occur.

4. The number of photoinductive cycles (short days and long nights for SDP or long days and short nights for LDP) required to induce flower initiation and continued flower development after transfer back to noninductive photoperiods varies greatly from species to species. Xanthium and a good many other SDP require only one photoinductive cycle, and the same is true of dill and a few other LDP. However, most species require from several

to a dozen or two photoinductive cycles and some, like soy beans (SDP), revert to vegetative growth as soon as transferred to long days after as many as 25 or more short days. Some plants, notably composites such as Rudbeckia and Cosmos, given a suboptimal number of photoinductive cycles will partially revert to vegetative growth and produce a variety of abnormal structures that are reproductive-vegetative intergrades. Even in species where the induced state is strong and lasts a long time, there are some interesting species differences. Thus, in Xanthium if only a single leaf is induced it can transmit the induced state to other leaves of the plant, whereas in Perilla a single induced leaf can continue to promote flower initiation (even if grafted on another plant) but cannot transmit the induced state to the other leaves.

5. One of the more important basic discoveries about photoperiodism was made by Hamner and Bonner, who reported in 1938 that interruption of the long night by a short period of light prevented flower initiation even though the total length of light was well below the critical length of a short day. This indicated that a long night, rather than a short day, was essential for flower initiation in Xanthium and suggested that some slow process or sequence of slow processes that could occur only in the dark were required for induction. The light interruption is not effective if given too near the beginning or end of the dark period, since the dark period after or before the interruption may still be long enough to be effective. For the light interruption to be effective it must provide a certain amount of light energy (100 to 1000 kiloergs/cm^2 depending on the species), the length of the required light break decreasing as the intensity increases. In any event, the quantity of light energy required is relatively low.

6. During the early 1950s Borthwick, Hendricks, and their coworkers at Beltsville determined the action spectrum of photoperiodism and found it to be

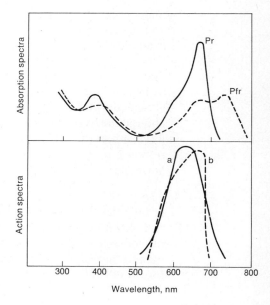

Figure 15-5. Above: Absorption spectra of red-absorbing (Pr) and far-red-absorbing forms of phytochrome. **Below:** Action spectra of inhibition of flower initiation in soybean (a) and promotion of lettuce seed germination (b). [Absorption spectra from data of Butler *et al.* Action spectrum (a) from data of Parker *et al.* and (b) from data of Hendricks]

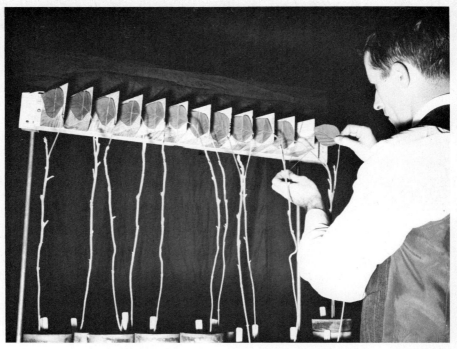

Figure 15-6. Soybean plants being set up for experimental determination of their photoperiodic action spectrum. The one leaf of each plant that was retained was later exposed to a narrow portion of a high-intensity spectrum projected on the leaves for a short period of time during the middle of the long dark period. The effectiveness of each waveband of light was determined by the degree to which it prevented flower initiation. [Courtesy of H. A. Borthwick, Agricultural Research Service, U. S. D. A., Beltsville, Md.]

essentially the same as that of the light promotion of seed germination (Figure 15-5). Exposure of plants to the different regions of the spectrum was facilitated by restricting the treatment to the short period of light needed to interrupt the long dark period, (Figure 15-6), thus determining the relative effectiveness of the different wavelengths in preventing flower formation in soy beans (SDP). As in seed germination, it was found that the effects of red or white light could be cancelled out by subsequent far-red light; thus it became apparent that photoperiodism was one of the numerous photomorphogenetic processes with phytochrome as the photoreceptor pigment. As in other processes involving phytochrome the far-red-absorbing form is the enzymatically active one, but in photoperiodism it evidently promotes processes that *prevent* the eventual production of the flowering hormone.

7. Although a certain minimum quantity of light energy is required for effective interruption of long dark periods, very dim light is just as effective as the same period of bright light for extending the length of the day. As little as 1 foot-candle (ft-c) or even somewhat less, is photoperiodically effective. Fortunately, the maximum intensity of moonlight is only 0.02 ft-c, so it does not disrupt the natural photoperiods. Of course, if the light intensity were never above 1 ft-c throughout the light period, plants could not carry on enough photosynthesis for survival.

It has been proposed that there is a high-intensity light requirement for plant photoperiodism, separate from the low-intensity light that is adequate for determining the effective length of the light period, and that this high-intensity light is required both before and after the inductive dark period by Xanthium. Hamner, and later Liverman and Bonner, conducted experiments with Xanthium that they believed demonstrated the need for the first high-intensity light period and also demonstrated the fact that this was necessary for photosynthesis. Among other evidence was the need for carbon dioxide during the light and the effectiveness of providing sucrose in place of the light. Lockhart and Hamner reached similar conclusions about the second high-intensity light requirement. However, Salisbury later claimed that his data showed no need for high-intensity light (unless the plants had been partially starved) but that a similar period of low-intensity light was just as effective. His explanation will be considered below. Further discussion of this matter can be found in Chapter 26 of Salisbury and Ross [38].

8. By maintaining experimental plants continuously in controlled-environment growth chambers it is possible to subject them to photoperiodic cycles other than the natural 24-hr one, for example, 12, 18, 36, 48, or 72. With cycles longer than 24 hr it is possible to provide plants with both longer days and longer nights than they would ever have in nature and, if desired, plants can be provided with nights of varying length without automatically varying the length of the light period simultaneously, or vice versa. The results of such experiments have been complex, quite diverse (depending on the experimental program and the species), sometimes contradictory, and usually rather surprising and difficult to evaluate. Only a few of the simpler examples can be given here, but further considerations can be found elsewhere [11, 21, 38].

For one thing, when the dark periods are unusually long, light interruptions are not effective in preventing flowering of SDP (or promoting that of LDP) unless given relatively near the beginning or end of the dark period. This suggests that the requirement for a dark period involves something more than a certain length that is essential for completion of slow processes inhibited by light. Hamner and his coworkers have found that cycles of 24, 48, or 72 hr are much more effective in inducing flowering of soybeans than cycles of 36 or 60 hr. They have also found that in Hyoscyamus (LDP) given a 10-hr light period (the critical length), flowering was most rapid with an 18-hr cycle (8 hr dark), slower or lacking with 24 or 30-hr cycles (14 or 20 hr dark), and again faster on a 42-hr cycle (32 hr dark). Salisbury found that the length of the light period between a barely inductive dark period of 8.5 hours and a subsequent 12-hr dark period had a marked effect on the stage of flowering of Xanthium.

The most successful attempts to explain these and many other results have involved implication of endogenous rhythms [8, 38]. These are rhythmical phenomena, such as the sleep movements of leaves, that are entrained by (adjusted to) some rhythmical environmental fluctuation (the daily light and dark periods in this case) but continue for at least some time if the organism is placed in a nonvarying environment, that is, continuous darkness with no variations in temperature or other known environmental factors. Thus, there appears to be within the organism some sort of clock or timing device independent of the environment but roughly geared to it. The sleep movements of bean leaves that have been in the dark since germination are random and uncoordinated but, if a group of such plants is given a brief flash of light to "set the clock" and are then kept in the dark, all exhibit coordinated sleep movements with a cycle of from 23.7 to 28.3 hr. Since the endogenous rhythms are not exactly 24 hr they are called circadian (about a day) rhythms. Some endogenous rhythms are attuned to the lunar cycle of 28 days, but most of them are circadian. Although endogenous rhythms were suspected before the turn of the century (at least in sleep movements of leaves), most of the research on them has been done since 1950 and numerous examples are now known in both plants and animals. Other examples in plants include rates of mitosis, rates of respiration in constant environments, and evidently the effect on photoperiodic response.

The role of circadian rhythms in photoperiodism is based predominantly on the theory of the German plant physiologist Erwin Bunning, the leading investigator of endogenous rhythms in plants. He has proposed that plants have an endogenous **photophil phase** (during which light promotes plant processes including flowering) alternating with a **skotophil phase** (during which light inhibits flowering and other processes). He and others believe that the degree to which each of these phases overlap any particular set of light and dark periods to which the plant is exposed greatly influences the photoperiodic response and provides the timing device which seems to be an essential feature of photoperiodism. Certainly, the photoperiodic behavior of plants under photoperiodic cycles other than 24 hr can be explained much more satisfactorily on the basis of this circadian rhythm than on any other basis. Salisbury has data that show the far-red-absorbing form of phytochrome promotes flowering during the light although, as long known, it inhibits flowering of SDP in the dark. He considers that this supports Bunning's theory. More detailed and documented discussions of this complex and still highly theoretical subject can be found elsewhere [8, 37].

9. The morphogenetic changes that result in the conversion of a vegetative bud primordium into a flower bud begin soon after florigen reaches the primordium from the leaves. According to Westmore and his coworkers, the first cytological evidence is increased cell division just below the apical initials. Then, flower primordia rather than leaf and lateral bud primordria begin developing. There is generally also an enlargement and change in shape of the apical meristem. If a solitary flower is being formed, the apex generally becomes broadened and flattened (Figure 15-7); however, if an inflorescence is developing, the apex generally becomes enlarged and globose or conical (Figure 15-8). The details vary considerably from species to species. Cocklebur is perhaps the most intensively studied plant, particularly as regards the developmental stages of the staminate inflorescence bud (Figure 15-9). Salisbury has identified a series of eight floral stages that can

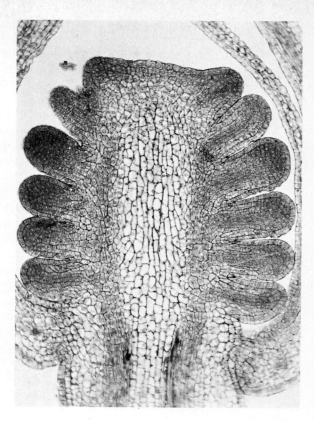

Figure 15-7. Photomicrograph of a longitudinal section of a Ranunculus (buttercup) flower primordium. Five stamen primordia are visible on each side and carpel primordia are beginning to develop on the sides of the apex. Portions of the sepal and petal primordia also show. [Courtesy of S. S. Tepfer]

be used as a semiquantitative measure of the degree of photoperiodic induction.

The morphogenetic changes involved in conversion to a flower bud are really substantial (Figure 15-10). In summary, they include

1. Cessation of indeterminate apical growth

2. Suppression of internode elongation and axillary bud formation, at least as regards individual flowers

3. The development of flower parts (sepals, petals, stamens, and carpels) instead of foliage leaves

4. A change in phyllotaxy.

A particularly striking and puzzling aspect of this marked morphogenesis is the shift, within a distance of a fraction of a millimeter, from the differentiation of sepals, to petals, then from petals to stamens and carpels, or as many of these structures as are prescribed by the genetic code of the individual. The fact that these develop in the proper sequence, with the numbers of each part characteristic of the species, and differentiate the particular structural details of the species as regards each part is one of the more amazing examples of the complexity and precision of differentiation (Figure 15-11).

The physiological processes involved in these morphogenetic changes are not well understood, but a number of interesting things have been discovered. Using histochemical techniques a number of investigators, particularly Gifford and Jensen and their students, have found that upon the arrival of the flowering stimulus there is a marked increase in RNA and proteins. Electronmicrographs have revealed an increase in the number of ribosomes and the complexity of the endoplasmic reticulum of the apical meristem cells [18]. Other investigators, including James Bonner and his students, [4] found that 5-fluorouracil (an inhibitor of RNA synthesis) applied to Xanthium buds prevents the flowering of photoperiodically induced plants. Others have found that actinomycin-D, another inhibitor of RNA synthesis, will prevent the flowering of induced plants of several species, as will various inhibitors of protein synthesis. Facts such as these suggest that florigen, like other hormones, may act by altering the kinds of mRNA synthesized, thus changing the enzyme complex of the meristematic cells and the processes leading to morphogenetic changes. However, this remains largely theoretical and does little to explain the shifts in differentiation as the flower parts develop.

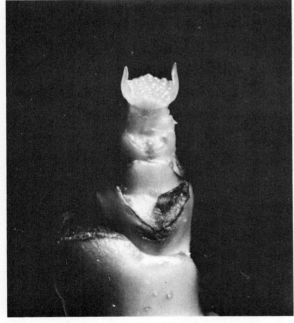

Figure 15-8. Photograph of an inflorescence primordium of Hydrangea, showing the primordia of the individual flowers. [Courtesy of Plant Science Research Division, Agricultural Research Service, U. S. D. A., Beltsville, Md.]

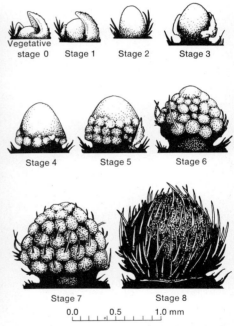

Figure 15-9. Stages in the development of a staminate inflorescence of Xanthium from a vegetative apex. Salisbury devised a system of using the stages of development attained to determine the effectiveness of various types of photoperiodic inductive treatments. [After F. B. Salisbury, *Plant Physiol.*, **30**:327 (1955)]

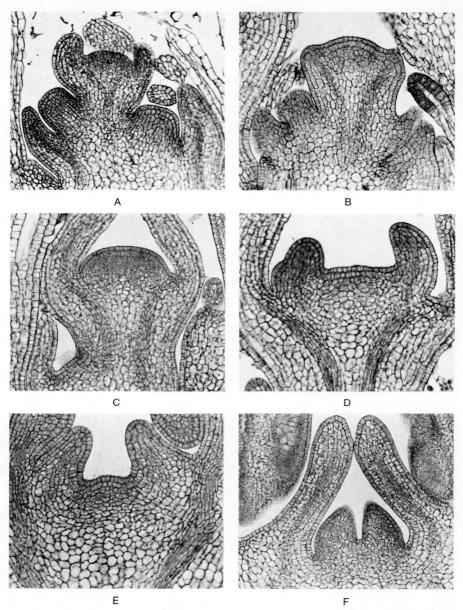

Figure 15-10. Photomicrographs of longitudinal sections of Vinca (periwinkle) flower primordia at different stages of development. **(A)** A young inflorescence. In the middle is the primordium of the first flower with two sepal primordia. To the left is the primordium of the second flower and to the right the vegetative apex ($\times 80$). **(B)** A flower primordium with sepal primordia just appearing ($\times 85$). **(C)** A flower primordium with larger sepals ($\times 95$). **(D)** A flower primordium with a stamen primordium at the left and a petal primordium at the right. Outside these are growing sepals ($\times 80$). **(E)** Carpel primordia on either side of the apical meristem, with stamen primordia in the upper corners ($\times 95$). **(F)** Later stage of carpel development. Two carpels show, the ventral margin of each being toward the center ($\times 95$). [Courtesy of N. H. Boke. From *Amer. J. Bot.,* **34:**433 (1947); **35:**413 (1948); **36:**535 (1949)]

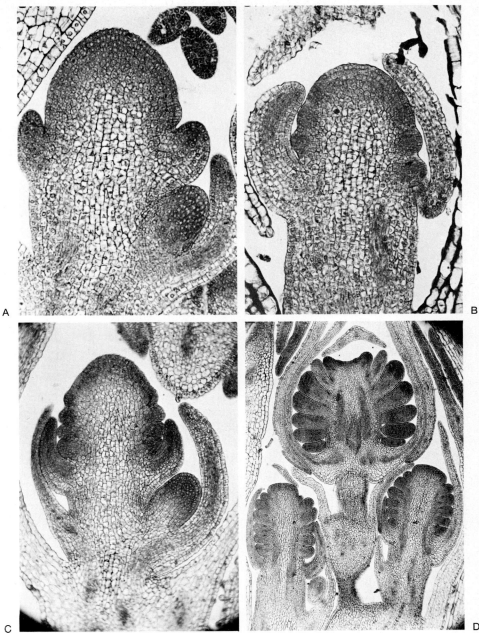

Figure 15-11. Photomicrographs of longitudinal sections of flower primordia of *Aquilegia formosa* (columbine). For section through vegetative apex see Figure 16-4. **(A)** Early primordium. At lower right is a bract, with an axillary bud above it and a sepal primordium above the bud. A stamen primordium is just beginning to form above the sepal. On the left side is a sepal primordium (×142). **(B)** Two developing sepals and stamen primordia above them (×142). **(C)** On right and left, from base up: bracteole, axillary bud, sepal, petal primordium, two stamen primordia (×95). **(D)** Primordia of three flowers. From base of older central flower up: sepals enclosing rest of the primordium, and on each side petal primordium, five stamen primordia, and carpel primordia developing from each side of the apex. The inverted V is a section through another sepal (×85). [Courtesy of S. S. Tepfer. From *Univ. Calif. Publ. Bot.,* **25:**513 (1953)]

HORMONES IN PHOTOPERIODISM. The Russian plant physiologist Chailakhyan postulated a flowering hormone called **florigen** in 1936, but florigen has never been identified and remains hypothetical. Despite substantial evidence that there is a hormone such as florigen, extensive efforts by many investigators to extract, isolate, and identify it have almost always failed. In 1961 Lincoln, Mayfield, and Cunningham [26] reported the extraction from flowering Xanthium plants of a crystalline, water-soluble organic acid that would cause noninduced Xanthium plants to bloom. The substance could not be extracted from vegetative plants. However, this extract induces blooming only when applied to the leaves and is ineffective when applied to the buds (whereas florigen should affect buds) so it may be a precursor of florigen or a substance necessary for florigen synthesis rather than florigen itself. Lincoln and his coworkers provided additional evidence in a series of papers, and in 1970 Hodson and Hamner [20] presented still more convincing evidence, for the effectiveness of the extract. It induced flowering in about three-fourths of the treated Xanthium plants and in about half of the treated duckweed (*Lemna perpusilla*), a more substantial response than previously reported. However, in Xanthium the extract was effective only when GA was added although GA by itself was ineffective. In duckweed GA was not needed to make the extract effective and, as a matter of fact, added GA made the extract totally ineffective. Further evidence is needed before the extract can be equated with florigen and besides the substance has not yet been identified chemically.

Despite the elusiveness of florigen, a number of investigators have measured its rate of translocation from a leaf to a bud by determing the time interval between completion of photoperiodic induction and the first evidence of morphogenetic changes in the bud. Salisbury found that in Xanthium the rate was influenced by temperature, with a Q_{10} of from 1.5 to 2.5, and possibly also by light intensity. The latter would tie in with the fact that the rate of translocation of sugars from photosynthesis out of leaves greatly influences the outward translocation of other solutes. Imamura and Takimoto [22] found that the rate of florigen translocation in Japanese morning glory was on the order of 2.5 to 3.0 mm/hr. This is much slower than the usual rate of sugar translocation through the phloem (often 200 mm/hr or more), but it is about the same as the rate of translocation of some viruses. In this connection, it is interesting to note that Bonner suspected that his success in getting the floral stimulus transmitted through a series of several plants by grafting might suggest that it was self-propagating like a virus, but all efforts to isolate such a viruslike substance failed. At any rate, the slow translocation of florigen suggests that it may be a substance of high molecular weight.

Since inductive photoperiods supplied to only one or a part of the leaves of a plant are often more effective (or effective only) if the leaves in the noninductive photoperiods are removed, it has been suggested that leaves in noninductive photoperiods produce a hormone that inhibits flowering. In some species the inhibition is stronger than in others, and the relative importance of inhibitors and promotors may differ with the species. However, natural flowering inhibitors are even more hypothetical than florigen; besides, at least some of the inhibition could be explained simply on the basis of massive sugar translocation from noninduced leaves to the buds preventing adequate translocation of florigen from the induced leaf or leaves.

Several of the established phytohormones have been found to have some influences on flowering, but none of these qualifies as being florigen nor do their effects go far in elucidating the hormonal aspects of photoperiodism. Synthetic auxins such as naphthaleneacetic acid (which are not destroyed by IAA oxidase) induce flowering of pineapples, apparently because they induce the formation of ethylene. However, this is a rather unique case and, in general, applied auxins either have no effect on flowering or inhibit it, particularly when applied to SDP during florigen synthesis or translocation. Auxin apparently is more abundant under long days than short ones, but this is true of both LDP and SDP so it does not help much in explaining photoperiodic flower induction. Wareing and others have found that abscisic acid will induce flowering in several SDP such as currants, Japanese morning glory, and pigweed under long days, and that it inhibits the flowering of a few LDP such as spinach under long days. However, it has no effect on the flowering of several SDP plants such as Xanthium, soybeans, and Maryland Mammouth tobacco.

Applied gibberellins have had the most spectacular effects on flowering, bring about flower initiation of most LDP on which they have been tried under short days (Chapter 13). However, they have no effect on flowering of SDP and of a few LDP and even inhibit flowering in a few species of each type [35]. Thus, despite the fact that increases in the gibberellin content of LDP have been reported following photoperiodic induction, gibberellins are evidently not florigen because florigen is evidently the same in all photoperiodic classes.

Chailakhyan, following the discovery of the influence of gibberellins on flower initiation in LDP, modified his florigen theory by proposing that florigen is two hormones rather than one, a gibberellin and a hypothetical hormone he called **anthesin.** He suggested that LDP could produce anthesin under any day length but gibberellin only under long days; that SDP could produce gibberellin under any day length but anthesin only under short days; and that DNP could produce both under any day length. Although this theory could explain most of the observed effects of gibberellins, it does not help much in clarifying the hormonal aspects of photoperiodism because anthesin is just as hypothetical and elusive as florigen has been.

THE INTERNAL PROCESSES AND CONDITIONS OF PHOTOPERIODISM. Despite the extensive research on plant photoperiodism during the past half century, our knowledge of the internal physiological, biochemical, and biophysical processes and conditions involved in photoperiodism is still extremely sketchy and incomplete. This is particularly true of LDP because most of the pertinent research has been on SDP, and there is still no satisfactory explanation for the different responses of these two and the other photoperiodic classes of plants to the length of day and night, although a variety of theories has been proposed.

What is known or at least strongly suggested, at least for SDP, can be outlined as follows.

1. Phytochrome is the photoreceptor pigment, and the conversion of P_{FR} to P_R (or at least a decrease in the P_{FR}/P_R ratio) during the early hours of the dark period is prerequisite to the dark processes leading to florigen synthesis in the leaves.

2. Although the phytochrome shift accounts for part of the long night requirement, it is apparently completed before the end of the critical dark period; therefore the timing device may also include the proposed photophil-skotophil circadian rhythm (which may also account for the first so-called high-intensity light process).

3. A sequence of unknown biochemical reactions requiring darkness presumably results in the eventual synthesis of florigen.

4. The second high-intensity light process may (or may not) involve either the stabilization of florigen from an unstable precursor or promote its translocation from the leaf.

5. Florigen is translocated rather slowly through the phloem from the leaves to the apical meristems.

6. In the buds florigen brings about the shift from vegetative primordia to flower primordia and the subsequent development of flower buds, perhaps by altering the kinds of mRNA and ribosomal RNA and thus, the kinds of enzymes synthesized by the cells of the apical meristem.

Although there are many gaps and uncertainties in the above sequence of events, the greatest gap is in the processes leading to florigen synthesis. For one thing, since florigen is still unidentified, it is impossible to propose even hypothetical pathways leading to its synthesis. A variety of metabolic inhibitors and antimetabolites have been applied to leaves during, before, and after photoperiodic induction in the hope of elucidating the processes involved. These experiments indicate respiration, RNA synthesis, and protein synthesis are essential to flowering. However, this is not surprising or unexpected and does not help much in elucidating the nature of the processes. Cobalt ions slow down the time-measuring process some, but their role is unknown.

One discovery that may be of more significance was the finding by Bonner, Heftmann, and Zeevart that SKF-7997, an inhibitor of steroid synthesis, inhibited flowering of Xanthium and Japanese morning glory when applied to the leaves. The investigation was conducted on the hunch that florigen, like the reproductive hormones of animals, might possibly be a steroid. However, this does not mean that florigen is necessarily a steroid because SKF-7997 may also inhibit other sytheses such as that of gibberellins, which like steroids are mevalonic acid derivatives. Much more extensive discussions of the photoperiodic processes are available elsewhere [6, 19, 35, 36, 37, 41, 42], and they make it even more evident that clarification of the photoperiodic reactions awaits more extensive future research.

Vernalization

The term vernalization [29, 32] was coined in 1928 by the Russian agronomist Lysenko who worked on this phenomenon, but there had been previous reports of vernalization by others, such as Gassner in Germany in 1918 and by investigators at the Ohio Agricultural Experiment Station before the turn of the century. Vernalization has been given several different definitions. In the most restricted sense it applies only to the technique of promoting the flowering of biennials or winter annuals by exposing soaked seeds to inductive low temperatures. The water content of the seeds is usually raised to about 40%, enough to make the embryos responsive to low temperatures but not enough to permit germination. However, we shall use the term in the broader and more generally used sense of promotion of flowering by low temperature preconditioning at any stage in the life of a plant.

In nature vernalization occurs during the winter and flower initiation generally does not occur until the warmer weather of the following spring, so the effect is usually inductive. However, the term vernalization can also be applied to the rather rare cases (as in Brussels sprouts) where flower initiation begins before the end of the cold period. Although the minimum length of the cold period and the effective temperature range vary from species to species, in general several weeks of temperatures ranging from around $-2°C$ to around $10°C$ are required. In some species even temperatures as high as $19°C$ are effective. In general, the optimal temperature range is quite broad ($10°C$ or so).

TYPES OF PLANTS AS REGARDS VERNALIZATION. Most of the plants requiring vernalization are winter annuals or biennials. The winter annuals include plants such as winter rye and winter wheat that are planted or naturally germinate and start growing in the fall, are vernalized during the winter, and bloom the following spring or summer. Spring rye and spring wheat are varieties that do not require vernalization and can be planted in the spring. In winter annuals the vernalization is quantitative, that is, flowering will usually occur eventually even without vernalization but after about twice the usual time and perhaps to a lesser degree. The seeds of biennials, such as henbane and many other weeds and celery and various other economic plants, generally germinate in the spring and grow into rosette plants, that is, plants with no internode elongation, no flowers, and with the leaves usually appressed close to the ground. In the spring of their second year, following vernalization during the winter, the plants bolt (the internodes elongate greatly) and bloom. Such true biennials have an absolute or quantitative requirement for vernalization and will remain as vegetative rosettes indefinitely if not provided with the required low temperature preconditioning (Figure 15-12).

Some perennials have either a qualitative or quantitative vernalization response (Figure 15-13). A number of species of perennial grasses have their flowering promoted by vernalization. Chrysanthemums require vernalization only once, and plants propagated vegetatively from a vernalized plant have no further vernalization requirement (Figure 15-14). Woody perennials that initiate their flowers may, in at least some species, have a vernalization requirement, but there has been little research done on this. Many trees and shrubs that bloom in the spring or early summer initiate their flowers the previous summer, and in such plants vernalization in the usual sense is obviously not necessary. However, most temperate zone trees and shrubs require low temperature preconditioning for breaking the dormancy of both vegetative buds and flower buds, and without it the buds do not open in the spring. If flowering is interpreted as including anthesis (opening of flower buds) as well

Figure 15-12. Celery plants of the same age. The group at the left, which have bolted and bloomed, were kept at temperatures between 10 and 15°C. The plants at the right were kept at temperatures between 15 and 21°C. [Photograph from H. C. Thompson.]

as flower initiation, this cold requirement for breaking flower bud dormancy could be considered as a type of vernalization. In some species such as lilac the cold requirement for breaking flower bud dormancy is greater than for breaking leaf bud dormancy, so in some of the southern states the plants develop full-grown leaves but no flowers. True annuals, in contrast with winter annuals (which actually can be considered to be a kind of biennial), are in general not influenced by vernalization, but there have been reports of somewhat earlier flowering in some species if they have been vernalized.

Some plants that require vernalization or have their flowering promoted by it also have a subsequent photoperiodism requirement, whereas others are day neutral. Most winter annuals also require subsequent long days, although it

Figure 15-13. Influence of 2 or 3 weeks of low temperature (10 to 15°C) on the blooming of Mathiola (stocks) plants. The plants were returned to higher temperatures after the cold treatment. Note that the plants not given cold treatment failed to bloom. [Data of K. Post, *Proc. Amer. Soc. Hort. Sci.,* **33:**649 (1935)]

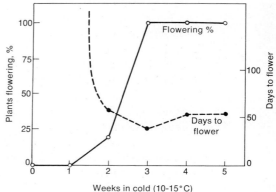

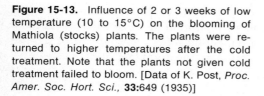

Figure 15-14. All these chrysanthemum plants were kept under photoinductive short days, but the two at the left were vernalized whereas the two at the right had been devernalized. [Courtesy of H. M. Cathey, Agricultural Research Service, U. S. D. A., Beltsville, Md.]

has been reported that short days may partially substitute for the vernalization requirement. Many biennials are also LDP, whereas chrysanthemum is a SDP.

Once a plant is vernalized the condition usually persists for a long period of time. However, vernalized plants may be devernalized by a period of high temperature (on the order of 30°C) if it comes within 4 or 5 days of completion of vernalization.

THE PHYSIOLOGY OF VERNALIZATION. Experiments involving localized low temperature treatments or the grafting of vernalized buds on unvernalized plants seem to make it clear that the meristems themselves are the site of vernalization. This is in contrast with the situation in photoperiodism where the leaves are the structures influenced by the environmental factor, and a hormone translocated to the meristems initiates the morphogenetic changes. However, once a plant has been vernalized it can cause an unvernalized plant to bolt and bloom when grafted to it. This indicated that vernalization results in the production of a hormone, which Melchers named **vernalin.** The efforts to isolate and identify vernalin were even less successful than the efforts to isolate and identify florigen. However, with the discovery that gibberellins would cause unvernalized biennials to bolt and bloom, it was suggested that vernalin was actually just a gibberellin. Gibberellins do come much closer to having the properties required of vernalin than they come to those required of florigen. The principal barrier to equating gibberellins with the hypothetical vernalin is that gibberellins induce bolting prior to any flower initiation, whereas vernalization induces flower initiation before bolting occurs.

Vernalization occurs only in the presence of an adequate supply of oxygen, food, and water, indicating that energy from respiration is required, but this is hardly surprising and provides little information about the biochemical processes involved. There has been some evidence that the meristematic cells must be actively dividing to be vernalized and that DNA and perhaps RNA may be involved, but the physiology of vernalization remains essentially obscure.

THERMOPERIODISM.   Thermoperiodism, the influence of daily temperature fluctuations on plant growth and development, has been elucidated largely by the investigations of Frits Went and his coworkers (Chapter 11). In contrast with vernalization, thermoperiodism is not an inductive phenomenon and the temperature affects current growth and development.

In most species the thermoperiod has more influence on the rate and vigor of vegetative growth and such aspects of reproduction as fruit set and development than it does on flower initiation. Even when there is an effect on flower initiation it is more quantitative rather than qualitative. For example, in tomatoes the optimal thermoperiods also increases the number of flowers per inflorescence. Favorable thermoperiods greatly promote both the vegetative growth and abundance of flowering of Browallia (Figure 15-15). African violets bloom most abundantly when the day temperature is around 16°C and the night temperature is about 22°C or more, one of the few exceptions to the

Figure 15-15. Two Browallia plants of the same age. Both were kept at day temperatures of about 21°C, but the one on the left was kept at 10°C at night whereas the one at the right was kept at 18°C at night. [Courtesy of the Department of Floriculture and Ornamental Horticulture, Cornell University]

usual requirement for warmer days than nights. Night temperatures below 20°C essentially inhibit flowering. The influence of thermoperiods on flowering is most obvious in DNP, but at least some of the modifying effects of temperature on photoperiodic responses appear to be thermoperiodic in nature.

Many temperate-zone flowering bulbs such as tulips and hyacinths have different optimal temperatures for various stages of their development, and in regions where they grow and bloom successfully the seasonal temperature changes closely approximate the changing optima. As might be expected, most of the research on this temperature effect has been done in Holland, mainly by Blaauw, Hartsema, and Luyten and their associates. Went calls this seasonal thermoperiodism, in contrast with the diurnal thermoperiodism we have described. However, seasonal thermoperiodism has almost as much in common with vernalization as it does with diurnal thermoperiodism and should perhaps be considered as a third type of temperature influence on flowering.

Tulips may be considered as an example of seasonal thermoperiodism. By the time the leaves have died in early summer several leaf primordia have already developed from the apical meristem of the bud. Then, within about three weeks flower primordia initiate and develop into flower buds, the optimal temperature being 20°C. However, completion of flower bud development requires a temperature of 9°C or so for a period of 13 to 14 weeks. For sprouting of the bulb, enlargement of the leaves, elongation of the peduncle, and growth and opening of the flower, gradually increasing temperatures up to 20°C or more are required. In regions where there is not enough low temperature for completion of flower bud development during the winter tulips fail to bloom even though they may survive, although bulbs brought from more suitable climates will do well the first year.

Flower Physiology

In the preceeding section we have been concerned primarily with the factors that bring about the initiation of flower primordia and to a lesser extent with the development of these into flower buds. In this section we will give further consideration to the development of flower buds, the opening of the buds **(anthesis)**, flower movements, pollen physiology, fertilization, and the senescence and abscission of flower parts.

Sex Expression

As applied to the sporophytes of angiosperms, sex expression refers to development of stamens or pistils in flowers. The majority of angiosperm species have perfect flowers with both stamens and pistils. In these sex expression results from a sudden and strong but subtle change in unknown morphogenetic factors in the minute distance between the primordia of the stamens and carpels. In monoecious species there may also be a shift in morphogenetic factors that influence sex expression, but over much greater distances. In many monoecious species, such as squash, staminate flowers are produced at the first (oldest) nodes, followed by a mixture of staminate and pistillate flowers, and finally at the youngest nodes pistillate flowers only. In squash these may then be followed by parthenocarpic pistillate flowers, the ultimate in what appears to be an increasing trend toward femaleness. In dioecious species there are, or course, staminate (male) and pistillate (female) plants, the ultimate degree of sex expression in the sporophyte generation.

Since the production of perfect (monoclinous) or imperfect (diclinous) flowers and the monoecious or dioecious condition are species characteristics, or in some cases varietal characteristics, they are obviously under basic hereditary control. Some species, notably various members of the cucumber family, have both monoecious and dioecious varieties. American holly trees are almost universally dioecious, but recently a monoecious tree was found in North Carolina. There have been reports of sex chromosomes in various dioecious species, but these reports have been questioned by some and, in any event, sex chromosomes do not appear to exert as strong control over sex expression in plants as in animals.

Despite the basic genetic nature of sex expression, it has been found that various factors such as the photoperiod, thermoperiod, hormones, and nutrition can greatly modify or even reverse the sex expression of many plants. A low carbon-to-nitrogen ratio has been found to increase the percentage of pistillate flowers developed by several monoecious species. In squash and cucumbers long days promote the development of staminate flowers, whereas short days promote the development pistillate flowers. In addition to this photoperiodic effect [31], there is a thermoperiodic effect at least in cucumbers: cool nights promote development of staminate flowers whereas warm nights favor pistillate flowers. Also, applied auxins promote the development of pistillate flowers by squashes and cucumbers, as do ethylene, acetylene, and even carbon monoxide, whereas gibberellins promote the development of staminate flowers.

Various factors such as those mentioned may induce the development of pistillate flowers as early as the ninth node or delay their formation until as late as node 100, but they do not obliterate the normal progression from staminate to pistillate flowers. However, the production of staminate flowers by pistillate plants of a dioecious variety of cucumber (economically desirable because of high yields) has been induced, making possible self-pollination of the plants for purposes of plant breeding. It is assumed that factors such as the photoperiod and thermoperiod influence sex expression by altering the hormonal or nutritional balance of the plants, but actually little is known about the physiology involved. In cucumber and at least some other monoecious plants, both stamen and carpel primordia are present initially but only one or another of these normally go on to develop into mature stamens or pistils.

The photoperiod has a marked influence on sex expression in a considerable number of species. One of the more spectacular examples is corn, in which short days bring about the development of pistillate flowers in the tassel. Generally the central spike of the tassel develops into a small but well-developed ear of corn, although it lacks the husks that enclose the usual axillary pistillate inflorescence (ear). Spinach is a dioecious LDP, but if short days are provided following induction the formation of staminate flowers on pistillate plants is promoted. John Heslop-Harrison found that the photoperiod may even influence sex expression of *Silene pendula,* a LDP with perfect flowers. If kept under long days continuously, the stamens are poorly developed and male sterility results whereas the pistils are unusually large. However, transfer to short days following induction results in normally developed perfect flowers.

One of the most extensively studied species from the standpoint of sex expression has been hemp (*Cannabis sativa*), a dioecious quantitative SDP. Usually staminate and pistillate plants are found in about a 1:1 ratio, and sex chromosomes have been reported. However, as long as a century ago occasional monoecious plants were reported, and in 1923 John H. Schaffner, who had been working on the sex expression of hemp, recognized that the photoperiod was an important factor. The literature on the subject was reviewed by Borthwick and Scully [7] in a paper in which they provided considerable additional information on the influence of the photoperiod on sex expression in hemp. In brief, very short photoperiods promote the development of staminate flowers on pistillate plants, whereas plants under photoperiods of 14 hr, just under the critical length, produce only pistillate flowers. Production of staminate flowers is further promoted by low temperatures just before or during induction. When female plants are pollinated with pollen from their staminate flowers, only female plants develop from the seeds under normal conditions, lending support to the sex chromosome concept. Production of pistillate flowers on staminate plants is less common and these are usually sterile, but Heslop-Harrison has reported that auxin applications will cause the change of stamen primordia into carpel primordia in male plants.

Flower Development

Flower initiation occurs with the formation of flower primordia in place of leaf and lateral bud primordia, and indeed the presence of the small flower primordia is a commonly used criterion indicating that photoperiodic induction has occured. However, the flower primordia do not necessarily develop into mature flowers. As we have noted previously, in many species once flower primordia have been induced photoperiodically they continue their development into flowers even if the plants are transferred to noninductive photoperiods, but in other species such as soy beans continuation of the inductive photoperiods is necessary for flower development. Also, in some species plants subjected to a minimum number of photoinductive cycles will only partially revert to vegetative development, resulting in a variety of reproductive-vegetative intergrades. In some species a shift from long to short days is required for development of primordia into flowers, whereas in others the reverse is true. Thus, the conditions suitable for flower initiation are not necessarily the same as for flower development. This is also shown by the different temperature requirements for flower initiation and flower development in tulips and other bulb plants.

Although there have been detailed descriptions of the morphological changes occurring during the development of flowers of a number of species, relatively little is known about the physiology of flower development. Among the more extensive physiological studies of flower development are those of Tepfer and his coworkers on columbine. They isolated immature flower buds or flower primordia from the plants and cultured them under sterile conditions. For initiation and initial development of all the flower parts, a culture medium containing sugar, minerals, vitamins, coconut milk, IAA, GA, and cytokinin was essential. The degree of development of the different flower parts could be altered by changes in the concentrations of the various hormones. In contrast with the natural situation, development of the sepal primordia inhibited the growth of the other flower parts, and it was necessary to excise them. Even if this was done stamens aborted despite the fact that they had developed anthers and pollen sacs, evidently because the medium did not provide all the necessary growth substances or the necessary concentrations of them. However, pistils did develop to mature size although mature ovules did not. Such experiments help elucidate the physiological requirements for flower development and, in particular, the importance of a certain hormonal balance that may be different for the different flower parts. Also, they indicate that hormones as well as foods from other parts of the plant are essential for flower development.

Peduncle and pedicel growth are generally quite rapid during flower development but slow down or stop when the flower reaches maturity, although there may be a second growth period after fruit development begins. Kaldewey found some degree of correlation between pedicel growth of Fritillaria and the diffusible auxin content of the flowers, and he suggested that a growth inhibitor might also be involved. The inhibitor appeared to be produced by the stamens. Another growth correlation was noted by Marré, who found that excission of the stamens of developing flowers of several species caused a decline in the mobilization of carbohydrate into the flower and cessation of mitosis in the developing ovulary. When flowers reach maturity there is a decrease in water translocation into them and generally a marked increase in the rate of respiration. Up to the time of anthesis there is generally a rapid accumulation of foods and mineral salts in a flower, but these then decline. In species that produce nectar part of the high sugar concentration of the petals is secreted in the nectar, and foods are also used in the development of the ovulary after fertilization.

There is scattered evidence that hormones play an essential role in the development of the sporangia (pollen sacs and ovules) and gametophytes (pollen and embryo sacs). Heslop-Harrison found that development of the megaspore of an orchid into an embryo sac did not occur unless auxin was supplied, either by pollination or by auxin applications on the ovulary. How general this auxin requirement is, is not known; however, in many species embryo sac development is at least not dependent on auxin from the pollen. Vasil has found that auxins, gibberellins, and cytokinins are all essential for maturation of the pollen sacs of onions.

Pollen Physiology

Research on pollen physiology has been predominantly in regard to the germination and growth of the pollen tube, which begins a short time after the pollen has been deposited on a stigma. The rate of pollen tube growth varies greatly with the species and environmental conditions but may be as high as 34 mm/hr. There is generally a peak optimal temperature for tube growth, which is 20°C in tomatoes. Tube growth occurs from the tip; both the cell wall and the plasma membrane are extended as the tube elongates.

The rather abundant starch usually present in pollen grains is hydrolyzed during tube growth and is probably used both in assimilation, including wall formation, and respiration. Pollen grains also secrete hydrolytic enzymes that apparently provide the pollen with additional food from the pistil and break down cell walls in the tissues of the pistil as the tube grows through them.

Pollen tubes readily grow in culture, either in a water solution or on an agar medium, and much that is known about pollen tube growth and physiology has been derived from culture experiments. Sugar is an essential component of the culture medium, both as a nutrient and for provision of a suitable osmotic potential. The optimal sugar concentration is surprisingly high: from 5 to 30%, depending on the species. This is related to the low water potentials in pollen grains, which generally develop such a high turgor in water that they burst open. Of course, too high a solute concentration in the medium results in plasmolysis of the pollen and failure of tube growth. The water potential of the stigmatic fluid is suitable for tube growth of pollen of the same species but may be either too high or too low for pollen of other species, thus causing either bursting or plasmolysis of the pollen.

Boron promotes pollen tube growth greatly in cultures, perhaps by influencing sugar utilization in the synthesis of cell wall components. Calcium ions also promote tube growth in culture, and in addition exert a chemotropic effect when unequally distributed through the agar. Calcium ions also play a chemotropic role in the pistils of some plants such as snapdragons, but other substances are the chemotropic agents in lilies and some other plants. Addition of auxins or gibberellins to the culture medium does not promote pollen tube growth, but auxin is quite abundant in pollen.

In cultures pollen generally does not develop tubes as long as it does in pistils, and pistil tissues or extracts promote pollen tube growth when added to culture media. Such results indicate that as pollen tubes grow through the pistil they obtain nutrients and perhaps growth substances from the pistil. Rosen and others have radioautographic evidence that substances from the pistil are incorporated into the elongating walls of pollen tubes. In pistils with solid styles the pollen tubes apparently secrete hydrolytic enzymes that digest cell walls in the style and provide a pathway through which the tube can grow, but in plants such as lilies that have hollow styles the tubes grow principally through the cavity. In some species pollen from the same plant fails to germinate on a stigma, or the tubes may grow much more slowly than those from other individuals of the species, thus providing one of the several means of promoting crossfertilization. Growth inhibitors with differential effects may be involved.

In recent years extensive work has been done on the ultrastructure and physiology of pollen, much of it by Rosen and his students. We cannot go into further detail regarding his work here except to mention that growing pollen tubes have a high level of metabolic activity, with substantial synthesis of proteins and RNA as well as cell wall materials. Rosen (34) has provided an extensive review of pollen tube structure and physiology.

Flower Movements

Flowers have a variety of movements that are of two principle types: movement of the entire flower or infloresence as a result of growth movements of

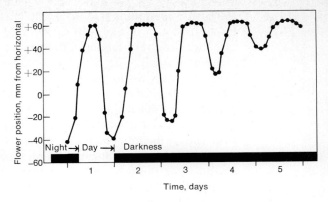

Figure 15-16. The elevation of Kalanchoë flowers by day and their drooping by night continues for a time as a circadian rhythm even after the plants have been placed in continuous darkness. [Data of R. Bünsow, *Biol. Zent.*, **72:**465 (1953)]

the pedicel or peduncle and opening and closing of flowers as a result of growth movements of the flower parts, particularly the petals. Turgor movements, which are responsible for the folding and opening of leaves, are apparently not involved in flower movements.

One well-known type of plant movement is the following of the sun by sunflower inflorescences, and this is also found in several other species. The east-west movement of the heads apparently results from a differential auxin concentration in the more and less brightly lighted sides of the peduncle, but a circadian rhythm is involved since the movement continues for several days in continuous darkness. Another and more common type of differential pedicel growth results in the elevation of flowers during the day and their drooping at night. At least in Fritillaria the growth movements result from differential concentrations of both auxin and inhibitors. These movements also have a circadian rhythm (Figure 15-16).

In most species flower opening occurs only once: at the time of anthesis. This results from the more rapid growth of the inner than the outer sides of the sepals and petals, and often also involves a rather rapid growth in size and unfolding of flower parts. However, some species have a daily opening and closing of the flowers, which also involves differential growth and a circadian rhythm (Figure 15-17). In most species with such movements the flowers are open during the day and closed at night, but in a number of plants

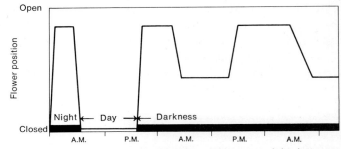

Figure 15-17. *Cestrum nocturnum* flowers are closed by day and open at night, but even in continuous darkness a circadian rhythm results in partial closing of the flowers. [Data of L. Overland, *Amer. J. Bot.*, **47:**378 (1960)]

such as evening primroses (Oenothera) and night-blooming jasmine (*Cestrum nocturnum*) the flowers are open at night and closed by day. At least in evening primrose, the synchronizing light is the time of dawn; opening occurs about 12 hr after dawn. In species with morning opening the time of dusk appears to provide the timing. In many species, including those with no daily opening and closing of flowers, the flowers close after pollination. In Portulaca the closing occurs 4 hr after pollination and apparently involves the stigma, since closing can be induced without pollination by excision of the stigma. However, in a larger number of species the flowers do not close after pollination although the petals and stamens generally become senescent and abscise.

Senescence and Abscission of Flower Parts

The senescence and abscission of petals and stamens following pollination, or in some cases fertilization, is a general if not universal aspect of flower ontogeny. Also, staminate flowers generally become senescent and abscise after they have shed their pollen. In some species, particularly ones like tulips where the sepals and petals are similar, the sepals also become senescent and abscise, whereas in other species the sepals are persistent and subtend the mature fruit. The stigma and style of the pistil also generally become senescent and either abscise or wither, leaving only the ovulary which begins developing into a fruit if fertilization has occurred or if there is natural or artificial parthenocarpy. Otherwise, the entire pistil abscises.

As in senescing leaves, there is generally extensive hydrolysis of proteins and other substances in senescing flower parts and substantial quantities of amino acids, sugars, mineral salts, and other solutes are translocated to the ovulary or other parts of the plant. There may also be a substantial loss of water, resulting in wilting and withering.

Fruit Physiology

Three aspects of fruit physiology have been identified. In chronological order these are fruit set, fruit growth, and fruit ripening. Most of the investigations of fruit physiology have been on fleshy fruits of economic importance. The physiology of fleshy fruits of wild plants is presumably similar, but the following discussion is applicable only in part to dry fruits of various types. Also, it should be noted that the details of fruit physiology may vary almost as much from species to species as fruit structure; therefore, it is not always safe to generalize too much from experiments on any particular species.

Fruit Set

Fruit set can be considered to have occurred if abscission of the ovulary has been prevented and the ovulary has begun its rapid growth into a fruit, although immature fruits sometimes abscise even after several weeks of growth. The usual prerequisits for fruit set is pollination and often also fertilization, but some plants such as bananas, pineapples, and seedless citrus fruits set

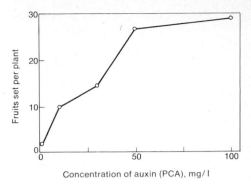

Figure 15-18. Influence of different concentrations of a synthetic auxin (*p*-chlorophenoxyacetic acid, PCA) on fruit set of tomatoes. [Data of H. B. Tukey, *Bot. Gaz.,* **94:**433 (1933)]

parthenocarpic fruits without either. A second type of parthenocarpy requires pollination but not fertilization, as in some orchids and in triploids such as some melon varieties. Failure of fertilization may result from slow pollen tube growth which may cause either a failure to reach the embryo sac or the loss of sperm viability before the tube reaches the embryo sac. In triploids fertilization fails because meiosis cannot occur, and there are also other types of genetic sterility that prevent fertilization. A third type of parthenocarpy occurs after both pollination and fertilization, but the embryos abort and there is no seed formed. This type occurs in a number of species including grapes and stone fruits. Lack of fertilization does not necessarily result in seedless fruits, since viable embryos and seeds may result from parthenogenesis or apomixis.

As was noted in Chapter 12, auxin applications induce artificial parthenocarpy in various species (Figure 15-18), including figs, hollies (Figure 15-19),

Figure 15-19. Artificial parthenocarpy of holly fruits induced by a synthetic auxin (methyl ester of NAA). Neither the controls (left) nor the treated plants were pollinated. [Courtesy of the Boyce Thompson Institute for Plant Research]

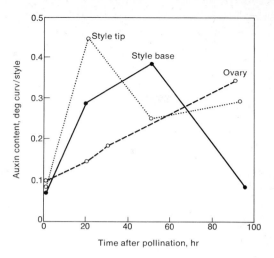

Figure 15-20. Increase in auxin content in tobacco pistils following pollination. (Auxin concentration was determined by the Avena curvature assay.) [Data of H. A. Lund, *Amer. J. Bot.,* **43**:562 (1956)]

tomatoes, melons, and other members of the cucumber family. Along with the fact that germinating pollen grains and developing embryos are rich sources of auxin (Figure 15-20), this suggests that auxin plays an essential role in fruit set and development. However, it has been impossible to induce fruit set and artificial parthenocarpy with auxins in about three-fourths of the species tested. Gibberellins do induce parthenocarpy in some of these species, including roses, apples, pears, stone fruits, and grapes, as well as in some species like tomato where auxin is effective. Gibberellins have no effect on fruit set in some species. At present it is still not clear to what extent auxin or gibberellins or both or other phytohormones are involved in natural fruit set in various species.

In nature the percentage of flowers setting fruit may be reduced by various factors such as premature flower abscission, limited pollination, limited fertilization, inadequate nutrition, embryo abortion, and premature abscission of young fruits. Premature flower abscission is common in a number of species including tomato and potato and can be reduced by auxin applications. Limited pollination can result from lack of pollinating agents, lack of staminate plants in dioecious species, a limited period of stigma receptivity to pollen, bursting of pollen grains in rain drops, and other factors. Limited fertilization may result from various factors such as pollen sterility, slow pollen tube growth, and failure of meiosis. Fruit set requires an adequate supply of foods and minerals and is greatly reduced by interference with photosynthesis. The presence of developing fruits on a plant frequently limits further fruit set, presumably because of the competition for food. Removal of the fruits already set generally removes the inhibition on further fruit setting.

The premature abscission of young fruits may also involve competition for foods. Premature abscission can often be reduced by auxin applications, but if applied too early the auxin may greatly increase fruit abscission. This apparently results from auxin-induced embryo and seed abortion (Figure 15-21). In some species, particularly with multiflowered inflorescences, only a small fraction of the flowers ever set fruit. A single panicle of a mango tree has about 6000 flowers, but normally only four or fewer set fruit.

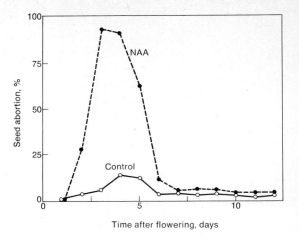

Figure 15-21. Effect of spraying apple flowers with a synthetic auxin (NAA, 40 mg/liter) on seed abortion and consequently premature abscission of fruits. [Data of L. C. Luckwill, *J. Hort. Sci.,* **28:**25 (1953)]

Limited fruit set, if not too severe, may be desirable from the standpoint of both the plant and the commercial producers of fruits. In species with numerous flowers per plant, the setting of fruit by every flower would result in such severe competition for food that all of the fruits would be small and might have underdeveloped seeds. As a matter of fact, commercial producers of fruits often spray their trees with synthetic growth substances to increase abscission when there has been a heavy fruit set. The remaining fruits will grow larger, increasing their cash value.

Fruit Growth

The conversion of an ovulary (and in accessory fruits certain adjacent structures) into a fruit is predominantly a matter of growth in size and involves little change in shape. Examination of an ovulary with a hand lens reveals that it has essentially the same shape as the mature fruit. The growth of fruits, particularly large ones such as pumpkins and watermelons, involves a marked and rapid increase in size. Sinnott found that, in only 18 days, a gourd ovulary 2.4 mm in diameter grew into a mature fruit 48 mm in diameter, a 20-fold increase in size.

The rate of fruit growth follows the usual sigmoid growth curve in most species, but there are some striking exceptions as in peaches, cherries, plums, and other stone fruits. Tukey and his coworkers found that these have a double sigmoid growth curve with a period of very slow growth separating the two grand periods of growth (Figure 15-22). During the first period of rapid growth the ovulary and the nucellus and integuments of the ovules are enlarging. During the period of slow ovulary growth the endosperm and embryos are developing rapidly. When these have completed their growth, the second period of rapid growth begins. Although the reasons for the double sigmoid growth rate of these fruits are not well understood, it appears that a competition for nutrients and perhaps water may be involved and that the developing endosperms and embryos are capable of mobilizing these from the ovularies.

Fruit growth always involves extensive cell enlargement. In some species there is also a period of active cell division preceding the period of cell enlargement, whereas in others there is little or no cell division beyond that which occurred during the development of the ovulary to its prepollination size. Differences in the sizes of fruits of different varieties of a species may result from cell size (as in cherries) or from cell numbers (as in apple). Fruit size may also be influenced by the number and sizes of the intercellular spaces formed. For example, the latter part of apple growth involves extensive intercellular space formation, and these spaces occupy about a quarter of the volume of the mature fruit. Thus, fruit volume increases more than fruit weight during this period.

The fact that auxins or gibberellins or both, depending on the species, are effective in bringing about fruit growth as well as fruit set in artificial parthenocarpy suggests that these hormones are involved in natural fruit growth. An auxin requirement would also seem logical because of the general necessity of auxin for growth. Although the auxin from the pollen tubes and the adjacent tissues of the pistil appear to be adequate for fruit set, developing seeds are essential for continued fruit growth in many species. For example, Nitsch induced the development of misshapen accessory strawberry fruits from the receptacles by preventing pollination in different patterns (Figure 15-23). If all the achenes were removed fruit growth ceased, but it could be promoted by auxin applications. The development of lopsided apples and other fruits has been correlated with the lack of normally developing seeds in the subnormal-sized segments. The fact that developing seeds produce abundant auxin, first in the endosperm and then in the embryo, suggests that the seeds promote fruit growth by providing auxin. Gustafson found that seedless varieties of oranges and grapes had up to seven times as much auxin in the ovularies of unopened flowers as seeded varieties, providing a possible explanation of their natural parthenocarpy.

However, despite rather extensive evidence of this type, some students of fruit growth are reluctant to draw any sweeping conclusions about the role of endogenous auxin (or at least IAA) in fruit growth. For one thing, peak

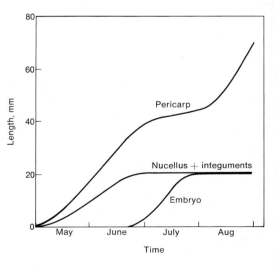

Figure 15-22. Double growth curve of the pericarp (fruit) of Elberta peach. Note that the growth plateau essentially coincides with the period when growth of the nucellus and integuments ceased and growth of the embryo occurred. [Data of H. B. Tukey, *Botan. Gaz.*, **94**:433 (1933)]

Figure 15-23. Influence of pollination and auxin application on the development of the accessory fruit (receptacle) of strawberry. **(A)** The fruit at the left was pollinated normally. Pollination was prevented in the other two, but the one at the right was treated with auxin. **(B)** In the fruit on the right pollination of the pistils on two sides of the receptacle was prevented, resulting in a flattened accessory fruit. **(C)** Only three of the many pistils of a strawberry flower were pollinated, resulting in localized development of the receptacle into accessory fruits. [Courtesy of J. P. Nitsch]

growth rates of fruits, particularly those with a double sigmoid growth curve, do not always correlate well with the peaks in auxin content. As an extreme example, in peaches the only appreciable increase in auxin content is during the period of slow growth. Also, chromatographic studies of growth-promoting substances from fruits have often revealed three or more substances, none of them IAA or other known phytohormones.

Substantial quantities of gibberellins have been found in the young fruits and developing seeds of a number of species. Gibberellins have also been found to increase fruit size in a number of species whether or not parthenocarpy was involved. Luckwell has pointed out that auxin is generally effective in species where fruit growth is predominantly by cell enlargement but ineffective if there is a substantial cell division component in growth. Since gibberellin is effective in some of the latter, it may act by promoting cell division as well as cell enlargement. Cytokinins might also be expected to play a role in these cell divisions, but there is little evidence for this aside from the fact that they have been found present in the young fruits of several species such as apple, tomato, and banana. Much remains to be clarified about the roles of phytohormones in fruit growth, not only as regards the relative importance of auxins, gibberellins, and cytokinins and the interactions among them but also as regards possible roles of unknown growth substances.

Fruit growth requires substantial quantities of foods, water, and mineral elements; the amounts required obviously increase with the size of the fruits. Most young fruits are green and produce some food by photosynthesis, but this is usually only a small fraction of the food used, and the rest must be translocated from the leaves (Figure 15-24). The foods are used not only in respiration and assimilation but also in the production of various substances that accumulate in quantity as the fruits are growing. Aside from the accumulation of starch, fat, and protein in seeds, the pericarps of most immature fleshy fruits accumulate much starch. The pericarps of a few fruits such as avocado accumulate fats, but this is not common. Many species accumulate organic acids in the vacuoles of the fruit cells, for example, citric acid in citrus fruits, malic acid in apples, tartaric acid in grapes, and isocitric acid in blackberries. These are derived principally from the dark fixation of carbon dioxide by pyruvic acid or phosphoenolpyruvic acid, rather than by drainage from the Krebs cycle. The dry weight of fleshy fruits generally ranges from 10 to 20% of the fresh weight, most of this being carbohydrates. There is usually little sugar accumulated in young fruits, but during the latter stages of fruit growth the sugar content of the vacuoles increases greatly in the fruits of many species. It has been suggested that the high turgor resulting from this increase in sugars is the principal factor in bringing about the second period of rapid cell enlargement in fruits with a double sigmoid growth rate.

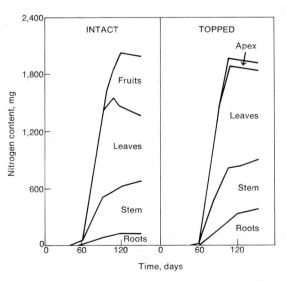

Figure 15-24. Topping of tobacco plants removes the developing fruits and so reduces the translocation of amino acids and other nitrogenous compounds from the leaves. Sugars and mineral elements are also translocated from leaves to developing fruits. [Data of R. Watson and A. H. K. Petrie, *Austr. J. Exp. Biol. Med. Sci.*, **18**:313 (1940)]

Fruit Ripening

With the completion of fruit growth and the attainment of the final fruit size, or in some cases even slightly earlier, fruits begin to ripen. In most species ripening is completed before the fruits abscise or are picked; however avocados do not ripen until after the fruits are picked, and picking promotes the ripening of other fruits such as apples and bananas. The ripening of fleshy fruits, with which this section is concerned, involves a complex of profound metabolic changes. The ripening of dry fruits is quite different, generally involving extensive secondary cell wall thickening and then death and desiccation of the cells. Dehiscence, if it occurs, can also be considered as part of the ripening process. The ripening of both dry and fleshy fruits is essentially a matter of senescence, even though in both cases it contributes toward seed dispersal.

The softening of fruits, which generally occurs as fleshy fruits ripen, results either from hydrolysis of pectic compounds or hydrolysis of starch or both. The extensive hydrolysis of hemicelluloses during ripening may also contribute some to softening, but there is generally no hydrolysis of cellulose. The principal factor in the softening of most fruits is the conversion of the protopectin and calcium pectate of the middle lamellae and primary walls to soluble pectins, which loosens the cells from one another. Also, in overripe fruits much of the pectin is converted to pectic acid. Since pectin forms gels, although neither pectic acid or protopectin do, fruits that are too ripe are not suitable for making jelly unless pectin is added.

The extensive hydrolysis of starch during ripening results in a greatly increased sugar content, and the completely ripened fruits of many species contain little or no starch. The sugars present are mostly fructose, glucose, and sucrose. Fats may also be hydrolyzed and converted to sugars. In most fruits there is a decline in organic acids during ripening, and some of these at least are converted to sugars. The decrease in sourness and increase in sweetness resulting from these processes contribute greatly toward making ripe fruits attractive to man and animals. Lemons are among the few fruits in which the organic acid content continues to increase during ripening.

Another set of changes involved in ripening is the synthesis of a wide variety of volatile substances that contribute greatly to the attractive odors and flavors of fruits. These substances include many different aldehydes, esters', ketones, hydrocarbons, terpenoid essential oils, and coumarin derivatives. Also contributing to desirable flavors of ripe fruits is the destruction of astringent substances, such as tannins and phenolics, that are present in some kinds of green fruits, notably persimmons.

Ripening also generally involves extensive changes in pigments. Usually chlorophyll breaks down, unmasking the chloroplast carotenoids, although chlorophyll is persistent in some ripe fruits such as watermelons and avocados. In some fruits such as bananas there is no additional carotenoid synthesis, whereas in others such as oranges carotenoid synthesis accompanies chlorophyll destruction. Most of the orange and yellow colors of ripe fruits result from carotenes or xanthophylls, and the red color of tomato flesh, red peppers, and rose fruits results from a carotenoid called lycopene. However, the red color

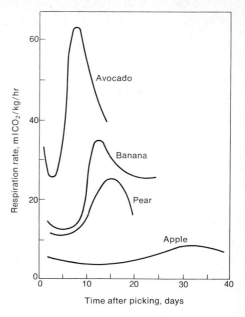

Figure 15-25. Rates of respiration of four ripening fruits, showing their respiratory climacterics. The rate of ripening correlates well with the intensity of the climacteric, ripening being fastest in avacado and slowest in apple of the four fruits shown. [After J. B. Biale, *Annu. Rev. Plant Physiol.,* **1:**183 (1950)]

of most fruits, and also the color of blue fruits, results from the presence of anthocyanins in the vacuoles of the cells. These are generally being synthesized about the time the chlorophyll is being broken down.

It has now been well established that energy from respiration is required for many of the processes involved in fruit ripening. Inhibitors of respiration have been found to delay or prevent the ripening of fruits. As fleshy fruits complete their growth there is generally a decrease in the rate of respiration, followed by a marked increase in respiration as ripening occurs. This is called the respiratory climacteric. The amount of increase in respiration during the climacteric varies greatly from species to species (Figure 15-25) and correlates positively with the rate of fruit ripening. Some fruits such as the citrus fruits do not have a respiratory climacteric. These are generally slowly ripening fruits, and Biale [3] has pointed out that in them ripening does not involve the extensive hydrolyses that occur during the ripening of fruits with a climacteric.

As might be expected from their influence on the rate of respiration, low temperature and low oxygen concentration reduce the climacteric and the rate of ripening. These environmental treatments have been used commercially to delay the ripening of fruits in storage to a desired degree. In contrast with the extensive hydrolyses of most substances during ripening, protein hydrolysis is limited and there may even be an increase in proteins. This fact may be related to the synthesis of new enzymes that catalyze the numerous metabolic processes involved in ripening. Protein synthesis appears to be particularly active during the climacteric, and it is possible that much of the energy from respiration is used in protein synthesis. However, mature fruits generally have a rather low protein content, on the order of 0.4 to 1.7% of the fresh weight.

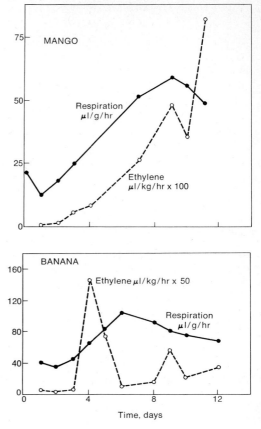

Figure 15-26. Relationship between rate of respiration and ethylene production of ripening fruits of mango and banana. Note that the patterns in the two species differ. [Data of S. P. Burg and E. A. Burg, *Plant Physiol.,* **37:**179 (1962)]

As far back as 1924 Denny found that ethylene promoted fruit ripening. It is now well established that ethylene is involved naturally in fruit ripening [3, 9] and that ripening fruits naturally contain and give off ethylene. Ripe fruits placed in a closed container with unripe ones, even of other species, will promote their ripening as a result of the diffusion of ethylene from the ripe fruits. Ethylene production correlates well with the rate of respiration (Figure 15-26), and there is considerable evidence that ethylene increases the rate of respiration. For example, ethylene will induce a climacteric even in fruits such as oranges that do not have one naturally.

Applications of auxins, particularly synthetic ones such as 2,4-D, can promote the ripening of green fruits, presumably because they increase ethylene synthesis. Dostal and Leopold have found that gibberellin retarded the development of the red color in ripening tomatoes but not the rate of respiration and suggested that ethylene operates in different ways on the various ripening processes. They found that ethylene would counteract the coloring inhibition induced by gibberellin. The literature on fruit ripening is extensive, largely because of its economic importance, and further discussions can be found elsewhere [25, 27].

Seed Physiology

The development and maturation of seeds is concurrent with the development and maturation of fruits. Seeds and fruits are indeed part of a single morphogenetic and physiological complex, particularly in view of the various correlative influences of one on the other. However, it has seemed best to include our brief consideration of the physiological aspects of seed growth and development in this section, along with other aspects of seed physiology—dormancy, viability, and germination[13].

Seed Development

Unless the embryo plant begins developing there is little or no growth of the ovule, the integuments fail to develop into seed coats, and no seeds form. The dependence of development of the seed as a whole on the development of the embryo is illustrated by the fact that, when embryos abort rather late in their development, development of the rest of the seed ceases; this results in small, partially developed, sterile seeds. Embryo abortion may also occur after only a few cell divisions, and the result is essentially the same as if there had been no fertilization. Abortion of the embryo usually occurs either because of genetic incompatabilities, failure of endosperm development, or inadequate nutrition of the embryo by the endosperm. The last situation may result from mutations or other genetic factors that do not permit the endosperm cells to synthesize all the essential growth substances.

When the zygote begins developing into an embryo the growth is steady and rapid; the increase in embryo volume is logarithmic. Jensen and others have found that young embryos are very active metabolically, having high rates of respiration, protein synthesis, and nucleic acid synthesis than older embryos. However, tissue culture studies of excised embryos suggest that very young embryos have extremely limited capabilities for synthesizing amino acids, vitamins, and phytohormones and that the ability to synthesize increases with age.

For example, van Overbeck found that when Datura embryos had attained their full size (5 mm long) they would grow in culture even if supplied only with sugar and the essential mineral elements. Younger embryos in the cotyledon stage (1 to 2 mm) also required vitamins and amino acids. Heart stage embryos (0.2 mm) grew only when coconut milk (liquid endosperm) was provided in addition, whereas proembryos (0.1 mm or less) rarely developed normally even with the added coconut milk. Young embryos have also been cultured successfully on completely synthetic media. For example, Rijven secured growth of heart-stage Capsella embryos on a medium containing salts, sugars, vitamins, amino acids, adenine, and several phytohormones.

The optimal osmotic concentration for embryo culture changes with age. Early heart stage Datura embryos grow best with 8 to 12% sucrose, late heart stages with 2% sucrose, and torpedo stage embryos with only 0.1% sucrose. It is interesting to note that high concentrations of sugars accumulate in the ensosperm until cell wall formation in the endosperm begins, after which the sugar concentration begins declining.

Substantial quantities of food are required during seed development, first for respiration and assimilation during the growth of the endosperm, then for development of the embryo and the seed coats, and later for accumulation in the endosperm, perisperm, or cotyledons as the case may be. Practically all of this food is translocated into the ovule from other parts of the plant, although some is recovered from the breakdown of first the nucellus and later the endosperm in species where they do not persist. Although the sugars, salts and water used in seed growth are translocated into the ovules, the cells of the endosperm and later on those of the embryo are capable of synthesizing most of the varied substances essential for growth, such as amino acids and the numerous vitamins, hormones and growth substances. Developing seeds are rich in auxin, cytokinins, and gibberellins and, as we have noted earlier, continued fruit growth is commonly dependent on hormones from the developing embryos.

There is at least one rather marked metabolic difference during the maturation and ripening of seeds as contrasted with fruits. Whereas in fruits there are extensive hydrolyses of starches, hemicelluloses, fats, and other substances, in seeds there is extensive synthesis and accumulation of these substances and also of proteins and, in some species, sucrose or inulin or other substances. The accumulation of substantial quantities of proteins, as well as starch and fats, notably in the seeds of grasses and legumes, make seeds an important and concentrated source of food for man and other animals. Accumulated proteins in seeds are generally crystallized and are quite different from those found elsewhere in the plant. This suggests that kinds of mRNA not found elsewhere in the plant are produced during seed maturation, and this may help explain the fact that RNA is ten times as abundant as DNA in mature wheat seeds.

Other changes that occur during seed ripening are a decline in the growth hormone content, synthesis of germination inhibitors in some species, progressive and extensive desiccation, and maturation of the seed coats. The latter usually involves considerable secondary cell wall deposition, hardening of the walls, synthesis of various pigments, and finally perhaps death of the cells. In some species the seed coats may also be impregnated with substances that make them impermeable to water, oxygen, or both. The desiccation of the maturing seed brings about a marked decrease in the rate of respiration and other processes. The low level of metabolism of embryos in dry seeds contributes toward survival for considerable periods of time.

Seed Viability

As long as the embryo in a mature seed remains alive, the seed is said to be viable. The period of viability varies greatly from species to species and is also influenced by environmental conditions. The seeds of most cultivated plants and of many wild species ordinarily remain viable for from 1 to 3 years. However, the seeds of some species such as silver maple remain viable only a few weeks. In this maple the embryo dies when the water content of the seed (about 58% at maturity) drops below about 30%, and this generally occurs within several weeks after the seed have been dispersed.

At the other extreme, Asian water lotus seeds about 200 years old and seeds of *Cassia multijuga* (a South American legume) 158 years old have been found to be viable. Accounts of discovery of viable wheat seeds 6000 years old or so in Egyptian tombs are fictitious, however. Except for the water lotus, all seeds known to have remained viable for over 100 years have been legumes with hard seed coats. However, many different species of weeds from a wide range of families have seeds that remain viable for many years.

In 1879 Dr. W. J. Beal of Lansing, Mich., mixed seeds of 20 different species of herbaceous weeds with sand, placed them in sealed jars, and buried the jars deep in the ground. Since then he and his successors have removed a jar, first every fifth year and later every tenth year, and tested the percentage of seed germination for each species. Four species had no viable seeds after only 5 years but, after 25 years, 11 species still had viable seeds. After 50 years there were viable seeds of 5 species. The species, with the percentage of viable seeds for each, were as follows: *Verbascum blattaria,* 62%; *Rumex crispus,* 52%; *Oenotheria biennis,* 38%; *Brassica nigra,* 8%; and *Polygonum hydropiper,* 4%. The first three still had viable seeds after 70 years, with germination percentages of 72, 8, and 14% respectively. After 80 years the germination percentages had dropped to 70, 2, and 10% [14].

Thus seeds of some species may remain viable for many years. The viability of stored seeds can be prolonged by cool temperatures, low humidity, and reduced oxygen. Of course, in nature seeds rarely attain their potential longevity. Seeds without dormancy can be expected to germinate within a year, and even seeds with dormancy are likely to have their dormancy broken and germinate within several years. Only rarely would seeds happen to be in an environment permitting continued viability but neither the breaking of dormancy nor germination.

Seed Dormancy

If a viable seed fails to germinate when supplied with adequate water, air with the usual oxygen content, and a suitable temperature for growth, it is considered to be **dormant.** If a seed does not germinate simply because of the lack of one or more of these three environmental factors it is said to be **quiescent.** A few botanists use the term dormancy in a broader sense, considering quiescence to be a type of dormancy, but we shall use the term in its more restricted and more generally used sense here.

A considerable number of species, including economic ones such as beans, peas, corn, and radish, have seeds that are never dormant and will germinate at any time they are in the proper environment. However, a considerable number of both cultivated and wild species have seeds with dormancy [24]. Five types of seed dormancy have been identified. Each type will be discussed briefly.

IMPERMEABLE SEED COATS. The impermeability of seed coats to water results from the impregnation of the seed coats with waxes or other waterproofing substances. Many legumes, including clovers, alfalfa, and locusts, and also other species such as water lotus and morning glories have this type of seed dormancy. These seed coats may also in some cases be impermeable to oxygen. The seeds of many grasses and composites have seed coats that are permeable to water but impermeable to oxygen. The two seeds present in a Xanthium (cocklebur) fruit provide a particularly interesting example of oxygen impermeability. The lower seed usually breaks dormancy and germinates the first spring, but the upper seed has a higher oxygen requirement and does not germinate until the second spring.

Seed coat dormancy may be overcome by abrasion of the seed coats by shaking the seeds (often in sand) or by the use of strong acids or by cracking or nicking the seed coats. Such procedures are referred to as **scarification.** Of course, removal of the seed coats is also effective. In nature seed coat dormancy is broken by gradual decay of the seed coats by bacteria and fungi, by cracking of the seed coats by alternate freezing and thawing during the winter and also sometimes by fire.

MECHANICALLY RESISTANT SEED COATS.   The seed coats of some species are permeable to both water and oxygen but are so hard and strong that they prevent the swelling of the seed and the growth of the embryo. The high imposed pressure greatly reduces the water potential gradient, thus limiting the absorption of water by both imbibition and osmosis. Plants with mechanically resistant seed coats include juniper and hazlenut and a considerable number of weeds including pigweed, shepherds's purse, mustard, and peppergrass. In pigweed, and also some of the others listed, dormancy persists for up to 30 years or more as long as the seed coats remain saturated with imbibed water. However, if the seed coats dry out, the dormancy is lost and germination occurs when the seeds are wet again. This type of seed coat dormancy may also be broken in the same ways as dormancy due to impermeable seed coats.

RUDIMENTARY EMBRYOS.   A number of species of plants have seeds that contain only partially developed embryos at the time of seed dispersal. These include ginkgo, European ash, and various species of hollies and orchids. In some cases the embryos are nearly mature but in others they are in early proembryo stages. The breaking of dormancy is simply a matter of the embryo continuing to develop until it reaches maturity, which is usually a rather slow process.

PHYSIOLOGICALLY IMMATURE EMBRYOS.   In a considerable number of species the embryos are morphologically mature but the embryos themselves are dormant, as evidenced by the fact that they fail to germinate even when removed from the seeds and placed under suitable cultural conditions. Amen [1, 2] describes these dormant embryos as being physiologically immature, that is, lacking some of the enzymes essential for catalyzing the processes involved in germination and growth. Among the species with physiologically immature or dormant embryos are lettuce, barley, iris, basswood, dogwood, various ashes, pines, and a good many members of the rose family including apples, peaches, pears, hawthornes, and Sorbus.

This type of dormancy is generally broken if moist seeds are subjected to 6 weeks or more of low temperatures, no more than a few degrees above freezing. This results in what has been called **after-ripening** of the embryo. In

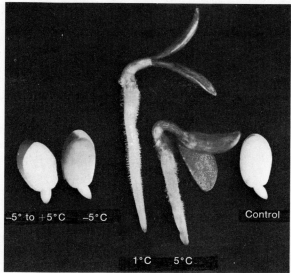

Figure 15-27. Germination of seeds of *Sorbus aucuparia* 2 days after they were placed on moist filter paper in petri dishes. The seeds had been stratified for 2 months at the indicated temperatures. The control seeds were not stratified (and so not after-ripened) and germinated only after 21 days. [Courtesy of the Boyce Thompson Institute for Plant Research]

nature after-ripening is achieved by the usual winter temperatures, whereas experimental and commercial after-ripening is accomplished by placing the seeds in a moist substrate and refrigerating them at 1 to 5°C for the required number of weeks (Figure 15-27). This is referred to as stratification. Pollock found that peach seeds that had not been after-ripened would germinate if about one-quarter of the seed coat and endosperm was removed from the hypocotyl end of each seed, but the temperature during the first week of germination was critical in determining whether the seedling was normal or a dwarf with small and abnormally shaped leaves (Figure 15-28). Germination at a temperature as high as 22°C resulted in essentially normal plants, but at 25°C and higher the plants were dwarfed.

GERMINATION INHIBITORS. The last type of seed dormancy results from the presence of germination inhibitors. These may be produced in the seed coats,

Figure 15-28. Elberta peach seedlings that had germinated 8 days at 27°C (right) and 19°C (left), and then both grew 4 weeks at 25°C under 16-hr days. Note the abnormal leaves as well as the dwarfing of the plant at the right. [Courtesy of B. M. Pollock. From *Plant Physiol.*, **37**:190 (1962)]

endosperm, or embryo itself, or they may diffuse into the seed from the fruit. They may even leach out of leaves and be absorbed by seeds after dispersal. The site of growth inhibitor synthesis varies from species to species, as does the chemical nature of the germination inhibitors. Evenari [16] has tabulated scores of substances that have been reported to act as germination inhibitors, and both the number and diversity of them are striking. Some germination inhibitors like abscisic acid are general growth inhibitors that are involved also in bud dormancy, but many are substances that are known to inhibit only germination. Among the many germination inhibitors listed by Evenari are mustard oils, coumarin and other unsaturated lactones, various organic acids, aldehydes, caffein and other alkaloids, essential oils, cyanide-releasing compounds such as amygdalin, ammonia-releasing compounds, and even sodium chloride and other salts! Some germination inhibitors are true hormonal growth inhibitors, but others apparently act osmotically or perhaps by providing an unfavorable pH. The mode of action is often not well understood. Thus, although it has long been known that tomato fruits (and tomato juice) contain germination inhibitors and most investigators regard them (caffeic and ferulic acids) as true growth inhibitors, some consider the effect to be simply osmotic.

In any event, dormancy resulting from germination inhibitors is generally broken by the eventual leaching of the inhibitors out of the seeds by rains in nature or by soaking and washing in the laboratory. However, some seeds with germination inhibitors can have their dormancy broken by a light exposure, and application of coumarin, a common germination inhibitor, to seeds without a light requirement will induce a light requirement. Also, low temperatures may be effective in breaking seed as well as bud dormancy, particularly in cases where the inhibitors involved are the same.

Amen [2] considers that the various types of seed dormancy, at least other than seed coat dormancy, have much more in common than is generally recognized. He points out that in all the types there is probably an interaction between growth promotors and growth inhibitors, and dormancy results from a lack of growth promotors or the presence of growth inhibitors or both. Future research should clarify the degree to which this is true and elucidate more details about the physiology of seed dormancy.

SIGNIFICANCE OF SEED DORMANCY. Seed dormancy has an important survival value, particularly in regions with marked seasonal changes in such environmental factors as temperature and rainfall. Dormancy keeps seeds from germinating the summer or fall they are produced, thus preventing the young plants from being frozen and killed before completing another life cycle. The seeds of many desert annuals remain dormant until there has been enough rain to leach out the germination inhibitors, and consequently enough to permit their rapid germination and growth to maturity during the rather brief rainy season. Annual weeds with dormant seeds can continue to flourish even though there might be a disasterous year with little or no seed produced, since the seeds may remain dormant for several years. The dormant seeds of various chaparral species are scarified by the periodic fires; thus, the burned plants are quite promptly replaced by their offspring.

Seed Germination

If a viable, nondormant seed is in a suitable environment it will germinate [28]. The essential environmental factors are an adequate supply of water, a suitable temperature, an adequate supply of oxygen, and in some species either the presence or absence of light. Seed germination is generally most rapid when the soil water content is near field capacity (FC), but seeds of some species will germinate even when the water content is near the permanent wilting percentage (PWP). In some cases seeds may even be able to imbibe enough water vapor from an atmosphere that is saturated, or nearly so, to initiate germination.

The temperature range suitable for germination is usually quite broad and varies considerably from species to species. For example, the range for wheat seeds is 1 to 35°C and for corn about 5 to 45°C. The optimum temperature is generally about midway between the two extremes and may be different for growth of the radicle and the plumule.

Although germination does not occur in the absence of oxygen, reduction of oxygen considerably below its atmospheric concentration does not inhibit germination of seeds of most species. Seeds will usually germinate even when completely submerged in water, and seeds of some hydrophytes such as cattails and rice germinate best at reduced oxygen pressures. The seed coats of peas and various other species are rather impermeable to oxygen even after they are hydrated and as a result, anaerobic respiration occurs until the seed coats are ruptured.

Although the seeds of most species will germinate either in the presence or absence of light, germination of the seeds of some species is influenced by light. Light prevents or retards the germination of seeds of tomato, onion, jimson-weed, some lilies, and a number of other plants. On the other hand, the seeds of beggar-ticks, Lepidium, tobacco, mistletoe, some varieties of lettuce, and various other plants will not germinate unless exposed to light at least briefly after they are hydrated (Figure 15-29). The role of phytochrome in light-induced germination was discussed in Chapter 11.

Even in species that do not have a light requirement for germination, phytochrome plays various roles: unfolding and enlargement of the young leaves of many dicotyledons, straightening of hypocotyl and plumule hooks, inhibition of hypocotyl elongation, inhibition of grass mesocotyl elongation, stimulation of coleoptile elongation, differentiation of xylem in hypocotyls, and stimulation of stomatal development. Mohr found that in white mustard seedlings phytochrome promotes protein synthesis in leucoplasts, the hydrolysis of accumulated proteins, fat hydrolysis, RNA synthesis, ascorbic acid synthesis, and protochlorophyll synthesis. The morphogenetic changes in seedlings that occur as they emerge from the ground are predominantly the result of phyto-chrome-mediated light responses.

The first event in germination is imbibition of water by the dry seed and the consequent hydration of the cell walls and protoplasts. Imbibition results in substantial swelling of the seed. The initial rapid imbibition is generally completed within a few hours (3 hr in lettuce seeds), and the subsequent

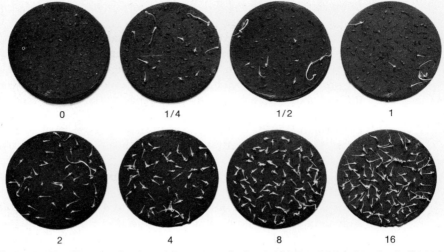

Figure 15-29. The germination of lettuce seeds that had been exposed to the indicated number of minutes of red light after soaking. [Courtesy of Agricultural Research Service, Plant Science Research Division, U. S. D. A., Beltsville, Md.]

absorption of water by imbition and osmosis is much slower. The hydration of the embryo and endosperm results in a marked increase in respiration and other metabolic processes.

The initial growth of the embryo is apparently controlled by preformed mRNA and enzymes, but soon new kinds of mRNA and enzymes are produced. It will be recalled that Varner and others (Chapter 13) found that gibberellin from the embryo induces the synthesis of several different hydrolytic enzymes

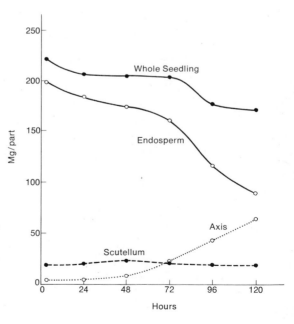

Figure 15-30. Changes in the dry weight of a corn seedling (and its various parts) germinating in the dark at 25°C. Note that only the axis (root and shoot) of the seedling increased in dry weight and that the dry weight of the scutellum (cotyledon) remained essentially constant. [Data of J. Ingle, L. Beevers, and R. H. Hageman, *Plant Physiol.,* **39**:735 (1964)]

in the aleurone layer of barley seeds. These enzymes then hydrolyze foods in the endosperm. Until a seedling emerges from the ground and begins carrying on photosynthesis, it is dependent for food upon the starch, lipids, proteins, or other substances accumulated in the endosperm, perisperm, or cotyledons of the seed. The accumulated foods are rapidly hydrolyzed to fatty acids, glycerol, sugars, and amino acids, which are then used in respiration or assimilation (Figure 15-30).

The first external evidence of germination is emergence of the growing radicle and its rapid development into the primary root. The plumule does not emerge until later—much later in some species. An extreme case is the snow trillium. Its seedlings develop only roots the first year. The plumule does not emerge and grow into a shoot until the second spring. Lela Barton found that two spaced cold periods were essential for breaking plumule dormancy, although one was enough to break radicle dormancy.

In most dicotyledons and some monocotyledons (including onions) the hypocotyl elongates and so raises the cotyledons above ground. However, in grasses and some dicotyledons, including peas and oaks, the hypocotyls do not elongate, so the cotyledons remain underground.

References

[1] Amen, R. D. "The concept of seed dormancy," *Amer. Sci.,* **51**:408–424 (1963).

[2] Amen, R. D. "A model of seed dormancy," *Bot. Rev.,* **34**:1–29 (1968).

[3] Biale, J. B. "Growth, maturation and senescence in fruits," *Science,* **146**:880–888 (1964).

[4] Bonner, James, and J. A. D. Zeevart. "Ribonucleic acid synthesis in the bud, an essential component of floral induction in *Xanthium,*" *Plant Physiol.,* **37**:43–49 (1962).

[5] Bopp, M. "Control of differentiation in fern allies and bryophytes," *Ann. Rev. Plant Physiol.,* **19**:361–380 (1968).

[6] Borthwick, H. A., and S. B. Hendricks. "Photoperiodism in plants," *Science,* **132**:1223–1228 (1960).

[7] Borthwick, H. A., and N. J. Scully. "Photoperiodic responses of hemp," *Bot. Gaz.,* **116**:14–29 (1954).

[8] Bunning, E. *The Physiological Clock,* Longmans, Springer-Verlag, New York, 1967.

[9] Burg, S. P., and E. A. Burg. "Ethylene action and the ripening of fruits," *Science,* **148**:1190–1195 (1965).

[10] Butler, W. L., and R. J. Downs. "Light and plant development," *Sci. Amer.,* **203**(6):56–63 (Dec. 1960).

[11] Chorney, W. *et al.* "Rhythmic flowering response in cocklebur," *BioScience,* **20**:31–32 (1970).

[12] Chailakhyan, M. K. "Flowering and photoperiodism of plants," *Plant Sci. Bull;* **16**(3):1–7 (Oct. 1970).

[13] Crocker, W., and L. V. Barton. *Physiology of Seeds,* Chronica Botanica, Waltham, Mass., 1957.

[14] Darlingtonm H. T., and G. P. Steinbauer. "The eighty-year period for Dr. Beal's seed viability experiment," *Amer. Jour. Bot.,* **48:**321–325 (1961).

[15] Evans, L. T. (ed.). *The Induction of Flowering,* Cornell University Press, Ithaca, N. Y., 1969.

[16] Evenari, M. "Germination inhibitors," *Bot. Rev.,* **15:**153–194 (1949).

[17] Garner, W. W., and H. A. Allard. "Effect of length of day on plant growth," *J. Agr. Res.,* **18:**553–606 (1920).

[18] Gifford, E. M., Jr., and K. D. Stewart. "Ultrastructure of vegetative and reproductive apices of *Chenopodium album,*" *Science,* **149:**75–77 (1965).

[19] Hillman, W. S. *The Physiology of Flowering,* Holt, Rinehart & Winston, New York, 1962.

[20] Hodson, H. K., and K. C. Hamner. "Floral inducing extract from Xanthium," *Science,* **167:**384–385 (1970).

[21] Hsu, J. C. S., and K. C. Hamner. "Studies on the involvement of an endogenous rhythm in the photoperiodic response of *Hyoscyamus niger,*" *Plant Physiol.,* **42:**725–730 (1967).

[22] Imamura, S., and A. Takimoto. "Transmission rate of photoperiodic stimulus in *Pharbitis nil,*" *Bot. Mag. (Tokyo),* **68:**260–266 (1955).

[23] Klebs, G. "Über die Blutenbildung von Sempervivum," *Flora* **11:**128–151 (1918).

[24] Koller, D. "Germination," *Sci. Amer.,* **200**(4):75–84 (April 1959).

[25] Leopold, A. C. *Plant Growth and Development,* McGraw-Hill, New York, 1964.

[26] Lincoln, R. G., D. L. Mayfield, and A. Cunningham. "Preparation of a floral initiating extract from Xanthium," *Science,* **133:**756 (1961).

[27] Mapson, L. W. "Biosynthesis of ethylene and the ripening of fruit," *Endeavour,* **29:**29–33 (1970).

[28] Mayer, A. M., and A. Poljakoff-Mayber. *The Germination of Seeds,* Permagon Press, Oxford, 1964.

[29] Murneek, A. E., and R. O. Whyte (eds.). *Vernalization and Photoperiodism, a Symposium,* Chronica Botanica, Waltham, Mass., 1948.

[30] Näf, U., J. Sullivan, and M. Cummins. "New antheridiogen from the fern *Onoclea sensibilis,*" *Science,* **163:**1357–1358 (1969).

[31] Nitsch, J. P., E. B. Kurtz, J. L. Liverman, and F. W. Went. "The development of sex expression in cucurbit flowers," *Amer. J. Bot.,* **39:**32–43 (1952).

[32] Purvis, O. N. "Vernalisation: a new method of hastening flowering," *Sci. Hort,* **4:**155–164 (1936).

[33] Raper, J. R. "Chemical regulation of the sexual processes in the thallophytes," *Bot. Rev.,* **18:**447-545 (1952).

[34] Rosen, W. C. "Ultrastructure and physiology of pollen," *Ann. Rev. Plant Physiol.,* **19:**435–462 (1968).

[35] Sachs, R. M., A. M. Kofranek, and S. Shyr. "Gibberellin-induced inhibition of floral initiation in Fuchsia," *Amer. J. Bot.,* **54:**921–929 (1967).

[36] Salisbury, F. B. *The Flowering Process,* Permagon Press, New York, 1963.

[37] Salisbury, F. B. "The initiation of flowering," *Endeavour,* **24:**74–80 (1965).

[38] Salisbury, F. B., and C. Ross. *Plant Physiology,* Wadsworth Pub. Co., Belmont, Calif., 1969.

[39] Voth, P. D., and K. C. Hamner. "Responses of *Marchantia polymorpha* to nutrient supply and photoperiod," *Bot. Gaz.,* **102:**169–205 (1940).

[40] Went, F. W. "The role of environment in plant growth," *Amer. Sci.,* **44:**378–398 (1956); "Climate and agriculture," *Sci. Amer.,* **196**(6):82–94 (June 1957).

[41] Wilkins, M. B. (ed.). *The Physiology of Plant Growth and Development,* McGraw-Hill, New York, 1969, Chaps. 16, 17, 18.

[42] Withrow, R. B. (ed.). *Photoperiodism,* AAAS, Washington, 1959.

[43] Zeevaart, J. A. D. "Physiology of flowering," *Science,* **137:**723–731 (1962).

Vegetative Development of Vascular Plants

16

We now come to a description of the differentiation of the cells and tissues of the vegetative organs of vascular plants as they develop from embryos into mature individuals. Space limitations prevent a similar consideration of vegetative development of the gametophytes of vascular plants and the great diversity of nonvascular plants, but discussions of these can be found elsewhere [5, 6, 20, 23, 24].

The growth and development of vascular sporophytes are characterized by at least two features that distinguish them from most other organisms. First, growth occurs from localized meristems once embryonic development is completed: the apical meristems of stems and roots and, in many species, the vascular cambium. Second, growth and development are indeterminate, that is, stems and roots continue to grow and develop throughout the life of the plant. As the plant grows older it also increases in size. The period of indeterminate growth ranges from less than a year in annuals to hundreds or even thousands of years in trees. However, indeterminate growth is not necessarily or even generally continuous. In temperate climates trees and shrubs usually become dormant and stop growing during the winter. Indeed, many species have active apical growth for only a few weeks in the spring and perhaps the early summer.

Thus, roots and stems have what amounts to a continuing embryogeny. This makes them particularly suitable for experimental studies of morphogenesis, especially since a single plant can provide a considerable number of meristems with identical heredity. The development of flowers, fruits, seeds, and most leaves is determinate, and growth ceases when a certain size has been attained.

It should be recalled that roots, stems, and leaves are the only vegetative organs of vascular plants, and all other plant structures are modifications of one or another of these. For example, stolons, rhizomes, tubers, corms, tendrils, thorns, and the leaflike cladophylls are all modified stems. The pitchers of pitcher plants, the insectivorous traps of Venus' fly-trap, spines, and some tendrils are modified leaves or parts of leaves. Vegetative buds are simply embryonic stems covered by young leaves and often by the modified leaves known as bud scales, and the bulbs of onions and tulips as well as the heads of cabbage and lettuce are essentially large buds.

Although the morphogenesis of roots and stems has much in common, the differences are great enough to warrant separate consideration. However, the development of stems and leaves is so interrelated that the morphogenesis of the shoot will be considered as a unit.

The Growth and Development of Shoots

The Shoot Apex

The extreme tip of the young stem consists of a generally hemispherical or conical tissue composed of meristematic cells (Figure 16-1, and Figure 2-2). This is referred to as the **apical meristem** or the **shoot apex,** and sometimes the portion of it distal to the youngest leaf primordium is called the **promeristem.** The shoot apex varies considerably in size and shape from species to species. In seed plants it is generally only a few hundred microns in diameter, but in such lower vascular plants as cycads and some ferns it may be several millimeters in diameter.

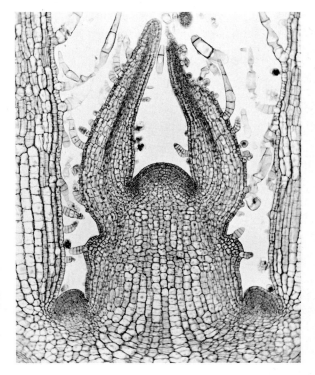

Figure 16-1. Photomicrograph of a shoot apex of Coleus. Note the youngest pair of leaves on either side of the apical meristem and the primordia of lateral buds in the axils of the older leaves. [Courtesy of Triarch Products]

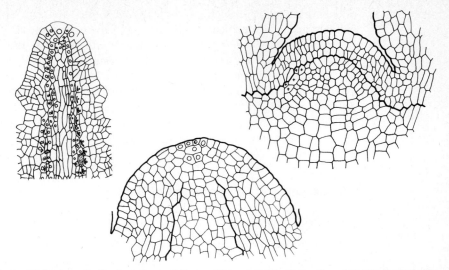

Figure 16-2. Longitudinal sections of three different types of shoot apex. **Left:** *Selaginella wildenovii,* showing the large apical cell characteristic of lower vascular plants. **Center:** *Torreya californica,* a gymnosperm, showing the cluster of apical initials and an almost complete lack of layering. **Right:** *Vinca minor,* an angiosperm, showing the three-layered tunica and below it the unlayered corpus. [Left to right: After B. D. Barclay, *Bot. Gaz.,* **91:**452 (1931); M. A. Johnson, *Phytomorphology,* **1:**188 (1951); A. Schmidt, *Bot. Arch.,* **8:**345 (1924)]

The shoot apex is of great morphogenetic significance and interest because the lineage of all cells of the shoot goes back to the apex and because the apex gives rise to differentiated stem tissues, leaf primordia, and lateral bud primordia. For well over a century botanists have studied the shoot apex intensively, first morphologically and then morphogenetically and physiologically, in an effort to elucidate its roles in shoot development. During the latter half of the 19th century there was great interest in describing the cellular organization of the shoot apex. Hofmeister, Naegli, and others found that, in ferns and other vascular cryptogams, there is a single tetrahedral apical cell which, through repeated divisions on its inner faces and subsequent divisions of its cell progeny, gives rise to all the cells of the shoot. Naegli proposed the theory that such apical cells were universally characteristic of apical meristems, but his theory soon became untenable when it was found that there are a least two other types of cell organization in the shoot apex (Figure 16-2). One, characteristic of conifers and some other gymnosperms, has a central cluster of several apical cells from which the other cells arise rather than a single cell. Another, characteristic of angiosperms, has neither a single apical cell nor a cluster of initial cells but rather several discrete layers of cells.

In 1868 Hanstein proposed that there were three distinct cell layers in the apex, each of which gave rise to certain tissues, thus providing a homology with the three germ layers of young animal embryos (Figure 16-3). He suggested that the outer single cell layer **(dermatogen)** produced the epidermis, the middle layer of several cells thickness **(periblem)** produced the cortex, and the inner core **(plerome)** produced the vascular cylinder and pith. However,

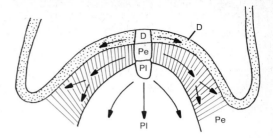

Figure 16-3. Diagram illustrating Hanstein's interpretation of the shoot apex. D, dermatogen; Pe, periblem; Pl, plerome.

his hypothesis proved to be untenable when it became evident that most angiosperms lacked these three distinct layers and that, even where they could be identified, there was no consistency in the particular tissues derived from them.

The more acceptable **tunica-corpus** theory was proposed by Schmidt in 1924. The outer tunica composed of one to four definite layers of cells covers the underlying corpus, an unlayered core of cells (Figure 16-4). The apical meristem does not have a fixed structure, and as it develops it may change in shape, size, and the number of layers in the tunica. Most proponents of the tunica-corpus theory consider it as primarily a description of apical organization and do not attempt to ascribe the universal origin of any specific tissue or organ to either the tunica or corpus. In general, however, the outer layer of the tunica appears to give rise to the epidermis (as would seem logical because of its location). The bulk of the stem tissues usually arises from the corpus, and the products of the inner tunica layers vary with the species.

Perhaps the best experimental evidence supporting the tunica-corpus theory of discrete layers and identifying the tissues derived from each region has been provided by the work of Satina, Blakeslee, and Avery [15]. They induced polyploidy in the embryos of germinating Datura seeds by the use of colchicine, and found that only certain cells of the shoot apex became polyploid (Figure 16-5). Since polyploid cells are readily identified under the microscope because of their larger size, larger nuclei, and more numerous chromosomes, it was possible to identify the progeny of the polyploid cells. In many cases polyploid

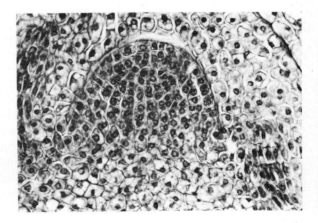

Figure 16-4. Photomicrograph of a vegetative shoot apex of *Aquilegia formosa* (×280). This species has a two-layered tunica. [Courtesy of S. S. Tepfer. From *Univ. Calif. Publ. Bot.*, **25**:513 (1953)]

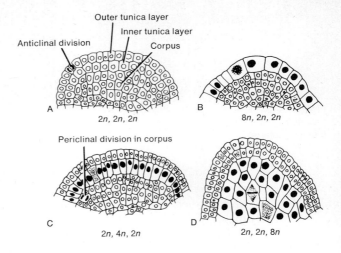

Figure 16-5. Longitudinal sections of shoot apices of Datura, showing periclinal chimeras induced by treatment with colchicine. **(A)** Untreated plant with all cells diploid. **(B)** Polyploidy in outer layer of the tunica. **(C)** Polyploidy in the inner layer of the tunica. **(D)** Polyploidy in the corpus. [Adapted from S. Satina, A. F. Blakeslee, and A. G. Avery, *Amer. J. Bot.*, **27**:895 (1940)]

cells were restricted to one or another of the tunica layers or to the corpus, indicating that each of these indeed had a common cell lineage distinct from that of the other regions. Also, it was determined that all three regions made contributions to the leaves as well as to the stem: the outer tunica layer formed the leaf epidermis, the inner tunica layer the mesophyll, and the corpus the vascular procambium.

The layered structure of the tunica results from the fact that its cells undergo **anticlinal** divisions, that is the new cell wall is at a right angle to the nearest outer surface of the apex. The cells of the corpus have **periclinal** divisions (new walls parallel with outer surface) as well as anticlinal divisions, so the corpus is not layered. However, anticlinal divisions tend to predominate in the periphery of the corpus. The corpus gives rise to longitudinal files of cells below it (the rib meristem).

Sinnott [16] thinks that the morphogenetic significance of layering in the shoot apex has been overestimated. Popham [14] and some other investigators have placed emphasis on zones of the apex, distinguished by differences in cell characteristics rather than by layering and planes of cell division. Popham identified seven types of shoot apex zonation. Newman [13] made a somewhat simpler classification of shoot apex types. His **monoplex** type has one apical initial cell and is characteristic of many ferns and other lower vascular plants. The **simplex** apex has several initials in one layer and is characteristic of most gymnosperms. The **duplex** type, characteristic of most angiosperms, has several initials in more than one layer. Further consideration of the organization of the shoot apex can be found elsewhere [7, 8, 21].

The shoot apex presents morphogenetic problems of substantial complexity and variety, most of which remain essentially unsolved. One problem is simply the matter of accounting for the different cellular patterns in the apex, another is how the apex simultaneously maintains its identity and gives rise to new tissues that will differentiate and mature. To what degree is the apex autonomous and independent of the tissues that have arisen from it in the control of development? What determines when and where the apex will give rise to leaf and lateral bud primordia in the pattern characteristic of the species? What role does the apex play in the differentiation of tissues in the subapical regions? These and other problems have been investigated extensively by a variety of techniques. A vast and complex literature has resulted, containing many hypotheses and theories as well as a substantial amount of factual information. We can review this in only a skeletonized way here, but more comprehensive discussions are available elsewhere [5, 6, 21, 22, 23].

During the two decades beginning about 1940 descriptive studies, which had predominated earlier, were replaced to a great extent by surgical experiments that involved cutting the apical meristems in various planes and patterns, thus partially isolating various parts of the apex and the leaf and bud primordia. Also, excised apical meristems and primordia were cultured in sterile tissue cultures, providing considerable information about their degree of autonomy. Such studies have been carried on by a considerable number of investigators, notably Wardlaw and the Snows in England and Wetmore and Ball in the United States, along with their students and coworkers. There have also been physiological and biochemical studies of the development from the shoot apex, using techniques such as phytohormone determinations and treatments, chromatography, histochemistry and cytochemistry.

Perhaps the earliest surgical experiment was conducted by Mary Pilkington in 1929. She cut a lupine shoot apex in half vertically and found that both halves developed into complete normal stems. In 1948 Ball found that even when an apex was quartered four normal stems resulted. Wardlaw isolated the shoot apex of the fern Dryopteris from the leaf primordia by four longitudinal cuts making a square. The apex was thus isolated from the vascular tissue and connected only through the pith (Figure 16-6); yet the apex continued to grow and produce leaf primordia and then leaves. Provascular tissues developed in the pith under the apex but did not connect with the existing vascular bundles. However, Ball and Wardlow both found later that in flowering plants the vascular tissue formed in the pith does connect with the vascular cylinder, thus reestablishing a vascular transport pathway. Ball and others found that isolated apical meristems in tissue culture (Figure 16-7) can develop into complete normal plants, both in seed plants and ferns. These and other experiments made it evident that the shoot apex is totipotent and autonomous, provided that it has access to essential substances such as water and food.

There has been much interest in the differentiation of the leaf and lateral bud primordia. Each species of vascular plant has a characteristic and quite

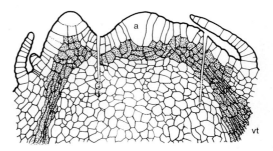

Figure 16-6. Longitudinal section through the shoot apex of Dryopteris (a fern) showing the isolation of the region around the apical cell (a) by cuts extending through the vascular tissue (vt). [After C. W. Wardlaw, *Phil. Trans. Roy. Soc. London,* **B232:**343 (1947)]

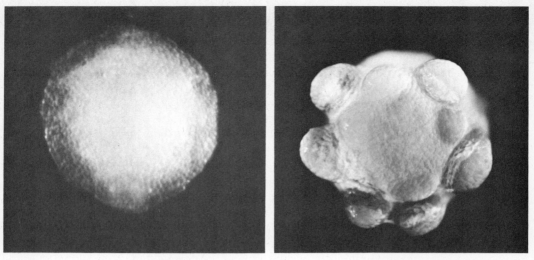

Figure 16–7. **Left:** Excised apical meristem of *Picea abies* (Norway spruce) in culture (×35). **Right:** After a week in culture an apex similar to the one shown on the left has developed at least 10 leaf primordia (×43). [Courtesy of J. A. Romberger, U. S. Forest Service. From J. A. Romberger, R. J. Varnell, and C. A. Tabor, *Tech. Bull. 1409,* U. S. D. A. Forest Service]

precise phyllotaxy (Chapter 17), the arrangement of leaves on the stem. In some species the leaves are opposite or whorled and in others they are alternate. If alternate, the leaves are arranged in a spiral, the steepness of the spiral being a species characteristic. The phyllotaxy of a plant is determined at the time the leaf primordia arise from the shoot apex (Figures 16-8 and 16-9). The question is what causes them to develop at just the right places and no place else, and the fact that phyllotaxy is under genetic control does not answer the question. Among the various theories the most acceptable one seems to be that the formation of a primordium inhibits the initiation of other primordia for some specific distance from it by the production of inhibitory substances or by physical space limitations, probably the latter.

Some investigators think that primordia arise because of more rapid cell division in the tunica than the corpus, which results in evagination of the tunica followed by the invasion of the corpus. Others think that the more rapid cell division is in the corpus, resulting in a protuberence that pushes out the tunica and places it under tension. The attempts to determine which occurs have been inconclusive.

Not long after a leaf primordium has differentiated, the primordium of its axillary bud begins to appear. Some have objected to calling these buds axillary rather than lateral because the bud primordia may arise some distance above the leaf primordium, even though subsequent development results in the location of the bud in the leaf axil. The usual arrangement of leaf and bud primordia may be altered by surgical manipulation. Wardlaw found that in Dryopteris (a fern), if the place where a new leaf primordium is to form is isolated from the apical cell by a deep tangential cut, a bud primordium rather than a leaf primordium developed. This indicates apical control over the type

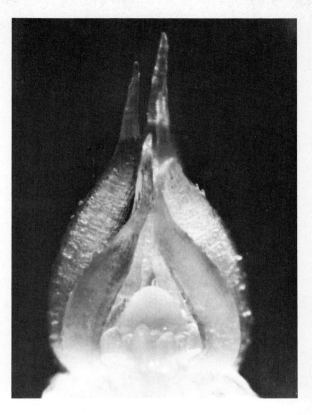

Figure 16-8. A partially dissected bud of a *Picea abies* seedling, showing the dome-shaped apical meristem with several leaf primordia, flanked by several developing leaves. [Courtesy of J. A. Romberger, U. S. Forest Service. From J. A. Romberger, R. J. Varnell, and C. A. Tabor, *Tech. Bull. 1409,* U. S. D. A. Forest Service]

Figure 16-9. The shoot apex of a dormant *Picea abies* bud, showing the apical meristem and numerous leaf primordia. Its height is about 2 mm. Usually the primordia of all the leaves that will develop the following summer are present in a dormant bud. [Courtesy J. A. Romberger, U. S. Forest Service. From J. A. Romberger, R. J. Varnell, and C. A. Tabor, *Tech. Bull. 1409,* U. S. D. A. Forest Service]

of primordium differentiated. Cutter later reported that only the three youngest primordia responded in this way. Steeves and Sussex found that when young leaf primordia of Osmunda (another fern) were excised and cultured they developed into buds and eventually entire plants, but older excised leaf primordia developed only into leaves. There appears to be a point in leaf primordium development before which it may differentiate into either a leaf or a bud, and after which its differentiation into a leaf is fixed. The critical point seems to be earlier in flowering plants than in ferns.

In recent decades the descriptive and surgical investigations of the shoot apex have been supplemented by histochemical and autoradiographic studies that have provided considerable information not otherwise available. These techniques promise to be even more productive in the future. Such studies have made it clear that some cells of the apex are more active in DNA, RNA, and protein synthesis and in division than others, the least active cells generally being the apical initials and the cells of the central zone. The phytohormones undoubtedly play roles in the growth and development of the shoot apex, but so far little has been accomplished in pinpointing any differential effects in various apical regions. However, it has been found that leaf primordia synthesize IAA and the quantity increases with the size of the primordium. The apical meristem apparently synthesizes little or none. Applications of various phytohormones to the apex may induce morphogenetic changes, but these are generally abnormal [6].

We now turn to the differentiation of the dermal, vascular, and fundamental tissue systems to which the shoot apex gives rise.

Differentiation of the Vascular Tissues

The first evidence of the differentiation of vascular tissues from cells produced by the shoot apex is the appearence of **procambial strands** [8] generally concurrent with the development of leaf primordia (Figure 16-10). The procambial strands extend into the developing leaves, where they will give rise to the vascular bundles of the petioles and the veins of the blade, and are continuous with the older vascular tissues of the stem below. The strands can be distinguished from the surrounding ground tissues by their more elongated and narrower cells. These differences in cell shape become more pronounced with time.

The intimate association between the initiation of leaves and provascular strands in plants with generally large leaves (the seed plants and ferns) has led many investigators to the conclusion that the primary vascular cylinder is composed principally if not entirely of leaf traces. However, other investigators consider that, at least in many species, the vascular cylinder contains many bundles that do not emerge as leaf traces. In plants such as the clubmosses and horsetails with small leaves, there is a solid vascular cylinder without leaf gaps from which relatively small leaf traces emerge (Figure 16-11). Here there is no doubt about the existence of a strictly cauline vascular cylinder.

The differentiation of the meristematic cells of the procambium into the characteristic cells of the xylem and phloem of the shoot follows a definite pattern in most species. The pattern is most clearly evident if observed in both

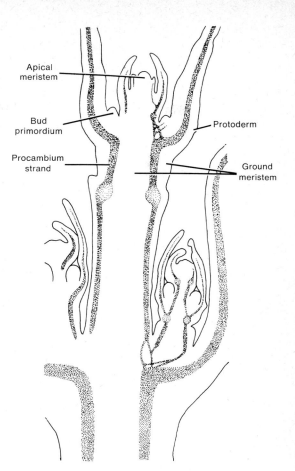

Apical meristem

Bud primordium

Procambium strand

Protoderm

Ground meristem

Figure 16-10. Diagrammatic drawing of a portion of a vegetative bud of Syringa (lilac), showing the developing primary tissues (×17). [After T. E. Weier, C. R. Stocking, and M. G. Barbour, *Botany,* 4th. ed., John Wiley & Sons, Inc., New York, 1970]

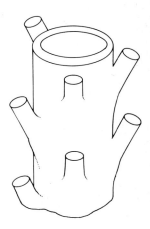

Figure 16-11. Diagrammatic drawing of a portion of the vascular cylinder of Selaginella (a clubmoss), a siphonostele without leaf gaps. Note the seven leaf traces.

longitudinal and transverse sections. Viewed longitudinally, the sieve-tube elements and other phloem cells differentiate acropetally from the previously differentiated phloem on up into the leaves. However, differentiation of the xylem generally begins near the base of the young leaf and, from there, extends acropetally into the leaf and basipetally into the stem, finally connecting with the previously differentiated xylem (Figure 16-12). Xylem differentiation [19] generally does not begin until after phloem differentiation is well under way. In an effort to explain the rather peculiar pattern of xylem differentiation, Wetmore suggested that differentiation begins at the point where an optimal concentration of auxin from the young leaf meets an optimal concentration of sugar from below.

That the apex plays a role in the differentiation of the underlying tissues is shown by the fact that differentiation ceases if the apex is removed. It is also evident that differentiation is under strong hereditary control, since the vascular tissues and others develop in a pattern characteristic of the species

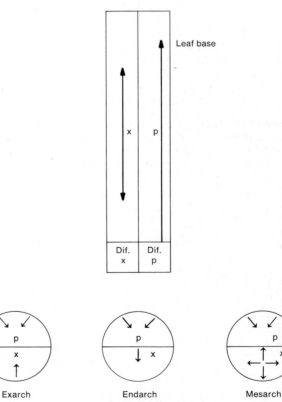

Figure 16-12. Diagrams illustrating the patterns of differentiation of xylem and phloem in procambium strands. **Above:** Longitudinal section of a strand. The phloem (p) differentiates upward from previously differentiated phloem, but the xylem (x) differentiates both upward and downward from a point just below a leaf base. **Below:** Three cross sections of procambial strands showing different patterns of xylem differentiation. Protoxylem develops at the base of the arrows and then metaxylem develops in the direction of the arrows. Phloem differentiation always proceeds inward from the periphery.

as regards both the sequence of events and the arrangement and structure of the mature tissues that result.

Considered from a transverse viewpoint differentiation of the vascular tissues may occur in one of three different patterns (Figure 16-12), but plants of a particular species display only one of them. In some species differentiation of the first xylem (protoxylem) begins at the edge of the procambial strand and differentiation of the metaxylem proceeds inward (the exarch type). In the second type (endarch) the protoxylem differentiates near the center of the procambial strand and differentiation continues toward the edge of the strand. In the third type (mesarch) differentiation of protoxylem begins within the xylem strand and differentiation of the metaxylem proceeds in a radial direction.

Regardless of the pattern of xylem differentiation, the differentiation of the protophloem begins at the outer periphery of the procambial strand and proceeds inward toward the xylem. However, in species with phloem internal to the xylem as well as external to it (bicollateral vascular bundles) the internal protophloem begins differentiating from the inner periphery of the procambial strand and proceeds outward toward the xylem.

The protoxylem and protophloem which differentiate early, while the internode is still elongating, constitute only a small part of the primary xylem and phloem and generally function only briefly in translocation. Because the differentiated cells of the protoxylem and protophloem are incapable of elongating, unlike the still undifferentiated cells of the adjacent ground meristem and the precambial strand, they tend to become stretched, ruptured, and perhaps crushed. Thus they become nonfunctional in translocation. In the meantime the more extensive and more permanent parts of the primary xylem and phloem (the metaxylem and metaphloem) have begun differentiating, their maturation being correlated with the cessation of internode elongation. The metaxylem and metaphloem continue to provide the sole pathway of vascular translocation through the stem until such a time as secondary xylem and phloem are laid down by the vascular cambium.

The cells of the protoxylem are generally smaller and have thinner walls than those of the metaxylem. The vessels of the protoxylem generally have annular or helical wall thickening, whereas those of the metaxylem are mostly reticulate, scalariform, or pitted (Chapter 2). The protoxylem has a higher percentage of parenchyma cells than the metaxylem. The cells of the protophloem are also generally smaller and thinner walled than those of the metaphloem. In the protophloem companion cells are often lacking and the sieve areas of the sieve-tube elements are less well defined than those in the metaphloem. In many dicotyledons the protophloem cells that remain after the disruption of the sieve tubes may differentiate into fibers.

The formation of fibers and tracheids in the primary vascular tissues involves substantial cell elongation, particularly in the case of phloem fibers. At first elongation occurs throughout the cell, but later only the ends retain their thin walls and the capacity for elongation. The median portion of the cell stops elongating and begins laying down secondary walls. In fibers there may be repeated nuclear divisions during elongation without laying down cross walls, resulting in a multinucleate cell. Once elongation and deposition of the secondary walls in the characteristic patterns have been completed the entire protoplast disintegrates and disappears, presumably as a result of the activity of a variety of hydrolytic enzymes. The disintegration of the protoplast, destructive as it may seem, is a normal and essential part of the differentiation of vessel elements as well as tracheids and fibers.

The cells that develop into vessel elements undergo a considerable amount of lateral enlargement, and usually also some elongation, before they begin laying a secondary walls in one of the characteristic patterns. The parts of the primary wall not covered by secondary wall, for example, pits, generally break down and disappear, leaving only the middle lamella. The differentiation of vessel elements involves not only dissolution of the protoplasts but also disintegration of most of the end walls, thus giving rise to the long multicellular vessels (Figure 11-3).

Differentiation of the sieve-tube elements involves substantial cell enlargement and formation of sieve plates and sieve areas as well as some secondary wall formation. The sieve areas arise from ordinary pits with plasmodesmata by the enlargement, and at times the fusion, of plasmodesmata. The differentiation of the protoplast begins with the appearance of proteinaceous slime bodies, which later lose their definite outlines and may fuse with one another to form the slime of the mature cell. At about the same time the nucleus disintegrates. In a number of cases nucleoli have been found to be extruded from the nucleus prior to its disintegration. These nucleoli may persist as long as the cell remains alive.

The roles of certain cell organelles and phytohormones in cell wall formation have been discussed in previous chapters. In recent decades various investigators including Esau, Evert, and Hepler and Newcomb have added considerably to the detailed knowledge of the differentiation of cells of the vascular tissues by electron microscope studies.

Differentiation of the Fundamental Tissues

Since the pith is composed of parenchyma cells, there is no differentiation other than that involved in the conversion of meristematic cells into parenchyma cells. However, the cells of the pith commonly enlarge to a considerably greater size than those of other tissues. In some species the growth of the pith fails to keep pace with the growth of the other tissues of the stem, resulting in the disruption of the pith and a stem with a hollow center. In a few species thin transverse layers of pith remain and alternate with cavities, producing what is known as a chambered pith.

The cortex of many species is composed predominantly of parenchyma cells, but chloroplasts commonly differentiate in these. In a considerable number of species the cortex also contains considerable collenchyma tissue, generally just inside the epidermis. Differentiation of collenchyma cells is principally a matter of increased wall thickening, especially in the corners of the cells. In some species collenchyma cells may also elongate greatly (to as much as 2 mm) and resemble fibers in size and shape. The wall thickenings are regarded as primary rather than secondary, since the thickening occurs before the cells reach their final size and the thickened walls consist principally of pectic compounds and cellulose. In some woody species the collenchyma cells may eventually lay down thick secondary walls that become lignified.

Differentiation of the Dermal Tissue

The epidermis differentiates from the outermost cell layer of the shoot apex, whether or not there is a definite tunica. The undifferentiated epidermis which extends for varying distances below the apex is sometimes referred to as the **protoderm.** The protodermal cells generally enlarge much less radially than they do tangentially and longitudinally, resulting in tabular epidermal cells that are proportionately much thinner than parenchyma cells. The epidermal cells of elongated structures such as stems, petioles, and the leaves of many monocotyledons tend to be greatly elongated, even though those covering other parts of the shoot may not be. The undulating shape of the epidermal cells of many species of dicotyledons as seen in face view has been variously ascribed to stresses that occurred during differentiation or to the pattern of solidification of the cuticle. The deposition of cutin in and on the outer walls of the epidermal cells begins early, even in the protodermal stage, but deposition continues throughout the period of differentiation.

As the cells of the protoderm divide, some of them at spaced intervals divide unequally. This process gives rise to a small cell with dense cytoplasm and a larger, more highly vacuolated cell. The smaller cell is a guard-cell mother cell (Figure 16-13). It then divides, thus giving rise to the two guard cells of a stomatal apparatus. At first the two guard cells are tightly joined, but then a lenticular swelling of the middle lamella between them occurs. This then disintegrates, forming the stomatal pore. Differentiation of the stomatal apparatus is completed as the guard cells assume their final size and shape, the substomatal cavity forms, and the guard cells are elevated or lowered in species that have raised or sunken stomata respectively. Stebbins and Shah [17] have found that the cytoplasm of the cell that gives rise to the mother cell polarizes before it divides, perhaps as early as the time that it was formed from its precursor cell. In plants of species that have subsidiary cells around the guard cells, these subsidiary cells may be either sister cells of the mother cell or may arise by the division of adjacent cells.

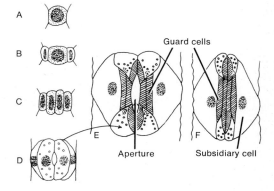

Figure 16-13. Stages in the development of stomata of sugarcane. **(A)** Stoma mother cell. **(B)** Stoma mother cell with the two subsidiary cells that were derived from cells adjacent to the stoma mother cell. **(C)** Division of the stoma mother cell into the two guard cells. **(D)** Immature stoma with the subsidiary cells. **(E)** Mature stoma, open. **(F)** Mature stoma, closed. The walls of the guard cells of the mature stomata are crosshatched. [After L. H. Flint and C. F. Morehead, *Amer. J. Bot.,* **33**:80 (1946)]

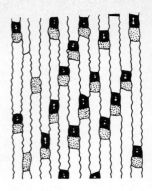

Figure 16-14. Surface view of epidermis of sugarcane stem showing the pairs of short cork cells (stippled) and silica cells (black) that alternate with the long cells. [After E. Artschwager, *J. Agr. Res.*, **30:**197 (1925)]

Differentiation of the stomata of a leaf is not simultaneous but occurs over a considerable period of time while the leaf is developing and enlarging. In netted-veined leaves the younger stomata are interspersed among older ones in various stages of development over the whole leaf surface. In the parallel-veined leaves of monocotyledons and a few dicotyledons, the differentiation of stomata begins at the leaf apex and progresses downward toward the leaf base.

In grasses and sedges there is often a greater diversity of epidermal cells than in most other plant families. The ordinary epidermal cells are usually greatly elongated and alternate in a regular sequence with short cells of two specialized types: silica cells and cork cells. Often a silica cell and a cork cell occur in pairs adjacent to one another (Figure 16-14). The silica cells contain massive deposits of silicon dioxide (SiO_2), whereas the cork cells have suberized walls and may contain solid organic deposits.

Stem Elongation

In buds and also in rosettes the stem internodes are very short—from a fraction of a millimeter to no more than a millimeter or so. In the plants of most species these internodes eventually attain a length of several centimeters to several decimeters. This extensive elongation is a product of both continued cell division and substantial cell elongation. Internode elongation is generally rapid as well as extensive, and an internode attains its final length in a matter of days or a few weeks. Once an internode has stopped elongating, any further stem elongation must occur in the younger internodes. Although the differentiation of the primary tissues of stems begins just below the stem apex before there has been much internode elongation, differentiation generally continues up to the time that the internode stops elongating. All growth in length of stems occurs in the primary tissues. Secondary growth from the cambium results in an increase in stem diameter but not in stem length.

It is often not recognized that cell division frequently continues, particularly in parenchyma cells, even after an internode has attained its final length. Division of elongated cells is not followed by elongation of the daughter cells, so there is a decrease in cell length and no increase in internode length. Treatment of plants with inhibitors of cell division such as maleic hydrazide results in some parenchyma cells twice or more the usual length in internodes no longer elongating, even though growth is severely inhibited [10].

There is some primary growth in diameter of stems, but it is much less than the primary growth in length of stems. Mature primary stems obviously have a greater diameter than the stem apex from which they developed and even than the young internodes. Large grasses such as corn, sugarcane, and bamboo have unusually marked primary growth in diameter.

Secondary Growth of Stems

The production of secondary xylem, phloem, and vascular rays by the cambium, the consequent disruption of the tissues external to the vascular cylinder, and the formation of a corky bark have been considered in Chapter 8. However, they deserve at least brief mention here since they are an important component in the total pattern of growth and development of woody plants. Even some species of herbaceous annuals have a cambium that may produce a substantial amount of secondary tissue during the one year of their life, although some species regarded as annuals in temperate regions are woody perennials in warmer regions where they can survive the winter.

It should be noted that whether a species is herbaceous or woody is determined by two rather simple basic factors, despite the rather marked structural differences between them. The first and more fundamental factor is whether all of the procambial cells differentiate into cells of the primary xylem and phloem or whether a layer of them between the xylem and phloem remain undifferentiated and meristematic, thus forming the cambium. The latter alternative also involves the emergence of interfascicular cambium between the vascular bundles. The second factor is the length of time the cambium remains active. It is, therefore, not surprising that many families and even genera of dicotyledons include both herbaceous and woody species.

The differentiation of the cells cut off from the cambium to either side begins promptly, so only a few layers of undifferentiated cells are evident in microscopic sections. Cell differentiation proceeds in much the same way as it does in the primary xylem and phloem, but the sieve-tube elements and vessel elements may have their characteristic features more highly developed. The vessel elements are reticulate, scalariform, or pitted with none of the annular or helical types.

The origin and differentiation of specialized cells from the cambium provides a number of interesting developmental problems that would throw considerable light on development in general—if they were solved. One problem is what keeps all the cells of the procambium, and later the cambium, from differentiating, thus making continued secondary growth possible. This is, of course, a species characteristic under genetic control. Another problem is what makes all the cells cut off to the inside develop into vessel elements, tracheids, or other cells of the xylem, whereas those cut off toward the outside of the same cambium, and only a few microns distant, develop into sieve-tube elements, companion cells, and other cells of the phloem. A similar problem relates to what causes a particular region of the cambium, that has been producing xylem and phloem, suddenly to begin giving rise to the parenchyma cells of a new vascular ray.

The Growth and Development of Leaves

Leaves, as well as stems, are an important part of the shoot. So far our consideration of shoot growth and development has dealt principally with stems, although we have noted the origin of leaf primordia from the apical meristem and much of the discussion of cell differentiation applies to leaves as well as to stems. We will now consider the development of leaf primordia into mature leaves.

Because of the diversity of leaf structure—the needle-shaped leaves of many conifers, the broad and thin leaves of most dicotyledons, and the narrow and elongated leaves of grasses and various other monocotyledons, to mention only a few types—there is also considerable diversity in the details of leaf development. Since the present discussion cannot cover all types, it will be devoted primarily to the development of typical dicotyledon leaves with broad, thin blades.

The first evidence of the differentiation of a leaf in the shoot apex is the appearance of a shoulderlike leaf buttress, which then elongates into a conical leaf primordium (Figure 16-15). The early primordium already contains a procambial strand in most cases. Most of the young leaf primordium will develop into the petiole (if one is present) and the midrib (Figure 16-16). Initial elongation of the primordium is mostly apical, but later on cell division and elongation begin to occur throughout it. Formation of the blade results from the activity of marginal and submarginal meristems. Repeated anticlinal divisions of the marginal cells give rise to the upper and lower epidermis. The submarginal cells give rise to the mesophyll by some periclinal in addition to anticlinal divisions.

At first development of the young leaf is primarily a matter of cell division, and its tissues are composed of small, compactly arranged cells without appreciable intercellular spaces. As time goes on, the number of cell divisions progressively decreases and finally all cell division stops. Thus, unlike stems and roots, leaves are determinate organs that reach a certain size and then stop growing. However, there are certain exceptions. The leaves of grasses and various other monocotyledons have a persistent basal meristem from which they continue to grow in length. Thus grass leaves being grazed by animals are not obliterated, and grasses make suitable lawn plants (although many

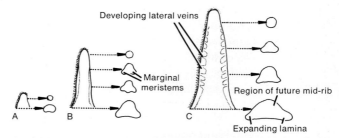

Figure 16-15. Stages in the development of leaf primordia of Nicotiana (tobacco) into young leaves. **(A)** Early primordium, 0.2 mm; **(B)** 1.1-mm primordium; **(C)** 1.8-mm primordium. Note the development of the lamina (blade) as a result of cell divisions of the marginal meristems. [After G. S. Avery, *Amer. J. Bot.,* **20:**566 (1933)]

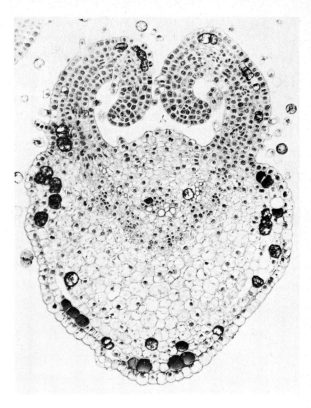

Figure 16-16. Section through a young leaf of *Sambucus canadensis* (elderberry). Note the small size of the developing lamina (blade) in comparison with the midrib. [Courtesy of Ripon Microslides]

home owners would appreciate less active meristems so that the interval between mowings could be extended). Also, apical growth of fern leaves may continue for some time after the rest of the leaf has matured and ceased growth.

Cell division ceases at different times in the various leaf tissues—first in the epidermis, next in the spongy mesophyll, and later in the palisade mesophyll. Cell division in the lateral veins may continue for some time after it has ceased elsewhere in the leaf. The differential cessation of cell division results in stresses within the leaf that, in turn, give rise to the differential pattern of intercellular spaces in the leaf tissues. Although the epidermal cells stop dividing first they generally continue to enlarge longer than the other cells, thus giving rise to the extensive intercellular spaces in the spongy mesophyll. Since the cells of the palisade layer continue to divide longer, the intercellular spaces between them are less marked. Another obvious difference in the development of the two layers of the mesophyll is that the palisade cells elongate more, whereas the spongy cells are more nearly isodiametric, although often quite irregular in shape. Marked development of intercellular spaces does not occur until the leaf has attained a fourth to a third of its final size.

Most of the leaf development that we have been describing occurs while the leaf is still in the bud. The very extensive and very rapid enlargement of leaves after they have emerged from the bud is largely a result of cell enlargement, although there may still be some cell division, particularly in the palisade layer and the veins. In annuals leaf development is generally a continuous process that proceeds from the leaf primordium to the mature leaf without interruption. However, most of the leaves of woody perennials in regions outside the tropics are formed the year before they emerge from the bud and are dormant during the winter along with the apical meristem.

That the shape and size of leaves are under strong hereditary control is evident from the fact that they are among the most widely used taxonomic criteria and that many species can be identified on the basis of leaf characteristics alone. Although usually there is little hereditary variation in leaf shape or size within a species, numerous single gene mutations that result in marked changes in leaf size and shape have been found in some species.

Despite the strong hereditary control of leaf shape and size, it is difficult to find two leaves of a plant that are exactly the same size and shape. A few species such as sassafras and mulberry have extremely varied leaves as regards the number, size, and position of the lobes, although they are all readily recognizable as characteristic of the species. In such species leaf variations are apparently random and are probably a result of either environmental differences or lack of extreme precision in the hereditary control of development. However, in many other species there are juvenile leaves that are quite different from the mature leaves (Chapter 18).

Environmental factors, particularly light, have a variety of marked effects on leaf development. The leaves of most dicotyledons will not enlarge until they have had at least a brief exposure to light, and light is also required for the unfolding of the leaves of species that have folded leaves in the buds. Both of these responses are mediated by phytochrome. Light intensity, and perhaps factors such as temperature and water availability, commonly has a marked influence on leaf development. For example the leaves of a tree that have developed in the shade are larger and thinner than those that have developed in full sunlight (Figure 16-17). The larger size of the shade leaves results mainly from the greater enlargement of the epidermal cells rather than from more cell divisions. However, the greater thickness of the sun leaves than the shade leaves results both from much greater elongation of the palisade cells and more cell divisions in the spongy mesophyll. There are also more cell divisions in the palisade, but this results in a more compact palisade layer rather than greater thickness. Formation of more layers of palisade cells may also contribute to the greater thickness of sun leaves. The photoperiod may also affect leaf development. Under short days Kalanchoë develops leaves that are sessile,

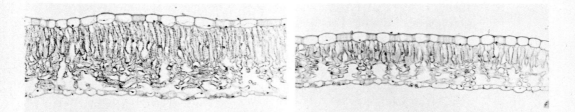

Figure 16-17. Sections through a sun leaf (left) and a shade leaf (right) of *Sambucus canadensis* (elderberry). [Courtesy of Ripon Microslides]

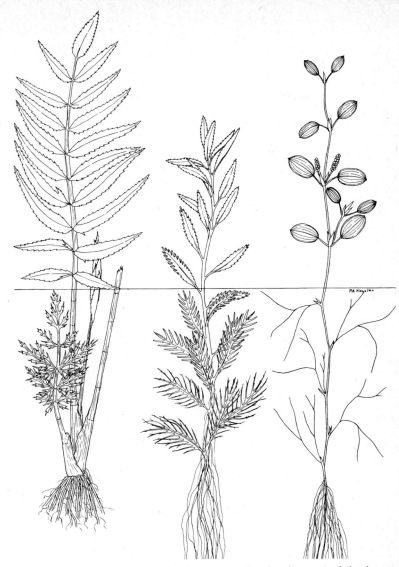

Figure 16-18. Influence of submergence under water on the development of the leaves of three aquatic angiosperms. [Drawing by Peggy-Ann Kessler Duke]

small, entire, and succulent in contrast with petiolate, large, notched, and relatively thin leaves under long days.

Temperature and humidity also may influence leaf development. English gorse has leafy shoots in moist habitats, but in dry habitats the leaves and branches are largely modified into spines and thorns. The various species of aquatic plants, with shoots that are partly submerged under water and partly aerial, generally have submerged leaves that are highly dissected and thin in contrast with the thicker and entire or only moderately lobed aerial leaves (Figure 16-18). Different investigators working with various amphibious species have ascribed the heterophylly to such factors as lower carbon dioxide concentration under water, lower under-water temperature, lower oxygen concentration under water, lower light intensity, and shorter under-water photoperiods (because the critical light intensity is attained later in the morning and earlier at night), rather than to mere submergence itself [2]. It has also been suggested that in some species the submerged leaves may merely be juvenile and that environmental factors are of secondary importance in bringing about the transition from juvenile to adult leaves.

In view of the quite marked influence of heredity and environment on leaf form and structure and of the accessibility of both developing and mature leaves to experimental manipulation, it would seem that leaves should provide unusually desirable materials for the study of many of the basic problems of development which transcend the study of leaf development *per se*. Yet, despite the rather numerous investigations of leaf development, there are still many unanswered questions. For example, in species with compound leaves, what causes the marginal and submarginal initials to divide only at spaced intervals, resulting in leaflets on a rachis? In species with lobed leaves, what causes the initials to stop dividing sooner in some places than others, resulting in the sinuses and lobes? What causes palisade cells to elongate much more than spongy mesophyll cells, and to elongate more in some environments than in others? What signals the cessation of cell divisions when there are enough cells for a size of leaf characteristic of the species, and why does the signal come sooner in elms than in catalpas?

Of course, we do have some information about leaf development, in particular as regards the effects of phytohormones. Auxin apparently is involved in the growth in length of petioles and veins but is not decisive in the growth in area of the blade, which is promoted by gibberellins and cytokinins. However, the latter promote cell enlargement in this case and, thus, are probably not involved in the promotion of leaf enlargement by light, which involves cell division. Fritz Went has suggested that the linear growth of leaves must be distinguished from growth in area if leaf development is to be understood, and also that the development of leaf parts such as the petioles, stipules, veins, and mesophyll are independent of one another even though the development of the leaf as a whole appears to be generally coordinated.

The Growth and Development of Roots

When a seed germinates, the root is generally the first part of the embryo plant to begin elongating and differentiating. The root may attain a considerable size and even form some branch roots before there is any very great growth of the hypocotyl, plumule, or coleoptile. The root meristem is commonly organized in the late heart or early torpedo stage of the embryo (Chapter 14). In some species the embryonic root consists of little more than the apical meristem, but in others it is a readily identifiable radicle quite distinct from the adjacent hypocotyl.

Our consideration of the growth and development of roots will be briefer than the preceding consideration of shoots, partly because several things, such as the differentiation of cells, apply to a considerable degree to roots as well as shoots and partly because the structure and development of roots is somewhat less complicated than that of shoots.

The Root Apex

The root apex (Figure 2-2) is, paradoxically, actually subapical because of the terminal root cap that it continues to lay down at the same time that it is adding to the root axis proper. However, use of the term apex is justified because of its homology with the stem apex as the meristematic region that gives rise to the primary tissues. The root apex, unlike the shoot apex, is not complicated by formation and presence of leaf primordia and lateral bud primordia. Root elongation is more uniform than stem elongation because there are no nodes and internodes that elongate differentially.

The undifferentiated embryonic tissues that lie just behind the root apex are similar to those just behind the shoot apex: a procambium that differentiates into the vascular tissues and the ground meristem that gives rise to the cortex and endodermis (Figure 16-19). In some species the protoderm is derived from the ground meristem and in others from the root cap. The pericycle has been described variously as arising from the procambium or the ground meristem. A meristem anterior to the root apex continues to produce root cap cells as the root grows, thus maintaining the root cap as its older cells are sloughed off or eroded as the cap penetrates the soil.

There is extensive and controversial literature dealing with the number of initial cells in the root apex and which of the initials, if more than one, gives rise to each of the embryonic tissues. The complexity and uncertainty are partly a result of extensive species differences and partly a result of conflicting conclusions from different experimental techniques or even conflicting interpretations of the same data.

Many lower vascular plants have a single apical initial in the root as well as in the shoot. Others have a tier of apical initials. Gymnosperms and dicotyledons have been described as having two sets of initials, one for the procambium and one for the other tissues including the root cap; monocotyledons have frequently been reported to have three sets of initials, one for the procambium, one for the ground meristem, and one for the root cap, the protoderm being derived from one of the last two. One thing about the apical initials of roots, regardless of their number, that appears to be well established is that they divide much less frequently than their cell progeny.

A number of those who have studied the root apex of angiosperms have proposed that there are only a few apical initial cells. For example, Von Gutenberg claims there is only one, whereas Brumfield suggests there are probably only three. Since such interpretations have been widely criticized they will not be considered here, but discussions of them can be found elsewhere [3, 7, 21]. The most generally accepted interpretation of the nature of the root apex is that of Clowes [4], based on his extensive investigations of the root apex by three different kinds of techniques. Other investigators, notably Jensen and his coworkers [12], have reached conclusions similar to those of Clowes.

Clowes traced cell lineages in fixed and stained median longitudinal sections of root tips (Figure 16-20) and confirmed the importance of T divisions in the

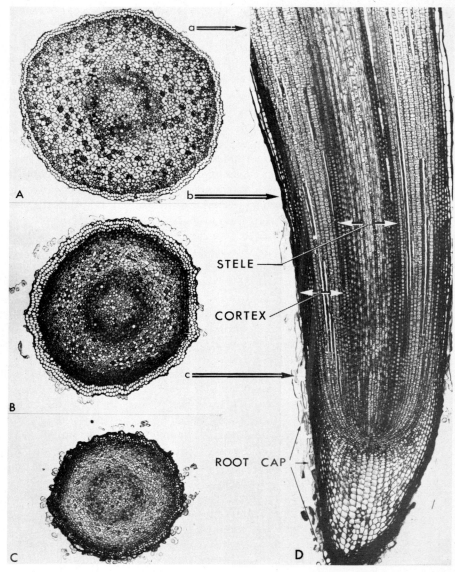

Figure 16-19. Photomicrographs of sections through the root tips of cultured Convolvulus (bindweed) roots. The cross sections **(A, B, C)** are from the indicated levels of a comparable root sectioned longitudinally **(D).** The open arrows in B enclose four sieve tubes, the first provascular cells to differentiate. The loose cells outside the epidermis in all four sections are cells detached from the root cap which cling to the epidermis under culture conditions. [Reprinted with permission of The Macmillan Company from *Plant Structure and Development* by T. P. O'Brien and Margaret E. McCully. Copyright © 1969 by The Macmillan Company]

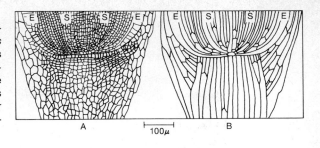

Figure 16-20. Diagrammatic drawings of longitudinal sections of the root tip of *Zea mays* (corn). **(A)** Outlines of the cells, with regions of the root shown by heavier lines. **(B)** Representation of the major T divisions that give rise to the different cell lineages, epidermis (E), stele (S), cortex, and root cap. [After F. A. L. Clowes, *Apical Meristems,* Blackwell Scientific Publications, 1961]

determination of cell lineages, originally proposed by Schuepp in 1917. In a T division a meristematic cell divides transversely and then one of the daughter cells divides longitudinally, the walls between these cells making a T-shaped configuration (Figure 16-21). The two cells formed by the longitudinal division then give rise to longitudinal rows of cells. In the cells giving rise to the body of the root the upper daughter cell divides longitudinally, forming an inverted T. In the cells giving rise to the root cap the lower daughter cell divides longitudinally, forming an upright T. From his analysis of T divisions Clowes concluded that the apical initials formed an inverted cuplike structure in the apex at the base of the root cap.

Clowes determined the frequency of cell divisions in the various regions of the root apex by supplying seedling roots with tritiated thymidine, fixing and sectioning them longitudinally after a day, and then making radioautographs of the root tips. Cells undergoing frequent divisions had incorporated much of the radioactive thymidine in their DNA, as evidenced by the radioautograph, whereas cells dividing little or not at all had synthesized little or no DNA containing the labeled thymidine. In all the sections checked Clowes found a group of 500 to over 1000 apical cells with essentially no labeling, which he called the quiescent center (Figure 16-22). The apical initial cells lie along the periphery of the quiescent center.

The third approach Clowes used was surgical destruction of various parts of the root apex and determination of its effect on root apex regeneration and growth. Data from his various investigations led Clowes to conclude that there are many initial cells, not just a few. Although the cells of the quiescent center divide little or not at all, Clowes found they began dividing after wounding or irradiation and regenerated the root apex. He found that the quiescent center

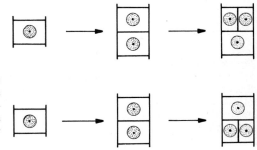

Figure 16-21. Diagram illustrating T divisions. **Above:** Inverted T division, which gives rise to all the tissues of the root except the root cap. **Below:** Upright T division, which gives rise to the root cap. The cell walls forming the T in each case are emphasized by heavier lines.

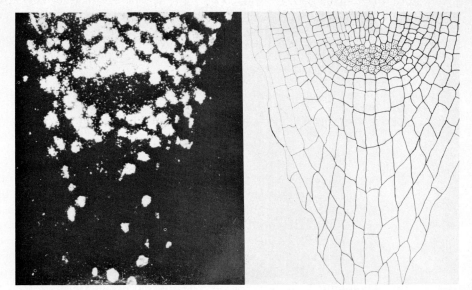

Figure 16-22. Left: Radioautograph of a section through the root apex of *Sinapsis alba* (white mustard) which had been supplied with ³H-thymidine, showing the location of cells (white spots) that had incorporated this tracer extensively in newly synthesized DNA (×320). Note the essential absence of labeling in the quiescent center. **Right:** Drawing of the cell outlines of the same section (×225), showing the quiescent center (stippled cells). [Courtesy of F. A. L. Clowes. Reprinted with the permission of The Macmillan Company from *Plant Structure and Development* by T. P. O'Brien and Margaret E. McCully. Copyright © 1969 by The Macmillan Company]

does not appear in seedlings until just after germination and in lateral roots until they emerge from the main root. Clowes suggests that the quiescent center may function by synthesizing hormones or by maintaining the geometry of the meristems.

Jensen and Kavaljian also investigated cell division in roots radioautographically and, like Clowes, found a definite quiescent center (Figure 16-23). Jensen used histochemical techniques to determine the nucleic acid and protein contents of root cells. He found that the cells of the quiescent center have the lowest content of DNA, RNA, and proteins in the whole root tip. Thus they appear to be relatively inactive metabolically as well as in cell division.

Figure 16-23. Diagrammatic longitudinal section of a root tip with the frequency of cell division indicated by the density of the stippling. Note the essential absence of cell division in the quiescent center. [After W. A. Jensen and L. G. Kavaljian, *Amer. J. Bot.,* **45**:365 (1958)]

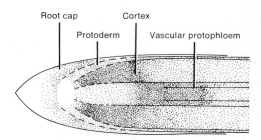

Root cap   Cortex
Protoderm   Vascular protophloem

Although the regions of cell division, cell enlargement, and cell maturation are not as precisely delimited as sometimes suggested in introductory textbooks, the region of maximum cell division usually extends 300 to 1500 μ behind the apex and the region of maximum cell elongation 1500 to 5200 μ from the apex. Heimsch found in barley that maturation of the cells of the vascular tissues occurs much farther back of the apex in the rapidly elongating main roots than in either the main roots near their final length or the small lateral roots. In the former, sieve tubes mature at about 750 μ, protoxylem at 8500 μ, and metaxylem at 10,500 μ. In the latter, the distances are 250, 400, and 550 μ respectively.

Although the cells of the phloem mature sooner and continue to divide longer than those of the xylem, the cells of the xylem begin elongating sooner. Differentiation of the procambium is acropetal and is not complicated by leaf traces as it is in stems. Both the xylem and phloem always differentiate centripetally.

A number of investigators including Erickson and Goddard, Brown and Broadbent, Brown and Robinson, and Jensen have reported greatly increased metabolic activity in elongating root cells. Based on their data, if the concentration or activity per cell in the root apex is taken as 1 in each case, the values for elongating cells are as follows: dry weight, 10; protein nitrogen, 9; oxygen consumption, 6; RNA, 5; DNA, 2; phosphatases, 3; and invertase, 25. Such data suggest the induction of new kinds of mRNA and enzymes, but most of the differences would be much less marked if calculated per unit of protoplasm than per cell.

There has long been a hypothesis that differentiated vascular tissues in some way, probably chemical, induce the differentiation of more tissue of their own kind from the undifferentiated cells produced by the apex. Since the differentiation of both the xylem and phloem in roots is acropetal and since the xylem and phloem occur in separate alternating strands, Torrey [18, 19] decided that development of the vascular tissues of roots provided ideal material for testing this hypothesis. The number of xylem strands (or later extensions from the central xylem cylinder) is a species characteristic and is generally consistent within a species. The number of strands may be two (diarch), three (triarch), four (tetrarch), five (pentarch), six (hexarch), or a larger number.

Torrey used the roots of pea plants, which are triarch. He excised 0.5 mm root tips from seedlings and cultured them for a week, by which time they were about 1 cm long. Most of the cultured roots retained the triarch pattern, but about a quarter of them were diarch and one was monarch. However, when diarch roots were continued in culture they reverted to triarch, and the monarch root changed to the diarch and then to the triarch pattern. The mature tissues evidently did not induce the preexisting pattern, the changes in pattern originating in the meristem. If the apical meristem was removed from cultured decapitated roots, it was regenerated within 3 days; the newly regenerated root was often triarch but sometimes the three strands were not connected with the old ones, and in some cases the new pattern was cylindrical rather than

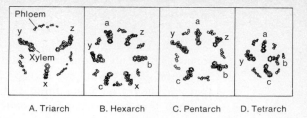

Figure 16-24. Influence of auxin on the pattern of the primary vascular tissue in a root of pea (*Pisum sativum*). **(A)** The triarch pattern of the root cut off and placed in a culture medium. **(B)** The new tip that developed in the presence of auxin in the medium formed a hexarch pattern (adding strands a, b, c). **(C)** After transfer to a medium lacking auxin a pentarch pattern developed. **(D)** Later the pattern became tetrarch. [After J. G. Torrey, *Amer. J. Bot.*, **44**:859 (1957)]

A. Triarch   B. Hexarch   C. Pentarch   D. Tetrarch

triarch. When $10^{-5} M$ IAA was added to the medium, the regenerated roots grew slowly and became hexarch (Figure 16-24). When the hexarch roots were transferred to a lower concentration of IAA, they started growing faster and slowly reverted to pentarch and then tetrarch patterns but never to the characteristic triarch pattern.

These experiments clearly show that acropetal induction does not control the pattern formation and also suggest that the natural IAA concentration in the roots of any species may play a role in determining the number of xylem and phloem strands. Torrey found that there was a correlation between the diameter of the procambial cylinder and the number of strands, and the IAA may act by increasing the diameter of the cylinder.

At the same time that the cells of the vascular tissues are differentiating, the Casparian strips are being laid down in the endodermal cells and the root hairs are developing in the epidermis, thus completing the differentiation of the primary tissues of the root. The differentiation of the root hairs is of considerable morphogenetic interest and has been studied extensively.

The first visual evidence of the differentiation of a root hair cell **(trichoblast)** is the assymetrical division of an epidermal cell, at about 150μ from the apex. The lower (more apical) daughter cell is shorter and becomes the trichoblast, whereas the longer cell will remain a hairless epidermal cell **(atrichoblast)** (Figure 16-25). The atrichoblast generally elongates more, and may divide, forming two atrichoblasts. Each species has a characteristic linear arrangement of trichoblasts, generally every second or every third cell bears a root hair,

Figure 16-25. Stages in the differentiation of root hair cells. **(A)** Early stages in the differentiation of trichoblasts and atrichoblasts in the epidermis of Trianea. Each trichoblast gives rise to a cell with a root hair, but the atrichoblasts may divide again and so give rise to two or more hairless cells. **(B)** Later stages of development in Phleum, where trichoblasts and atrichoblasts alternate. Note that the atrichoblasts are much longer. [A after L. Geitler, B after E. W. Sinnott and R. Bloch. Reprinted with permission of The Macmillan Company from *Development in Flowering Plants* by John G. Torrey. Copyright © 1967 by John G. Torrey]

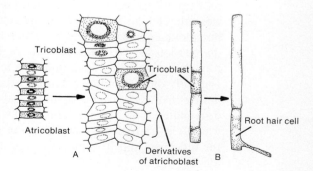

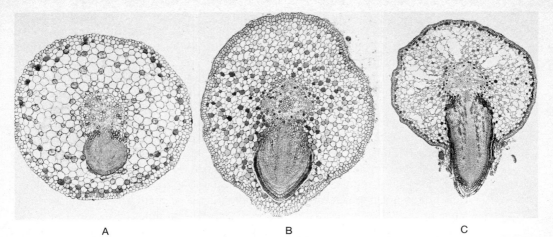

Figure 16-26. Three successive stages **(A, B, C)** in the development of a branch root of Salix (willow). [Courtesy of Ripon Microslides]

although environmental factors may modify the usual pattern. Once the root hair evaginates from the surface of the trichoblast, it continues to grow in length for some time. Growth of a hair occurs at the tip, and the nucleus of the trichoblast is generally near the tip. Because of their continued elongation, the hairs of a root present a conical appearance (Figure 16-26), tapering from the longer, older hairs at the top of the root hair zone to the shorter, younger ones nearer the tip. Root hairs generally range from 5 to 17 μ in diameter and from 80 to 1500 μ in final length. Root hairs generally live and function for only a few days, but because of the continued growth of the root the width of the root hair zone remains more or less constant and the hairs continue to be located at about the same distance from the apex.

Despite the smaller size of the trichoblasts, they have a nuclear volume two to three times that of the atrichoblasts, a substantially higher RNA content, and subsequently a marked increase in enzymes not present in atrichoblasts. Using cytochemical techniques, Avers [1] was able to identify trichoblasts by the presence of enzymes such as acid phosphatase and peroxidase prior to the time that the hairs developed. Thus, trichoblasts apparently provide a good example of the relationship of differential mRNA and enzyme production to cell differentiation.

The Development of Lateral Roots

The development of branch roots differs markedly from that of branch stems, which originate superficially from axillary buds whose primordia arise from the stem apex soon after the leaf primordia. In contrast, branch roots originate deep within the tissues of the main root. The lateral root primordia of gymnosperms and angiosperms arise in the pericycle, whereas those of the lower vascular plants generally originate in the endodermis. However, the endodermis may also participate in the lateral root primordia of higher vascular plants when the branch root arises in undifferentiated tissue. Once the apical meristem has been established it begins growing through the cortex, soon penetrates the epidermis (Figure 16-27), and continues its growth externally. There has been a difference of opinion as to whether penetration of the cortex and epidermis is purely mechanical or involves also the action of hydrolytic enzymes, but the latter seems probable. The death of cells adjacent to the penetrating root supports the enzymatic theory. The endodermis originally covers the laterals, but is shed or dies just before or soon after emergence.

Branch stems have the same precise phyllotaxic arrangement as the leaves that subtended the buds from which they arose, although this is often not obvious because many of the lateral buds may remain dormant and never develop into branches. Branch roots obviously do not arise at nodes nor do they have any regular arrangement comparable with phyllotaxy, but their location is not haphazard. There is some degree of regularity in their vertical spacing, and they generally arise opposite the xylem poles. However, in diarch roots they generally arise opposite the phloem poles, and this also may be the case in some families such as the grasses and sedges. It is frequently possible to determine the type of vascular cylinder (triarch, tetrarch, and so on) of a

Figure 16-27. A root tip of a radish seedling (×15). The continuing elongation of the root hairs until they reach their final length gives the root hair zone a conical profile. [Reprinted with the permission of The Macmillan Company from *Plant Structure and Development* by T. P. O'Brien and Margaret E. McCully. Copyright © 1969 by The Macmillan Company]

species merely by observing the external arrangement of the lateral roots. In species that have secondary root growth the xylem of the branch becomes embedded in the secondary xylem of the main root, just as is the case with the xylem of branch stems.

The origin and development of adventitious roots is similar to that of lateral roots. Adventitious roots are generally regarded as all those that do not arise from the root pole or radicle of the embryo, that is, roots that develop on stems, leaves, calluses, or perhaps even on older roots that have passed the stage when they would normally produce branches. The primordia of adventitious roots that develop in young or herbaceous stems generally form in the interfascicular parenchyma near the periphery of the vascular tissue. In older stems with secondary growth, the primordia arise near the vascular cambium from the cells of a vascular ray or in some cases even from the cambium itself.

Secondary Growth of Roots

Woody dicotyledons and gymnosperms have extensive secondary root growth from vascular cambia, except in small branch roots and, like woody stems, they become thick and woody with a corky bark. Many herbaceous dicotyledons have no secondary growth, but a substantial number have vestiges of secondary vascular tissues. Still others, particularly biennials and herbaceous perennials, have extensive secondary root growth.

Secondary growth of roots was discussed in Chapter 8, but will be reviewed here briefly. At first the cambium is restricted to the regions between phloem bundles and the adjacent bays of the xylem. As a result of the activity of the cambium the bays are filled with secondary xylem so that the outer edge of the xylem becomes cylindrical. Then cambium develops adjacent to the projections of the xylem, connecting the previously isolated strips of cambium and forming a cylinder of cambium. From this point on the production of secondary xylem and phloem by the cambium proceeds as it does in stems. Also as in stems, the tissues external to the cambium are compressed and then split as the vascular cylinder increases in diameter, phellogens develop in the cortex and later in the phloem, and eventually the woody root consists only of bark (cork and phloem), the cambium, and the xylem.

Environmental factors acting on the shoot may have marked effects on the secondary growth of roots. For example, short days promote the growth of radish roots. Excised roots of species that have secondary growth generally fail to initiate a cambium even though the culture medium is suitable for continued primary growth. However, secondary growth of excised pea roots will occur if 8% sucrose and $10^{-5}\,M$ IAA are provided to the root via its basal end. Secondary growth of radish roots requires application of a cytokinin in addition to the sucrose and IAA. These and other similar data lead to the conclusion that nutrients and phytohormones translocated from the shoot to the root are essential for initiation of the cambium and secondary growth. Thiamin and other B vitamins from the shoot are essential for primary root growth and probably also for secondary root growth. A higher concentration of sugars is evidently required for initiation of the cambium and secondary growth than for primary growth.

References

[1] Avers, C. J. "Biochemical localization of enzyme activities in root meristem cells," *Amer. J. Bot.,* **48:**137–143 (1961).

[2] Bostrack, J. M., and W. F. Millington. "On the determination of leaf form in an aquatic heterophyllous species of Ranunculus," *Bull. Torrey Bot. Club,* **89:**1–20 (1962); see also Davis, G. J. "Proserpinaca: photoperiodic and chemical differentiation of leaf development and flowering," *Plant Physiol.,* **42:**667–668 (1967).

[3] Brumfield, R. T. "Cell-lineage studies in root meristems by means of chromosome rearrangements induced by X-rays," *Amer. J. Bot.,* **30:**101–110 (1943).

[4] Clowes, F. A. L. *Apical Meristems,* Blackwell, Oxford, 1961.

[5] Cutter, Elizabeth G. (ed.). *Trends in Plant Morphology,* Longmans, Green & Co., London, 1966.

[6] Cutter, Elizabeth G. *Plant Anatomy: Experiment and Interpretation,* "Part 1: Cells and Tissues," "Part 2: Organs," Addison-Wesley, Reading, Mass., 1971.

[7] Esau, Kathryn. *Plant Anatomy,* 2nd. ed., John Wiley & Sons, New York, 1965.

[8] Esau, Kathryn. *Vascular Tissue Differentiation,* Holt, Rinehart & Winston, New York, 1965.

[9] Gemmell, A. R. *Developmental Plant Anatomy,* St. Martin's Press, New York, 1969.

[10] Greulach, V. A., and J. G. Haesloop. "Some effects of maleic hydrazide on internode elongation, cell enlargement, and stem anatomy," *Amer. J. Bot.,* **41:**44–50 (1954).

[11] Jacobs, W. P., and I. B. Morrow. "Quantitative relations between stages of leaf development and differentiation of sieve tubes," *Science,* **128:**1084–1085 (1958).

[12] Jensen, W. A., and L. G. Kavaljian. "An analysis of cell morphology and the periodicity of division in the root tip of *Allium cepa, Amer. J. Bot.,* **45:**365 (1958).

[13] Newman, I. V. "Pattern in meristems of vascular plants. I. Cell participation in living apices and in the cambial zone in relation to the concepts of initial cells and apical cells." *Phytomorphology,* **6:**1–19 (1956).

[14] Popham, R. A. "Principal types of vegetative shoot organization in vascular plants," *Ohio J. Sci.,* **51:**249–270 (1951); "Cytogenetics and zonation in the shoot apex of *Chrysanthemum morifolium,*" *Amer. J. Bot.,* **45:**198–206 (1958).

[15] Satina, S., A. F. Blakeslee, and A. G. Avery. "Demonstration of the three germ layers in the shoot apex of Datura by means of induced polyploidy in periclinal chimeras," *Amer. J. Bot.,* **27:**895–905 (1940).

[16] Sinnott, E. W. *Plant Morphogenesis,* McGraw-Hill, New York, 1960.

[17] Stebbins, C. L., and S. S. Shah. "Developmental studies of cell differentiation in the epidermis of monocotyledons. II. Cytological features of stomatal development in the Gramineae," *Develop. Biol.,* **2:**477–500 (1960).

[18] Torrey, J. G. "Auxin control of vascular pattern formation in regenerating pea root meristems grown *in vitro*," *Amer. J. Bot.*, **44:**859–870 (1957).

[19] Torrey, J. G., D. E. Fosket, and P. K. Hepler. "Xylem formation: a paradigm of cytodifferentiation in higher plants," *Amer. Sci.*, **59:**338–352 (1971).

[20] Voeller, B. "Developmental physiology of fern gametophytes: relevance for biology," *BioScience*, **21:**266–275 (1971).

[21] Wardlaw, C. W. *Morphogenesis in Plants*, Methuen, London, 1968.

[22] Wardlaw, C. W. *Essays of Form in Plants*, Barns & Noble, New York, 1968.

[23] Wardlaw, C. W. *Cellular Differentiation in Plants and Other Essays*, Barnes & Noble, New York, 1970.

[24] Wetmore, R. H. "Morphogenesis in plants—a new approach," *Amer. Sci.*, **47:**326–340 (1959).

Some Developmental Phenomena 17

This chapter deals with several aspects of plant development that have received little or no consideration in the previous chapters. First, we shall consider such morphogenetic phenomena such as polarity, symmetry, spirality, differentiation, correlation, and regeneration. We will then turn to the development of cells, tissues, and organs detached from plants and cultured in sterile media. Although their development is often markedly different from what it would have been in the intact plant, these excised plant parts have contributed much to our understanding of the development of intact plants. Finally, we will consider various types of abnormal plant development.

Some Aspects of Plant Morphogenesis

Perhaps the most striking and significant thing about the development of an organism, yet one that is often simply taken for granted, is the precision and high degree of organization of both the morphogenetic processes and the resulting structures. Several aspects of this organization will be considered in this section.

Polarity

The vast majority of all organisms are polar, that is, they have an axial structure in which the two ends or portions differ from one another physiologically as well as morphologically. In vascular plants polarity is characteristically expressed as a lower root portion and an upper shoot portion. Both the gametophytes and sporophytes of mosses and liverworts are also polar, and most algae and fungi are polar although the polarity may be less marked both morphologically and physiologically and more readily subject to reversal.

Vertebrate animals, and the great majority of invertebrates have an axis with head and tail ends. Even relatively simple invertebrates such as hydras exhibit definite polarity. All the red and brown algae, including the coenocytic genera (Chapter 2), have definite polarity as do many green algae (notably those with holdfasts) and most fungi. Even many unicellular organisms are polar, as evidenced by the location of their flagella and various internal organelles. However, many filamentous green algae such as Spirogyra lack morphological polarity, even though their structural organization is axial, as do spherical organisms like Volvox and many unicellular algae, although they may have some degree of physiological polarity. The bacteria, blue-green algae, amebae, and the vegetative stages of slime molds are characteristically nonpolar.

Among the physiological aspects of polarity in vascular plants are such things as the polar translocation of some substances such as auxin and calcium ions, metabolic gradients from the base to the apex of both roots and stems, and possibly the generally unidirectional mobilization pathways. Roots and shoots differ from one another physiologically as well as morphologically. For example, the optimal auxin concentration for root growth is much lower than for stem growth. Aside from the obvious case of photosynthesis, roots and shoots differ greatly in their metabolic capabilities. To mention just a few examples, only the roots of tobacco plants can synthesize nicotine, plants in general can synthesize thiamin and other B vitamins only in their shoots, and roots and shoots often differ in their capacity for synthesizing the various amino acids. There are definite electrical polarities in plants with small but consistent potential differences between any two points on an organ, or even from one end of a cell to the other. These bioelectrical fields have been extensively investigated, by Lund [12] and his associates and more recently by others, including Jaffe.

Among the more striking manifestations of polarity is the development of adventitious roots and buds on cuttings, a phenomenon investigated as far back as 1871 by Pfeffer and 1878 by Vochting. With only a few exceptions, stem cuttings develop adventitious roots at their morphologically lower ends, even though they are inverted, whereas the buds at the morphologically upper ends develop into branches (Figure 17-1). A long cutting develops branches and roots only toward the ends, but if it had been cut in half each piece would have developed branches and roots, a situation comparable with the presence of both positive and negative poles in a bar magnet cut in half.

Root cuttings generally exhibit a similar polarity, developing adventitious buds at their morphologically upper ends and roots at their lower ends, even when inverted (Figure 17-2). However, the polarity of roots is often weaker than that of stems, and inverted roots sometimes develop roots at their morphologically upper ends and buds at their morphologically lower ends. Half slices of fleshy roots and tubers may show transverse polarity, forming shoots along the diameter and roots along the circumference of the half-disc (Figure 17-3).

Leaves also have regenerative polarity, as shown by the fact that detached leaves develop adventitious roots only near the cut end of the petiole. However, if the leaves are those of a species capable of forming adventitious buds, these also develop near the end of the petiole but above the roots. Leaf polarity is relatively weak, and may even be lacking. When pieces of leaf blade are

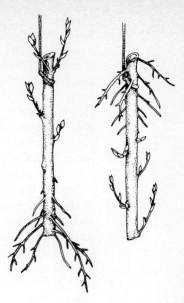

Figure 17-1. Polarity of regeneration in cuttings of willow stems suspended in moist air. **Left:** a cutting suspended in its normal position. **Right:** a cutting suspended in an inverted position. [After W. Pfeffer. From E. W. Sinnott, *Plant Morphogenesis,* McGraw-Hill Book Company, New York, 1960]

used in propagation, as in begonia, there is little or no evidence of polarity in the formation of buds and roots. In general, cuttings of the lower vascular plants and nonvascular plants have a weaker polarity that is more readily reversed than those of the vascular plants.

Initially spores and zygotes have no obvious polarity, and the induction of polarity in these nonpolar cells has been a matter of great interest to various

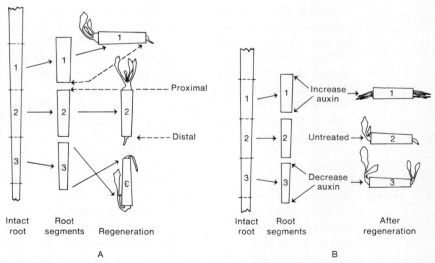

Figure 17-2. Polarity of regeneration in segments of dandelion (Taraxacum) root. **(A)** Note that shoots are produced on the proximal (upper) end and roots on the distal (lower) end regardless of the orientation of a segment. **(B)** If auxin is applied to a segment placed horizontally roots develop at both ends, but if auxin is decreased shoots develop at both ends. [After H. E. Warmke and G. L. Warmke, *Amer. J. Bot.,* **37:**272 (1950)]

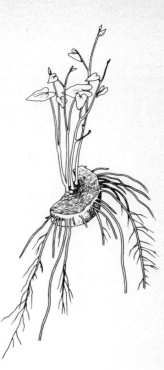

Figure 17-3. Transverse polarity of regeneration from a half slice from a tuber of Dioscorea (yam). [After K. Goebel. From E. W. Sinnott, *Plant Morphogenesis,* McGraw-Hill Book Company, New York, 1960]

investigators. Polarity may develop even before the first division as indicated by a rearrangement of the cell organelles (Figure 17-4) or a change in the shape of the cell. In seed plants polarity of the zygote seems to be determined in some way by the polarity of the embryo sac, the root pole (radicle) developing toward the micropyle of the ovule. Not only is the embryo sac or female gametophyte polar but the megaspore mother cell from which it develops is polarized to a certain extent. It is always the megaspore nearest the micropyle that is functional and develops into the female gametophyte whereas the other three microspores disintegrate. In Isoetes and at least some other ferns the polarity of the zygote is apparently determined by the archegonium, and the first division is always at right angles to the axis of the archegonium.

In contrast to the above examples, induction of polarity in spores and zygotes that develop outside a parental structure appears to be determined primarily by environmental factors. In general, light appears to be the most important polarizing factor for the spores of the lower vascular plants, bryophytes, and algae and also for the zygotes of algae. The darker side of the cell gives rise

Figure 17-4. At the left is an unpolarized spore of Equisetum. Polarization of the spore results in concentration of the chloroplasts (chl) on the upper side and movement of the nucleus (k) toward the lower side. Note position of the equator in the first mitotic division. [After W. Nienburg, *Ber. Deut. Bot. Ges.,* **42:**95 (1924)]

Figure 17-5. Polarization of embryo of Fucus (a brown alga). **(A)** Zygote before division. **(B)** Polarization of the two-cell embryo. **(C)** The lower cell gives rise to the rhizoid and the upper cell to the rest of the thallus. **(D)** An embryo about 12 days old with developing rhizoid and thallus, magnified less than A–C. [After J. G. Torrey, *Development of Flowering Plants,* The Macmillan Company, New York, 1967]

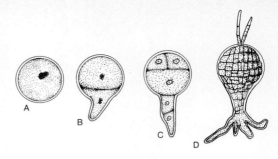

to the rhizoids and the more brightly lighted side develops into the apical organs. Blue light is most effective, whereas red light is ineffective. The polarizing effect of light is rapid. Haupt found that the polarity of Equisetum spores could be reversed within a brief period by light from the direction opposite to that from which it was originally supplied. Perhaps surprisingly, gravity plays a minor role, if any, in the induction of the polarity, although sometimes polarity can be reversed by centrifugation. Although light seems to be the principal natural polarizing factor, polarity has been induced, reversed, or eliminated by a wide variety of other factors such as auxin, thiamin, electrolytes, and electrical fields.

The most extensive studies of the induction of polarity have been on the zygotes of the brown alga Fucus and other genera of the order Fucales. The unfertilized egg is nonpolar, but polarity soon develops in the zygote. Its first division is unequal; the smaller cell develops a protuberance and grows into the rhizoid, and the larger cell grows into the remainder of the thallus (Figure 17-5). Gravity does not influence this polarity.

Whitaker [27] found that the rhizoid will develop on the side toward a higher temperature, a lower pH, or a higher auxin concentration. Lund [11] found that the rhizoid developed toward the + pole in an electrical field. Whitaker, as well as others, noted that in a cluster of Fucus zygotes the rhizoids develop toward the center of the group. It has been suggested that, in the absence of any effective environmental gradients, the point of entry of the sperm into the egg may determine polarity. Polarity is evident even before the first division of the zygote since the assumption of a pear shape occurs before the division. Whitaker attempted to explain the influences of the various factors on polarity on the basis that all of them generated a carbon dioxide-pH gradient, but this theory has been placed in serious doubt by the extensive work of Jaffe [9].

For one thing, Jaffe found that the group effect was not lost or changed by a 500-fold change in the buffer capacity of the system nor by changes from emission of carbon dioxide by respiration or its uptake by photosynthesis. Jaffe has also shown that in plane-polarized white light the protuberance grows out at approximately right angles to the incident light, and many zygotes form two protuberances at 180° from one another. He also found that at times zygotes in a cluster develop rhizoids away from, rather than toward, one another and suggested that these group effects result because of the diffusion from zygotes of a rhizoid-stimulating substance (rhizin) or a rhizoid-inhibiting substance (antirhizin). Jaffe proposed that the variety of polarity-inducing factors may all operate basically by influencing membrane potentials and generating a transcellular electrical potential. He suggested that this might result in electrophoresis within the cell, in generating ion gradients, or in a popped balloon effect. One hole in a balloon keeps others from forming because of the reduced air pressure, and rhizoidal initiation could prevent others from forming by keeping membrane potential down. Jaffe found that the development of a transcellular electrical potential and the first evidences of polarization are almost simultaneous.

Quatrano [15] implicated the production of new kinds of RNA and proteins in the polarization of Fucus zygotes. He reported that the RNA was produced several hours before the new proteins (enzymes?) and that the period of protein synthesis coincided with the time of irreversible polarization.

Symmetry

A second marked developmental aspect of plant (as well as animal) structure is symmetry. The great majority of all organisms are symmetrical as well as polar, and individual cells, tissues, and organs are also often symmetrical. Only a few organisms such as amebae and the plasmodia of slime molds lack symmetry, but the sporangia of slime molds are symmetrical. Although the leaves of some plants such as elms and begonias are asymmetrical, they are components of a larger symmetry. There are four basic types of symmetry: spherical, radial, bilateral, and dorsiventral (Figure 17-6).

Spherical symmetry is very rare among organisms. It is restricted to spherical cells or colonies such as coccus bacteria, some spherical unicellular algae, some spores, eggs, and zygotes, and Volvox colonies. For perfect spherical symmetry two identical halves must result if a sphere is cut along a diameter in any plane. Even cells with superficial spherical symmetry may have asymetrical distribution of their organelles. Although a fruit such as an orange may be spherical, it does not have spherical symmetry, because its stem and blossom ends differ and its internal structure is radial.

Radially symmetrical structures, in contrast with spherically symmetrical ones, are polar and have only one axis of rotation. They can be cut around this axis along a diameter into two identical halves. Strictly radial structures are symmetrical along an infinite number of diameters, but some radial structures such as stems bearing opposite leaves are symmetrical only along one or two diameters. Roots and vertical stems are usually radially symmetrical, as are many mushrooms and mosses and also the tubular leaves of rushes. Most regular (actinomorphic) flowers are radially symmetrical, as are many inflorescences such as the heads of most composites and the umbels of members of the carrot family. Filamentous algae generally have radial symmetry, as do many unicellular algae. Thus, radial symmetry is very common in the plant kingdom.

In contrast, **bilateral symmetry** is quite rare. Bilaterally symmetrical structures are flattened and their front and back as well as their right and left sides are similar, so there are two planes of symmetry. The stems of some cacti such as Opuntia, the leaves of iris, the phylloclads of various species, the thalli of various algae such as Fucus and Laminaria, and the coenocytic alga Bryopsis

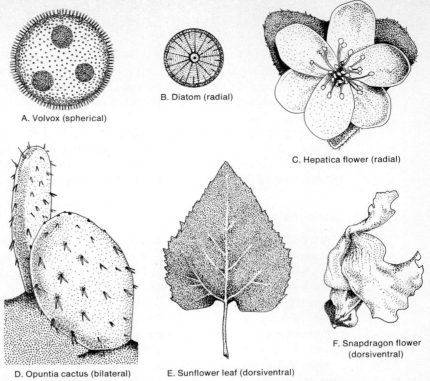

A. Volvox (spherical)

B. Diatom (radial)

C. Hepatica flower (radial)

D. Opuntia cactus (bilateral)

E. Sunflower leaf (dorsiventral)

F. Snapdragon flower (dorsiventral)

Figure 17-6. Drawings illustrating examples of different types of symmetry plants. **(A)** Spherical symmetry; **(B, C)** radial symmetry; **(D)** bilateral symmetry; **(E, F)** dorsiventral symmetry. (All are drawn to different scales.)

are bilaterally symmetrical. The flowers of the members of the mustard family appear to be radially symmetrical but are actually bilaterally symmetrical because two of the six stamens are shorter than the other four and opposite one another.

Dorsiventral symmetry differs from bilateral symmetry in that the front and back of the structure are different so there is only one plane of symmetry, that is, only the right and left sides are similar. Dorsiventral symmetry is sometimes incorrectly referred to as bilateral symmetry. Like radial symmetry, dorsiventral symmetry is widespread in the plant kingdom. Leaves typically have dorsiventral symmetry, not only as regards external morphology but also as regards the arrangement of their tissues. The phloem is on the under side and the xylem on the upper side of the veins, the spongy mesophyll is usually below the palisade mesophyll, and the number of stomata in the upper and lower epidermis usually differs. Fern prothallia and liverwort gametophytes are typically dorsiventral. Horizontal stems, such as rhizomes, stolons, and the trailing stems of various plants as diverse as ivy and clubmosses, are usually dorsiventral rather than radial, although horizontal roots commonly retain radial symmetry. Irregular (zygomorphic) flowers, such as those of legumes, orchids, and snapdragons have dorsiventral symmetry.

The development of symmetrical structures presents interesting morphogenetic problems that are still essentially unsolved. The symmetry of a plant is obviously basically hereditary, and the development of symmetrical structures is one of the more complex aspects of genetic control. However, in some cases internal or external environment can alter or even prevent symmetrical development. To take a common and simple example, trees develop a symmetrical shoot or crown (Figure 17-7) except when exposed to substantial environmental gradients. Thus, a tree growing at the edge of a forest is likely to be asymmetrical in the branches on the lighted and shaded sides and also in the thickness of the annual rings in the trunk, whereas mountain trees exposed to strong prevailing winds are notoriously asymmetrical. Näf and others have found that young fern prothallia growing in continuously shaken culture media under conditions that eliminate environmental gradients fail to develop both

Figure 17-7. A tree that has grown in the open. Although the individual branches are not symmetrical, the crown as a whole approaches radial symmetry. Pines and other conifers that have grown in the open are generally even more symmetrical. [Courtesy of the Chicago Natural History Museum]

polarity and symmetry and grow into calluslike masses. The development of isolated tissues in culture into asymmetrical calluses illustrates the importance of both internal and external environment in the development of organized symmetrical structures and also stresses the importance of symmetry as a normal component of plant organization.

Goebel found that when Lycopodium was placed in the dark the shoots became radially symmetrical in contrast to their usual dorsiventral symmetry, showing that light rather than gravity influenced the symmetry. However, gravity may also influence symmetry as Sinnott found to be the case in maple leaves (Figure 11-7). The influence of internal environment on the development of symmetry is illustrated by the experiment of Sussex, who made an incision in a stem apex above the point where a leaf primordium would develop. The primordium grew into a radially symmetrical bud rather than a normal dorsiventral leaf. The apical meristem evidently plays a role in the induction of normal leaf symmetry. Many other examples of the morphogenetic aspects of symmetry, and of the other morphogenetic phenomena discussed in this section, can be found in the books by Sinnott [19], Wardlaw [25], and Thompson [23].

Spirality

A third aspect of plant organization and development (in addition to polarity and symmetry) is spirality. The extensive occurrence of spiral structures in plants, at the molecular, cellular, tissue, and organ levels, is striking and rather amazing. The morphogenetic or general biological significance of spirality, if any, is not clear. Apparently, spirality at one organizational level has little to do with that at higher levels. At the molecular level we find spiral molecules such as the helices of starch, DNA, and proteins. At the cellular level there are spiral thickenings of the walls of protoxylem vessel elements, the helical chloroplasts of Spirogyra, the spiral orientation of cellulose fibers in secondary cell walls, and the spiral streaming of protoplasm.

Castle found that the sporangiophore of the fungus Phycomyces grew in a spiral fashion, probably because of the spiral orientation of the chitin fibers in the wall, and Green found similar spiral growth of the long cells of the alga Nitella. At the tissue level the cells of the xylem are generally oriented spirally, rather than vertically, although it is interesting that this is not the case in roots as it is in stems. At the organ and organism levels we find spirally twining vines and tendrils and the spiral arrangement of alternate leaves on stems (phyllotaxy). The growth of stems occurs in a spiral pattern, resulting in spiral growth movements **(nutations)** that are strikingly independent of environmental gradients. Nutations are clearly evident in time lapse moving pictures of growing stems. However, despite the widespread occurrence of spirality in plants, it should be noted that many structures at most levels of organization lack it.

Perhaps the most extensively studied case of spirality is **phyllotaxy** (Figure 17-8). The spiral arrangement of alternate leaves on stems may not be obvious, but it can easily be made so by tying a string to the petiole of the lowest leaf on a branch and then winding the string around the stem so as to touch the petioles of all the successive leaves in order. The string now describes a spiral.

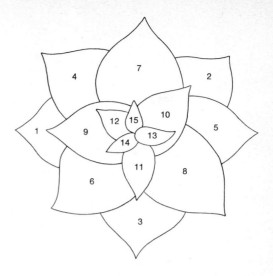

Figure 17-8. Diagram of a shoot with 3/8 phyllotaxy viewed from above. Note that every eighth leaf is directly above another leaf (for example leaf 9 is over leaf 1 and 10 over 2) and that three turns around the stem are required to reach the eighth leaf. [After E. W. Sinnott and K. S. Wilson, *Botany: Principles and Problems,* 6th ed., McGraw-Hill Book Company, New York, 1963]

The simplest phyllotaxy is the one in which successive leaves occur at an angle of 180°. Thus, to go from a leaf to the next one directly above it requires one circuit of the stem and passage of two leaves. This kind of phyllotaxy is described as 1/2. If the third leaf is directly above the first one, then one circuit of the stem and three leaves are involved, and the phyllotaxy is 1/3. More common than these phyllotaxies is one in which the fifth leaf is directly above the first and two circuits of the stem are required to reach it: 2/5 phyllotaxy.

The number of different phyllotaxies is definitely limited: the series is 1/2, 1/3, 2/5, 3/8, 5/13, 8/21, 13/34, and 21/55. There are, for example, no phyllotaxies of 1/4, 3/5, or 5/8. The intriguing thing is that from the third fraction of the series on the numerator is the sum of the two previous numerators and the denominator is also the sum of the two preceding denominators. This is the Fibonacci series of fractions long known to mathematicians, and because of this mathematical basis of phyllotaxy it has been of interest to mathematicians and philosophers as well as botanists. The most complex phyllotaxy of foliage leaves is 5/13; the higher ones such as 8/21, 13/34, and 21/55 are found in cones and other compact axes with greatly reduced internodes (Figure 17-9). Although flower parts are so close together that they appear whorled, monocotyledon flowers generally have a 1/3 spiral and dicotyledon flowers a 2/5 spiral.

Although there has been a great deal of speculation about the significance of phyllotaxy, most of it has been philosophical rather than scientific. From the standpoint of morphogenesis, the principal interest in phyllotaxy is what causes leaf primordia to arise at just the right place on the stem apex so as to provide the kind of phyllotaxy specified by the genetic code of the species. Some of the theories regarding this have been mentioned earlier (Chapter 16). It should be made clear that each species of vascular plant has a certain phyllotaxy (for example, 2/5) and that this is remarkably resistant to alteration by environmental factors. However, many plants slowly shift from one phyllotaxy to another during their development. It has been suggested that the

Figure 17-9. Drawing of a pine cone showing the tight spirals in which the cone scales are arranged. Note that there are both clockwise and counter-clockwise spirals.

spiral arrangement of leaves has survival value because it reduces the shading of one leaf by another to a minimum, but this is dubious.

Differentiation

Since differentiation [8] has already been discussed in various places in the six preceding chapters, there will be no detailed consideration of it here. However, differentiation is one of the key aspects of morphogenesis and central to the problems of development, so at least a brief review of its general nature is essential at this point. The importance of differentiation in the development of all but the simplest plants becomes evident if one visualizes an organism that has at least rudimentary polarity, symmetry, and spirality and yet has no differentiated cells, tissues, or organs.

Although differentiation is evident at all levels of organization, it occurs basically at the cellular level. Cell differentiation is usually thought of in terms of cell morphology—changes in cell size and shape, changes in cell walls such as characteristic patterns of secondary wall deposition, and changes in the numbers, kinds, and sizes of cell organelles. However, cells that are morphologically similar may differ in their physiological capabilities, as in the case of parenchyma cells in stems and roots. Also, differences in the plane, number, and duration of cell divisions are important in determining the characteristic sizes and shapes of the various tissues and organs. Even death of the cell is an essential component in the differentiation of cells such as vessel elements, fibers, and cork cells.

The intriguing thing about differentiation is what causes the various changes in cells to occur at just the right times and right places during normal development so as to produce the kinds of tissues and organs characteristic of the species. Despite the extensive investigations of development that have been conducted, we still have no really satisfactory explanation of this.

Of course, genetic control is obviously involved, since the resulting structures and physiological capabilities are characteristic of the species. During the past century several biologists postulated that differentiation might be explained by the differential distribution of genes to cells during cell division, but it is now clear that all cells produced by mitosis normally have the full complement of hereditary potentialities. However, as we have pointed out previously, in recent years there has been a major breakthrough in understanding the differential expression of hereditary potentialities by the various cells of an organism. In any particular cell only a fraction of the kinds of mRNA for which there are DNA codes are actually being produced; the kinds of mRNA produced change from time to time in any particular cell; and such changes in the mRNA present precede cell differentiation. Thus, the changes in kinds of mRNA would result in changes in the kinds of enzymes produced, and this could explain the physiological and morphological changes involved in differentiation. There is also evidence that in some cases a specific enzyme may not be synthesized by a cell even though the mRNA coding for it is present.

That hormones play a role in differentiation has been pointed out previously, including the relatively simple example of the induction of vessel element formation in otherwise undifferentiated calluses by localized auxin applications. It has long been recognized that the mere position of a cell in the developing organism can influence its differentiation. This seems to be, at least partially, a result of the movement into the cell of such substances as hormones and foods from other cells, but it may also involve gradients of other factors of the internal environment such as oxygen, carbon dioxide, or even electrical potentials.

Much more extensive discussions of differentiation are available elsewhere, as in Sinnott [19, Chap. 8] and Salisbury and Ross [18, Chap. 22]. The latter authors draw interesting parallels between plant differentiation and the implements of modern technology such as transducers, feedback systems, logic circuits, and computers.

Correlation

The growth and development of organisms is integrated and controlled in such a way that they attain a size and shape characteristic of the species with quite characteristic ratios between the sizes of the various organs. Such integration of growth of the organism as a whole is referred to as correlation or correlative growth. Correlations involve growth more and development less than the other aspects of morphogenesis we have considered. In general, correlations involve the influence of one organ or structure on the growth of another. Sinnott [19] compares the normal growth and development of an organism with the operation of an organized army in contrast with an unruly mob.

The results of correlative growth are obvious in human beings, where the appendages and head are normally proportional in size to the trunk; the rare exceptions only emphasize the importance of correlation. Correlations are perhaps less obvious in plants because of their more diffuse and less precisely organized bodies, but still each species has a characteristic shape, size, and pattern of branching. Contrast, for example, a spruce tree, an elm tree, and a sunflower plant. Sinnott classifies correlations as physiological and genetic; the latter are generally less subject to environmental modification, but he admits all correlations have both genetic and physiological components and that the physiological aspects of genetic correlations are probably just more basic and complex and not as clearly evident. The physiological correlations are mostly either nutritional or hormonal.

One rather simple nutritional correlation is the ratio between the size of fruits and the total area of leaf surface supplying them with foods. The size of fruits developed on a branch can be increased by a reduction in their number (which sometimes occurs naturally as a result of premature abscission of some of the fruits) or by girdling the branch, thus preventing translocation of foods out of the branch. The correlation between the leaf area of a tree and the rate and amount of trunk growth is also obviously nutritional. The mobilization of foods and mineral nutrients by young and rapidly growing organs plays a role in many nutritional correlations. The presence of growing fruits on squash plants may inhibit the normally continuing development of flowers, as shown by renewed flower development from the primordia when the fruits are removed. The inhibition of flower development is probably a result of mobilization by the fruits, although there may also be a hormonal factor. Potato plants rarely develop fruits but, if the environmental factors are unfavorable for tuber growth, fruits will develop in abundance. Tukey found an interesting case in certain hybrid cherries where the embryos died after an initial period of normal development but grew to maturity if excised and cultured in a suitable medium. These embryos were evidently not capable of normal mobilization of nutrients in competition with the developing fruits.

The ratio between the size and weight of the shoot and roots of a plant (S/R ratio) is to a large degree a matter of nutritional correlation, although hormonal correlation is also involved.

A good many examples of hormonal correlations were given in Chapters 12 and 13. One of these is the influence of auxin (and possibly also gibberellin) from pollen and the developing embryos on the growth of fruits. In contrast to such stimulatory hormonal correlations are inhibitory ones like apical dominance. As pointed out in Chapter 12, apical dominance plays an important role in the development of a shoot system that is not hopelessly cluttered, and the varying degrees of apical dominance in different species are responsible in a large part for the characteristic branching patterns of each species. Correlative inhibitions in intact plants generally prevent the development of adventitious roots and buds that may be formed on cuttings and also the division and growth of cells that occurs if they are isolated and cultured in suitable media. Thus, both stimulatory and inhibitory correlations are important in the development of a normally organized and coordinated plant body.

Regeneration

Regeneration is the restoration of tissues or organs of an organism that have been mechanically severed or physiologically isolated, thus reconstituting the organism as a whole. Unlike the morphogenetic phenomena discussed above, regeneration is not primarily a component of normal development. On the whole, the regenerative capacities of plants are much more extensive than those of animals and are widespread among both nonvascular and vascular plants, although some species of vascular plants in particular have rather limited capabilities for regeneration. Among animals the regeneration of organs is limited principally to invertebrates such as planaria, starfish, and certain crustaceans, and in general the regenerative abilities of vertebrates are limited to the restoration of tissues in the healing of wounds.

Under natural conditions regeneration may follow such disruptions of plants as the breaking off of branches or whole shoots by strong winds, the death of parts of the plant as a result of parasites, insect attacks, freezing, fire, or other factors, or the removal of shoots by grazing animals. The deliberate use of cuttings of stems, leaves, or roots by man for vegetative propagation is dependent on the regenerative capacity of the cuttings (Chapter 14). One of the principal experimental approaches to plant morphogenesis has involved the study of regeneration after the surgical isolation or removal of various tissues or organs. Similarly, the study of the development of isolated cells or tissues in culture involves consideration of their regenerative capacities. The most extreme case of regeneration is the development of entire plants from isolated cells or cell clusters in culture.

Among the common examples of regeneration in plants are the development of adventitious roots in stem cuttings, the formation of adventitious buds on roots following removal of the shoot, and the development of both adventitious roots and adventitious buds on detached leaves or leaf cuttings (Figure 17-10). In all these cases there is restoration of a complete plant with a full complement

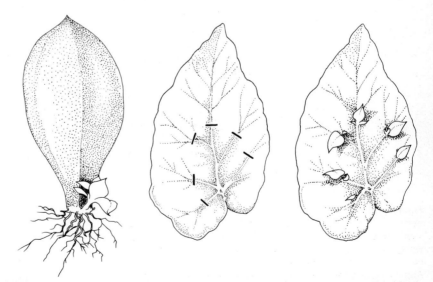

Figure 17-10. **Left:** Development of roots and a shoot from the base of a detached leaf of *Echeveria peacocki*. **Right:** If the veins of a *Rex begonia* are severed, adventitious shoots and roots develop at the cuts. Similar regeneration occurs if a leaf is cut into pieces. [After J. P. Mahlstede and E. S. Haber, *Plant Propagation,* John Wiley & Sons, Inc., New York, 1957]

of organs. The growth of water sprouts from the stumps or roots of trees that have been cut down is a well-known case of regeneration. In this, as in some other examples of regeneration, callus tissue generally develops first and the adventitious buds develop from it.

There are marked species differences in regenerative capacity. For example, the formation of water sprouts is generally restricted to woody angiosperms. There are species differences in ability to develop adventitious roots on cuttings. Some species will do so only after auxin applications, some such as coleus and willow root readily in water, whereas still others will not root even after auxin treatments. The roots of many herbaceous species have little or no regenerative capacity, whereas at the other extreme there are some plants such as bindweed with roots that readily form both adventitious buds and roots. Even rather short pieces of the brittle bindweed roots can regenerate whole plants, making it very difficult to eradicate this weed. There are marked species differences in regeneration by leaves [29]. Some plants such as begonia and African violet readily develop both adventitious roots and buds from leaves, and these can be used in propagation. However, the leaves of most species are incapable of regenerating either buds or roots. The capacity of leaves for developing adventitious roots only is found in more species than the capacity for forming buds as well as roots.

Plants are generally capable of regenerating parts of organs or individual tissues as well as entire organs, particularly in meristematic tissues or in tissues that can revert to a meristematic state and then redifferentiate. When apical meristems are severed vertically into two or more segments, each segment generally reconstitutes a cylindrical apex that can grow into a shoot. Root tips are commonly reconstituted if only the extreme tip (500 to 750 μ) is removed; however, because of the complication of leaf primordia, normal reconstitution of the stem apex will occur only if excision is limited to the terminal dome less than 100 μ back of the tip. Loss of a portion of the epidermis of a stem often results in the redifferentiation of cortical cells into typical epidermis, including cutinization and the presence of guard cells. If one or more vascular bundles are severed by cutting a notch in the stem, the severed bundles become reconnected by the differentiation of parenchyma into xylem elements (Figure 17-11). Phloem may be similarly regenerated.

There are many interesting examples of regeneration in nonvascular plants, only a few of which will be noted here. Among the algae even individual cells

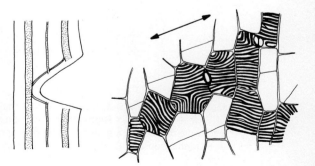

Figure 17-11. Left: Regeneration of vascular tissue in a Coleus stem after some of the vascular bundles were severed by cutting a wedge out of the stem. The regenerated connecting strands are at angles with the vertical. **Right:** Drawing showing the short regenerated vessel elements, which differentiated from parenchyma cells. The arrow indicates the direction of the differentiation. [After E. W. Sinnott and R. Bloch, *Amer. J. Bot.*, **32**:151 (1945)]

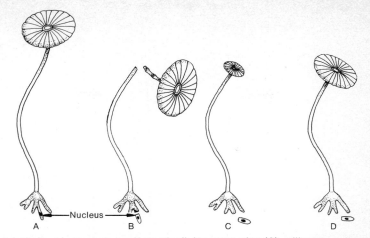

Figure 17-12. Acetabularia, a large and complex unicellular green alga **(A)**, will regenerate a new cap **(C** and **D)** after its cap has been cut off **(B)**, even if the portion of the cell containing the nucleus is also removed. [After T. E. Weier, C. R. Stocking, and M. G. Barbour, *Botany,* 4th. ed., John Wiley & Sons, Inc., New York, 1970]

(especially large ones) and coenocytes may be reconstituted if severed. Acetabularia (Figure 17-12) is a well-known example. If the apical cell of a filamentous alga is destroyed, the cell just below it may redifferentiate as an apical cell from which the filament continues to grow. Hofler described an interesting case of regeneration in the filamentous alga Griffithsia. If a cell in the filament dies, the cell above it grows through or around the dead cell and connects with the cell below, thus reconstituting the filament. Fungi have extensive regenerative capacities, among the most striking being the reconstitution of normally shaped sporophores such as mushrooms and brackets after even substantial portions of them have been cut away.

Bryophytes, and in particular liverworts, have marked regenerative capabilities. The regeneration of moss gametophytes generally involves an initial protonematal stage. Protonemata may even regenerate on moss sporophytes. The diploid gemetophytes then give rise to tetraploid sporophytes when they reproduce, although in some cases the protononal buds have developed into diploid sporophytes instead of the usual leafy gametophytes. The significance of such unusual development from the standpoint of the alternation of generations as well as of morphogenesis is just one of the things that have made bryophytes favored organisms for investigations of development.

The Development of Isolated Plant Parts

In its broadest sense the development of isolated plant parts includes such things as the regenerative development of cuttings, but in this section we are concerned with the development of isolated plant cells, tissues, and organs in tissue culture (Figure 17-13). Although the results of investigations involving tissue culture have been mentioned a number of times previously, they deserve

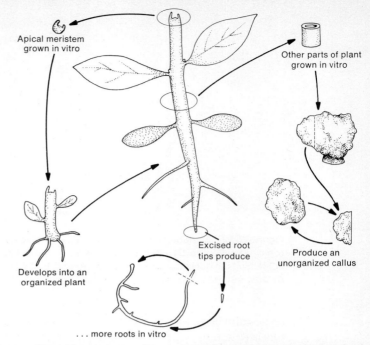

Figure 17-13. Diagram illustrating various patterns of growth and development of excised plant parts cultured in sterile nutrient media. [Adapted from a drawing by R. H. Wetmore. From J. G. Torrey, *Development in Flowering Plants,* The Macmillan Company, New York, 1967]

a somewhat more extensive and unified treatment here because of their increasing importance as an experimental approach in clarifying the problems of plant growth and development.

Haberlandt pointed out the potential value of the culture of isolated plant cells as an approach to the study of development at the turn of the century. However, his own efforts were unsuccessful, partly owing to a poor choice of experimental material and partly to the lack of knowledge of plant growth substances at that time. The successful culture of isolated plant parts over indefinite periods of time was first reported by Gautheret, Nobécourt, and White in 1939, all of them working independently. Since then numerous investigators have used tissue cultures experimentally with increasing success.

Figure 17-14. Photograph of an unorganized callus that grew in a tissue culture.

The first successful organ cultures were of isolated roots. The first cultures of plant tissues resulted only in the growth of unorganized and undifferentiated calluses (Figure 17-14), but later means were found for inducing the differentiation of specialized cells such as vessel elements and even adventitious buds and roots on a callus. More recently, the successful culture of isolated cells and small cell clusters that can develop into embryoids and then mature normal plants has been achieved.

The basic or minimal media used in plant tissue culture are like the medium originally prepared by White for the culture of isolated tomato roots or some modification of it. These media contain salts of the essential mineral elements, a carbon source such as sugar, and generally thiamin. Such basic media are generally supplemented with other substances including auxin, cytokinin, other hormones, various vitamins, and reduced nitrogen (ammonium compounds or amino acids), either as a means of determining the requirements for various types of development or for securing such development. Such defined media may also be supplemented by the addition of complex mixtures of inorganic and organic substances from plant sources such as coconut water (a liquid endosperm), the milky endosperm of young corn, walnut, or horse chestnut seeds, or yeast extract. The chemical composition of these is incompletely known. Sterile bacteriological culture techniques must be used to prevent contamination by microorganisms.

Organ Culture

Root tips were the first plant organs to be cultured successfully, and those of many different species have been used, principally herbaceous dicotyledons. Monocotyledon root tips are more difficult to culture. The growth of excised roots is usually normal as regards both morphology and anatomy, and branch roots are formed freely. Cultured roots of most species do not produce adventitious buds, but there are a few exceptions such as bindweed which was studied by Torrey. Tomato roots originally isolated by Philip White have been passed through over a thousand subcultures for more than a third of a century with undiminished vigor, growing at an average of about 2 cm per day.

Isolated stem apices usually develop normally in culture and produce adventitious roots, thus giving rise to an entire organized plant. Excised leaf primordia in culture generally develop into small but normally shaped leaves. Isolated flower primordia attain varying degrees of normal development in culture, as do young stamens and other isolated flower parts. The ovularies of pollinated excised flowers frequently develop into normal, though often small, fruits in culture. Thus, all the various plant organs can develop in culture media of suitable composition.

Although organ cultures are interesting and sometimes rather spectacular, they are of limited value in the clarification of differentiation because they generally include previously organized meristematic tissues. However, they have been useful in determining the substances necessary for the growth and development of various organs. For example, they revealed the necessity of thiamin, and in some species other B vitamins, for root growth.

The culture of excised embryos involves, not just organs, but an entire organism. Excision of embryos at various stages of development has made it clear that the youngest embryos are essentially heterotrophic, with very limited synthetic capabilities, but that they become progressively more autotrophic. Raghavan found that the early globular embryos of Capsella required suitable concentrations of auxin, cytokinin, adenine, and various vitamins in addition to the basic salts and sugar; amino acids are also generally needed for the development of young embryos. As development proceeds the embryos are capable of synthesizing more and more of the necessary organic compounds themselves, until excised mature embryos require nothing but mineral salts, water, light, carbon dioxide and oxygen.

Tissue Culture

Isolated differentiated tissues, unlike isolated organs, generally do not develop normally in culture but rather grow into an unorganized and undifferentiated callus, even if the medium is supplemented with auxin and a variety of vitamins. However, further supplements to the medium (notably of cytokinins), alteration of the physical conditions of the culture, and other manipulations have resulted in the differentiation of tissues and even bud and root primordia in the calluses, thus providing considerable morphogenetic information. The differentiation of xylem in callus tissue after localized application of auxin (Chapter 12) and the differentiation of buds, roots, or both by tobacco pith callus, depending on the ratio of auxin to cytokinin, (Chapter 13) are examples. The buds and roots that develop on callus cultures are often not connected by vascular tissue, but if they are plantlets that can grow into mature plants develop. Winton and others [28] have used these for the vegetative propagation of quaking aspen trees.

Cell Cultures

The most exciting development in tissue culture has been the culture of isolated cells and small clusters of cells that are usually detached from calluses by mechanical agitation (rotation or shaking) of callus cultures (Figure 17-15). The isolated cells may divide and separate, particularly if agitation is continued, thus producing what essentially amounts to a culture of unicellular organisms. More commonly, especially if plated on an agar medium, these cells may bud like yeast cells, form filaments or clusters of cells, or enlarge into giant cells without cell division or sometimes with nuclear divisions resulting in coenocytes. Such coenocytes may later divide internally by the formation of walls.

The most striking aspect of cell culture is the development, under suitable conditions, from cells in culture of embryoids similar in structure to the embryos that normally develop in the seeds of the species. Even more remarkable is the fact that the developing embryoids usually pass through stages comparable to those of developing embryos, including proembryos, and the globular, heart, and torpedo stages (Figure 17-16). The embryoids usually develop readily into normal plants that can be transplanted to soil and mature in the usual way. Steward has carried carrot plants through several generations, each one derived

Figure 17-15. Some of the rotors used by F. C. Steward and his associates in the culture of isolated plant cells. [Courtesy of F. C. Steward]

from suspensions of isolated cells in culture, without any loss of vigor or normal development.

Carrot plants, both wild and cultivated, have been used most widely, but the development of embryoids has also been secured from cell cultures of other species such as Cuscuta, tobacco [24], endive, parsley, and asparagus. The work of Steward [21, 22] and his coworkers with carrot is the best known, but Reinert [16] found plantlets in carrot cultures as early as 1959 and many other investigators have worked with this species. The isolated cells have been derived from various tissues including secondary phloem of the root, normal embryos, and petioles. Steward has reported the development of as many as 100,000 embryoids from the isolated cells of a single normal embryo. Efforts to induce embryoid or plantlet formation in cultures of cells from a number of species have been unsuccessful, either because they lack the regenerative capacity of species such as carrot or because of a lack of all the necessary substances or conditions in the culture medium.

Steward has proposed that carrot embryoids will develop only from isolated single cells freed from the inhibiting influences of adjacent cells and also that coconut water (milk) or some other liquid endosperm is an essential component of the culture medium. However, Halperin [7] and other investigators have questioned both these statements. Halperin claims that embryoids develop from

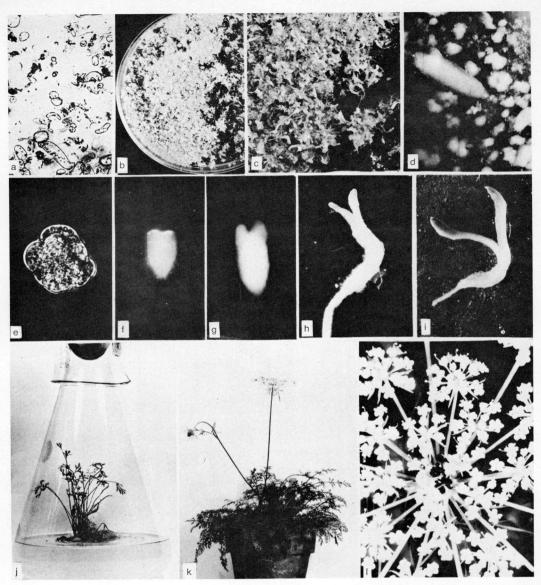

Figure 17-16. Development of isolated cells of carrot (*Daucus carota*) in culture. **(a)** Photomicrograph of cells isolated from carrot embryos. **(b)** A petri dish containing about 100,000 embryoids, all derived from the isolated cells of one embryo. **(c)** Higher magnification of b. **(d)** Embryoids and cell clusters that developed from isolated cells. **(e–i)** Successive stages in embryoid development. **(j–k)** Stages in the development of carrot plants derived from isolated cells. **(l)** Detail of an inflorescence of a mature plant derived from cultured cells. [From F. C. Steward *et al.*, *Science*, **143**:20–27 (Jan. 3, 1964) with the permission of the authors and publisher. Copyright 1964 by the American Association for the Advancement of Science]

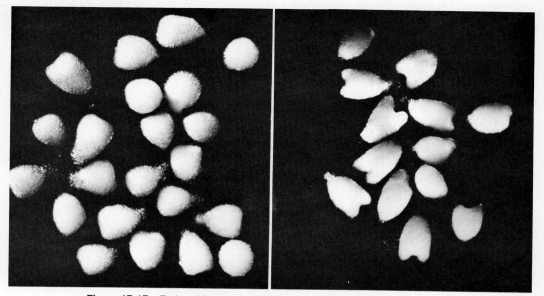

Figure 17-17. Embryoids that developed from cells isolated from petioles of *Daucus carota* (wild carrot or Queen Anne's Lace) (×110). **Left:** Cotyledon initiation stage. There is a single tubular cotyledon primordium. **Right:** Heart stage. Halperin obtained such normal embryoid development in culture media containing only mineral salts and sucrose. [Courtesy of Walter G. Halperin. From W. G. Halperin, *Amer. J. Bot.,* **53:**443 (1966)]

clusters of cells. He also points out that he and other investigators have secured embryoid development from cell clusters of carrots and other plants in culture using simple defined media without coconut milk and that in cultures of wild carrot cells coconut water has even been found to be inhibitory. Halperin secured wild carrot embryoids in a medium containing only minerals, sucrose, vitamins and auxin (Figure 17-17).

Steward noted that cells that develop into embryoids are relatively small and have a high starch content. Later Thorpe and Murashige found that the cells of tobacco callus that form buds also have a high starch content and that gibberellin inhibited bud formation by inducing α-amylase production. Steward and his coworkers found that large, nondividing carrot cells in basal medium had a much higher DNA and RNA content than the smaller dividing cells with coconut water supplement. He suggested that the protein synthesis prerequisite to cell division was not dependent upon new nucleic acid synthesis. Cells in culture frequently develop various degrees of polyploidy and rarely undergo reduction division, thus producing haploid cells. Although there is controversy as to whether embryoids can develop from single isolated cells, sequential photomicrographs have made it clear that a single cell can give rise to a small callus that later develops buds and roots and so an entire plant.

One of the most interesting and significant cases of embryoid development in cultures was reported by Nitsch and Nitsch [13], who secured embryoid development from young tobacco pollen grains (Figure 17-18). The stage of pollen development is critical, since the cells must have separated from the

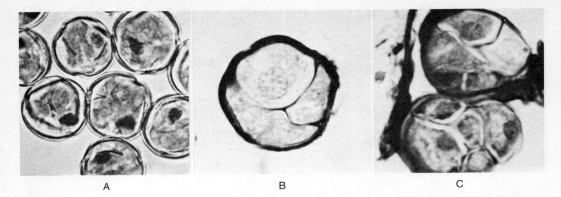

A   B   C

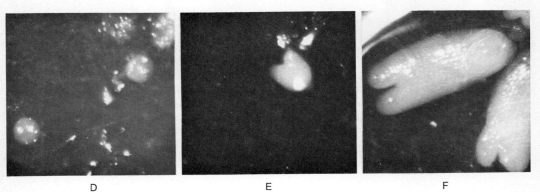

D   E   F

Figure 17-18. Stages in the development of sporophyte plants from pollen grains of *Nicotiana tabacum* (tobacco). **(A)** Microspores ($\times$1050). **(B)** Cell division within wall of pollen grain ($\times$650). **(C)** Young proembryos breaking out of the pollen wall ($\times$1000). **(D)** Globular stage of embryo ($\times$43). **(E)** Heart stage of embryo ($\times$43). **(F)** Torpedo stage of embryo ($\times$43). **(G)** Plantlets emerging from an anther after 1 month in culture. [A–C from J. P. Nitsch, *Phytomorphology*, **19:**389 (1970); D–G from J. P. Nitsch and C. Nitsch, *Science*, **163:**85–87 (3 Jan. 1969). Copyright 1969 by the American Association for the Advancement of Science]

G

tetrad and still be uninucleate. The culture procedures involved placing entire young anthers on a completely defined agar culture medium.

These findings have a number of interesting implications. They conclusively demonstrate that embryoids can develop from single cells. The development of a haploid sporophyte from a cell that normally develops into a male gametophyte provides further evidence that ploidy is not a determining factor in the development of gametophyte and sporophyte generations. The development of the haploid sporophytes could be regarded as a case of male parthenogenesis, even though the sperm and tube nucleus of the pollen had apparently not yet been produced. A means is provided whereby numerous haploid sporophytes can be provided. If any of these were treated with colchicine, completely homozygous diploids could be produced. Both of these kinds of plants could provide valuable experimental material and might even prove to be valuable from a commercial standpoint.

In general, the development of cultured isolated cells into normal mature plants has demonstrated beyond doubt that even mature cells are totipotent; that is, they contain the complete genome of the zygote from which the plant they were derived from developed. Cell cultures provide an excellent means of studying both the physiological and morphogenetic aspects of plant development. Furthermore, cell cultures provide a potentially valuable means of propagating large numbers of plants with identical genomes for experimental or even economic use.

Abnormal Plant Development

In view of the complexity of plant development, the remarkable thing is not that plants develop abnormally at times but that, in the vast majority of cases, development results in the production of an organism with the precise characteristics specified by its hereditary potentialities. It is this precision of development that makes possible the taxonomic classification of organisms and the construction of keys for their identification. In plants abnormal development may result in the formation of calluses, tumors, galls, or fasciation. Alterations resulting from somatic mutations, ionizing radiation, and chimeras are also commonly referred to as abnormal development. Each of these will be discussed briefly. Alterations in development such as the growth of stunted and chlorotic plants lacking one or more essential mineral elements, the etiolation of plants growing in the absence of light, and the malformation of diseased plants could equally well be considered as abnormal, although they are usually not.

Strictly speaking, none of the above should really be designated as abnormal. In each case the development is actually normal for the particular set of environmental factors and alterations in heredity and internal processes involved. What we are really dealing with is development that differs from plants with the usual range of hereditary potentialities that are characteristic of the species and are growing in an environment optimal for development of members of the species. The various types of unusual development commonly referred to as abnormal have a number of interesting aspects, and their study can contribute even more than it has in the past to an understanding of the usual pattern of development.

Calluses and Tumors

Calluses and tumors are both amorphous structures with little or no cell differentiation, without organized tissues, lacking polarity, and composed principally or entirely of rather large and often misshapen parenchyma cells. They are often polyploid. They usually result from resumption of mitosis in tissues where it had ceased, and the planes of cell division are random, resulting in the amorphous structure and the lack of symmetry. The usage of the terms callus and tumor is variable, and they are partially but not entirely interchangeable.

The calluses that develop in tissue cultures are never called tumors. Their growth generally requires the presence of auxin and a cytokinin in the culture medium and growth ceases if they are transferred to a medium lacking phytohormones. The calluses may differentiate buds or roots or both under certain conditions and, as has been noted previously, the ratio of auxin to cytokinin plays an important role in such differentiation. Those calluses that develop in plants as a result of wounding, application of substances with auxin activity, virus infection, or certain parasites including insects, nematodes, mites, bacteria, and fungi, are sometimes referred to as tumors (Figure 17-19).

As in animals, some plant tumors are localized and self-limiting. They are dependent on the normal tissues of the plant for phytohormones and vitamins. Other plant tumors are autonomous; that is, they synthesize their own phytohormones and vitamins. They are capable of continued growth and may spread from one region of the plant to another. If inoculated in other parts of the plant or other plants of the same species, these autonomous tumors readily take hold and grow. They are generally induced by viruses or parasites, particularly bacteria. Such tumors are never referred to as calluses, but they are frequently called plant cancers [2]. Their spread through the plant is more limited than the spread of animal cancer, probably because the vascular tissues provide less effective dispersal channels than the blood streams of animals.

The most common and widespread, and also the most extensively investi-

Figure 17-19. **Left:** Tumor on cut off stem of a Kalanchoë plant. **Right:** Differentiation of abnormal shoots from a similar tumor. [Courtesy of Arman C. Braun]

Figure 17-20. Tumors caused by crown gall bacteria on Kalanchoë plants. [Courtesy of Armin C. Braun]

gated, autonomous tumors of plants are the crown galls (Figure 17-20) of about 150 genera of plants which are induced by the bacterium *Agrobacterium tumefaciens*. The bacteria enter through a wound and produce tumor-inducing substances. Wound substances produced by the plant may also play a role in conditioning and inducing the cells. Once the tumor cells are autonomous, further participation by the bacteria is not essential and the secondary crown galls formed elsewhere in the plant are generally bacteria-free. Because they produce their own phytohormones and vitamins, crown galls grow readily on simple culture media when exercised.

Crown gall tumors have altered synthetic capabilities other than those for producing auxin and cytokinin. These include the ability to synthesize unusual amino acids such as lysopine (N-α-propionyl-L-lysine) and octopine (N-carboxyethylargenine) that are absent or present in very low concentration in normal plant tissues [3]. The altered synthetic capabilities are probably the result of gene derepressions. Crown galls produce a cytokinin that is a nicotinamide derivative rather than a purine derivative. Braun [3] has suggested that possibly the usual cytokinins act by inducing the synthesis of this substance rather than by promoting cell division directly.

Galls

In contrast with crown galls, which perhaps should really not be designated as galls, plant galls are highly organized structures with a specific and characteristic morphology [10]. Aside from certain tumors, which may be called galls, there are two types of plant galls. One type, designated by Küster as **prosoplasmatic galls,** includes galls that are always induced by a parasite and have a characteristic structure differing from that of any plant organ, often in the nature of their cells and tissues as well as their external morphology. Many of these galls are symmetrical and have a highly complex structure (Figure 17-21). The second type are designated as **organoid galls.** These are all recognizable plant organs that have developed in a greatly modified and abnormal pattern or are essentially normal in structure but develop in places where they do not normally occur. Organoid galls are also generally induced by parasites, but in some cases they result from other things such as altered phytohormone

Figure 17-21. Spiny rose galls on a blackberry plant. [Photograph by Ross E. Hutchins]

balances, other physiological disturbances, environmental factors, and perhaps even mutations.

Most of the prosoplasmatic galls are induced by insects (Figure 17-22) that lay their eggs in plant tissues. The developing larvae, or other immature stages, secrete substances that induce differentiation of the gall and use the tissues as food and, when the immature insect reaches maturity, it emerges. In some cases the gall-inducing substances may be injected by the female when the eggs are laid, or in other cases the substances are secreted by the eggs. The gall-inducing substances have not been clearly identified, but they are probably hormones, DNA, RNA, enzymes, or perhaps some combination of these. The latter appears most probable, since the insect secretions are rich in nucleic acids and proteins and since each kind of gall has a highly species-specific structure. Of course, the plant itself does the metabolic work of gall differentiation, and its enzymes and phytohormones are undoubtedly also involved. Almost every order of insects includes some gall-inducing species, but three-fourths of the species are gall wasps and gall midges. There are some 1500 species of these in the United States alone. Each species of gall insect parasitises only a single species of plant, but the plants of a species may be parasitized by a considerable number of different gall insects. If this is the case each species of insect induces galls with a different structure, and this fact indicates clearly that the insect secretions are the dominant factor in determining gall structure. Some galls are induced by mites and others by fungi (Figure 17-23).

Perhaps the most common and conspicuous kind of organoid gall is the **witches' broom** (Figure 17-24). Witches' brooms are induced by a variety of parasites including fungi, bacteria, mistletoe, and mites. They have been found on some 200 species of plants, mostly woody plants. A witches' broom is a dense cluster of small highly branched shoots with short internodes and reduced leaves. It results from the loss of apical dominance and, in some cases also, from the formation of adventitious buds in the internodes. The leaf and stem tissues are generally less highly differentiated than they normally are. Sachs and Thimann [17] found that a bacterium that induces witches' brooms produces a cytokinin that is evidently involved in the induction. Samuels reported that witches' brooms caused by a bacterium (*Corynebacterium fascians*) can be duplicated by kinetin applications. However, the inducing substances produced by most species of parasites have not been identified and some may not be cytokinins.

Other examples of organoid galls induced by parasites are modified leaves. In a sedge (Juncus) the leaf sheath enlarges abnormally whereas the blade is unusually small. In poplar the stipules enlarge greatly, becoming bladelike. Mites are known to cause flowers or cones to give rise to vegetative shoots. Among the examples of misplaced organs are ovaries in staminate flowers, stamens on ovularies, and ovules on the surface of an ovulary. Mites sometimes cause small leaves to develop from a leaf blade.

Figure 17-22. Two of the many types of galls caused by insects. [**Left:** Courtesy of the Carolina Biological Supply Company. **Right:** Photograph by Ross E. Hutchins]

Figure 17-23. Galls on jack pine (*Pinus banksiana*) caused by the pin-oak rust fungus (*Cronartium quercum*). [Courtesy of the U. S. Forest Service]

Figure 17-24. A globose witches' broom on a spruce tree. [Courtesy of the U. S. Forest Service]

Other Abnormal Organs

The organoid galls intergrade with other cases of abnormal organ development that are not induced by parasites, the causative factors often being obscure or unknown. Flowers are particularly subject to such abnormal development. Sepals may develop into what are essentially foliage leaves or stamens and pistils may become petallike. Occasionally the receptacle of a flower regains indeterminate growth and develops into a vegetative shoot, and thus the flower parts appear to be arranged around a stem. This is particularly common in roses. A number of species of plants with dorsiventral flowers, including mints, snapdragons, and larkspurs, sometimes develop radially symmetrical flowers. Usually these occur primarily toward the base of an inflorescence, but in the mints they generally occur toward the top of the inflorescence. The cause of such **pelory,** as it is called, is generally not known, although in larkspur gravity appears to be a factor. The flower heads of composites developed in response to a minimum number of inductive photoperiods may revert to vegetative development in varying degrees. This gives rise to such abnormalities as vegetative flowers in Rudbeckia and divided, leafy bracts in Cosmos or, even to various degrees of reversion to abnormal vegetative shoots.

Fasciation is a relatively common type of abnormal organ development characterized by the development of stems, and less commonly roots, into flattened or elliptical structures that are ribbonlike or ringlike rather than the usual radially symmetrical shape (Figure 17-25). Fasciation may result from the expansion of the apex principally in one direction or from the lateral fusion of several adjacent apeces. The causes of fasciation are obscure, although it has been induced by high concentrations of growth substances and by pruning or wounding. For example, when the epicotyls of *Phaseolus multiflorus* seedlings are removed the cotyledonary buds develop into fasciated stems. These stems generally revert to the usual cylindrical form later on. Short days have been reported to favor fasciation in strawberries. In some cases, such as the flattened inflorescences of cockscomb, fasciation is hereditary. Hereditary fasciation has also been found in tobacco, and large tomato fruits with more than two carpels are considered to be fasciations. It is doubtful whether such hereditary fasciation, which has become a standard characteristic of a species or variety, is any more abnormal than the flattened stems of certain cacti or the cladophylls that constitute the normal photosynthetic organs of some species.

Abnormalities Induced by Ionizing Radiation

Among the most extreme examples of abnormal plant development are those induced by sublethal doses of ionizing radiation (Figure 17-26). The abnormalities are so numerous and varied [5, 6] that we can mention only a few of them such as the curving, twisting, and fasciation of stems, reduced internode length, the formation of surplus axillary buds resulting in witches' brooms, and

Figure 17-25. A fasciated branch of *Rhus glabra* (smooth sumac). Note the flattening of the stem, which is normally round. [Photograph by H. H. Lyon, Department of Plant Pathology, Cornell University]

Figure 17-26. Maryland Mammouth tobacco plants exposed to the indicated dosages of γ-radiation by placing them different distances from the radiation source. See also Figure 11-12. [Courtesy of the Boyce Thompson Institute for Plant Research]

the development of adventitious aerial roots. Leaves may become thickened, asymetrical, puckered, curled, cupped, or fused and the veination may be distorted. Flowering may be delayed or reduced, inflorescences may proliferate abnormally, and the flower parts may be variously modified and malformed (Figure 17-27), as by the development of embryo sacs in anthers. It will be

Figure 17-27. Morphogenetic changes of tulip flowers induced by the indicated amounts of radiation. [Courtesy of the Brookhaven National Laboratory]

noted that many of these abnormalities may occur in nonirradiated plants, although not in such abundance or to such an extreme degree.

Some of the abnormalities induced by ionizing radiation are apparently the result of somatic mutations, but more commonly they are not, as indicated by cases of reversion to normal development particularly in cuttings. In such cases the abnormal development may involve alterations in such substances as mRNA, enzymes, hormones, or other metabolites. In particular, a number of the abnormalities resemble effects of abnormally low or high hormone concentrations.

Abnormalities Induced by Applications of Growth Substances

Like abnormalities induced by ionizing radiation, the abnormalities resulting from application of excessive quantities of growth substances are products of human manipulation in most cases. The formation of callus on stems following localized auxin application has already been mentioned. Excessive auxin may not only inhibit growth but also cause distorted and malformed development of leaves and stems, both in young plants and in plantlets in tissue culture. The synthetic auxins are more likely to induce abnormalities than IAA since they are not destroyed by IAA oxidase. The herbicide 2,4-D generally causes some degree of malformation before it kills a plant, and sublethal doses may induce marked morphogenetic alterations (Figure 12-29). The abnormally tall growth of rice plants infected by *Gibberella fujikuroi* provides an example of natural abnormal growth induced by a hormone (gibberellin). Growth substances may also induce fasciation and other abnormalities.

References

[1] Brachet, J. L. A. "Acetabularia," *Endeavour,* **24**:155–161 (1965).

[2] Braun, A. C. "Plant cancer," *Sci. Amer.,* **186**(6):66–72 (June 1952); "The reversal of tumor growth," *Sci. Amer.,* **213**(5):75–83 (Nov. 1965).

[3] Braun, A. C. *The Cancer Problem,* Columbia University Press, New York, 1969.

[4] Brown, R. "The growth of the isolated root in culture," *J. Exp. Bot.,* **10**:169–177 (1959).

[5] Gunckel, J. E. "The effects of ionizing radiation on plants: morphological effects," *Quart. Rev. Biol.,* **32**:46–56 (1957).

[6] Gunckel, J. E. *et al.* "Vegetative and floral morphology of irradiated and nonirradiated plants of *Tradescantia paludosa,*" *Amer. J. Bot.,* **40**:317–332 (1953).

[7] Halperin, W. "Single cells, coconut milk, and embryogenesis *in vitro,*" *Science,* **153**:1287–1288 (1966).

[8] Heslop-Harrison, J. "Differentiation," *Ann. Rev. Plant Physiol.,* **18**:325–348 (1967).

[9] Jaffe, L. F. "On the centripetal course of development of the Fucus egg, and self-electrophoresis," *28th. Symposium of the Society for Developmental Biology,* Academic Press, New York, 1970, pp. 83–111.

[10] Hovanitz, W. "Insects and plant galls," *Sci. Amer.,* **201**(5):152–162 (Nov. 1959).

[11] Lund, E. J. "Electrical control of organic polarity of the egg of Fucus," *Bot. Gaz.,* **76**:288–301 (1923).

[12] Lund, E. J. *et al. Bioelectric Fields and Growth,* Univ. of Texas Press, Austin, Tex., 1947.

[13] Nitsch, J. P., and C. Nitsch. "Haploid plants from pollen grains," *Science,* **163**:85–86 (1969).

[14] Nojoku, E. "Effect of gibberellic acid on leaf form," *Nature,* **182**:1097–1098 (1958).

[15] Quatrano, R. S. "Rhizoid formation in Fucus zygotes: dependence on protein and ribonucleic acid synthesis," *Science,* **162**:468–470 (1968).

[16] Reinert, J. "Morphogenesis in plant tissue cultures," *Endeavour,* **21**:85–90 (1962).

[17] Sachs, T., and K. V. Thimann. "Release of lateral buds from dominance," *Nature,* **201**:939–940 (1964).

[18] Salisbury, F. B., and C. Ross. *Plant Physiology,* Wadsworth Pub. Co., Belmont, Calif., 1969.

[19] Sinnott, E. W. *Plant Morphogenesis,* McGraw-Hill, New York, 1960.

[20] Skoog, F., and C. G. Miller. "Chemical regulation of growth and organ formation in plant tissues cultured *in vitro,*" *Symp. Soc. Exp. Biol.,* **11**:118–131 (1957).

[21] Steward, F. C. "The control of growth in plant cells," *Sci. Amer.,* **209**(4):104–113 (Oct. 1963); "Totipotency, variation and clonal development of cultured cells," *Endeavour,* **29**:117–124 (1970).

[22] Steward, F. C. *et al.* "Growth and development of cultured plant cells," *Science,* **143**:20–27 (1964).

[23] Thompson, D. *On Growth and Form* (abridged edition, J. T. Bonner, ed.), Cambridge University Press, Cambridge, 1966.

[24] Vasil, V., and A. C. Hildebrandt. "Growth and tissue formation from single, isolated tobacco cells in microculture," *Science,* **147**:1454–1455 (1965).

[25] Wardlaw, C. W. *Morphogenesis in Plants,* 2nd. ed., Meuthen, London, 1968.

[26] Wetmore, R. H. "Morphogenesis in plants—a new approach," *Amer. Sci.,* **47**:326–340 (1959).

[27] Whitaker, D. M. "Determination of polarity by centrifuging eggs of *Fucus furcatus,*" *Biol. Bull.,* **73**:249–260 (1937).

[28] Winton, L. L. "Plantlets from aspen tissue cultures," *Science,* **160**:1234–1235 (1968).

[29] Yarwood, C. E. "Detached leaf culture," *Bot. Rev.,* **12**:1–56 (1946).

Phases in the Life of Plants

18

The growth and development of a zygote into an embryo and subsequently a seedling, young plant, mature plant, and finally a senescent plant that eventually dies is a continuum. However, each of these phases of plant life is characterized by physiological, developmental, and morphogenetic features that differ in part from those of the other phases. Since the embryonic and seedling phases have already received some consideration, this chapter will deal with the three subsequent phases: juvenility, maturity, and senescence.

The juvenile phase begins with the young seedling and lasts until the plant begins reproductive development. Juvenile plants have high rates of metabolic activity and growth and commonly differ from mature plants of the species in such respects as the size and shape of the leaves, the growth patterns of the stems, and the general absence of abscission layers. Mature plants generally have a reduced rate of vegetative growth, active reproductive development, and alterations in the rates and pathways of some metabolic processes. In senescent plants catabolic processes predominate over anabolic processes, growth ceases, various organs may atrophy or abscise, and finally the plant dies.

The duration of each of the three phases varies greatly from species to species. Annuals complete all three phases within one to several months. A biennial is juvenile the first year of its life, and maturity and senescence occur within several months of the second year. In perennials each of the three stages may last for many years. The life span of trees often extends through hundreds or even thousands of years. The production of flowers, fruits, and seeds, which in most perennials occurs over a period of many years, does not lead to senescence of the plant as it does in annuals and biennials. In a few perennials such as English ivy the juvenile stage may last indefinitely, maturity rarely being attained.

Some organs of a plant may become senescent while the rest of the plant retains vigorous juvenility or maturity. The leaves of deciduous woody perennials become senescent and abscise every autumn, and those of evergreens become senescent after several years. The shoots of herbaceous perennials become senescent and die every autumn, but their underground rhizomes and roots remain vigorous for years. The petals and stamens of flowers generally become senescent and abscise not long after anthesis, as do entire flowers that have not been pollinated, and fruits become senescent sooner or later.

Phase changes are frequently initiated by environmental factors, including day length and low temperature preconditioning, but continuation of the phase is generally independent of such factors. Phase change is basically an aspect of the expression of hereditary potentialities, but the mechanisms involved are not clear. Brink [1] pointed out that there is limited evidence that the chromosomes are the site of such discontinuous and potentially reversible ontogenetic changes. He proposed the hypothesis that, in addition to the stable genes, chromosomes also contain self-perpetuating accessory substances, which undergo paramutation in an orderly way in somatic cells, as an essential aspect of a nucleocytoplasmic system of morphogenetic determination. It appears likely that the less hypothetical gene repressions and derepressions may also be involved. At a different level, altered hormonal balances apparently play a role.

Note that animals also undergo more or less marked phasic changes. Perhaps the most marked and spectacular examples are those insects that pass through zygote, larval, pupal, and adult phases.

Juvenility

Juvenility [6] is characterized not only by a high level of metabolic activity and rapid growth but also by the differentiation of structures that are often quite unlike those produced after the plant reaches maturity or even a later stage of juvenility. Among the more striking morphogenetic differences are those of leaves. The juvenile leaves of the garden bean plant (*Phaseolus vulgaris*) develop only at the first node and are opposite and simple rather than alternate and compound like all the subsequently developed leaves (Figure 18-1). At the other extreme are leaves of various plants such as cotton and morning glories that gradually change from the juvenile to the mature form through a series of intermediate forms at successive nodes. Morning glories and a number of other plants also have an eventual gradual reversion of leaf shape to the juvenile form (Figure 18-2). Various species of cucurbits exhibit a similar pattern of change in leaf size, from small juvenile leaves through progressively larger mature leaves to a maximum size, followed by progressively smaller leaves.

Many species have transitions from juvenile to mature leaves intermediate between the extremes mentioned above. Some, like peas, have several juvenile leaves with a rather sharp shift to the mature form. Arbor vitae (Thuja) and several other conifers form needle shaped juvenile leaves, often for several years, followed by a sudden shift to mature leaves, which are appressed scales. English ivy commonly retains its trailing juvenile form and lobed juvenile

Figure 18-1. Young bean (*Phaseolus vulgaris*) plants showing the single pair of opposite, simple, cordate juvenile leaves and the alternate, trifoliate, pinnately compound adult leaves.

leaves year after year, only rarely converting to its bushy mature form with entire leaves (Figure 18-3).

Acacia trees generally have pinnately compound leaves, but in some species only the juvenile leaves are pinnately compound. At successive nodes the pinnate blades become progressively reduced and the petioles become broadened into bladelike phyllodes which are simple and entire (Figure 18-4). Most of the leaves have no pinnate blades at all. The juvenile leaves of Eucalyptus are distinctly different from the mature leaves (Figure 18-5).

Juvenile leaves may also differ from mature leaves as regards the formation of abscission layers, particularly in a number of woody species such as beech and some oaks. The juvenile leaves fail to form effective abscission layers and remain on the stems until growth is resumed in the spring, whereas the leaves

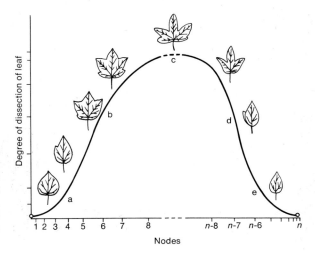

Figure 18-2. Progressive changes in the shape of *Ipomoea caerula* (morning glory) leaves from the base to the apex of the stem. On the left the nodes are numbered from the base and on the right from the apex. [Data of E. Ashby, *New Phytol.*, **49**:375 (1950)]

Figure 18-3. Mature (left) and juvenile shoots of *Hedera helix* (English ivy). Note the differences in leaf shape and the erect shrubby stems and inflorescence of the mature form. The juvenile form is a trailing or climbing vine and never blooms. [Courtesy of V. T. Stoutemyer]

Figure 18-4. *Acacia melanoxylon* (black acacia) showing the transition from the pinnate juvenile leaves to the mature leaves that consist entirely of phyllodes (broadened petioles). Note the intermediate forms with both a phyllode and compound blade. [Courtesy of V. T. Stoutemyer]

Figure 18-5. Mature (left) and juvenile (right) leaves of *Eucalyptus globulus*. [Courtesy of V. T. Stoutemyer]

Figure 18-6. A beech tree in the winter, showing the delayed leaf abscission on the juvenile branches. Note, however, that abscission has occurred on the ends of the lower branches, which have attained the mature condition. [Courtesy of A. C. Leopold]

on mature branches abscise in the autumn (Figure 18-6). There is generally a rather abrupt transition from the lower juvenile branches and leaves to the higher mature ones that developed later, and the juvenile branches may retain their characteristics indefinitely. However, as the juvenile branches continue to grow they too generally convert to the mature state. A similar pattern of juvenile and mature branches occurs in many other trees, including fruit trees such as apples, even though there may be no differences in leaf abscission. In general, flower formation is limited to the mature branches. Cuttings from juvenile branches generally develop adventitious roots much more readily than those from mature branches.

In many woody species, particularly shrubs such as Eleagnus, the juvenile branches are long and whiplike, have unusually long internodes, grow rapidly, are strongly geotropic (and so vertical), and branch at a sharp angle. Many species with thorns bear them exclusively or largely on juvenile branches. One of the more striking changes in stem morphology with conversion from the juvenile to the mature form is in English ivy (Figure 18-3). In the common juvenile form the stems are flexible trailing or climbing vines. In the rare mature form the branches become much more woody and rigid, more highly geotropic, and more highly branched, resulting in a shrubby growth pattern. As noted previously, the mature leaves are entire instead of lobed, and flowers are born only on the mature branches. The juvenile stem of Brussels sprouts is thin and pointed with a small apex, but when the stem attains maturity it becomes very thick with a broad flat apex. A similar, if not quite so extreme, change is found in many other herbaceous species.

Juvenile organs may differ from mature ones in internal as well as external structure. For example, young fern sporophytes have relatively simple protosteles or siphonosteles in contrast with the more complex vascular systems of the mature plants. The number of bundles in the leaf traces of many angiosperm seedlings is less than in the mature plants, and in general the secondary vascular tissues are less complex in the juvenile than the mature stems.

Many more examples of the differences between juvenile and mature organs could be given but, despite the large amount of descriptive information available, little is known about the causes of these differences between the juvenile and mature states. Robbins [8] and others have demonstrated that application of gibberellin can cause at least temporary reversion of mature English ivy plants to the juvenile form, while Njoku [7] found that gibberellin prolonged the formation of juvenile leaves by morning glories. Other characteristics of juvenile stems, such as rapid growth and long internodes, also suggest that they may have a higher gibberellin content than mature stems. Characteristics of juvenile stems, such as their strong geotropic responses, apical dominance, and capacity for forming adventitious roots, suggest that they may also have a higher auxin content than mature stems. However, de Zeeuw and Leopold [3] found that auxin applications hastened transition of Brussels sprouts stems to the mature condition.

Thus, it appears that phytohormones are involved in juvenility and maturity, but much more experimental evidence is needed to clarify their roles and their importance in relation to other factors. At a more basic level, we can hypothesize that there may be a shift in the kinds of mRNA and enzymes produced prior to the shift from the juvenile to the mature stage.

Environmental factors may also influence the change from juvenility to maturity, although no type of environmental manipulation has ever been found to influence the juvenile and mature leaves of bean plants. Long-day and short-day plants may be kept juvenile for long periods of time, if not indefinitely, by keeping them under noninductive photoperiods. Biennials remain juvenile until they have been subjected to sufficient low-temperature preconditioning (or have been treated with gibberellin). If the lower leaves of amphibious plants are indeed juvenile forms, the conversion to the upper mature leaves is evidently influenced by the shift from a submerged to aerial environment.

Although ivy generally retains its juvenile form indefinitely even when it has climbed high on a tree or wall, with only rare shifts to the mature phase, a colony of ivy in a Gothland pine forest was found to shift consistently to the mature phase whenever it climbed a tree trunk to a height of more than 2 m. In ivy, as in most other species, cuttings taken from juvenile plants develop into juvenile plants, whereas cuttings taken from mature plants develop into mature plants. A similar persistence of form in plants from cuttings, although not a matter of juvenile and mature phases, is found in the gymnosperm *Araucaria excelsa,* which has horizontal branches. Cuttings from branches grow into plants that are entirely horizontal, whereas cuttings from the main stem have the usual vertical stem with horizontal branches. Seedlings are always juvenile, even though the seeds were produced on mature plants. This applies also to seedlings that develop from nucellar embryos, at least in citrus plants.

In 1954 Doorenboos conducted some interesting grafting experiments in which he grafted adult scions to juvenile stocks and *vice versa.* If the stock was juvenile and the scion mature, the scion reverted to the juvenile form. If the scion was juvenile and the stock was mature, the stock reverted to the juvenile form. These results suggest that juvenile plants contain translocatable substances that promote juvenility.

Maturity

The discussion of maturity will be brief, not because it is an unimportant phase but because it is well defined and delimited by juvenility on the one hand and senescence on the other. Maturity is characterized by a reduction in the rate of vegetative growth and the development of flowers or other reproductive structures. The decreased rate of vegetative growth results in part from the mobilization of foods, minerals, and other metabolites into the reproductive organs, particularly developing seeds and fruits. Vegetative growth can be promoted in many cases by removal of the flowers. There is also the possibility that substances produced by reproductive structures act as inhibitors of vegetative growth. In some species, particularly certain annuals and biennials, most or all the buds are converted to flower buds, thus reducing or preventing apical growth.

In mature plants, as in juvenile ones, anabolic processes proceed more rapidly than catabolic processes. Notably, the synthesis of proteins and nucleic acids exceeds their degradation. There is a high photosynthetic/respiratory ratio and in some species extensive lipid synthesis. There is commonly extensive accumulation of carbohydrates and lipids, and sometimes proteins, although much of the food is used in the development of the reproductive organs. A considerable amount of the food that accumulates is found in the seeds and fruits. There is evidence that auxin levels decline as a plant progresses from juvenility to maturity, and it has been suggested that reduced auxin concentration is a factor in the initiation of reproductive development. There is also some evidence for reduced levels of gibberellin with the onset of maturity. Some of the morphological differences between juvenile and mature plants have been discussed in the previous section.

There is a marked difference in the duration of maturity from species to species. In annuals and biennials the mature stage lasts only a few weeks or at most a few months, whereas in perennials it is a matter of years. Trees, in particular, may remain in the mature condition for hundreds or even thousands of years with no signs of senescence, although eventually they too become senescent. The transition from maturity to senescence is more gradual and less marked in woody plants than herbaceous ones.

Senescence

Senescence [5] is characterized by deteriorative processes that eventually result in the death of one or more organs or of the entire plant. The pattern of senescence varies from species to species. Overall senescence is characteristic of annuals and biennials and generally occurs soon after the seeds and fruits are mature. There is evidently a causal relationship between the two because removal of flowers commonly prevents the onset of senescence. Herbaceous perennials have another pattern called top senescence. Although the aerial shoots senesce and die annually, the roots and crowns (stem tissues at the upper ends of the roots) retain maturity and often continue to live for many years. In woody perennials both the stems and roots remain in the active mature condition for years, but the leaves and reproductive organs become senescent and die. In deciduous species the leaves become senescent, abscise, and die every autumn. In evergreens, on the other hand, the leaves become senescent only after several years, a progressive type of senescence that results in the presence of leaves at all times.

Each of the various flower structures has a characteristic pattern of senescence. The petals (and in some species the sepals) are generally the first flower parts to become senescent, abscise, and die. Stamens generally become scenescent and abscise soon after they have shed their pollen, and the same is true of entire staminate flowers. Pistillate or perfect flowers become senescent and abscise if there has been a failure of pollination and fertilization, except in naturally parthenocarpic species. The ripening of fruits is also a matter of senescence, desirable as it may be from a commercial standpoint. Only the embryos of the seeds, and for a time the endosperms, escape the fate of senescence. The carpellate cones of gymnosperms become senescent and die soon after the seeds mature, although they may remain on the trees for some time after the seeds are shed. The pollen cones, like staminate flowers, senesce and abscise soon after the pollen is shed.

During senescence there is a marked preponderence of catabolic over anabolic processes. There is a progressive decrease in the rate of photosynthesis as a leaf becomes more senescent and, except in evergreen leaves, this generally begins soon after a leaf has attained its final size. There is also a decline in the rate of respiration, although it is generally not as rapid as the decline in photosynthesis, and senescent leaves commonly decrease in dry weight because of this and because of the outward translocation of solutes. There is a decline in proteins (including enzymes) as the rate of synthesis drops below the rate of degradation, and the rate of hydrolysis of other substances such as fats and starch generally exceeds their rate of synthesis.

Senescence also involves a marked decrease in RNA, and this may well be basic to the decline in protein synthesis and the general metabolic decline. Chlorophyll breakdown exceeds chlorophyll synthesis, resulting in the complete loss of chlorophyll. The carotenoids persist longer than the chlorophylls, as evidenced by the yellow color of autumn leaves, but eventually they too are broken down (in this case into volatile terpenoids). Similar, although not entirely the same, degradative processes occur in the senescence of organs other than leaves. There is generally extensive translocation of both foods and minerals out of senescing leaves prior to their abscission. As has already been noted, the formation of abscission layers is a common aspect of senescence of both leaves and flower parts.

It is interesting to note that, although the transition from juvenility to maturity may involve numerous morphological changes, the transition from maturity to senescence is primarily a shift in metabolism with relatively few morphological changes and those are largely degenerative. Various botanists have attempted to provide explanations for senescence without any great success. One of the first was Molisch, who suggested that senescence resulted from the mobilization of foods and minerals to the developing fruits and seeds and consequent starvation of the vegetative organs. Mobilization of foods to vegetative storage organs may also promote senescence (Figure 18-7).

Removal of young fruits will, indeed, prevent senescence in many cases, but subsequent investigators have found that senescence can be initiated even by the presence of young flowers (even staminate ones), and these do not mobilize nutrients to a degree that would bring about starvation of the vegetative organs (Figure 18-8). Therefore it has been suggested by Leopold and others that the reproductive organs may produce inhibitors of a hormonal nature, which induce senescence, although such substances have not been identified. It has also been found that the senescence of leaves and cotyledons (the first plant structures to become senescent and abscise) can be prevented or delayed by removal of the buds. Although buds do mobilize substances from leaves, again the quantity does not seem great enough to produce leaf senescence by itself,

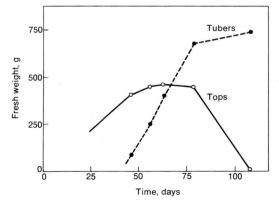

Figure 18-7. Influence of potato tuber growth in suppressing growth of the shoot and then in promoting senescence. [Data of E. C. Wassink and J. A. J. Stolwijk, *Meded. Landbouwhogeschool Wageningen* **53**:99 (1953)]

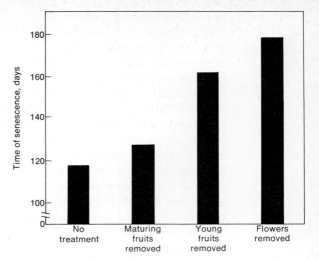

Figure 18-8. Influence of removal of flowers or fruits of soybean plants in delaying senescence. [Data of A. C. Leopold, E. Nidergang-Kamien, and J. Janick, *Plant Physiol.*, **34:**570 (1959)]

and the transport of senescence-inducing substances from buds to leaves has been suggested.

Brown [2] pointed out that, when cultures of excised tomato roots developed mature tissues, there was a decrease in growth rate and other symptoms of senescence, and he suggested that the mature tissues produced toxic (inhibitory?) substances. Root tips cut off and transferred to fresh medium resumed normal growth. Varner [9] and others have suggested that senescence results from loss of membrane integrity including the previous degree of differential permeability. This could be a factor in any of the other proposed mechanisms and could help explain the extensive translocation from senescent organs and perhaps also the altered functioning of various organelles, but neither this nor any other of the proposals is an explanation at a basic level.

There is considerable evidence that various phytohormones are present in different concentrations in senescent and mature tissues and also that application of phytohormones to senescing organs may either speed senescence or result in rejuvenation, depending on the substance and concentration. As was pointed out in Chapter 13, the cytokinins can promote mobilization of substances into leaves and greatly delay senescence. Several investigators have found that there is a marked decline (tenfold in sunflowers) in the quantity of cytokinins translocated from roots to shoots at the time that growth ceases and senescence begins.

Applications of auxin or gibberellins have also been found to delay senescence in some cases, including some in which cytokinins are ineffective. As noted previously auxin plays a role in the prevention of abscission, and auxin or gibberellin applications to unpollinated flowers can prevent their senescence and abscission and bring about the development of the ovulary into a fruit. On the other hand, abscisic acid not only promotes abscission and dormancy but also at least some of the symptoms of senescence. Ethylene may also promote abscission, and it has long been known to be a factor in fruit ripening, a senescence phenomenon. These are just a few examples of the known roles of phytohormones in senescence, but their position in the series of cause and effect relationships leading to senescence is by no means clear.

A number of environmental factors are known to promote, if not initiate, senescence. Among these factors are high temperatures, water deficits, deficiencies of nitrogen or various mineral elements, ionizing radiation, pathogens, and short days. Any of these might affect phytohormone levels or might influence more directly the various changes in metabolism involved in senescence. Short days appear to be one of the more important environmental factors involved in senescence. Aside from the fact that they initiate reproductive development of short-day plants, they have long been known to play a role in the autumnal senescence and abscission of the leaves of deciduous trees and shrubs. Also, Krizek *et. al.* [4] found that short days promoted the senescence of Xanthium plants even though they were kept from developing flowers by debudding. This fact suggests that mobilization may not be important and that short days may induce the production of senescence-inducing substances entirely aside from their effect on flower initiation.

Of course, neither the influence of environmental factors, a description of the metabolic changes that occur, the influence of phytohormones, nor the explanations so far attempted really explain why senescence occurs. One thing that is sometimes ignored is that senescence, like all phenomena of life, is under hereditary control. Just one evidence of this is the different patterns of senescence found in annuals, biennials, herbaceous perennials, and the deciduous and evergreen woody perennials—all of these patterns are hereditary species characteristics.

One can hypothesize, but not much more at this time, that the basic factor in senescence is a change in the kinds and perhaps of the quantities of mRNA produced. This could explain the general shift from predominantly anabolic to predominantly catabolic processes and also such things as the synthesis of anthocyanins only by the senescent leaves (or fruits) in many species. It could also explain such things as a change in membrane integrity and permeability. Perhaps phytohormones, or environmental factors operating through an effect on phytohormones or otherwise, are responsible for the changes in mRNA production. Perhaps the sequence of change in the kinds of mRNA produced in an organism from its initiation through to its senescence is in some way programmed by a genetic mechanism. Much still remains to be learned about senescence through future research.

It should be noted that senescence of cells, tissues, and organs can be prevented by isolating them from a plant and culturing them in suitable media, although transfers to fresh media may be needed from time to time. For example, tomato root tips isolated well over a quarter of a century ago and passed through a series of subcultures are still actively growing, whereas they would long ago have senesced and died if left on the plant. Also, the meristematic cells of the apical meristems and cambia of trees may be considered to avoid senescence for years, even though the leaves and other organs continue to become senescent.

Although senescence is deteriorative, it should not be considered as an undesirable aspect of plant development. Basically, death is an essential partner of reproduction because, if reproduction continued without death, suitable habitats for any species would soon be exhausted. Of course, in temperate and arctic climates most annuals would be killed by freezing even if they did not senesce and die earlier, and many perennials would also be killed by freezing if leaf senescence, abscission, and dormancy did not occur.

However, most annuals and herbaceous perennials are active during only a part of the growing season. For example, tulips are active during the spring and their tops senesce before June. The spaced periods of activity and senescence of various species is important ecologically since it reduces competition and enables a habitat to support a wider variety of species. Senescence is commonly attuned to ecological adaptation and generally occurs when environmental factors such as temperature, water availability, and photoperiod become unfavorable for active metabolism and growth.

Before senescent leaves abscise there is extensive translocation of minerals, hydrolized foods, and other substances from them into other parts of the plant where they can be utilized. The abscission of the older, senescent leaves reduces the total transpiration from the plant, and such minerals as remain in the fallen leaves may leach into the soil where they can be reabsorbed. The senescence, or ripening, of fruits may contribute toward the dissemination of the seeds by animals, facilitate the release of the seeds, or result in the formation of germination inhibitors. These and other aspects of the positive contributions of senescence to survival have been discussed in greater detail by Leopold [5].

References

[1] Brink, R. A. "Phase change in higher plants and somatic cell heredity," *Quart. Rev. Biol.,* **37:**1–22 (1962).
[2] Brown, R. "The growth of isolated roots in culture," *J. Exp. Bot.,* **10:**169–177 (1959).
[3] de Zeeuw, D., and A. C. Leopold. "Altering juvenility with auxin," *Science,* **122:**925–926 (1955).
[4] Krizek, D. T., W. J. McIlrath, and B. S. Vergara. "Photoperiodic induction of senescence in Xanthium plants," *Science,* **151:**95–96 (1966).
[5] Leopold, A. C. "Senescence in plant development," *Science,* **134:**1727–1732 (1961).
[6] Leopold, A. C. *Plant Growth and Development,* McGraw-Hill, New York, 1964.
[7] Njoku, E. "Effect of gibberellic acid on leaf form," *Nature,* **182:**1097–1098 (1958).
[8] Robbins, J. "Gibberellic acid and the reversal of adult Hedera to a juvenile state," *Amer. J. Bot.,* **44:**743–746 (1957).
[9] Varner, J. E. "Biochemistry of senescence," *Ann. Rev. Plant Physiol.,* **12:**245–264 (1961).

Appendix A
Metric Units and Equivalents

Basic Measurements

Weight, Length. See the table on the following page.

Area. The square of the units of length, for example, m^2, cm^2, μm^2. The hecatre is a unit of area used in the metric system in place of the acre used in the English system. 1 hecatre = 1 hm^2 = 2.471 acres.

Volume. The liter and its multiples are used primarily for liquid volume. Solid volume is the cube of the units of length, for example, m^3, cm^3 or cc, μm^3. In the Système International d'Unités (SI units) adopted in 1960 the liter and its multiples are abandoned and volume is expressed entirely as the cubes of the units of length, with the cubic meter (m^3) as the basic unit. However, the liter is still widely used as a unit of volume.

Force

Force is mass $\times$ acceleration ($F = ma$). The metric unit of force is the dyne. 1 dyne = 1 g $\times$ 1 cm/sec^2.

Energy

Energy is the capacity for doing work. Work = force $\times$ distance ($W = Fd$). The basic metric unit of energy is the erg. 1 erg = 1 dyne-cm. In the SI units the joule has replaced the dyne as the basic unit. 1 joule = 10^7 ergs.

The calorie is commonly used as a measure of heat energy and chemical bond energy. 1 calorie (cal) is the quantity of heat required to raise the temperature of 1 g of water 1°C, more specifically from 14.5 to 15.5°C. 1 cal = 4.185 $\times$ 10^7 erg = 4.185 joule. The large or kilogram calorie (Cal or kcal) = 1000 calories. In biology the kcal is most generally used.

Table A-1 Metric units of weight, length, and volume with equivalents.

Weight

Basic unit, prefixes, and abbreviations		Magnitude	Equivalents
mega-	Mg	10^6 g	metric ton (1.1 US ton)
kilo-	kg	10^3 g	2.205 lb
hecto-	hg	10^2 g	
deka-	Dg	10 g	
gram	g	1 g	0.0353 oz
deci-	dg	10^{-1} g	
centi-	cg	10^{-2} g	
milli-	mg	10^{-3} g	
micro-	μg	10^{-6} g	1 gamma (γ)

Length

Basic unit, prefixes, and abbreviations		Magnitude	Equivalents
kilo-	km	10^3 m	0.6214 mile
hecto-	hm	10^2 m	
deka-	Dm	10 m	
meter	m	1 m	3.281 ft 1.094 yd
deci-	dm	10^{-1} m	
centi-	cm	10^{-2} m	0.3937 in
milli-	mm	10^{-3} m	
micro-	μm	10^{-6} m	1 micron (μ)
nano-	nm	10^{-9} m	1 millimicron (mμ) 10 Ångstroms (Å)

Volume

Basic unit, prefixes, and abbreviations		Magnitude	Equivalents
kilo-	kl	10^3 liter	1 m^3
hecto-	hl	10^2 liter	2.838 bu
deka-	Dl	10 liter	
liter	l	1 liter	1.0567 qu
deci-	dl	10^{-1} liter	
centi-	cl	10^{-2} liter	3.38 fl oz
milli-	ml	10^{-3} liter	1 cc or 1 cm^3
micro-	μl	10^{-6} liter	1 lambda (λ)

The energy in a photon is the product of Planck's constant (h) and the frequency of the radiation (ν). $E = h\nu$. Planck's constant is 6.624×10^{-27} erg-sec or 1.58×10^{-34} cal-sec. If a mole of a substance is excited by radiation Avogadro's number (6.02×10^{23}/mole) of photons will be required. This is 1 einstein.

Pressure

The basic unit of pressure in the SI units is Newtons per square meter (N/m^2). $1 N = 1$ kg-m/sec^2, a unit of force. However, in biology the most commonly used metric unit of pressure is the bar. 1 bar $= 10^6$ dynes/cm$^2 = 10^6$ ergs/cm^3. 1 bar $= 0.987$ atmosphere (atm) $= 14.504$ lb/in$^2 = 75$ cm Hg. 1 atm $= 14.696$ lb/in$^2 = 76$ cm Hg.

Light Quantity

The unit of luminous flux is the lumen, which is the amount of light emitted by 1 candella (unit of light intensity) and falling per second on a unit area (m^2 in the metric system) at a unit distance (m) from the source. The metric unit of illumination is the lux. 1 lux $= 1$ lumen/m$^2 = 1$ meter candle (m-c). The English unit of illumination is the foot candle (ft-c). 1 ft-c $= 1$ lumen/ft$^2 = 0.133$ lux.

Temperature

The metric temperature unit is degrees centigrade (Celsius), °C. However in SI units the standard is degrees Kelvin (°K). $1°C = 1°K$, but °K is absolute zero, so $0°C = 273°K$. Degrees farenheit (F) $= 9/5°C + 32$. Degrees centigrade $= 5/9 \, (°F - 32)$.

Concentration

A 1 molar (M) solution contains 1 mole of solute per liter of solution. A 1 molal (m) solution contains 1 mole of solute per 1 liter of water. A 1 normal (N) solution of an acid contains 1.008 g of ionizable hydrogen per liter of solution. A 1 N solution of a base contains 17.008 g of ionizable hydroxide per liter of solution.

A weight percentage solution is the grams solute per grams solvent $\times$ 100. A volume percentage solution is the milliliters solute per milliliters solvent $\times$ 100. Used for low concentrations of solutes are parts per thousand (ppt) and parts per million (ppm). Solute concentrations are also expressed in such units as g/liter, μg/liter, and μg/g.

Appendix B
Major Groups of Plants

The following outline of the major groups of plants is included here primarily to provide a guide to the various categories of plants referred to in this textbook and in various references. Except for Subkingdoms (if they are used in a particular system of classification), the major taxonomic subdivisions of the Plant Kingdom are the Divisions. Divisions are coordinate with Phyla in the Animal Kingdom, and are sometimes called Phyla, but Phyla are not recognized as a category by the International Code of Botanical Nomenclature. The Subkingdoms and Divisions listed on the right side are those used by Bold in his system of classification [See Bold, H. C., *Morphology of Plants,* Harper & Row, New York, 1967.]

A number of systems of classifying the major groups of plants has been proposed, and in the more modern ones the tendency has been to increase the number of Divisions and, thus, decrease the diversity of plants in a Division. However, Bold has set up more Divisions than are found in most other systems of classification. The names applied to the Divisions vary from system to system. The Classes into which Divisions are subdivided are not given, but it should be noted that the dicotyledons and monocotyledons are the two classes (or subclasses in some systems) of flowering plants (Anthophyta).

In some classifications of the Plant Kingdom the seed plants are divided into two classes: the Gymnosperms, which have naked seeds not enclosed in fruits; and the Angiosperms, which have seeds enclosed in fruits. The Gymnosperms include Divisions 21 through 24 in Bold's system of classification. The Angiosperms are the same plants as those in Bold's Division 25 (Anthophyta). In the rather widely-used classification of the Plant Kingdom proposed by Oswald Tippo in 1942 all vascular plants (Bold's Divisions 17 through 25) are placed in a single Division, the Tracheophyta.

The categories to the left of the Subkingdoms have been abandoned in modern systems of classification of the Plant Kingdom (except for the Tracheophyta in some systems) but are still used as convenient common names for various large groups of plants.

The following outline of the Plant Kingdom is based on the classification of all organisms into two kingdoms: the Plant Kingdom and the Animal Kingdom. However, some biologists prefer systems in which there are three or even four kingdoms: Monera (bacteria and blue-green algae), Protista, Plantae, Fungi, and Animalia. [See Whittaker, R. H., *Science,* **163:**150–160 (1969).] The organisms included in the Kingdom Protista vary greatly from system to system. The protozoa are always included, as are many unicellular and colonial eucaryotic algae, but in some systems all algae and fungi are classified in the Kingdom Protista.

If all this seems confusing, recall that taxonomic categories do not exist in nature. Classifications are human artifacts designed to separate the great diversity of organisms into groups that show their evolutionary relationships to one another as accurately as possible on the basis of available information.

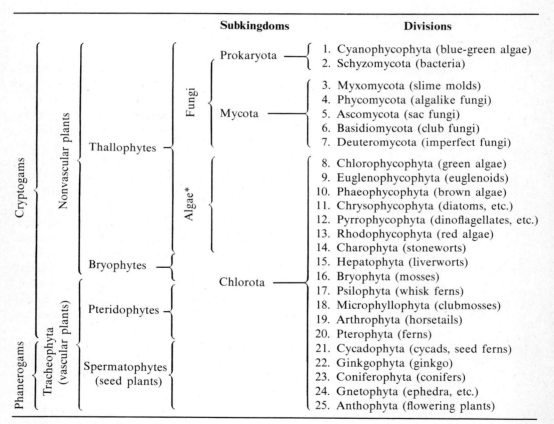

*The algae include also the blue-green algae.

Index

Abnormal development, 543–551
Abscisic acid (ABA), 375–381
Abscission, 344–345, 379,
    467–469, 555–557, 560
Abscission zone, 344–345
Absolute humidity, 239
Absorption, of mineral salts,
    179–183
  of water, 256–265
Absorption spectra, of
    photosynthetic pigments,
    68–71
  of phytochrome, 313, 446
Action spectrum, of
    photoperiodism, 446
  of photosynthesis, 71
  of phototropism, 337–338
  of seed germination, 446
Activated diffusion, in phloem,
    284
Adenine, 145
Adenosine diphosphate (ADP),
    67. See also Adenosine
    triphosphate.
Adenosine triphosphate, 67,
    74–80, 96–101, 149
Adventitious buds, 371–372, 391,
    533–534
Adventitious roots, 346–347, 372,
    388–389, 517, 533–534
After-ripening, 480–481
Alcoholic fermentation, 104–105
Aleuroplasts, 40
Alkaloids, 152–155
Amino acids, 142–144, 149–150
Ammonia, 137, 141–142, 144
Amo-1618, 376–377
Amylases, 126, 361, 367, 484
Amylopectin, 123–124
Amyloplasts, 40
Amylose, 123–124
Angiosperms, life cycle, 411–421
  reproductive development,
    435–485
  reproductive structures,
    423–433
  vascular system, 198–210,
    212–223
  vegetative development,
    488–519

Annual rings, 206–207, 210–211
Antheridia, 396, 398–399, 404
Anthers, 412, 415
Anthesin, 455
Anthocyanin, 156–158
Anticlinal divisions, 492
Apex, root, 9–10, 509–512
  shoot, 9–10, 489–495
Apical dominance, 342–344, 372,
    378, 532
Apical meristems, 9–10, 489–495,
    509–512
Apomixis, 391–392
Arabanose, 117–118
Arabans, 118, 123
Archegonia, 398–399, 402, 405,
    409
Arils, 432
Ascorbic acid, 106–107, 356
Ascorbic acid oxidase, 106–107
Asexual reproduction, 385–392
ATP. See Adenosine
    triphosphate.
Auxin, 325–352, 362, 462,
    468–476, 513–514, 552, 558,
    562

Bacteria, auxin synthesis by, 332
  cell ultrastructure, 54–56
  gall induction by, 547
  in nitrogen cycle, 136–140
  photosynthetic, 54–56, 64
  reproduction, 385
Bark, 205, 209–210, 214–216, 503,
    517
Beta inhibitor, 376
Beta oxidation, 134
Betacyanin, 158
Bicollateral bundles, 203–204
Binary fission, 385
Bleeding, 226
Blue-green algae, cell
    ultrastructure, 54–56
  in nitrogen fixation, 136, 139
  photosynthetic pigments, 70–71
  reproduction, 385
Boron, 164–165, 171
Bracts, 424, 429
Branch gaps, 216
Brassin, 356

Bud dormancy, 322, 372, 438
Bud traces, 216
Budding, in plant propagation,
    390
  of yeasts, 390
Bulliform cells, 235–236

Calcium, 164–165, 169
Callose, 280
Callus, 536, 543–545
Calvin–Benson cycle, 77–80
Carbohydrate metabolism,
    115–129
Carbon dioxide, influence on rate
    of photosynthesis, 86–89
  influence on rate of respiration,
    109
Carboxylation, 64–65
Carotenes. See Caroteneoids.
Carotenoids, 41, 68–71, 377–378
Carpels, 412, 415–416
Catalase, 106–107
Cell culture, 538–543
Cell differentiation, 298–300,
    335–336, 370–371
Cell division, 10–11, 296–298,
    334–335, 355–356, 370–371
Cell elongation (enlargement),
    298, 334, 366, 370–371
Cell sap, 26
Cell walls, 48–54
Cellobiose, 120, 128
Cells, 6–56
  chlorenchyma, 13
  collenchyma, 13
  companion, 17, 201
  cork, 24–25, 209–210, 215–216
  endodermal, 15–16, 212–214,
    262
  epidermal, 13–15, 501–502
  guard, 15, 227–228, 234,
    501–502
  meristematic, 10–11, 489–495,
    502–512
  parenchyma, 11–13
  procaryotic, 54–56
  sclerenchyma, 18–19
  sieve, 16–17, 212
  sieve tube elements, 16–17, 201,
    500

Cells (*cont.*)
    size range, 6
    subsidiary, 15, 234, 501
    tracheids, 20–21, 210–212,
        499–500
    vessel elements, 19–20, 22–23,
        201, 208, 300, 499–500
Cellulose, 38, 48–52, 123, 127–128
Cephalin, 131
Chlorenchyma, 13
Chlorophyll, 68–72
Chloroplasts, 41–44, 46–47
Chromatin, 28
Chromoplasts, 41, 46–47
Chromosomes, 11, 28
Circidian rhythms, 448–449
Circulation of solutes, 291–292
Cisternae, 36
Citric acid cycle, 97–98
Clubmosses, life cycle, 403–406
Cobalt, 162–164
Coconut water (milk), 356, 367,
        432, 539–541
Coenzyme A (CoA), 97–99,
        132–135
Coenzyme Q, 100
Cohesion mechanism, 252–256
Coleoptile, 325–327, 330, 432–433
Collenchyma, 13
Companion cells, 17, 201
Conidiospores, 386
Conifers, life cycle, 406–410
    stem structure, 210–212
Copper, 164–170
Cork, 24–25, 209–210, 215–216
Correlations, 531–532
Culture, of cells, 538–543
    of organs, 537–538
    of tissues, 538
Cuticle, 13, 52, 226–227, 234–235,
        501
Cutin, 136. *See also* Cuticle.
Cuttings, 389–391
Cytochromes, 74–76, 100, 152
Cytokinins, 367–375. *See also*
    Kinetin.
Cytoplasmic streaming (cyclosis),
        31, 281
Cytosine, 145

2,4-D. *See*
    2,4-Dichlorophenoxyacetic
    acid.
Day-neutral plants, 438, 441, 444
Denitrification, 142
D-enzyme, 125
Deoxyribonucleic acid, 28, 38, 40,
        145–147, 149–150
Deoxyribose, 117, 145–146
Dermal system, 7–9, 501–502
Development, 294–315, 520–551
    abnormal, 543–551

environmental influences, 299,
        303–314
    of flowers, 449–453
    of fruits, 421, 470–473
    hereditary influences, 299–303
    of isolated plant parts, 535–543
    of leaves, 504–508
    reproductive, 435–487
    of roots, 508–517
    of seeds, 410, 419–421, 477–478
    of stems, 489–503
    vegetative, 488–519
2,4-Dichlorophenoxyacetic acid,
        350–352
Dictyosomes, 36–38
Differentiation, 298–299, 530–531
    auxin effects on, 335
    cytokinin effects on, 370–371
    of dermal tissues, 501–502
    of flower primordia, 449–453
    of fundamental tissues, 500–501
    gibberellin effects on, 366
    of leaf and bud primordia,
        492–496
    of vascular tissues, 496–500,
        513–515
Diffusion, of gases, 176–177
    of solutes, 178–179, 183
    through stomata, 231–233
    of water, 183–188
Diffusion pressure deficit, 184
Diffusion shells, 231–232
Digestion, 121–122, 126–128, 134,
        151
Dihydroxyacetone, 117
Dihydroxyacetone phosphate, 78,
        96, 103, 132
Disaccharides, 120–122
DNA. *See* Deoxyribonucleic acid.
DNP. *See* Day-neutral plants.
Dormancy, bud, 322, 365, 372,
        378, 438
    seed, 365, 372, 378, 479–482
Dormin, 377
Double fertilization, 411, 418–419

Eggs, 396 *ff.*
Einstein, 72
Elioplasts, 40
Embryo sac, 416–419
Embryoids, 538–543
Embryos, 405, 410, 419–420,
        431–433, 477–478
Emerson effect, 73
Endodermis, 9–10, 15–16,
        212–214, 262
Endogenous rhythms, 448–449
Endoplasmic reticulum (ER),
        35–36
Endosperm, 419–421, 432–433
Epidermis, 13–15, 226–229,
        501–502
Erythrose, 117

Essential oils, 156
Ethylene, 330, 352–353, 475–476
Exine, 413

FAD, $FADH_2$. *See* Flavin
    adenine dinucleotide.
Fasciation, 549
Fats, 130–135
Fern, life cycle, 399–403
Ferredoxin, 73–75
Fertilizers, 163
Fibers, 18–19
Field capacity, 258–261
Filament, 412
Flavin adenine dinucleotide, 100,
        133–135, 151
Flavonols, 157–158
Florigen, 454–456
Flower bud differentiation,
        449–453
Flower development, 463–464
Flower initiation, 346–347, 365,
        372, 435–461
Flower movements, 465–467
Flower physiology, 461–467
Flower sex expression, 461–463
Flowers, 411–413, 423–428
Fluorescence, 72
Fructose, 119–123, 128
Fructose-1,6-diphosphate, 77–79,
        96, 101–103, 119
Fructose-6-phosphate, 96, 103
Fruit abscission, 469
Fruit development, 347, 365,
        419–420
Fruit growth, 470–473
Fruit physiology, 467–476
Fruit ripening, 352, 474–476
Fruit set, 467–470
Fruit types, 429–431
Fundamental tissues,
    differentiation of, 500–501
Funiculus, 416

GA. *See* Gibberellins.
Galactose, 119
Galls, 545–549
Gametes, 392–419
Gametophytes, 395–419
Gemmae, 386
Germination, 378, 483–485,
        361–362
Germination inhibitors, 481–482
Geotropism, 340–342
Gibberellins (gibberellic acid),
        356–367, 469, 471–473, 558
Girdling, 269, 272–273
Glucose, 96–97, 103–105, 119,
        124, 127
Glucose-1-phosphate, 119, 121,
        125
Glucose-6-phosphate, 96, 101–103
Glyceraldehyde, 117

Glycerol, 131–132
Glycolysis, 96–97, 104
Glyoxysomes, 47
Golgi apparatus, 36–38
Grafting, 390
Gravity, influences on plants,
    306, 340–342
Growth, 294–323
    auxin influences, 335–347
    cellular, 296–298, 334–335
    determinate and indeterminate,
        318–319
    differential, 319–320
    environmental influences, 219,
        303–314
    of fruits, 470–473
    hereditary influences, 299–303
    kinetics of, 315–318
    periodicity of, 205–207,
        320–323
    rates of, 318–333
    of roots, 214–216
    of stems, 204–207, 209–210
Growth inhibitors, 375–379,
    481–482
Growth substances. See
    Phytohormones.
Guanine, 145
Guard cells, 15, 227–228, 234,
    501–502
Guttation, 226

Herbicides, 350–352
Heterocysts, 385–386
Heterogametes, 396
Heterophylly, 507–508, 554–558
Heterospory, 401
Heterothallism, 397
Hexosans, 122–129
Hexose monophosphate shunt,
    101–102
Hexoses, 118–120
Hill reaction, 64
Hilum, 432–433
Homothallism, 397
Hormones. See Phytohormones.
Hydathodes, 226
Hydroponics, 164

Imbibition, 193–196
Indole-3-acetic acid (IAA),
    328–329, 332. See also
    Auxin.
Inflorescences, 423–429
Inhibitors, of germination,
    481–482
    of growth, 352–353, 375–381
Inine, 413
Interfacial flow, 284–285
Inulin, 123, 128
Invertase, 121
Ion accumulation, 178, 180–182
Ion exchange, 179–180

Ionizing radiation, morphogenetic
    effects, 310–311, 549–551
Iron, 164, 169
Isogametes, 393–394
Isoprene, 155

Juvenility, 553–559

Kinetin, 287, 367–369, 547. See
    also Cytokinins.
Krebs cycle, 97–98

Laticifers, 18
Layering, 389
LDP. See Long-day plants.
Leaf abscission, 344–345, 379,
    555–557, 561, 563
Leaf gaps, 216
Leaf growth and development,
    306, 313–314, 363–364, 371,
    504–508
Leaf structure, 7–8, 216–221,
    226–228, 235–236, 506–507
Leaf traces, 202, 216–217,
    496–497
Leaves, juvenile, 554–559
    modified, 488–489
    sun and shade, 506
    vascular system, 216–221
Lecithin, 131
Lenticels, 24, 205
Leucoplasts, 40–41, 46–47
Life, origin of, 59
Life cycles, 392–422
Light, influence on growth and
    development, 310–315,
    506–507
    influence on rate of
        photosynthesis, 89–90
    influence on rate of
        transpiration, 233
    influence on seed germination,
        483–484
    influence on stomatal opening,
        229–230
    in photoperiodism, 437–448
    in photorespiration, 83
    in photosynthesis, 68–73
    in phototropism, 337–340
Light spectrum, 310
Lignin, 159
Limiting factors, 84–86
Lipid metabolism, 130–136
Lipids, as products of
    photosynthesis, 80
Lomasomes, 47–48
Long-day plants, 438, 440–444
Low-temperature preconditioning,
    309–310
Lutein, 69
Lysosomes, 47

Magnesium, 164–165, 169–170
Maleic hydrazide, 317, 376–377

Maltose, 120
Manganese, 165, 171
Mannose, 119
Mass flow, 180, 281–284
Matric potential, 184, 189,
    193–194
Maturity, 559–560
Megaspores, 401–418
Megasporophylls, 404–412
Meiosis, 393–400, 404–409,
    411–416
Meiospores, 386, 393–413
Membrane permeability, 166,
    173–175
Membranes, 26–27, 29, 32–48,
    54–55, 173–175
Meristems, 9–11, 319, 489–496,
    503, 509–512, 517
Metaphloem, 499
Metaxylem, 499
Metric units, 565–567
Mevalonic acid, 155, 358
Micropyle, 408, 418, 433
Microspores, 401–415
Microsporophylls, 404–412
Microtubules, 31–32
Middle lamella, 49
Mineral deficiency symptoms,
    166–172
Mineral elements, roles of,
    164–166
Mineral nutrition, 161–172
Mitochondria, 38–40
Mitosis, 11, 26, 28, 297
Mobilization, 285–291, 373, 473,
    560–564
Molybdenum, 164, 171–172
Monocarpic perennials, 436–437
Monosaccharides, 116–120
Morphogenesis, 489–517, 520–551
Moss, life cycle, 379–399
Münch hypothesis, 281–284
Mycorrhizae, 264–265, 303

α-Naphthalene acetic acid
    (NAA), 350–351
Nicotinamide adenine
    dinucleotide (NAD$^+$,
    NADH), 67, 96–97, 100–103,
    132–135, 139–141, 151
Nicotinamide adenine
    dinucleotide phosphate
    (NADP$^+$, NADPH), 67,
    74–80, 101–103
Nitrite oxidation, 142
Nitrate reduction, 140–141
Nitrogen, 136–140, 164, 167–168
Nitrogen cycle, 136–142
Nitrogen fixation, 136–140
Nitrogen metabolism, 136–155
Nitrogenase, 138
Nucellus, 408, 416, 420
Nuclear envelope, 29–30

Nucleic acids, 145–147
Nucleolus, 27–28
Nucleus, 26–29
Nutation, 528

Oedogonium, chloroplasts, 42
    life cycle, 396–397
Oogonia, 396, 398
Organ culture, 537–538
Organelles, cell, 25–54
Osmosis, 183–193
Osmotic potential, 184–185, 192
Osmotic pressure, 184
Ovularies (ovaries), 412, 415, 421,
    429–430, 467, 470
Ovules, 412, 415–417, 477
Oxygen, influence on ion
    accummulation, 180–181
    influence on phloem
        translocation, 279
    influence on photosynthetic
        rate, 92
    influence on respiratory rate,
        111–112
    influence on root growth, 304
    influence on seed germination,
        483
    photosynthetic production of,
        4, 61–63, 74–75

Papain, 151
Parenchyma, 9, 11–13
Parthenocarpy, 347, 365, 372,
    467–468
Parthenogenesis, 391
Pectic compounds, 129
Pelory, 549
Pentosans, 118, 122
Pentose phosphate pathway,
    101–103
Pentoses, 116–117
Periclinal divisions, 492
Perisperm, 420, 432
Permanent wilting percentage,
    258–261
Permeability, 173–175
Peroxidase, 106–107
Petals, 412, 423–425
Phellem, 24, 209, 215–216
Phelloderm, 24, 209
Phellogen, 24, 209
Phloem, cellular components, 9,
    16–17, 212
    primary, 199–204, 212–221
    secondary, 204–206, 209,
        215–216
    solute translocation, 269–285
Phosphatidic acid, 131
Phosphoenol pyruvate pathway,
    80–81
Phosphoglyceric acid (PGA), 66,
    77–81, 96–97
Phospholipids, 32–33, 135
Phosphon-D, 376–377

Phosphorus, 164, 168
Phosphorylation, oxidative,
    98–100
    photosynthetic, 73–76
Photons, 70, 72
Photoperiodic classes of plants,
    438, 441–444
Photoperiodism, 437–456
    action spectrum, 446–447
    hormones in, 454–455
    morphogenetic influences,
        449–453
    role in senescence, 563
    role in sex expression, 462–463
Photophil phase, 449
Photorespiration, 83, 92
Photosynthesis, 58–93
    action spectrum, 63, 71
    annual productivity, 4–5, 59–60
    apparent, 82–83
    bacterial, 1, 54–56, 64
    carbon fixation and reduction,
        76–81
    energy efficiency, 81–82
    net, 82–83
    photochemical reactions, 68–76
    quantum efficiency, 76
    rates of, 2, 82–92
    sugarcane type, 80–81
    true, 82–83
Photosynthetic/respiratory ratio,
    83–84
Phototropism, 312–313, 337–340
Phycocyanin, 68–71
Phycoerythrin, 68–71
Phyllotaxy, 494, 528–530
Phytochrome, 313–314, 446–447,
    455–456, 483–484
Phytohormones, 325–383,
    454–455, 459, 551, 562–563
Pigments, photoreceptor, 311–314
Pigment systems, photosynthetic,
    44–45, 72–75
Pine, life cycle, 406–410
    vascular system, 210–212
Pinocytosis, 32
Pistils, 412, 415–416, 423–426
Pits, 20–23
Placenta, 416
Plasmalemma, 33
Plasmodesmata, 33–34, 175
Plastids, 40–47
Plastocyanin, 74–76
Plastoquinone, 74
Polarity, 520–525
Pollen, 406–408, 413–415, 464–465
Pollen tube, 410, 418
Pollination, 410, 418
Polyphenol oxidase, 106–107
Polyribosomes, 31
Polysaccharides, 122–129
Porphyrin, 152–153
Potassium, 165, 169
Pressure flow, in phloem, 281–284

Pressure potential, 184–185, 192
Primary tissues, 7–10, 488–517
Procambial strands, 496–499
Propagation, vegetative, 387–392,
    533, 538–543
Proplastids, 44–47
Proteins, 148–151
Prothallia, 400
Protoderm, 501
Protophloem, 498–499
Protoxylem, 498–499
P/R ratio. See
    Photosynthetic/respiratory
    ratio.
Purines, 144–145
PWP. See Permanent wilting
    percentage.
Pyrenoids, 41–42
Pyrimidines, 144–145
Pyrrole, 152–153
Pyruvic acid, 80–81, 96–97

Q-enzyme, 124, 126
Quantasomes, 44
Quantum, 70, 72
Quiescent center, 511–512

Raffinose, 122, 268, 277
Reduction division. See Meiosis.
Regeneration, 532–535
Relative humidity, 238–239
R-enzyme, 126
Reproduction, 384–487
    asexual, 384–392
    sexual, 392–434
Reproductive development,
    435–487
Reproductive organs,
    mobilization into, 288–290
    vascular system, 221–223
Respiration, 83, 92, 94–113,
    134–135
    aerobic, 95–100
    anaerobic, 104–106
    rates of, 108–113
Respiratory climacteric, in fruits,
    475–476
Respiratory energy, uses of, 101
Respiratory quotient (RQ),
    107–108
Riboflavin, 337–338
Ribonucleic acid, 28–31, 38,
    55–56, 147, 149–150
    messenger, 149–150, 301–303,
        348–349, 367
    ribosomal, 28–31
    transfer, 149–150
Ribose, 117
Ribose-5-phosphate, 103, 145–147
Ribosomes, 29–31, 36, 149–150
Ribulose, 117
Ribulose-1,5-diphosphate, 77–80
Ripeness to flower, 436–437
RNA. See Ribonucleic acid.

Root apex, 7, 9–10, 509–512
Root growth and development, 336–337, 340–341, 346–347, 508–517
Root hairs, 13–15, 256–257, 514–516
Root pressure, 250–251
Root structure 212–216
Roots, absorption by, 261–265

Scarification, 480
Sclereids, 18–20
Sclerenchyma, 18–20
Scutellum, 432–433
SDP. *See* Short-day plants.
Secondary growth, of roots, 214–216, 517
of stems, 204–210, 503
Seed development, 410, 419–421, 477–478
Seed dormancy, 372, 479–482
Seed germination, 361–362, 378, 483–485
Seed physiology, 477–485
Seed structure, 431–433
Seed viability, 478–479
Senescence, 373–374, 379, 467, 560–564
Sepals, 412, 424
Sex expression, in flowers, 365, 372, 461–463
Shoot apex, 489–495
Shoot tension, 250, 252–256
Short-day plants, 438, 441–443
Sieve cells, 16–17
Sieve tube elements, 16–17
Silicon, 162
Skotophil phase, 449
Slime bodies, 280
Sodium, 162, 165, 182–183
Soil water, 257–261
Solutes, absorption of, 177–183
translocation of, 267–285
Soredia, 386
Sori, 400
Sound waves, influence on growth, 305–306
SPAC, 224–225
Spawn, 387
Spectrum, electromagnetic, 310
Sperm, 396 *ff.*
Spherosomes, 47
Spirality, 528–530
Sporangia, 385–386, 398–401, 404–409, 411–418
Spores, 385–386 *ff.*
Sporophylls, 399–412
Sporophytes, 394–396, 398–406, 411–412, 422
Sporopollenin, 52–53
Sporulation, 386
Stachyose, 122, 268, 277
Stamens, 412, 423–425
Starch, 40, 62, 123–127, 229, 485

Starch phosphorylase, 125, 229
Statoliths, 341
Stem elongation, 336, 362–363, 502–503
Stem structure, 7–10, 199–212
Stems, modified, 488–489
primary growth, 488–500
secondary growth, 204–210, 503
vascular system, 199–212
Sterols, 158–159
Subsisiary cells, 14–15, 234, 501
Sucrase, 121
Sucrose, 120–121, 268, 277
Sucrose phosphorylase, 121
Sucrose synthetase, 121
Stomata, 14–15, 227–234, 501–502
Strobili, 403–404, 406–407
Sugars, 116–122
Sulfur, 164, 169
Suspensors, 410, 419–420
Symmetry, 525–528

2,4,5-T. *See* 2,4,5-Trichlorophenoxyacetic acid.
T divisions, 511
Tannins, 158
Tapetum, 408
Temperature, influence on growth and development, 308–310, 456–461
influence on photosynthetic rate, 90–92
influence on respiratory rate, 112–113
influence on seed germination, 483
influence on transpiration rate, 239–240
Terpenes, 155–156
Tetrapyrroles, 152–153
Tetroses, 116–117
Thermoperiodicity, 308–309, 460–461
Thiamin, 355
Thigmotropism, 337
Thylakoids, 43
Thymine, 145
Tissue culture, 538
Tissue systems, 7–9
Tonoplast, 33, 35
Toxic elements, 166
Tracheids, 19–22, 210–212
Translocation, of solutes, 267–285
of water, 250–256
Transpiration, 226–250
Transpiration stream, 250
Traumatin, 355
2,4,5-Trichlorophenoxyacetic acid, 350–352
Trichoblasts, 514
Triglycerides, 130–135
Triose phosphates, 77-80, 96, 131–132

Trioses, 116–117
Tropisms, 337–342
Tumors, 543–545
Tunica-corpus theory, 491–493
Tyloses, 209

UDP-glucose-transglycolase, 125
Ulothrix, chloroplasts, 42
life cycle, 393–394
Ulva, life cycle, 394–396
Uracil, 145

Vacuoles, 11–12, 26, 33, 35
Vapor pressure, 238–240
Vascular system, 7–9, 198–223
Vascular tissues, differentiation of, 496–500
Vegetative development, 488–519
Vegetative propagation, 387–392
Veins, 218–221
Verbascose, 122, 268, 277
Vernalin, 459
Vernalization, 456–460
Vessel elements, 19–23, 208, 299–300
Vessels, 119–209
Vitamins, 151–152, 355–356

Water, absorption of, 256–265
diffusion of, 183-193
influence on photosynthetic rate, 88–89
influence on respiratory rate, 111
loss of (transpiration), 226–250
translocation of, 250–256
Water potential, 184–196, 224–225, 257–259
Waxes, 130, 135–136
Wilting, 241–242
Witches' brooms, 547–548

Xylans, 118, 123
Xylem, cellular components, 9, 16–23, 207–212
differentiation of, 496–500, 513–514
primary, 199–204, 212–214, 216–223
secondary, 204–212, 214–216
solute translocation through, 269–274
water translocation through, 250–256
Xylose, 117–118
Xylulose, 117

Zeatin, 368
Zinc, 164, 169
Zoospores, 386, 393
Zygospores, 386
Zygotes, 392–393, 395–399, 405, 409–410, 418